T0325820

HYALURONAN IN CANCER BIOLOGY

HYALURONAN IN CANCER BIOLOGY

Edited by

ROBERT STERN

Department of Pathology
Al Quds University
Abu-Dies, East Jerusalem 20002
Palestinian Territory

AMSTERDAM • BOSTON • HEIDELBERG • LONDON
NEW YORK • OXFORD • PARIS • SAN DIEGO
SAN FRANCISCO • SINGAPORE • SYDNEY • TOKYO

Academic Press is an imprint of Elsevier

Academic Press is an imprint of Elsevier

525 B Street, Suite 1900, San Diego, CA 92101-4495, USA
30 Corporate Drive, Suite 400, Burlington, MA 01803, USA
32 Jamestown Road, London, NW1 7BY, UK
Radarweg 29, PO Box 211, 1000 AE Amsterdam, The Netherlands

First edition 2009

Library of Congress Cataloging-in-Publication Data

Hyaluronan in cancer biology/edited by Robert Stern. – 1st ed.
 p. ; cm.
 Includes bibliographical references and index.
 ISBN 978-0-12-374178-3
 1. Hyaluronic acid–Pathophysiology. 2. Carcinogenesis. I. Stern,
Robert, M.D.
 [DNLM: 1. Hyaluronic Acid–metabolism. 2. Neoplasms–metabolism. 3.
Antigens, CD44–metabolism. 4. Disease Progression. 5. Hyaluronic
Acid–therapeutic use. QZ 200 H992 2009]
 RC268.5.H93 2009
 616.99′4071–dc22

                    2008054964
British Library Cataloguing-in-Publication Data
A catalogue record for this book is available from the British Library

ISBN: 978-0-12-374178-3

For information on all Academic Press publications
visit our website at elsevierdirect.com

Printed and bound in USA
Transferred to Digital Printing, 2011

To the memory of my father and my mother,
and for Tali, Aaron & David

Short Contents

I

HISTORICAL OVERVIEW

II

CELL BIOLOGY OF HYALURONAN IN CANCER

III

HYALURONAN RECEPTORS AND SIGNAL TRANSDUCTION PATHWAYS

IV

HYALURONAN SYNTHESIS

V

HYALURONAN DEGRADATION, THE HYALURONIDASES, AND THE PRODUCTS OF DEGRADATION

VI

HYALURONAN IN CANCER EPITHELIAL-STROMAL INTERACTIONS

VII

HYALURONAN AND INDIVIDUAL CANCERS

VIII

CLINICAL USES OF HYALURONAN-RELATED BIOMATERIALS AS ANTI-CANCER AGENTS

IX

A NEW PERSPECTIVE

Full Contents

III

HYALURONAN RECEPTORS AND SIGNAL TRANSDUCTION PATHWAYS

6. Hyaluronan-Mediated CD44 Interaction With Receptor and Non-Receptor Kinases Promotes Oncogenic Signaling, Cytoskeleton Activation and Tumor Progression

LILLY Y.W. BOURGUIGNON

7. Adhesion and Penetration: Two Sides of CD44 Signal Transduction Cascades in the Context of Cancer Cell Metastasis

DAVID J.J. WAUGH, ASHLEIGH McCLATCHEY, NICOLA MONTGOMERY, AND SUZANNE McFARLANE

8. Involvement of CD44, a Molecule with a Thousand Faces, in Cancer Dissemination

DAVID NAOR, SHULAMIT B. WALLACH-DAYAN, MUAYAD A. ZAHALKA, AND RONIT VOGT SIONOV

9. RHAMM/HMMR: An Itinerant and Multifunctional Hyaluronan Binding Protein that Modifies CD44 Signaling and Mitotic Spindle Formation

JAMES B. McCARTHY AND EVA A. TURLEY

IV

HYALURONAN SYNTHESIS

10. Altered Hyaluronan Biosynthesis in Cancer Progression
NAOKI ITANO AND KOJI KIMATA

V

HYALURONAN DEGRADATION, THE HYALURONIDASES, AND THE PRODUCTS OF DEGRADATION

11. Hyaluronidase: Both a Tumor Promoter and Suppressor
VINATA B. LOKESHWAR AND MARIE G. SELZER

VI

HYALURONAN IN CANCER EPITHELIAL–STROMAL INTERACTIONS

VII

HYALURONAN AND INDIVIDUAL CANCERS

VIII

CLINICAL USES OF HYALURONAN-RELATED BIOMATERIALS AS ANTI-CANCER AGENTS

19. Clinical Use of Hyaluronidase in Combination Cancer Chemotherapy: A Historic Perspective

GERHARD BAUMGARTNER AND GERHARD HAMILTON

20. Exploiting the Hyaluronan–CD44 Interaction for Cancer Therapy

VIRGINIA M. PLATT AND FRANCIS C. SZOKA, JR.

IX

A NEW PERSPECTIVE

Color Plate Section at the End of the Book

Preface

This volume is devoted entirely to the subject of hyaluronan (HA) and cancer. The importance of HA in malignancy is very well documented. However, the subject has been largely neglected. This volume attempts to bring attention to the critical role of HA in cancer biology, in its initiation, progression, and spread. An excellent monograph on cancer, published in 2007 (R.A. Weinberg, The Biology of Cancer, Garland Science, New York) does not cite HA a single time in the index. Reference to the HA receptor, CD44 is given once. However, that citation does not mention that CD44 is the receptor for HA.

The current volume attempts to redress such oversights, to draw attention to this important molecule to colleagues in the cancer field, and to provide students that are entering the area of HA cancer biology with a comprehensive overview.

Hyaluronan is a ubiquitous high molecular size unbranched carbohydrate polymer that is prominent in vertebrate extracellular matrix during embryogenesis, inflammation, in wound healing, whenever there is rapid tissue turnover and repair, but particularly, in neoplasia. Although HA is a simple disaccharide that repeats thousands of times, reaching a molecular mass of several million Daltons, it has a remarkable array of biological functions. This is unusual because among all the glycosaminoglycans, it is the only one that is not sulfated nor modified in any other way throughout its length.

Preparations of HA are well known in the commercial world. It is the filler used by ophthalmologists following cataract surgery, known as Healon.® Lightly cross-linked forms of HA are used as cosmetic fillers, as in Restylane,® and as a visco-supplement for synovial fluid, used by orthopedic surgeons, known as Synvisc.® It can be found in many cosmetic preparations, as a "feel-good" for facial creams and ointments.

Proteins associated with HA metabolism are also finding increasing commercial use. Hyaluronidases, the enzymes that degrade HA, are used to enhance sperm penetration in the process of *in vitro* fertilization (Cumulase®), as an aid in dispersing *i.v.* solutions that have accumulated subcutaneously, as in the newborn nursery, and when caustic chemotherapy agents accumulate in local tissues. They are used for enhancing drug absorption of small and large molecules, as Hylenex® and Enhanze® respectively.

A stabilized pegylated (polyethylene glycol cross-linked) version of a hyaluronidase is now in early clinical trials, as an adjunct in cancer chemotherapy, acting to promoting drug uptake and penetration. However, as HA and its associated molecules achieve increasing commercial visibility, it remains an obscure entity in the life sciences.

The cancer community is beginning to realize the importance of HA, now that its involvement has been documented in stem cells, the stem cell niche, and particularly its involvement in cancer stem cells. It makes this volume all the more germane for enhancing our understanding of HA in the malignant process, and for highlighting how it functions as a critical tool for cancer research.

A historic overview is given in Section I (Stern, Israel/Palestine), tracing the history of HA through its several metamorphoses and baptisms, from ground substance, to acid mucopolysaccharide, to hyaluronic acid, and to hyaluronan, presumably its final incarnation.

A general context is then provided for this interesting molecule in Section II, by Toole and Slomiany (USA). Heldin and her co-workers (Sweden) outline the growth factors that modulate HA deposition, while Datta (India) describes the role of an HA-binding protein in cancer biology.

There are several receptors for HA. An overview of this area is provided in Section III. The predominant receptor for HA, CD44, is one of the most complex molecules in all biology. It has a variety of isoforms, derived from combinations of ten alternatively spliced exons. Vast numbers of post-translational modifications of CD44 increase dramatically the multiple forms in which the receptor occurs. Four chapters, by Stamenkovic (Switzerland) and Yu (USA), another by Bourguignon (USA), by Naor and his colleagues (Israel), and Waugh and colleagues (U.K.) provide overviews of CD44 in its multiple guises, and their involvement in cancer biology. Another HA receptor, Rhamm (Hmmr) interacts with CD44, and modulates its malignant involvement, as shown by McCarthy (USA) and Turley (Canada).

Hyaluronan has an extremely rapid rate of turnover, providing controls at multiple levels for its net deposition. The several enzymes that synthesize HA are the HA synthases, or HASs, that sequentially add sugars to the reducing termini. These are described by two pioneers in the field, Kimata and Itano (Japan) in Section IV.

The hyaluronidase enzymes are the endoglycosidases that degrade HA, described in Section V. Known as the HYALs, they have been controversial, since both increases and decreases in enzyme levels are associated with cancer progression. Two chapters summarize that area, by Lokeshwar and Selzer (USA), and another by Stern (Israel/Palestine). The HA polymer takes on different biological properties as it becomes cleaved, as outlined by Sugahara (Japan and USA). The hyaluronidases are presumably involved in size-specific cleavage reactions. Binding proteins and hyaluronidase inhibitors are presumed to be involved in generating fragments, as well as maintaining polymers at a particular fragment length. But how this occurs is entirely unknown.

The stroma surrounding malignancies is highly abnormal, and is the subject of Section VI. Hyaluronan is intimately involved in the cross-talk between cancers and the host peritumor stromal. The scirrhous reaction or desmoplasia of carcinomas has long been recognized by Pathologists. The extent of that reaction, and the prominence of HA in the reaction are often utilized clinically as prognostic indicators. However, as outlined by the group headed by Raija and Markku Tammi (Finland), the dynamic reciprocity between cancers and stroma, and the role of HA therein have many subtleties. The laboratory of Seth and Ana Schor (U.K.), another husband and wife team, was among the first to document the striking similarity between fetal fibroblasts and peritumor fibroblasts. This is an association that parallels the biology of oncofetal proteins.

In Section VII, focus is placed on site-specific cancers, on prostate, malignant melanoma, and breast cancers. These are summarized by three leading laboratories in their fields, respectively, headed by Simpson (USA), Simon (Germany), and Brown (Australia).

Translational research is now being emphasized by granting agencies. The importance of HA-related molecules has finally begun to be realized in the clinic, as outlined in Section VIII. A historic overview is provided by Baumgartner and Hamilton (Austria). In Europe, hyaluronidase has a history of being used in combination chemotherapy regimens, something that has, until recently, not been permitted in the United States. Certain aggressive lymphoblastic lymphomas, resistant to chemotherapies, became sensitive when hyaluronidase was included in the protocol. Dr. Baumgartner was the first oncologist to incorporate that strategy into cancer treatment. Platt and Szoka (USA) explore various strategies for targeting cancer chemotherapies using the high-affinity binding of HA to its CD44 receptor.

And finally, in Section IX, as in much of cancer research, we have learned to expect the unexpected, an entirely unanticipated dimension has appeared. One of the hyaluronidases, HYAL-2, is a cell surface receptor for a class of animal tumor viruses, as described by Miller (USA).

Hyaluronan does not give up its secrets easily. But recent rapid progress makes this volume all the more timely. Within a few years, I predict it will not be possible to summarize the field again within a single volume.

Robert Stern
Jerusalem, Palestinian Territory
August, 2008

Foreword

Karl Meyer described a polysaccharide in the vitreous body of the eye in 1934 and gave it the name hyaluronic acid (now hyaluronan). He and his collaborators subsequently showed its presence in many other tissues, and determined its chemical structure as a linear chain of alternating units of glucuronic acid and N-acetylglucosamine linked by $\beta(1\text{-}3)$ and $\beta(1\text{-}4)$ linkages. Around 1950, Alexander G. Ogston and his collaborators in Oxford characterized a hyaluronan–protein complex from synovial fluid, found that it had a molecular weight in the order of millions, and extended over a large volume. Duran-Reynals described the so called "spreading factor" in 1928, which turned out to be hyaluronidase.

This was the background that existed when I started to work on hyaluronan in 1949 under the tutorship of Endre A. Balazs. It was at that time commonly believed that hyaluronan was an inert filling material between cells without any specific biological activities. Much work on the polymer during the 1950s and 1960s was therefore directed towards understanding the macromolecular properties of the compound and their importance for the physical state of the cell environment. However, it is notable that by about 1950, Endre Balazs had already begun studying the effects of hyaluronan on cell growth in tissue culture. Notably, he together with my classmate in medical school, Jan von Euler, investigated the connection between hyaluronan and cancer.

A breakthrough in hyaluronan research came in 1972 when Hardingham and Muir found that cartilage proteoglycans specifically bind to hyaluronan. Subsequently a number of extracellular proteins and cell surface receptors have been discovered that interact with the polymer. Suddenly hyaluronan was found to directly and specifically regulate many cellular functions.

The development of the hyaluronan field has accelerated in the last few decades. It is apparent that hyaluronan plays an important role in such fields as mitosis, embryological development, cellular motility, pathological reactions such as inflammation and many other basic functions. Of special interest in recent years has been the discovery of specific biological effects of different size fragments of hyaluronan. The number of researchers working in the field has increased rapidly, and international conferences on the specific subject of hyaluronan are now held every third year.

In parallel with the discoveries of the basic functions of hyaluronan, the substance has become a tool in clinical medicine, much of that due to Endre Balazs. It is used for example in eye surgery, in treatment of arthrosis, and as a space filler in tissues. It is also used as a moisturizer in skin creams and has become a commercial success.

Robert Stern has been a leading scientist in the hyaluronan field in the last decades during its period of very rapid development, and I have admired his work. He has now edited a volume on hyaluronan that focuses entirely on cancer biology, in order to make researchers in the cancer field aware of the importance of this unique polymer. I sincerely hope that this will become a successful endeavor. I also wish that, had I been younger, I could have helped him in this important task.

Torvard Laurent
Uppsala, Sweden
October, 2008

List of Contributors

Ulf Anderegg (Chapter 17) Department of Dermatology, Venerology and Allergology, University of Leipzig, Leipzig, Germany

Päivi Auvinen (Chapter 14) Department of Oncology, Kuopio University Hospital, Kuopio, Finland

Marco Averbeck (Chapter 17) Department of Dermatology, Venerology and Allergology, University of Leipzig, Leipzig, Germany

Gerhard Baumgartner (Chapter 19) Ludwig Boltzmann Institute for Clinical Oncology and Photodynamic Therapy, KH Hietzing, Vienna, Austria

Lilly Y.W. Bourguignon (Chapter 6) Department of Medicine, University of California-San Francisco and Veterans Affairs Medical Center, San Francisco, CA, USA

Tracey J. Brown (Chapter 18) Laborartory for Hyaluronan Research, Department of Biochemistry and Molecular Biology, Monash University, Melbourne, Australia

Anindya Roy Chowdhury (Chapter 4) School of Environmental Sciences, Jawaharlal Nehru University, New Delhi, India

Jacqueline Cox (Chapter 15) Unit of Cell and Molecular Biology, The Dental School, University of Dundee, Dundee, Scotland, UK

Kasturi Datta (Chapter 4) School of Envirnomental Sciences, Jawaharlal Nehru University, New Delhi, India

Ian R. Ellis (Chapter 15) Unit of Cell and Molecular Biology, The Dental School, University of Dundee, Dundee, Scotland, UK

Margaret Florence (Chapter 15) Unit of Cell and Molecular Biology, The Dental School, University of Dundee, Dundee, Scotland, UK

Carl Gebhardt (Chapter 17) Department of Dermatology, Venerology and Allergology, University of Leipzig, Leipzig, Germany

Ilora Ghosh (Chapter 4) Environmental Toxicology, School of Environmental Sciences, Jawaharlal Nehru University, New Delhi-110067, India

Gerhard Hamilton (Chapter 19) Ludwig Boltzmann Institute for Clinical Oncology and Photodynamic Therapy, KH Hietzing, Vienna, Austria

Paraskevi Heldin (Chapter 3) Matrix Biology Group, Ludwig Institute for Cancer Research, Uppsala University, Biomedical Center, Uppsala, Sweden

Naoki Itano (Chapter 10) Department of Molecular Oncology, Institute on Aging and Adaptation, Shinshu University Graduate School of Medicine, Nagano, Japan

Sarah J. Jones (Chapter 15) Unit of Cell and Molecular Biology, The Dental School, University of Dundee, Dundee, Scotland, UK

Anupama Kamal (Chapter 4) School of Environmental Sciences, Jawaharlal Nehru University, New Delhi, India

Eugenia Karousou (Chapter 3) Ludwig Institute for Cancer Research, Uppsala University, Biomedical Center, Uppsala, Sweden

Koji Kimata (Chapter 10) Research Complex for the Medicine Frontiers, Aichi Medical University, Aichi, Japan

Veli-Matti Kosma (Chapter 14) Institute of Clinical Medicine, Pathology and Forensic Medicine, University of Kuopio, and Department of Pathology, Kuopio University Hospital, Kuopio, Finland

Anne H. Kultti (Chapter 14) Department of Anatomy, University of Kuopio, Kuopio, Finland

Vinata B. Lokeshwar (Chapter 11) Department of Urology, Miller School of Medicine, University of Miami, Miami, FL, USA

James B. McCarthy (Chapter 9) Department of Laboratory Medicine and Pathology, Masonic Cancer Center, University of Minnesota, Minneapolis, MN, USA

Ashleigh McClatchey (Chapter 7) Centre for Cancer Research and Cell Biology, Queen's University Belfast, Belfast, Northern Ireland

Suzanne McFarlane (Chapter 7) Centre for Cancer Research and Cell Biology, Queen's University Belfast, Belfast, Northern Ireland

A. Dusty Miller (Chapter 21) Fred Hutchinson Cancer Research Center, Seattle, WA, USA

Nicola Montgomery (Chapter 7) Centre for Cancer Research and Cell Biology, Queen's University Belfast, Belfast, Northern Ireland

David Naor (Chapter 8) The Lautenberg Center for General and Tumor Immunology, The Hebrew University, Hadassah Medical School, Jerusalem, Israel

Risto Pirinen (Chapter 14) Institute of Clinical Medicine, Pathology and Forensic Medicine, University of Kuopio, and Department of Pathology, Kuopio University Hospital, Kuopio, Finland

Virginia M. Platt (Chapter 20) Joint Graduate Group in Bioengineering at the University of California Berkeley and the University of California San Francisco, San Francisco, CA, USA

Ana M. Schor (Chapter 15) Unit of Cell and Molecular Biology, The Dental School, University of Dundee, Dundee, Scotland, UK

Seth L. Schor (Chapter 15) Unit of Cell and Molecular Biology, The Dental School, University of Dundee, Dundee, Scotland, UK

Marie G. Selzer (Chapter 11) Department of Urology, Miller School of Medicine, University of Miami, Miami, FL, USA

Jan C. Simon (Chapter 17) Department of Dermatology, Venerology and Allergology, University of Leipzig, Leipzig, Germany

Melanie A. Simpson (Chapter 16) Department of Biochemistry, University of Nebraska, Lincoln, NE, USA

Ronit Vogt Sionov (Chapter 8) The Lautenberg Center for General and Tumor Immunology, The Hebrew University- Hadassah Medical School, Jerusalem, Israel

Spyros S. Skandalis (Chapter 3) Ludwig Institute for Cancer Research, Uppsala University, Biomedical Center, Uppsala, Sweden

Mark G. Slomiany (Chapter 2) Department of Cell Biology and Anatomy, College of Medicine, Medical University of South Carolina, Charleston, SC, USA

Ivan Stamenkovic (Chapter 5) Division of Experimental Pathology, Institute of Pathology, University of Lausanne and CHUV, Lausanne, Switzerland

Robert Stern (Chapters 1 and 12) Department of Pathology, Faculty of Medicine, Al Quds University, Abu-Dies, East Jerusalem 20002, Palestinian Territory

Kazuki N. Sugahara (Chapter 13) Vascular Mapping Center, Burnham Institute for Medical Research at UCSB, University of California, Santa Barbara, CA, USA

Francis C. Szoka, Jr. (Chapter 20) Department of Biopharmaceutical Sciences and Pharmaceutical Chemistry, University of California San Francisco, San Francisco, CA, USA

Markku I. Tammi (Chapter 14) Department of Anatomy, University of Kuopio, Kuopio, Finland

Raija H. Tammi (Chapter 14) Department of Anatomy, University of Kuopio, Kuopio, Finland

Natalie K. Thomas (Chapter 18) Department of Biochemistry and Molecular Biology, Monash University, Melbourne, Australia

Bryan P. Toole (Chapter 2) Department of Cell Biology and Anatomy, College of Medicine, Medical University of South Carolina, Charleston, SC, USA

Eva A. Turley (Chapter 9) Department of Biochemistry, London Regional Cancer Program, University of Western Ontario, London, ON, Canada

Shulamit B. Wallach-Dayan (Chapter 8) The Lautenberg Center for General and Tumor Immunology, The Hebrew University- Hadassah Medical School, Jerusalem, Israel

David J.J. Waugh (Chapter 7) Reader, Centre for Cancer Research and Cell Biology, Queen's University Belfast, Belfast, Northern Ireland

Anne-Marie Woolston (Chapter 15) Unit of Cell and Molecular Biology, The Dental School, University of Dundee, Dundee, Scotland, UK

Qin Yu (Chapter 5) Department of Oncological Sciences, Mount Sinai School of Medicine, New York, NY, USA

Muayad A. Zahalka (Chapter 8) The Lautenberg Center for General and Tumor Immunology, The Hebrew University- Hadassah Medical School, Jerusalem, Israel

SECTION I

HISTORICAL OVERVIEW

CHAPTER

1

Association Between Cancer and "Acid Mucopolysaccharides": An Old Concept Comes of Age, Finally

Robert Stern

INTRODUCTION

The influence of hyaluronan (HA) on cancer progression has been exceedingly well described (Toole, 2002; Toole et al., 2002; Toole and Hascall, 2002; Stern, 2005). However, recognition of this important phenomenon has lagged, and inexplicably, continues to be neglected by most cancer biologists. Knowledge in this area has advanced extremely rapidly, and has taken on additional significance, now that it is documented that the major receptor for HA, CD44, is expressed on the surface of virtually all stem cells, including cancer stem cells (e.g., Al Hajj et al., 2003). This volume aims to bring attention to the field of HA and its role in cancer initiation, progression, and spread.

Assembly of these reviews is now particularly timely. It is the first volume ever to appear dedicated entirely to the role of HA in cancer biology. A recent textbook on basic oncology, widely recognized to be of superior quality, does not have a single citation in the index for HA (Weinberg, 2006). CD44 is given one citation, without mentioning that it is the predominant receptor for HA. Ironically, even the Weinberg laboratory has since then become aware of the significance of HA and CD44 in cancer progression (Godar et al., 2008).

Our purpose here is to draw attention to a critical molecule that that has been neglected, and up until now, poorly understood by most cancer scientists. The time has come, finally, to bring HA, previously known as hyaluronic acid (Balazs et al., 1986), and before that, as simply an acid mucopolysaccharide, to the attention of a wider audience.

HYALURONAN

Historical Perspective

The term "ground substance" was first applied to the amorphous material between cells by the German anatomist, Henle, in 1841 (Henle, 1841). It is a mistranslation of the German "Grundsubstanz," which would be better translated as "basic," "fundamental," or "primordial" substance. By 1852, sufficient information had accrued for the inclusion of "Grundsubstanz" in a textbook of human histology (Koellicker, 1852).

The modern era of ground substance research began in 1928 with the discovery of a "spreading factor" by Francisco Duran-Reynals. Testicular extracts stimulated rapid spread of materials injected subcutaneously on the backs of shaved rabbits, while simultaneously causing dissolution of the ground substance (Duran-Reynals, 1928; 1929; Duran-Reynals and Suner Pi, 1929; Duran-Reynals and Stewart, 1933). The active principal of these extracts was later shown to be the enzyme, hyaluronidase (Chain and Duthrie, 1940; Hobby et al., 1941), the class of enzymes that degrade HA. Interestingly, in one of the studies by Duran-Reynals, hyaluronidase-like activity was demonstrated in extracts of human malignancies, particularly from breast cancers and malignant melanoma (Duran-Reynals et al., 1929).

"Ground substance" was subsequently renamed "acid mucopolysaccharides," a term first proposed by Karl Meyer (1938), who first described HA (Meyer and Palmer, 1934; 1936). This was the term to designate the hexosamine-containing sugar polymers that occurred in animal tissues alone, as well as when bound to proteins. Chondroitin sulfate is the major GAG of the matrix of such tissues as cartilage, tendon, and scar. However, it is now well established that HA is by far the predominant "acid mucopolysaccharide" that constitutes true "ground substance," though heparan sulfate is the most abundant GAG at the cell surface.

Overview

Hyaluronan is a high-molar-mass linear glycosaminoglycan (GAG) found intracellularly, on the surface of cells, but predominantly in the extracellular matrix (ECM) between cells. This linear polysaccharide can reach a size of 6 to 8 MDa. It is a ubiquitous polymer with the repeating disaccharide structure of $(-\beta1,3-N$-acetyl-D-glucosamine-$\beta1,4$-D-glucuronic acid-$)_n$. It has one carboxyl group per disaccharide repeating unit, and is therefore a polyelectrolyte with a negative charge at neutral pH. It is near perfect in chemical repeats, with no known deviations in its simple disaccharide structure with the possible exception of occasional deacetylated glucosamine residues.

Hyaluronan, at low concentrations, is ubiquitous. However, it is found in high concentrations during embryogenesis, and whenever rapid tissue turnover and repair are occurring. It occurs in particularly high concentrations in fetal tissues, in amniotic fluid, is the major constituent of fetal structures such as Wharton's jelly of the umbilical cord, but also in malignancies. Over 50% of total body HA occurs in the skin (Reed et al., 1988).

At the cellular level, a burst of HA synthesis occurs just prior to mitosis, enabling some cells to become dissociated from neighboring cells and to lose the adhesion from their surrounding ECM in preparation for division (Toole et al., 1972; Tomida et al., 1974; Mian, 1986; Brecht et al., 1986). It is during this short period within the cell cycle that normal cells most closely resemble transformed cells. The deposition of HA preceding mitosis promotes detachment, and also confers motility directly upon cells (Turley and Torrance, 1984; Turley et al., 1985), correlating possibly with the movement of metastatic tumor cells.

Cancer cells do not do unusual things, but do usual things at unusual times. The formulation can be posited that cancer cells emulate that point in the cell cycle when cells synthesize increased levels of HA, round up, detach from their substratum, and leave temporarily the social contract in order to divide. Normal cells then degrade that HA in order to reattach to the substratum and to carry on the business of being normal tissue components. Cancer cells have learned to eliminate this step, to retain their HA coat, enabling them instead, to continue to divide endlessly (Itano et al., 2002).

Hyaluronan Can Influence Cell Fate: Studies from Embryology

Classical studies in embryogenesis document that HA is ubiquitous in developmental processes and in tissue modeling. Hyaluronan is particularly prevalent when undifferentiated cells are proliferating rapidly and move from their stem cell niche to the site of organ development. This stage of cell proliferation and movement ends when cells commit to a program of differentiation. In fact, the HA environment actively inhibits differentiation, creating instead an environment that promotes proliferation (Ozzello et al., 1960). Cells must lose their HA-rich environment in order for that commitment to differentiation to occur (Toole, 1991). Such a series of events were demonstrated for limb development, as well as cornea, the neural tube, cartilage and muscle development, and branching morphogenesis of parenchymal organs (Bernfield and Banerjee, 1972; Gakunga et al., 1997). Neuroectoderm pinches off to become neural crest elements, which then wander through the vertebrate body in an HA-rich environment. Such movement ceases just as HA becomes degraded (Pratt et al., 1975).

Again, parallels can be drawn between this window of normal tissue development and the onset of tumor growth, when cancer cells move and proliferate. Normal proliferating cells shed their HA through hyaluronidase activity. In most cases, it may be the failure to remove the HA coat, or the continuous turnover and replacement that promotes, malignant cell growth and the development of cancer.

Early studies of the influence of an HA environment on cell fate were from the laboratory of Arnold Caplan (Kujawa et al., 1986b). Primitive myoblasts derived from chick embryo skeletal muscle plated on plastic will proliferate, fuse to form a syncytium, and will begin to synthesize actin and myosin, and even begin to have contractile activity. However, the same cells grown on an HA-covered dish will grow and proliferate, but will not fuse, will not express skeletal muscle actin or myosin, nor show contractile behavior.

An effect chain length was demonstrated for this phenomenon (Kujawa et al., 1986a), one of the first demonstration of size dependency for HA polymers. Similar results have been shown for chondrogenesis; addition of small amounts of HA inhibit formation of cartilage nodules (Toole et al., 1972).

Cancer Is a Price Paid for Metazoan Evolution

Hyaluronan synthesis occurred relatively late in metazoan evolution. The primitive nematode *C. elegans* contains chondroitin, and no HA (Yamada et al., 1999; Toyoda et al., 2000; Hwang et al., 2003). It can be postulated that HA developed late in evolution, at a time when stem cells had to move from their original niche, and travel some distance to another body site for growth, proliferation, and differentiation. It is precisely this fragment of metazoan biology that may have been commandeered by malignant cells. Emergence of HA may parallel the step in evolution when malignancies first arose.

The difference between chondroitin and HA is the epimerization of one 4-hydroxyl group, resulting in an *N*-acetylglucosamine from the original *N*-acetylgalactosamine. The galactose moiety is widely utilized in immune recognition in higher organisms. The axial hydroxyl group when it is epimerized becomes covert and unavailable for recognition. The HA chain is thus able to avoid the primitive immune-like surveillance system, enabling stem cells to move through the metazoan organism without recognition. This may explain how HA became a "stealth" molecule (Lee and Spicer, 2000), and why it was necessary in evolution.

STROMAL–EPITHELIAL INTERACTION IN CANCER

Extracellular Matrix of Normal Cells

The ECM is a heterogeneous mixture of proteins and proteoglycans that surrounds and separates cells, supports their structure and their organization in tissues. It contains myriads of smaller molecules including growth factors, adhesion molecules and a host of other small moieties, and controls their presentation to cells. Hyaluronan and other GAGs, most of which are covalently bound to proteins, influence the behavior of malignant cells by virtue of their expansive configuration, regulate basic processes such as proliferation, recognition, modulation of adhesion, and cell–cell communication. In addition to being supportive structures, they create links intracellular and extracellular environment, are involved in transduction of key intracellular biological signals. They act as receptors, co-receptors, and catalyze profound changes that lead to the malignant phenotype.

The Stroma Around Tumors Is Highly Abnormal, but Tends to Resemble Embryonic Mesenchyme

The nature of the tumors' abilities to commandeer stromal cells to their own agenda is just now beginning to be understood. An association between the stromal elements surrounding malignant tumors, and unusual histochemical features has been noted for over a century (Kuru, 1909). These include a marked increase in the deposition of "acid mucopolysaccharides," and a hyaluronidase-sensitive metachromasia. Such observations have been routinely made by surgical pathologists over the decades during examination of cancerous tissues but without sufficient fanfare. This enriched acid mucopolysaccharide deposition in the stroma surrounding malignant tumors, was long ago predicted to be the product of the peritumor fibroblasts, rather than from the cancer cells themselves (Grossfeld et al., 1955; Ozello and Speer, 1958).

It soon became apparent that the stroma that surrounded cancers was not entirely abnormal. The peritumor stroma tended to resemble fetal or embryonic fibroblasts, more than normal adult fibroblasts, as documented in the pioneering work of Seth and Ana Schor (Schor et al., 1989; Chen et al., 1989; Gray et al., 1989; Schor and Schor, this volume). This again underlines the concept that cancers do not always do unusual things. Sometimes, they do usual things at unusual times.

Clinically, striking increases of HA in the serum of many cancer patients has also been well documented (Manley and Warren, 1987; Wilkinson et al., 1996). This suggests that the increased deposition of "acid mucopolysaccharides" or HA in cancerous tissue is not a local phenomenon, but may have had wide spread consequences.

Mechanisms for Peritumor Stromal Abnormalities

There are a number of mechanisms that can be invoked for the differences observed between normal and peritumor stroma, or how it was that malignant cells were able to influence their surrounding stroma. Some of these mechanisms could not have been conceivable when such differences were first documented. Tumor cells may commandeer a small subpopulation of stromal cells to expand, and to become the predominant population. Tumor cells may be able to recruit cells from the bone marrow, to take up residence in and around the tumor cell population. The purpose of such stromal cells is to provide growth factors, and perhaps those very growth factors that are provided in a fetal-like environment, providing an environment conducive to angiogenesis, growth and remodeling. It is likely that a combination of these two scenarios attend human malignancies in their Darwinian drive to survive, grow, and spread.

HYALURONAN IN CANCER

Malignancies Have Increased Hyaluronan

It is now widely recognized that HA is dramatically increased in most malignancies. This increase in HA correlates with tumor virulence, and is often used as a prognostic indicator. But such observations were made in number of experimental systems, long before it was appreciated in human cancers.

Among the earliest observations on HA in animal tumors was by Elvin Kabat in 1939 (Kabat, 1939), who later went on to make major contributions to immunology and to the chemistry of blood group substances. In those early studies, he demonstrated that the "mucinous substance" associated with Rous sarcomas in the chicken was identical to the material that had been characterized by Karl Meyer. The same material was then demonstrated to be produced in cultures of Rous sarcoma cells (Grossfeld, 1962).

Following infection of avian cells with the sarcoma virus, there is a five-fold increase in the HA-synthases, the enzymes that synthesize HA (Ishimoto et al., 1966), causing a dramatic stimulation of HA deposition. Infection with other oncogenic viruses also caused enormous increases in rates of HA production, as well as abnormal acceleration of cell growth (Hamerman et al., 1965). Treatment with tumor promoters stimulated HA synthesis as well (Ulrich and Hawkes, 1983).

The HA isolated from such transformed cells has the ability to stimulate proliferation of growth-retarded, non-transformed cells (Henrich and Hawkes, 1989). The constitutive HA synthesized by non-transformed cells does not possess this property, an ability attributed to the size difference between the two classes of HA polymers (Stern et al., 2006; Sugahara et al., 2006).

Hyaluronan was demonstrated in a number of other experimental animal model tumors, including the rat Walker carcinoma (Fiszer-Szafarz and Gallino, 1970). Another was the rabbit V_2 carcinoma (Toole et al., 1979), one of the earliest studies to demonstrate a direct relationship between HA and invasive tumor growth. Aggressiveness of other murine tumors were subsequently shown to correlate with HA content (Knudson et al., 1984). Increased levels of HA correlate with high metastatic potential in variants of mouse mammary carcinoma cells (Angello et al., 1982a; b; Kimata et al., 1983).

In human mammary cell culture systems, highly aggressive breast cancer cell lines such as MDA-MB-231 synthesize greater amounts of HA than the much less virulent cell line MCF-7. But, in addition, the HA synthesized by the breast cancer lines remains cell-associated while normal breast epithelial cells secrete most of their HA into the medium (Chandrasekaran and Davis, 1979).

Hyaluronan is produced not only by cancer cells, but production can be induced by the tumor cells in their surrounding stromal cells. In human cancers, levels of HA often correlate inversely with prognosis (Ropponen, 1998; Auvinen et al., 2000; Anttila et al., 2000). But, as a practicing anatomic pathologist, when staining for HA in human breast cancers, it is apparent that patterns differ widely (Stern, R., unpublished). Some tumors have abundant HA within the tumor, some within the surrounding stroma, and some with pronounced HA deposition in both tumor and stroma, while some breast malignancies show little HA deposition in either tumor or stroma. In pursuing prognostic indicators, it may be important to separate such patterns. As indicated in the opening line of Tolstoy's *Anna Karenina*, "All happy families are happy in the same way, but all unhappy families are unhappy each in their own way." The same can be said of malignancies.

Mechanisms for the Increased Hyaluronan in Malignancies

Cancer cell culture systems facilitated identification of "factors" that modulated expression of HA. Among the earliest of such observations included the ability of 17-β-estradiol and of growth hormone to stimulate production of acid mucopolysaccharides in fetal fibroblasts (Ozello, 1964). Tumor cells also secrete factors that can induce increased synthesis of HA in fibroblasts (Knudson et al., 1984a; Knudson and Pauli, 1987; Asplund et al., 1993). A similar factor occurs in both fetal serum and the serum of cancer patients (Decker et al., 1989). Some of these tumor-derived factors have become defined (Suzuki et al., 1985), while others have defied explication, despite intense efforts (Decker et al., 1989). Some of these are soluble factors, while others require cell–cell contact (Knudson et al., 1984b).

Do stromal cells become abnormal by direct contact with cancer cells? Are there soluble factors that influence such a conversion, and are such putative factors similar to fetal-derived growth factors? Do such factors influence HA production in the stromal populations induced to expand, or induced to migrate from the bone marrow by tumor cells? Again, it is likely that all of these scenarios participate in cancer growth and spread.

Cancers Are Resilient in Utilizing Hyaluronan Metabolism for Their Own Promotion

Not surprisingly, examples abound demonstrating that cancer cells have commandeered every aspect of the metabolism of HA in promoting their Darwinian quest to survive. For HA, examples of its synthesis (Kimata and Itano, this volume), receptors and related signal transduction networks (Bourguignon, this volume), fragmentation (Sugahara, this volume), degradation (Lokeshwar, this volume), and the various other strategies that cancer cells have achieved utilizing HA for their own promotion (Toole, this volume).

Anomalously, Hyaluronan Oligomers Can Inhibit Tumor Growth

An anomaly of HA cancer biology is that HA oligomers injected into cancer sites markedly inhibit tumor growth. This observation was made *in vivo* using visible skin tumors, and would be attractive for the control of tumors such as malignant melanoma (Zeng et al., 1998). *In vitro*, the HA oligomers inhibit anchorage-independent growth of several tumor cell types. They induce apoptosis and stimulate caspase-3 activity through the phosphoinositide 3-kinase/Akt cell survival pathway (Ghata et al., 2002). A possible mechanism for HA oligomers' ability to thwart tumor growth may be by competing with high molecular weight chains for HA receptors.

These observations are best understood in the context that high molecular weight HA is a reflection of intact healthy tissues, and that HA oligomers are distress signals indicating that tissue injury has occurred. The range of activities and biological functions of variably sized HA fragments have recently been reviewed (Stern et al., 2006). Some of the recent observations that hyaluronidase treatment can suppress cancer growth may well be a reflection of the HA fragments generated, rather than a direct effect of the hyaluronidase itself (Shuster et al., 2002).

The physiological effects of HA and its associated water of hydration on tumor interstitial fluid pressure, and the ability of hyaluronidase treatment to relieve such pressure adds another intriguing element to their relationship.

ABNORMALITIES IN OTHER GLYCOSAMINOGLYCANS OCCUR IN MALIGNANCY

Abnormal forms and concentrations of glycosaminoglycans other than HA have also been reported for a variety of cancers. These proteoglycans play an important role in neoplasia. One of the earliest reports described the stimulating properties of chondroitin sulfate on the growth of Ehrlich ascites tumor cells (Takeuchi, 1965).

Early studies were performed using radiolabeled [^{35}S]sulfate, comparing normal and cancerous tissues. Such experiments would not have detected changes in HA metabolism. Twelve-fold increases in the deposition of chondroitin-4 and -6 sulfate were documented in colon tumors (Iozzo et al., 1982). Histochemically, this increase occurred in the intercellular matrix of the connective tissue adjacent to the tumor. The leucine- and chondroitin sulfate-rich proteoglycan, decorin, functions as a paradigm for the profound changes that tumor matrix can exert (Iozzo et al., 1989; Iozzo and Cohen, 1994).

Heparan sulfate (HS) proteoglycans have also been implicated in tumor pathogenesis in a widespread and convincing manner (Sanderson et al., 2004). The HS proteoglycans actually comprise a wide variety but closely related family of GAGs derived from a common precursor, but varying in their glycan sequence and composition, particularly in relation to their sulfate composition. Other families of proteoglycans rich in HS side chains are characterized by having entirely different proteoglycan core proteins. These include syndecans, glypicans, and perlecan. Each of these HS proteoglycans is associated with tumor progression or suppression, or both. The reason that proteoglycan and HA biology has not been studied more carefully by cancer biologists is precisely because of the complexity of their structure and functions. Nevertheless, the HS proteoglycans and HA constitute major targets for potential anti-cancer therapeutics.

CONCLUSIONS

Hyaluronan has long been recognized to be concentrated in the areas around cancer cells. The purpose of this volume is to bring attention to this relatively neglected area of cancer cell biology. Stromal influence on malignancies is a related concept that also has not achieved the attention it deserves. And, lastly, entirely new and unpredicted directions are identified that further widen the role of hyaluronan in cancer biology (Miller, A.D., in this volume).

Hyaluronan has long been recognized as essential to cancer biology. However, critical recognition has been lacking. Investigations of the ECM have up to now been at the periphery of cell biology. A dynamic reciprocity is becoming apparent between extra- and intracellular events, the ability of the ECM to control and orchestrate that dialogue, the ability of HA and its size-specific fragments to induce signal transduction pathways by engagement of HA-specific receptors, and the recognition that HA is prominent in the peritumor ECM, and is at the core of such interactions in malignancy.

References

Al-Hajj, M., Wicha, M. S., Benito-Hernandez, A., Morrison, S. J., and Clarke, M. F. (2003). Prospective identification of tumorigenic breast cancer cells. *Proc Natl Acad Sci USA* **100**, 3983–3988.

Angello, J. C., Danielson, K. G., Anderson, L. W., and Hosick, H. L. (1982a). Glycosaminoglycan synthesis by subpopulations of epithelial cells from a mammary adenocarcinoma. *Cancer Res* **42**, 2207–2210.

Angello, J. C., Hosick, H. L., and Anderson, L. W. (1982b). Glycosaminoglycan synthesis by a cell line (C1–S1) established from a preneoplastic mouse mammary outgrowth. *Cancer Res* **42**, 4975–4978.

Anttila, M. A., Tammi, R. H., Tammi, M. I., Syrjanen, K. J., Saarikoski, S. V., and Kosma, V. M. (2000). High levels of stromal hyaluronan predict poor disease outcome in epithelial ovarian cancer. *Cancer Res* **60**, 150–155.

Auvinen, P., Tammi, R., Parkkinen, J. et al. (2000). Hyaluronan in peritumoral stroma and malignant cells associates with breast cancer spreading and predicts survival. *Am J Pathol* **156**, 529–536.

Asplund, T., Versnel, M. A., Laurent, T. C., and Heldin, P. (1993). Human mesothelioma cells produce factors that stimulate the production of hyaluronan by mesothelial cells and fibroblasts. *Cancer Res* **53**, 388–392.

Balazs, E. A., Laurent, T. C., and Jeanloz, R. W. (1986). Nomenclature of hyaluronic acid. *Biochem J* **235**, 903.

Bernfield, M. R. and Banerjee, S. D. (1972). Acid mucopolysaccharide (glycosaminoglycan) at the epithelial–mesenchymal interface of mouse embryo salivary glands. *J Cell Biol* **52**, 664–673.

Brecht, M., Mayer, U., Schlosser, E., and Prehm, P. (1986). Increased hyaluronate synthesis is required for fibroblast detachment and mitosis. *Biochem J* **239**, 445–450.

Chain, E. and Duthie, E. S. (1940). Identity of hyaluronidase and spreading factor. *Br J Exp Path* **21**, 324–338.

Chandrasekaran, E. V. and Davidson, E. A. (1979). Glycosaminoglycans of normal and malignant cultured human mammary cells. *Cancer Res* **39**, 870–880.

Chen, W. Y., Grant, M. E., Schor, A. M., and Schor, S. L. (1989). Differences between adult and foetal fibroblasts in the regulation of hyaluronate synthesis: correlation with migratory activity. *J Cell Sci* **94**, 577–584.

Decker, M., Chiu, E. S., Dollbaum, C., et al. (1989). Hyaluronic acid-stimulating activity in sera from the bovine fetus and from breast cancer patients. *Cancer Res* **49**, 3499–3505.

Duran-Reynals, F. (1928). Exaltation de l'activité du virus vaccinal par les extraits de certains organs. *CR Soc Biol* **99**, 6–7.

Duran-Reynals, F. (1929). The effects of extracts of certain organs from normal and immunized animals on the infecting power of vaccine virus. *J Exp Med* **50**, 327–340.

Duran-Reynals, F. (1933). Studies on a certain spreading factor existing in bacteria and its signficance for bacterial invasiveness. *J Exp Med* **58**, 161–181.

Duran-Reynals, F. and Stewart, F. W. (1933). The action of tumor extracts on the spread of experimental vaccinia of the rabbit. *Am J Cancer* **15**, 2790–2797.

Duran-Reynals, F. and Suner Pi, J. (1929). Exaltation de l'activité du Staphylocoque par les extraits testiculaires. *CR Soc Biol* **99**, 1908–1911.

Fiszer-Szafarz, B. and Gullino, P. M. (1970). Hyaluronic acid content of the interstitial fluid of Walker carcinoma 256. *Proc Soc Exp Biol Med* **133**, 597–600.

Gakunga, P., Frost, G., Shuster, S., Cunha, G., Formby, B., and Stern, R. (1997). Hyaluronan is a prerequisite for ductal branching morphogenesis. *Development* **124**, 3987–3997.

Ghatak, S., Misra, S., and Toole, B. P. (2002). Hyaluronan oligosaccharides inhibit anchorage-independent growth of tumor cells by suppressing the phosphoinositide 3-kinase/Akt cell survival pathway. *J Biol Chem* **277**, 38013–38020.

Godar, S., Ince, T. A., Bell, G. W., et al. (2008). Growth-inhibitory and tumor-suppressive functions of p53 depend on its repression of CD44 expression. *Cell* **134**, 62–73.

Gowda, D. C., Bhavanandan, V. P., and Davidson, E. A. (1986). Isolation and characterization of proteoglycans secreted by normal and malignant human mammary epithelial cells. *J Biol Chem* **261**, 4926–4934.

Grey, A. M., Schor, A. M., Rushton, G., Ellis, I., and Schor, S. L. (1989). Purification of the migration stimulating factor produced by fetal and breast cancer patient fibroblasts. *Proc Natl Acad Sci USA* **86**, 2438–2442.

Grossfeld, H. (1962). Production of hyaluronic acid in tissue culture of Rous sarcoma. *Nature* **196**, 782–783.

Grossfeld, H., Meyer, K., and Godman, G. (1955). Differentiation of fibroblasts in tissue culture as determined by mucopolysaccharide production. *Proc Soc Exp Biol Med* **88**, 31–35.

Hamerman, D., Todaro, G. J., and Green, H. (1965). The production of hyaluronate by spontaneously established cell lines and viral transformed lines of fibroblastic origin. *Biochim Biophys Acta* **101**, 343–351.

Henle, F. (1841). Vom Knorpelgewebe. Allgemeine Anatomielehre, von den Mischungs und Formbestandteilen des menschlichen Koerpers. Leopold Voss Verlag, Leipzig, pp. 791–799.

Henrich, C. J. and Hawkes, S. P. (1989). Molecular weight dependence of hyaluronic acid produced during oncogenic transformation. *Cancer Biochem Biophys* **10**, 257–267.

Hobby, G. L., Dawson, M. H., Meyer, K., and Chaffee, E. (1941). The relationship between spreading factor and hyaluronidase. *J Exp Med* **73**, 109–123.

Hopwood, J. J. and Dorfman, A. (1977). Glycosaminoglycan synthesis by cultured human skin fibroblasts after transformation with simian virus 40. *J Biol Chem* **252**, 4777–4785.

Hwang, H. Y., Olson, S. K., Esko, J. D., and Horvitz, H. R. (2003). *Caenorhabditis elegans* early embryogenesis and vulval morphogenesis require chondroitin biosynthesis. *Nature* **423**, 439–443.

Ishimoto, N., Temin, H. M., and Strominger, J. L. (1966). Studies of carcinogenesis by avian sarcoma viruses. II. Virus-induced increase in hyaluronic acid synthetase in chicken fibroblasts. *J Biol Chem* **241**, 2052–2057.

Iozzo, R. V., Bolender, R. P., and Wight, T. N. (1982). Proteoglycan changes in the intercellular matrix of human colon carcinoma: an integrated biochemical and stereological analysis. *Lab Invest* **47**, 124–128.

Iozzo, R. V. and Cohen, I. (1993). Altered proteoglycan gene expression and the tumor stroma. *Experientia* **49**, 447–455.

Iozzo, R. V., Sampson, P. M., and Schmitt, G. K. (1989). Neoplastic modulation of extracellular matrix: stimulation of chondroitin sulfate proteoglycan and hyaluronic acid synthesis in co-cultures of human colon carcinoma and smooth muscle cells. *J Cell Biochem* **39**, 355–378.

Iozzo, R. V. and Wight, T. N. (1982). Isolation and characterization of proteoglycans synthesized by human colon and colon carcinoma. *J Biol Chem* **257**, 11135–11144.

Itano, N., Atsumi, F., Sawai, T., et al. (2002). Abnormal accumulation of hyaluronan matrix diminishes contact inhibition of cell growth and promotes cell migration. *Proc Natl Acad Sci USA* **99**, 3609–3614.

Kabat, E. A. (1939). A polysaccharide in tumors due to a virus of leucosis and sarcoma of fowls. *J Biol Chem* **130**, 143–147.

Kimata, K., Honma, Y., Okayama, M., Oguri, K., Hozumi, M., and Suzuki, S. (1983). Increased synthesis of hyaluronic acid by mouse mammary carcinoma cell variants with high metastatic potential. *Cancer Res* **43**, 1347–1354.

Knudson, W., Biswas, C., and Toole, B. P. (1984a). Stimulation of glycosaminoglycan production in murine tumors. *J Cell Biochem* **25**, 183–196.

Knudson, W., Biswas, C., and Toole, B. P. (1984b). Interactions between human tumor cells and fibroblasts stimulate hyaluronate synthesis. *Proc Natl Acad Sci USA* **81**, 6767–6771.

Koellicker, A. (1852). Von den Geweben. Handbuch der Gewebelehre des Menschen. Wilhelm Engelmann Verlag, Leipzig, pp. 51–89.

Kujawa, M. J., Carrino, D. A., and Caplan, A. I. (1986a). Substrate-bonded hyaluronic acid exhibits a size-dependent stimulation of chondrogenic differentiation of stage 24 limb mesenchymal cells in culture. *Dev Biol* **114**, 519–528.

Kujawa, M. J., Pechak, D. G., Fiszman, M. Y., and Caplan, A. I. (1986b). Hyaluronic acid bonded to cell culture surfaces inhibits the program of myogenesis. *Dev Biol* **113**, 10–16.

Kuru, H. (1909). Studies in tumorigenesis: examples from breast tumor pathology, with a particular emphasis on the malignant transformation of fibroadenomas. *Deutsche Ztschr Chir* **98**, 415–463.

Lee, J. Y. and Spicer, A. P. (2000). Hyaluronan: a multifunctional, megaDalton, stealth molecule. *Curr Opin Cell Biol* **12**, 581–586.

Manley, G. and Warren, C. (1987). Serum hyaluronic acid in patients with disseminated neoplasm. *J Clin Pathol* **40**, 626–630.

Meyer, K. (1938). The chemistry and biology of mucopolysaccharides and glycoproteins. *Sympos Quant Biol* **6**, 91–118.

Meyer, K. and Palmer, J. W. (1934). The polysaccharide of the vitreous humor. *J Biol Chem* **107**, 629–634.

Meyer, K. and Palmer, J. W. (1936). On glycoproteins II. The polysaccharides of vitreous humor and of umbilical cord. *J Biol Chem* **114**, 689–703.

Mian, N. (1986). Analysis of cell-growth-phase-related variations in hyaluronate synthase activity of isolated plasma-membrane fractions of cultured human skin fibroblasts. *Biochem J* **237**, 333–342.

Ozello, L. (1964). Effect of 17-beta-estradiol and growth hormone on the production of acid mucopolysaccharides by human embryonal fibroblasts in vitro. *J Cell Biol* **21**, 283–286.

Ozzello, L., Lasfargues, E. Y., and Murray, M. R. (1960). Growth-promoting activity of acid mucopolysaccharides on a strain of human mammary carcinoma cells. *Cancer Res* **20**, 600–604.

Ozzello, L. and Speer, F. D. (1958). The mucopolysaccharides in the normal and diseased breast: their distribution and significance. *Am J Path* **34**, 993–1009.

Pauli, B. U. and Knudson, W. (1988). Tumor invasion: a consequence of destructive and compositional matrix alterations. *Hum Pathol* **19**, 628–639.

Pratt, R. M., Larsen, M. A., and Johnston, M. C. (1975). Migration of cranial neural crest cells in a cell-free hyaluronate-rich matrix. *Dev Biol* **44**, 298–305.

Reed, R. K., Lilja, K., and Laurent, T. C. (1988). Hyaluronan in the rat with special reference to the skin. *Acta Physiol Scand* **134**, 405–411.

Ropponen, K., Tammi, M., Parkkinen, J., et al. (1998). Tumor cell-associated hyaluronan as an unfavorable prognostic factor in colorectal cancer. *Cancer Res* **58**, 342–347.

Sanderson, R. D., Yang, Y., Suva, L. J., and Kelly, T. (2004). Heparan sulfate proteoglycans and heparanase – partners in osteolytic tumor growth and metastasis. *Matrix Biol* **23**, 341–352.

Schor, S. L., Schor, A. M., Grey, A. M., Chen, J., Rushton, G., Grant, M. E., and Ellis, I. (1989). Mechanism of action of the migration stimulating factor produced by fetal and cancer patient fibroblasts: effect on hyaluronic and synthesis. *Vitro Cell Dev Biol* **25**, 737–746.

Shuster, S., Frost, G. I., Csoka, A. B., Formby, B., and Stern, R. (2002). Hyaluronidase reduces human breast cancer xenografts in SCID mice. *Int J Cancer* **102**, 192–197.

Stern, R. (2005). Hyaluronan metabolism: a major paradox in cancer biology. *Pathol Biol (Paris)* **53**, 372–382.

Stern, R., Asari, A. A., and Sugahara, K. N. (2006). Hyaluronan fragments: an information-rich system. *Eur J Cell Biol* **85**, 699–715.

Sugahara, K. N., Hirata, T., Hayasaka, H., Stern, R., Murai, T., and Miyasaka, M. (2006). Tumor cells enhance their own CD44 cleavage and motility by generating hyaluronan fragments. *J Biol Chem* **281**, 5861–5868.

Suzuki, M., Asplund, T., Yamashita, H., Heldin, C. H., and Heldin, P. (1995). Stimulation of hyaluronan biosynthesis by platelet-derived growth factor-BB and transforming growth factor-beta 1 involves activation of protein kinase C. *Biochem J* **307**, 817–821.

Takeuchi, J. (1965). Growth-promoting effect of chondroitin sulphate on solid Ehrlich ascites tumour. *Nature* **207**, 537–538.

Takeuchi, J. (1966). Growth-promoting effect of acid mucopolysaccharides on Ehrlich ascites tumor. *Cancer Res* **26**, 797–802.

Tomida, M., Koyama, H., and Ono, T. (1974). Hyaluronate synthetase in cultured mammalian cells producing hyaluronic acid: oscillatory changes during growth phase and suppression by 5-bromodeoxyuridine. *Biochim Biophys Acta* **338**, 352–363.

Toole, B.P. (1991). Proteoglycans and hyaluronan in morphogenesis and differentiation. In: E.D. Hay (Ed.), Cell Biology of Extracellular Matrix, 2nd edn., Plenum Press, New York, pp. 305–341.

Toole, B. P. (2002). Hyaluronan promotes the malignant phenotype. *Glycobiology* **12**, 37R–42R.

Toole, B. P., Biswas, C., and Gross, J. (1979). Hyaluronate and invasiveness of the rabbit V2 carcinoma. *Proc Natl Acad Sci USA* **76**, 6299–6303.

Toole, B. P. and Hascall, V. C. (2002). Hyaluronan and tumor growth. *Am J Pathol* **161**, 745–747.

Toole, B. P., Jackson, G., and Gross, J. (1972). Hyaluronate in morphogenesis: inhibition of chondrogenesis in vitro. *Proc Natl Acad Sci USA* **69**, 1384–1386.

Toole, B. P., Wight, T. N., and Tammi, M. I. (2002). Hyaluronan–cell interactions in cancer and vascular disease. *J Biol Chem* **277**, 4593–4596.

Toyoda, H., Kinoshita-Toyoda, A., and Selleck, S. B. (2000). Structural analysis of glycosaminoglycans in *Drosophila* and *Caenorhabditis elegans* and demonstration that tout-velu, a *Drosophila* gene related to EXT tumor suppressors, affects heparan sulfate in vivo. *J Biol Chem* **275**, 2269–2275.

Turley, E. A., Bowman, P., and Kytryk, M. A. (1985). Effects of hyaluronate and hyaluronate binding proteins on cell motile and contact behaviour. *J Cell Sci* **78**, 1331–1345.

Turley, E. A. and Torrance, J. (1985). Localization of hyaluronate and hyaluronate-binding protein on motile and non-motile fibroblasts. *Exp Cell Res* **161**, 17–28.

Ullrich, S. J. and Hawkes, S. P. (1983). The effect of the tumor promoter, phorbol myristate acetate (PMA), on hyaluronic acid (HA) synthesis by chicken embryo fibroblasts. *Exp Cell Res* **148**, 377–386.

Weinberg, R. A. (2007). The Biology of Cancer. Garland Science, New York.

Wilkinson, C. R., Bower, L. M., and Warren, C. (1996). The relationship between hyaluronidase activity and hyaluronic acid concentration in sera from normal controls and from patients with disseminated neoplasm. *Clin Chim Acta* **256**, 165–173.

Yamada, S., Van Die, I., Van den Eijnden, D. H., Yokota, A., Kitagawa, H., and Sugahara, K. (1999). Demonstration of glycosaminoglycans in *Caenorhabditis elegans*. *FEBS Lett* **459**, 327–331.

Zeng, C., Toole, B. P., Kinney, S. D., Kuo, J. W., and Stamenkovic, I. (1998). Inhibition of tumor growth in vivo by hyaluronan oligomers. *Int J Cancer* **77**, 396–401.

Zhang, L., Underhill, C. B., and Chen, L. (1995). Hyaluronan on the surface of tumor cells is correlated with metastatic behavior. *Cancer Res* **55**, 428–433.

CELL BIOLOGY OF HYALURONAN IN CANCER

CHAPTER

2

Hyaluronan–CD44 Interactions and Chemoresistance in Cancer Cells

Mark G. Slomiany and Bryan P. Toole

INTRODUCTION

The association of high levels of hyaluronan with malignant tumors has been known for some time (Knudson et al., 1989). Hyaluronan is often enriched preferentially in the stroma that surrounds tumors rather than in parenchymal regions (Bertrand et al., 1992; Koyama et al., 2007; Tammi et al., 2008; Toole et al., 1979), most likely as a result of stromal cell–tumor cell interactions (Asplund et al., 1993; Edward et al., 2005; Knudson et al., 1984). However, hyaluronan synthesis is also increased in many malignant tumor cells themselves (Calabro et al., 2002; Kimata et al., 1983; Zhang et al., 1995) and is frequently found in direct association with tumor cells *in vivo* (Tammi et al., 2008).

Resistance of cancers to various chemotherapeutic agents, i.e. multi-drug resistance, can arise in numerous ways, e.g. decreased uptake of drugs due to cell and tissue barriers, activation of repair and detoxification mechanisms, altered metabolic phenotype, increased activities of cell survival/anti-apoptotic signaling pathways, or enhanced drug efflux via cell membrane transporters of the ATP-binding cassette (ABC) family (Cheng et al., 2005; Dai and Grant, 2007; Gottesman et al., 2002; Li and Dalton, 2006; Tredan et al., 2007). A relatively new paradigm is the likely contribution of cancer stem-like cells to chemoresistance since these cells are highly enriched in ABC-family drug transporters (Dean et al., 2005).

In recent years, hyaluronan and CD44 have been shown to influence drug resistance at several of these different levels, namely, cell survival signaling pathways, drug transporter expression and activity, glycolytic phenotype, and cancer stem-like cell characteristics. These advances are discussed in this review.

HYALURONAN, CD44, AND DRUG RESISTANCE

The possibility that hyaluronan might influence drug resistance was suggested in the findings that local hyaluronidase treatment enhances the action of various chemotherapeutic agents *in vivo* (Baumgartner et al., 1998), and that hyaluronidase-induced dispersion of drug-resistant, multicellular, tumor cell spheroids reverses their drug resistance (Kerbel et al., 1996; St Croix et al., 1998). The mechanistic action of hyaluronidase on drug resistance was not understood at the time of these studies, but was usually explained in terms of possible effects on cell adhesion barriers (Kerbel et al., 1996) or drug penetration (Baumgartner et al., 1998; Desoize and Jardillier, 2000) rather than hyaluronan-specific effects on signaling pathways. Initial studies by our laboratory showed that calcium-independent aggregation of transformed cells can be due to hyaluronan-mediated, multivalent cross-bridging of receptors on adjacent cells (Underhill and Toole, 1981). This observation and the finding that hyaluronan–receptor interactions regulate cell survival signaling path-ways known to be important in drug resistance (Ghatak et al., 2002) led our group and others to further investigate the possible role of hyaluronan in multidrug resistance.

Employing a drug-resistant human carcinoma cell line, we demon-strated that disruption of endogenous hyaluronan-induced signaling by treatment with small hyaluronan oligomers suppresses resistance to several drugs, including doxorubicin, taxol, vincristine, and methotrexate

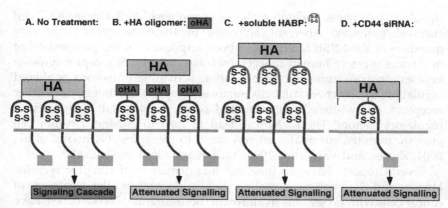

FIGURE 2.1 Antagonists of hyaluronan–CD44 interactions. Other antagonists include blocking antibodies against CD44 (Guo et al., 1994) and inhibitors of hyaluronan synthesis (Simpson et al., 2002). From Toole and Slomiany (2008). HA, hyaluronan; HABP, HA binding protein.

(Misra et al., 2003). Other antagonists of hyaluronan–CD44 signaling (Fig. 2.1) had similar effects. In addition, we showed that increased hyaluronan production, induced by over-expression of the hyaluronan synthase, HAS-2, caused increased drug resistance in the relatively chemosensitive MCF-7 breast cancer cell line. This increased resistance was reversed by treatment with hyaluronan oligomers or other antagonists of hyaluronan–CD44 signaling (Misra et al., 2003; 2005). It should be noted that the resistant cell line used in these studies was the so-called MCF-7/Adr human breast cancer cell. However, it has now been shown that this cell line is actually a drug-resistant ovarian carcinoma line, specifically OVCAR-9 (Liscovitch and Ravid, 2007). Subsequently, studies from other laboratories have similarly shown that hyaluronan promotes resistance to cisplatin, methotrexate, doxorubicin and etoposide in head and neck squamous carcinoma cells (Wang and Bourguignon, 2006; Wang et al., 2007), to cisplatin in non-small cell lung cancer cells (Ohashi et al., 2007), and to vincristine in lymphoma cells (Cordo Russo et al., 2008).

CELL-AUTONOMOUS REGULATION OF CELL SURVIVAL SIGNALING PATHWAYS BY HYALURONAN AND CD44

Elevated levels of cell survival/anti-apoptotic pathways, a common occurrence in cancer cells, are major factors contributing to drug resistance (Cheng et al., 2005; Dai and Grant, 2007). Receptor tyrosine kinases are a class of plasma membrane receptors that bind various regulatory factors, such as EGF, IGF, HGF, and PDGF, and activate several intracellular signaling pathways, including the phosphoinositide 3-kinase/AKT cell survival pathway. Aberrant activities of these receptors, especially members of the ERBB family, have been implicated in the progression of numerous types of human cancers. Increased activity of receptor tyrosine kinases can arise from gene amplification, activating mutations or altered regulation, e.g. by cross-talk between these receptors and integrins or other receptors, or by altered autocrine and paracrine stimulation by various regulatory factors. These changes lead in turn to enhanced tumor cell growth, motility, survival, and resistance to therapies (Gschwind et al., 2004; Krause and Van Etten, 2005; Yarden and Sliwkowski, 2001).

Several reports have documented augmentation of receptor tyrosine kinases and downstream signaling pathway activities after treatment of cancer cells with exogenous hyaluronan (Bourguignon, 2008a; b). We have demonstrated that manipulation of constitutive hyaluronan production and interactions in cancer cell themselves has profound effects on these pathways. Consequently, we have shown that constitutively high levels of active, i.e. autophosphorylated, ERBB2 in carcinoma cells are dependent on endogenous hyaluronan–CD44 interaction and that experimentally increased hyaluronan production causes sustained, elevated ERBB2 phosphorylation in cells that normally exhibit low levels of ERBB2 activity (Ghatak et al., 2005). Furthermore, stimulation of hyaluronan production induces assembly of a constitutive, lipid raft-associated, signaling complex containing phosphorylated ERBB2, CD44, ezrin, phosphoinositide 3-kinase, and the chaperone molecules, HSP90 and CDC37; inhibition of endogenous hyaluronan–receptor interactions causes disassembly of this complex. Antagonists of hyaluronan interactions used in these studies included hyaluronan oligomers, soluble hyaluronan binding proteins and siRNA against CD44 (see Fig. 2.1), all of which caused disassembly of this complex and inactivation of ERBB2 (Ghatak et al., 2005). Recent work in our lab (unpublished results) shows that hyaluronan oligomers also cause rapid internalization of ERBB2 and CD44. In addition, similar influences of constitutive hyaluronan–CD44 interaction may occur with other receptor tyrosine kinases, i.e. EGFR, IGF-1R, PDGFR, and c-MET (Misra et al., 2006), and corresponding effects have been shown for downstream anti-apoptotic and proliferation pathways that are known to be regulated by these receptor kinases and to be important oncogenic pathways in several cancers. For example, increased hyaluronan production stimulates the phosphoinositide 3-kinase, MAP kinase and COX-2 pathways whereas antagonists of hyaluronan interactions suppress these pathways (Ghatak et al., 2002; Misra et al., 2003; Misra et al., 2008). Importantly, interactions between CD44 and receptor tyrosine kinases may lead to very different outcomes in normal versus cancer cells and at different pericellular hyaluronan concentrations (Li et al., 2006; Li et al., 2007c).

HYALURONAN, CD44, AND DRUG TRANSPORTERS

In addition to the oncogenic, anti-apoptotic activities of the phospho-inositide 3-kinase/AKT signaling pathway, mentioned above, this pathway also induces increased expression of ABC family multidrug transporters, such as P-glycoprotein (MDR1/ABCB1), multidrug resistance-associated protein-1 (MRP-1/ABCC1) and breast cancer resistance protein (BCRP/ABCG2) (Lee et al., 2004; Misra et al., 2005; Mogi et al., 2003). We have demonstrated that constitutive hyaluronan–CD44 interaction regulates expression of the ABC family drug transporters, P-glycoprotein and BCRP, in carcinoma and glioma cells, respectively, most likely via receptor tyrosine kinase-mediated activation of the phos-phoinositide 3-kinase/AKT pathway (Gilg et al., 2008; Misra et al., 2005).

Hyaluronan–CD44 interaction may also stabilize drug transporters at the plasma membrane. We have found that CD44 co-localizes with P-glycoprotein and BCRP in the plasma membrane of cancer cells and that treatment of these cells with an antagonist of hyaluronan interactions, viz. hyaluronan oligomers, rapidly induces internalization of CD44 and the transporters into the cell and inhibits drug efflux (Fig. 2.2) (unpublished results). Others have also shown that hyaluronan and CD44 influence transporter expression and activity, as well as malignant cell properties (Cordo Russo et al., 2008; Miletti-Gonzalez et al., 2005; Ohashi et al., 2007). In a study comparing multi-drug resistant cell lines of breast, oral, and ovarian origin that express elevated levels of P-glycoprotein with their respective P-glycoprotein-negative, drug-sensitive, parental cell lines, a positive correlation was demonstrated between the expression of CD44 and P-glycoprotein. The two proteins were found to co-immunoprecipi-tate, and drugs or siRNA that interfere with the function of P-glycoprotein were shown to inhibit cell motility and invasion (Miletti-Gonzalez et al., 2005), which are properties strongly related to CD44 receptor activity (Hill et al., 2006; Tzircotis et al., 2005). In a similar way, co-immunoprecipitation and co-localization of P-glycoprotein and CD44 have been demonstrated in drug resistant melanoma cells, and the two molecules were found to cooperate in promoting invasive behavior (Colone et al., 2008). These and other observations (Dean et al., 2005; Raguz et al., 2004; Yang et al., 2003) suggest a close relationship between malignant cell properties and resistance to therapy, and the likely involvement of hyaluronan and CD44 as major factors mediating this relationship.

Confocal microscopic co-localization and fluorescence resonance energy transfer studies in NIH3T3 cells have shown that P-glycoprotein is closely associated with CD44 and other components of plasma membrane lipid microdomains, commonly known as lipid rafts (Bacso et al., 2004). It was also shown in this study that P-glycoprotein is anchored to the

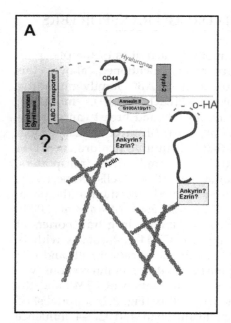
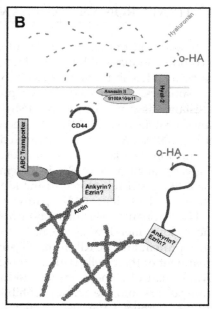

FIGURE 2.2 Internalization of drug transporters by treatment with hyaluronan oligomers. (A) In the absence of treatment, hyaluronan is tethered by CD44 at the plasma membrane whereby it stabilizes actin-linked CD44–transporter complexes in lipid microdomains. Hyaluronan is cleaved by hyaluronidase (most likely HYAL-2) and internalized via CD44 in an orderly manner (Knudson et al., 2002; Tammi et al., 2001). (B) Oligomers of hyaluronan (o-HA) stimulate CD44 internalization en masse, destabilizing transporter complexes. Inhibitors of actin assembly inhibit this process. From Toole and Slomiany (2008).

cytoskeleton. CD44 binds to the actin cytokeleton through ERM-family proteins (Tsukita et al., 1994) or ankyrin (Singleton and Bourguignon, 2004). Thus, these results suggest that CD44 resides in close molecular vicinity to P-glycoprotein and may be one of the proteins responsible for the cytoskeletal association of this transporter. Furthermore, raft localization of P-glycoprotein seems to be of functional importance since cholesterol depletion results in inhibition of transporter activity (Bacso et al., 2004). It has also been noted that drugs that interfere with P-glycoprotein can also affect localization of CD44 on the cell membrane and promote CD44 capping, and therefore might act via inhibition of actin polymerization (Miletti-Gonzalez et al., 2005). Similarly, we have seen that hyaluronan oligomer-induced internalization of CD44 and transporters is inhibited if the cells are co-treated with an inhibitor of actin polymerization, latrunculin, thus suggesting that the transporters and CD44 are anchored to actin filaments (Fig. 2.2) (unpublished results).

Hyaluronan synthesis and secretion may be directly related to drug transport since recent work suggests that hyaluronan might be secreted through multidrug transporters in vertebrate cells (Prehm and Schumacher, 2004; Schulz et al., 2007). Studies employing a battery of inhibitors as well as siRNA to sort out possible transporters involved in hyaluronan export led to the conclusion that MRP5 is the most likely hyaluronan transporter in human fibroblasts, but other transporters are probably involved in other cells (Prehm and Schumacher, 2004). Although this evidence supports a role for drug transporters in hyaluronan secretion, other studies strongly suggest that constitutive export of hyaluronan requires only the hyaluronan synthases themselves (Weigel and Deangelis, 2007). Moreover, definitive direct evidence for hyaluronan export through ABC transporters, rather than regulation by transporter activity, is lacking. Nevertheless it is likely that such export does occur at least under certain circumstances. Our findings support a close relationship between hyaluronan and drug transporters in that treatment with hyaluronan oligomers inhibits hyaluronan production or export (Misra et al., 2008) and induces rapid internalization of the drug transporters, BCRP and P-glycoprotein (unpublished results).

HYALURONAN, EMMPRIN, AND THE GLYCOLYTIC PHENOTYPE

An almost universal property of malignant cancers is increased glycolysis – the "Warburg effect." Increased glycolysis in cancer is associated with various conditions such as hypoxia, acidosis, and mitochondrial defects, which result in enhanced drug resistance and malignancy. An outcome of increased glycolysis is lactate production and extrusion across the plasma membrane via proton-coupled monocarboxylate transporters (MCTs). Lactate efflux and resultant pericellular acidification stimulate cell invasion, metastasis, and drug resistance (Gatenby and Gillies, 2004; Martinez-Zaguilan et al., 1996; Pelicano et al., 2006; Tredan et al., 2007). Two sets of observations indicate a possible relationship of hyaluronan to the glycolytic phenotype.

First, an essential partner in the activity of MCTs is emmprin (CD147) (Halestrap and Meredith, 2004), a tumor cell surface glycoprotein that also stimulates hyaluronan production (Marieb et al., 2004). Emmprin was originally identified as a factor that is expressed at high levels on the surface of malignant tumor cells and induces production of matrix metalloproteinases via cell–cell interactions (Biswas et al., 1995; Yan et al., 2005). More recent work from our lab revealed that emmprin also stimulates hyaluronan production in cancer cells and, as a consequence, induces anchorage-independent cell proliferation (Marieb et al., 2004) and drug resistance (Misra et al., 2003). Emmprin is crucial for the proper function of several monocarboxylate transporters, specifically MCT1, MCT3, and

MCT4. These MCTs require association with emmprin in the endoplasmic reticulum for trafficking to the plasma membrane and, in the absence of emmprin, are targeted for degradation (Gallagher et al., 2007; Kirk et al., 2000; Wilson et al., 2005). It has been shown that emmprin and MCT4 trafficking to the plasma membrane of breast cancer cells are mutually interdependent and that suppressed expression of MCT4 results in decreased migratory capacity in these cells, most likely due to inhibition of emmprin function (Gallagher et al., 2007).

Second, lactate stimulates hyaluronan synthesis and expression of CD44 variants in fibroblasts (Stern et al., 2002) and melanoma cells (Rudrabhatla et al., 2006), and lactate response elements are present in several hyaluronan-related genes, e.g. CD44 and the hyaluronidase, HYAL-1 (Formby and Stern, 2003).

Since emmprin–MCT interaction is required for lactate secretion and both emmprin and lactate stimulate hyaluronan production, we have examined the relationship of hyaluronan and CD44 to MCT1 and MCT4, the most commonly expressed MCTs in cancer cells. We find that CD44 interacts and co-localizes with these MCTs in the plasma membrane of breast cancer cells (Slomiany et al., 2008). Furthermore, treatment of the cells with hyaluronan oligomers leads to their internalization and to attenuation of their function, in similar fashion to our findings with RTKs and drug transporters mentioned above. Other investigators have documented a similar relationship between hyaluronan–CD44 interaction and the Na(+)-H(+) exchanger 1 (Bourguignon et al., 2004).

HYALURONAN, CANCER STEM CELLS, AND DRUG RESISTANCE

Recent evidence suggests that malignant tumors contain sub-populations of stem-like cells, termed: "cancer stem cells," "cancer progenitor cells", or "tumor-initiating cells," that are highly malignant and resistant to therapies. These cells are able to rapidly regenerate a fully grown tumor that recapitulates the heterogeneous cellular composition of the tumor of origin when implanted in small numbers in an animal host (Dalerba et al., 2007a; Hill and Perris, 2007; Vescovi et al., 2006). These cells may also comprise the metastatic sub-population of tumors (Brabletz et al., 2005; Li et al., 2007b). A striking property of these cells is their resistance to treatment with cytotoxic chemotherapeutic agents or radiation, possibly underlying the tendency of many malignant cancers to recur after treatment (Ailles and Weissman, 2007; Dean et al., 2005; Neuzil et al., 2007).

However, the cancer stem cell hypothesis remains controversial since the precise nature and origin of cancer stem-like cells have yet to be elucidated (Hill and Perris, 2007; Patrawala et al., 2005; Shipitsin et al.,

2007; Wang et al., 2008). These cells may comprise the metastatic sub-population of tumors (Brabletz et al., 2005; Li et al., 2007b) or, stated differently, may merely reflect heterogeneity within tumors with respect to various oncogenic and metastatic properties (Shipitsin et al., 2007). Nevertheless, the presence of highly malignant, therapy-resistant sub-populations within human tumors is well-established and our increased understanding of the properties of these cells is likely to yield more effective therapeutic strategies.

The hyaluronan receptor, CD44, is one of the most common markers used for isolation of cancer stem-like cells from carcinomas (Al-Hajj et al., 2003; Dalerba et al., 2007b; Li et al., 2007a; Prince et al., 2007). Recent studies indicate that CD44 is functionally important in leukemia stem cells (Jin et al., 2006; Krause et al., 2006). Other studies point to a possible role for another hyaluronan binding protein, Rhamm, in myeloma progenitors (Crainie et al., 1999; Maxwell et al., 2004); hyaluronan synthases are also altered in myeloma progenitors (Adamia et al., 2005; Calabro et al., 2002). Moreover, hyaluronan may have an important role in normal stem cell behavior (Avigdor et al., 2004; Matrosova et al., 2004; Nilsson et al., 2003; Pilarski et al., 1999). However, virtually nothing has been published on the potential role of hyaluronan–CD44 interactions in cancer stem-like cells.

Recently we have begun to examine the effects of perturbing hyaluronan interactions with hyaluronan oligomers on the malignant and therapy-resistant properties of stem-like cells isolated from cancer cell lines and from patient-derived tumors. These oligomers most likely displace constitutively bound hyaluronan polymer from their receptors, resulting in attenuation of hyaluronan-induced signaling (Ghatak et al., 2002; Ghatak et al., 2005; Misra et al., 2006). We find that hyaluronan oligomers inhibit the growth of a very aggressive stem-like sub-population of C6 glioma cells in a novel spinal cord engraftment model that replicates invasive behaviors of human gliomas in the central nervous system (Gilg et al., 2008). Furthermore, the oligomers cause increased apoptosis and decreased proliferation in these tumors. The C6 stem-like cells show elevated activation of EGFR and AKT, expression of the BCRP drug transporter and resistance to treatment with methotrexate, when compared with the C6 parental cells; these parameters were also reduced by treatment with the hyaluronan oligomers (Gilg et al., 2008), indicating the potential importance of hyaluronan in the properties of these cells.

Cancer stem-like cell sub-populations can be enriched using Hoechst dye exclusion (the "side-population": due to efflux by drug transporters, especially BCRP), cell surface markers, or spheroid formation. However, the efficacy of these different methods varies among different cancer types, and in each case the enriched preparations are heterogeneous. Depending on the method of separation and the cancer type, these cell preparations express various subsets of markers and exhibit varying degrees of malignancy and resistance to drug treatment. For example, in a study of lung cancer the side-population cells were found to exhibit stem cell-like properties such as multidrug resistance, high telomerase activity, and tumor-repopulating capacity, and consistently expressed high levels of ABC transporters, in comparison to non-side population cells (Ho et al., 2007). However, CD44 and other commonly used cancer stem cell markers were present in both side-population and non-side population cells, pointing to lack of exclusivity of the markers to the side-population in these lung cancer cells (Ho et al., 2007). In another study it was found that the side-populations from various carcinoma cells were enriched in stem-like and malignant properties, yet cancer cells lacking BCRP expression were similar in tumorigenicity to BCRP-positive cells (Patrawala et al., 2005). A common marker for identification of cancer stem-like cells is CD133 (Mizrak et al., 2008; Neuzil et al., 2007) but it has been shown recently that CD133-negative cells can also express stem-like properties (Beier et al., 2007; Wang et al., 2008). Thus, a challenge to understanding the relationship of multidrug resistance to cancer stem-like cells will be isolation of homogenous populations that can be compared and analyzed with respect to transporter expression and function, anti-apoptotic signaling pathways and their specific relationships to hyaluronan and CD44.

Despite these caveats, it is reasonable to expect that antagonists of hyaluronan–CD44 interaction, e.g. small hyaluronan oligomers, may be useful in therapeutic strategies aimed at preventing tumor recurrence from therapy-resistant sub-populations within malignant cancers.

CONCLUSIONS

The findings discussed in this review imply that hyaluronan–CD44 interactions stabilize several types of cell surface complexes that lead to numerous downstream effects, including multidrug resistance (summarized in Fig. 2.3). It is not yet clear whether these effects are due to unique involvements of specific sub-fractions of CD44 variants, to general interactions with membrane compartments such as lipid rafts, or to global effects of a hyaluronan-based pericellular matrix that influences mechano-cellular signaling interactions. With respect to the latter, it should be pointed out that constitutive cell surface hyaluronan not only tethers the pericellular matrix to the cell surface but also forms complexes with numerous other factors that may play a role in stabilizing membrane-bound complexes (Evanko et al., 2007). For example, PDGF induces formation of a pericellular matrix around smooth muscle cells that contains aggregates of hyaluronan, versican, and link protein (Evanko et al., 2001) and that is required for proliferation and migration (Evanko et al., 1999). Other components of these matrices may be TSG-6, inter-alpha trypsin

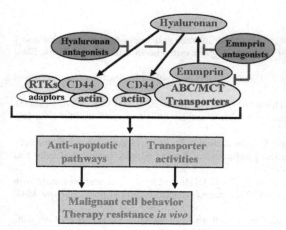

FIGURE 2.3 HA, CD44, and emmprin interactions. A scheme showing possible interactions of hyaluronan, CD44, and emmprin with receptor tyrosine kinases (RTKs), ATP-binding cassette (ABC) multidrug transporters, and monocarboxylate (lactate) transporters (MCT). The potential effects of antagonists of hyaluronan–CD44 and emmprin interactions are also shown. From Toole and Slomiany (2008).

inhibitor, tenascin, and thrombospondin-1; also, versican can be replaced by other hyaluronan binding proteoglycans such as aggrecan, neurocan, or brevican in different tissues (Evanko et al., 2007). Perturbation of pericellular matrices, either with hyaluronan oligomers or by other means, may lead to destabilization and internalization of matrix components themselves as well as associated membrane complexes. For example, protease-generated fragments of aggrecan that are complexed with hyaluronan, but not intact aggrecan–hyaluronan complexes, are internalized via hyaluronan–CD44 interaction (Embry Flory et al., 2006). Further investigation of the relationships of pericellular matrix composition and perturbation to plasma membrane-bound complex formation and stability is needed to clarify these issues.

These interactions of pericellular hyaluronan with the plasma membrane and with other pericellular macromolecules may serve both structural and signaling functions that influence malignancy and resistance to therapies. Irrespective of the mechanisms involved, it is clear that antagonists of hyaluronan, CD44, and emmprin-induced events (Fig. 2.3) are promising candidates for therapeutic strategies aimed at preventing tumor refractoriness or recurrence due to drug-resistant sub-populations within malignant cancers.

ACKNOWLEDGMENTS

Recent work from our lab that is described herein was supported by grants to B.P.T. from the National Institutes of Health (CA073839 and CA082867), the Department of Defense (OC050368), and The Charlotte Geyer Foundation.

References

Adamia, S., Reiman, T., Crainie, M., et al. (2005). Intronic splicing of hyaluronan synthase 1 (HAS-1): a biologically relevant indicator of poor outcome in multiple myeloma. *Blood* **105**, 4836–4844.

Ailles, L. E. and Weissman, I. L. (2007). Cancer stem cells in solid tumors. *Curr Opin Biotechnol* **18**, 460–466.

Al-Hajj, M., Wicha, M. S., Benito-Hernandez, A., Morrison, S. J., and Clarke, M. F. (2003). Prospective identification of tumorigenic breast cancer cells. *Proc Natl Acad Sci USA* **100**, 3983–3988.

Asplund, T., Versnel, M. A., Laurent, T. C., and Heldin, P. (1993). Human mesothelioma cells produce factors that stimulate the production of hyaluronan by mesothelial cells and fibroblasts. *Cancer Res* **53**, 388–392.

Avigdor, A., Goichberg, P., Shivtiel, S., et al. (2004). CD44 and hyaluronic acid cooperate with SDF-1 in the trafficking of human CD34+ stem/progenitor cells to bone marrow. *Blood* **103**, 2981–2989.

Bacso, Z., Nagy, H., Goda, K., et al. (2004). Raft and cytoskeleton associations of an ABC transporter: P-glycoprotein. *Cytometry A* **61**, 105–106.

Baumgartner, G., Gomar-Hoss, C., Sakr, L., Ulsperger, E., and Wogritsch, C. (1998). The impact of extracellular matrix on the chemoresistance of solid tumors – experimental and clinical results of hyaluronidase as additive to cytostatic chemotherapy. *Cancer Lett* **131**, 85–99.

Beier, D., Hau, P., Proescholdt, M., et al. (2007). CD133(+) and CD133(−) glioblastoma-derived cancer stem cells show differential growth characteristics and molecular profiles. *Cancer Res* **67**, 4010–4015.

Bertrand, P., Girard, N., Delpech, B., Duval, C., d'Anjou, J., and Dauce, J. P. (1992). Hyaluronan (hyaluronic acid) and hyaluronectin in the extracellular matrix of human breast carcinomas: comparison between invasive and non-invasive areas. *Int J Cancer* **52**, 1–6.

Biswas, C., Zhang, Y., DeCastro, R., et al. (1995). The human tumor cell-derived collagenase stimulatory factor (renamed EMMPRIN) is a member of the immunoglobulin super-family. *Cancer Res* **55**, 434–439.

Bourguignon, L. Y. (2008a). This volume.

Bourguignon, L. Y. (2008b). Hyaluronan-mediated CD44 activation of RhoGTPase signaling and cytoskeleton function promotes tumor progression. *Semin Cancer Biol* **18**, 251–259.

Bourguignon, L. Y., Singleton, P. A., Diedrich, F., Stern, R., and Gilad, E. (2004). CD44 interaction with Na+-H+ exchanger (NHE1) creates acidic microenvironments leading to hyaluronidase-2 and cathepsin B activation and breast tumor cell invasion. *J Biol Chem* **279**, 26991–27007.

Brabletz, T., Jung, A., Spaderna, S., Hlubek, F., and Kirchner, T. (2005). Opinion: migrating cancer stem cells – an integrated concept of malignant tumour progression. *Nat Rev Cancer* **5**, 744–749.

Calabro, A., Oken, M. M., Hascall, V. C., and Masellis, A. M. (2002). Characterization of hyaluronan synthase expression and hyaluronan synthesis in bone marrow mesenchymal progenitor cells: predominant expression of HAS-1 mRNA and up-regulated hyaluronan synthesis in bone marrow cells derived from multiple myeloma patients. *Blood* **100**, 2578–2585.

Cheng, J. Q., Lindsley, C. W., Cheng, G. Z., Yang, H., and Nicosia, S. V. (2005). The Akt/PKB pathway: molecular target for cancer drug discovery. *Oncogene* **24**, 7482–7492.

Colone, M., Calcabrini, A., Toccacieli, L., et al. (2008). The multidrug transporter P-glyco-protein: a mediator of melanoma invasion? *J Invest Dermatol* **128**, 957–971.

Cordo Russo, R. I., Garcia, M. G., Alaniz, L., Blanco, G., Alvarez, E., and Hajos, S. E. (2008). Hyaluronan oligosaccharides sensitize lymphoma resistant cell lines to vincristine by modulating P-glycoprotein activity and PI3K/Akt pathway. *Int J Cancer* **122**, 1012–1018.

Crainie, M., Belch, A. R., Mant, M. J., and Pilarski, L. M. (1999). Overexpression of the receptor for hyaluronan-mediated motility (RHAMM) characterizes the malignant clone in multiple myeloma: identification of three distinct RHAMM variants. *Blood* **93**, 1684–1696.

Dai, Y. and Grant, S. (2007). Targeting multiple arms of the apoptotic regulatory machinery. *Cancer Res* **67**, 2908–2911.

Dalerba, P., Cho, R. W., and Clarke, M. F. (2007a). Cancer stem cells: models and concepts. *Annu Rev Med* **58**, 267–284.

Dalerba, P., Dylla, S. J., Park, I. K., et al. (2007b). Phenotypic characterization of human colorectal cancer stem cells. *Proc Natl Acad Sci USA* **104**, 10158–10163.

Dean, M., Fojo, T., and Bates, S. (2005). Tumour stem cells and drug resistance. *Nat Rev Cancer* **5**, 275–284.

Desoize, B. and Jardillier, J. (2000). Multicellular resistance: a paradigm for clinical resistance? *Crit Rev Oncol Hematol* **36**, 193–207.

Edward, M., Gillan, C., Micha, D., and Tammi, R. H. (2005). Tumour regulation of fibroblast hyaluronan expression: a mechanism to facilitate tumour growth and invasion. *Carcinogenesis* **26**, 1215–1223.

Embry Flory, J. J., Fosang, A. J., and Knudson, W. (2006). The accumulation of intracellular ITEGE and DIPEN neoepitopes in bovine articular chondrocytes is mediated by CD44 internalization of hyaluronan. *Arthritis Rheum* **54**, 443–454.

Evanko, S. P., Angello, J. C., and Wight, T. N. (1999). Formation of hyaluronan- and versican-rich pericellular matrix is required for proliferation and migration of vascular smooth muscle cells. *Arterioscler Thromb Vasc Biol* **19**, 1004–1013.

Evanko, S. P., Johnson, P. Y., Braun, K. R., et al. (2001). Platelet-derived growth factor stimulates the formation of versican–hyaluronan aggregates and pericellular matrix expansion in arterial smooth muscle cells. *Arch Biochem Biophys* **394**, 29–38.

Evanko, S. P., Tammi, M. I., Tammi, R. H., and Wight, T. N. (2007). Hyaluronan-dependent pericellular matrix. *Adv Drug Deliv Rev* **59**, 1351–1365.

Formby, B. and Stern, R. (2003). Lactate-sensitive response elements in genes involved in hyaluronan catabolism. *Biochem Biophys Res Commun* **305**, 203–208.

Gallagher, S. M., Castorino, J. J., Wang, D., and Philp, N. J. (2007). Monocarboxylate transporter 4 regulates maturation and trafficking of CD147 to the plasma membrane in the metastatic breast cancer cell line MDA-MB-231. *Cancer Res* **67**, 4182–4189.

Gatenby, R. A. and Gillies, R. J. (2004). Why do cancers have high aerobic glycolysis? *Nat Rev Cancer* **4**, 891–899.

Ghatak, S., Misra, S., and Toole, B. P. (2002). Hyaluronan oligosaccharides inhibit anchorage-independent growth of tumor cells by suppressing the phosphoinositide 3-kinase/Akt cell survival pathway. *J Biol Chem* **277**, 38013–38020.

Ghatak, S., Misra, S., and Toole, B. P. (2005). Hyaluronan regulates constitutive ErbB2 phosphorylation and signal complex formation in carcinoma cells. *J Biol Chem* **280**, 8875–8883.

Gilg, A. G., Tye, S. L., Tolliver, L. B., et al. (2008). Targeting hyaluronan interactions in malignant gliomas and their drug-resistant multipotent progenitors. *Clin Cancer Res* **14**, 1804–1813.

Gottesman, M. M., Fojo, T., and Bates, S. E. (2002). Multidrug resistance in cancer: role of ATP-dependent transporters. *Nature Rev Cancer* **2**, 48–58.

Gschwind, A., Fischer, O. M., and Ullrich, A. (2004). The discovery of receptor tyrosine kinases: targets for cancer therapy. *Nat Rev Cancer* **4**, 361–370.

Guo, Y., Ma, J., Wang, J., et al. (1994). Inhibition of human melanoma growth and metastasis in vivo by anti-CD44 monoclonal antibody. *Cancer Res* **54**, 1561–1565.

Halestrap, A. P. and Meredith, D. (2004). The SLC16 gene family – from monocarboxylate transporters (MCTs) to aromatic amino acid transporters and beyond. *Pflugers Arch* **447**, 619–628.

II. CELL BIOLOGY OF HYALURONAN IN CANCER

Hill, A., McFarlane, S., Mulligan, K., et al. (2006). Cortactin underpins CD44-promoted invasion and adhesion of breast cancer cells to bone marrow endothelial cells. *Oncogene* **25**, 6079–6091.

Hill, R. P. and Perris, R. (2007). "Destemming" cancer stem cells. *J Natl Cancer Inst* **99**, 1435–1440.

Ho, M. M., Ng, A. V., Lam, S., and Hung, J. Y. (2007). Side population in human lung cancer cell lines and tumors is enriched with stem-like cancer cells. *Cancer Res* **67**, 4827–4833.

Jin, L., Hope, K. J., Zhai, Q., Smadja-Joffe, F., and Dick, J. E. (2006). Targeting of CD44 eradicates human acute myeloid leukemic stem cells. *Nat Med* **12**, 1167–1174.

Kerbel, R. S., St Croix, B., Florenes, V. A., and Rak, J. (1996). Induction and reversal of cell adhesion-dependent multicellular drug resistance in solid breast tumors. *Hum Cell* **9**, 257–264.

Kimata, K., Honma, Y., Okayama, M., Oguri, K., Hozumi, M., and Suzuki, S. (1983). Increased synthesis of hyaluronic acid by mouse mammary carcinoma cell variants with high metastatic potential. *Cancer Res* **43**, 1347–1354.

Kirk, P., Wilson, M. C., Heddle, C., Brown, M. H., Barclay, A. N., and Halestrap, A. P. (2000). CD147 is tightly associated with lactate transporters MCT1 and MCT4 and facilitates their cell surface expression. *Embo J* **19**, 3896–3904.

Knudson, W., Biswas, C., Li, X. Q., Nemec, R. E., and Toole, B. P. (1989). The role and regulation of tumour-associated hyaluronan. *Ciba Found Symp* **143**, 150–159.

Knudson, W., Biswas, C., and Toole, B. P. (1984). Interactions between human tumor cells and fibroblasts stimulate hyaluronate synthesis. *Proc Natl Acad Sci USA* **81**, 6767–6771.

Knudson, W., Chow, G., and Knudson, C. B. (2002). CD44-mediated uptake and degradation of hyaluronan. *Matrix Biol* **21**, 15–23.

Koyama, H., Hibi, T., Isogai, Z., et al. (2007). Hyperproduction of hyaluronan in neu-induced mammary tumor accelerates angiogenesis through stromal cell recruitment: possible involvement of versican/PG-M. *Am J Pathol* **170**, 1086–1099.

Krause, D. S., Lazarides, K., von Andrian, U. H., and Van Etten, R. A. (2006). Requirement for CD44 in homing and engraftment of BCR-ABL-expressing leukemic stem cells. *Nat Med* **12**, 1175–1180.

Krause, D. S. and Van Etten, R. A. (2005). Tyrosine kinases as targets for cancer therapy. *N Engl J Med* **353**, 172–187.

Lee, J. T., Jr., Steelman, L. S., and McCubrey, J. A. (2004). Phosphatidylinositol 3'-kinase activation leads to multidrug resistance protein-1 expression and subsequent chemo-resistance in advanced prostate cancer cells. *Cancer Res* **64**, 8397–8404.

Li, C., Heidt, D. G., Dalerba, P., Burant, C. F., et al. (2007a). Identification of pancreatic cancer stem cells. *Cancer Res* **67**, 1030–1037.

Li, F., Tiede, B., Massague, J., and Kang, Y. (2007b). Beyond tumorigenesis: cancer stem cells in metastasis. *Cell Res* **17**, 3–14.

Li, L., Asteriou, T., Bernert, B., Heldin, C.H., and Heldin, P. (2007c). Growth factor regulation of hyaluronan synthesis and degradation in human dermal fibroblasts: importance of hyaluronan for the mitogenic response of PDGF-BB. *Biochem J* **404**, 327–336.

Li, L., Heldin, C. H., and Heldin, P. (2006). Inhibition of platelet-derived growth factor-BB-induced receptor activation and fibroblast migration by hyaluronan activation of CD44. *J Biol Chem* **281**, 26512–26519.

Li, Z. W. and Dalton, W. S. (2006). Tumor microenvironment and drug resistance in hema-tologic malignancies. *Blood Rev* **20**, 333–342.

Liscovitch, M. and Ravid, D. (2007). A case study in misidentification of cancer cell lines: MCF-7/AdrR cells (re-designated NCI/ADR-RES) are derived from OVCAR-8 human ovarian carcinoma cells. *Cancer Lett* **245**, 350–352.

Marieb, E. A., Zoltan-Jones, A., Li, R., et al. (2004). Emmprin promotes anchorage-independent growth in human mammary carcinoma cells by stimulating hyaluronan production. *Cancer Res* **64**, 1229–1232.

Martinez-Zaguilan, R., Seftor, E. A., Seftor, R. E., Chu, Y. W., Gillies, R. J., and Hendrix, M. J. (1996). Acidic pH enhances the invasive behavior of human melanoma cells. *Clin Exp Metastasis* **14**, 176–186.

Matrosova, V. Y., Orlovskaya, I. A., Serobyan, N., and Khaldoyanidi, S. K. (2004). Hyaluronic acid facilitates the recovery of hematopoiesis following 5-fluorouracil administration. *Stem Cells* **22**, 544–555.

Maxwell, C. A., Rasmussen, E., Zhan, F., et al. (2004). RHAMM expression and isoform balance predict aggressive disease and poor survival in multiple myeloma. *Blood* **104**, 1151–1158.

Miletti-Gonzalez, K. E., Chen, S., Muthukumaran, N., et al. (2005). The CD44 receptor interacts with P-glycoprotein to promote cell migration and invasion in cancer. *Cancer Res* **65**, 6660–6667.

Misra, S., Ghatak, S., and Toole, B. P. (2005). Regulation of MDR1 expression and drug resistance by a positive feedback loop involving hyaluronan, phosphoinositide 3-kinase, and ErbB2. *J Biol Chem* **280**, 20310–20315.

Misra, S., Ghatak, S., Zoltan-Jones, A., and Toole, B. P. (2003). Regulation of multi-drug resistance in cancer cells by hyaluronan. *J Biol Chem* **278**, 25285–25288.

Misra, S., Obeid, L. M., Hannun, Y. A., et al. (2008). Hyaluronan constitutively regulates activation of COX-2-mediated cell survival activity in intestinal epithelial and colon carcinoma cells. *J Biol Chem* **283**, 14335–14344.

Misra, S., Toole, B. P., and Ghatak, S. (2006). Hyaluronan constitutively regulates activation of multiple receptor tyrosine kinases in epithelial and carcinoma cells. *J Biol Chem* **281**, 34936–34941.

Mizrak, D., Brittan, M., and Alison, M. R. (2008). CD133: molecule of the moment. *J Pathol* **214**, 3–9.

Mogi, M., Yang, J., Lambert, J. F., et al. (2003). Akt signaling regulates side population cell phenotype via Bcrp1 translocation. *J Biol Chem* **278**, 39068–39075.

Neuzil, J., Stantic, M., Zobalova, R., et al. (2007). Tumour-initiating cells vs. cancer 'stem' cells and CD133: what's in the name? *Biochem Biophys Res Commun* **355**, 855–859.

Nilsson, S. K., Haylock, D. N., Johnston, H. M., et al. (2003). Hyaluronan is synthesized by primitive hemopoietic cells, participates in their lodgment at the endosteum following transplantation, and is involved in the regulation of their proliferation and differentiation in vitro. *Blood* **101**, 856–862.

Ohashi, R., Takahashi, F., Cui, R., et al. (2007). Interaction between CD44 and hyaluronate induces chemoresistance in non-small cell lung cancer cell. *Cancer Lett* **252**, 225–234.

Patrawala, L., Calhoun, T., Schneider-Broussard, R., Zhou, J., Claypool, K., and Tang, D. G. (2005). Side population is enriched in tumorigenic, stem-like cancer cells, whereas ABCG2+ and ABCG2-cancer cells are similarly tumorigenic. *Cancer Res* **65**, 6207–6219.

Pelicano, H., Martin, D. S., Xu, R. H., and Huang, P. (2006). Glycolysis inhibition for anti-cancer treatment. *Oncogene* **25**, 4633–4646.

Pilarski, L. M., Pruski, E., Wizniak, J., et al. (1999). Potential role for hyaluronan and the hyaluronan receptor RHAMM in mobilization and trafficking of hematopoietic progenitor cells. *Blood* **93**, 2918–2927.

Prehm, P. and Schumacher, U. (2004). Inhibition of hyaluronan export from human fibroblasts by inhibitors of multidrug resistance transporters. *Biochem Pharmacol* **68**, 1401–1410.

Prince, M. E., Sivanandan, R., Kaczorowski, A., et al. (2007). Identification of a subpopulation of cells with cancer stem cell properties in head and neck squamous cell carcinoma. *Proc Natl Acad Sci USA* **104**, 973–978.

Raguz, S., Tamburo De Bella, M., Tripuraneni, G., et al. (2004). Activation of the MDR1 upstream promoter in breast carcinoma as a surrogate for metastatic invasion. *Clin Cancer Res* **10**, 2776–2783.

Rudrabhatla, S. R., Mahaffey, C. L., and Mummert, M. E. (2006). Tumor microenvironment modulates hyaluronan expression: the lactate effect. *J Invest Dermatol* **126**, 1378–1387.

Schulz, T., Schumacher, U., and Prehm, P. (2007). Hyaluronan export by the ABC transporter MRP5 and its modulation by intracellular cGMP. *J Biol Chem* **282**, 20999–21004.

Shipitsin, M., Campbell, L. L., Argani, P., et al. (2007). Molecular definition of breast tumor heterogeneity. *Cancer Cell* **11**, 259–273.

Simpson, M. A., Wilson, C. M., and McCarthy, J. B. (2002). Inhibition of prostate tumor cell hyaluronan synthesis impairs subcutaneous growth and vascularization in immuno-compromised mice. *Am J Pathol* **161**, 849–857.

Singleton, P. A. and Bourguignon, L. Y. (2004). CD44 interaction with ankyrin and IP(3) receptor in lipid rafts promotes hyaluronan-mediated Ca(2+) signaling leading to nitric oxide production and endothelial cell adhesion and proliferation. *Exp Cell Res* **295**, 102–118.

Slomiany, M. G., Grass, G. D., Robertson, A. D., et al. (2008) Hyaluronan, CD44 and emmprin regulate lactate efflux and membrane localization of monocarboxylate transporters in human breast carcinoma cells. *Cancer Res.*, in press.

St Croix, B., Man, S., and Kerbel, R. S. (1998). Reversal of intrinsic and acquired forms of drug resistance by hyaluronidase treatment of solid tumors. *Cancer Lett* **131**, 35–44.

Stern, R., Shuster, S., Neudecker, B. A., and Formby, B. (2002). Lactate stimulates fibroblast expression of hyaluronan and CD44: the Warburg effect revisited. *Exp Cell Res* **276**, 24–31.

Tammi, R., Rilla, K., Pienimaki, J. P., et al. (2001). Hyaluronan enters keratinocytes by a novel endocytic route for catabolism. *J Biol Chem* **276**, 35111–35122.

Tammi, R. H., Kultti, A., Kosma, V. M., Pirinen, R., Auvinen, P., and Tammi, M. I. (2008). Hyaluronan in human tumors: pathobiological and prognostic messages from cell-associated and stromal hyaluronan. *Semin Cancer Biol* **18**, 288–295.

Toole, B. P., Biswas, C., and Gross, J. (1979). Hyaluronate and invasiveness of the rabbit V2 carcinoma. *Proc Natl Acad Sci USA* **76**, 6299–6303.

Toole, B. P. and Slomiany, M. G. (2008). Hyaluronan, CD44 and emmprin: partners in cancer cell chemoresistance. *Drug Resist Updat* **11**, 110–121.

Tredan, O., Galmarini, C. M., Patel, K., and Tannock, I. F. (2007). Drug resistance and the solid tumor microenvironment. *J Natl Cancer Inst* **99**, 1441–1454.

Tsukita, S., Oishi, K., Sato, N., Sagara, J., and Kawai, A. (1994). ERM family members as molecular linkers between the cell surface glycoprotein CD44 and actin-based cytoskel-etons. *J Cell Biol* **126**, 391–401.

Tzircotis, G., Thorne, R. F., and Isacke, C. M. (2005). Chemotaxis towards hyaluronan is dependent on CD44 expression and modulated by cell type variation in CD44–hyaluronan binding. *J Cell Sci* **118**, 5119–5128.

Underhill, C. B. and Toole, B. P. (1981). Receptors for hyaluronate on the surface of parent and virus-transformed cell lines: binding and aggregation studies. *Exp Cell Res* **131**, 419–423.

Vescovi, A. L., Galli, R., and Reynolds, B. A. (2006). Brain tumour stem cells. *Nat Rev Cancer* **6**, 425–436.

Wang, J., Sakariassen, P. O., Tsinkalovsky, O., et al. (2008). CD133 negative glioma cells form tumors in nude rats and give rise to CD133 positive cells. *Int J Cancer* **122**, 761–768.

Wang, S. J. and Bourguignon, L. Y. (2006). Hyaluronan and the interaction between CD44 and epidermal growth factor receptor in oncogenic signaling and chemotherapy resistance in head and neck cancer. *Arch Otolaryngol Head Neck Surg* **132**, 771–778.

Wang, S. J., Peyrollier, K., and Bourguignon, L. Y. (2007). The influence of hyaluronan–CD44 interaction on topoisomerase II activity and etoposide cytotoxicity in head and neck cancer. *Arch Otolaryngol Head Neck Surg* **133**, 281–288.

Weigel, P. H. and DeAngelis, P. L. (2007). Hyaluronan synthases: a decade-plus of novel glycosyltransferases. *J Biol Chem* **282**, 36777–36781.

Wilson, M. C., Meredith, D., Fox, J. E., Manoharan, C., Davies, A. J., and Halestrap, A. P. (2005). Basigin (CD147) is the target for organomercurial inhibition of monocarboxylate transporter isoforms 1 and 4: the ancillary protein for the insensitive MCT2 is EMBIGIN (gp70). *J Biol Chem* **280**, 27213–27221.

Yan, L., Zucker, S., and Toole, B. P. (2005). Roles of the multifunctional glycoprotein, emmprin (basigin; CD147), in tumour progression. *Thromb Haemost* **93**, 199–204.

Yang, J. M., Xu, Z., Wu, H., Zhu, H., Wu, X., and Hait, W. N. (2003). Overexpression of extracellular matrix metalloproteinase inducer in multidrug resistant cancer cells. *Mol Cancer Res* **1**, 420–427.

Yarden, Y. and Sliwkowski, M. X. (2001). Untangling the ErbB signalling network. *Nat Rev Mol Cell Biol* **2**, 127–137.

Zhang, L., Underhill, C. B., and Chen, L. (1995). Hyaluronan on the surface of tumor cells is correlated with metastatic behavior. *Cancer Res* **55**, 428–433.

CHAPTER

3

Growth Factor Regulation of Hyaluronan Deposition in Malignancies

Paraskevi Heldin, Eugenia Karousou, and Spyros S. Skandalis

INTRODUCTION

The link between the stromal microenvironment and the promotion of cancer was first described in 1889 by Stephen Paget (Paget, 1889), who predicted that the interactions between tumor cells (the "seed," including secreted growth factors and cell surface proteins) and the host microenvironment ("the soil") determine the metastatic outcome. In recent years it has become accepted that the microenvironment of local host tissue provides tumor cells with a scaffold that promotes their attachment and serves as reservoir for regulatory signals and thereby actively participates in tumor progression and metastasis (Schor and Schor, 2001).

In this review we focus on the extracellular molecule hyaluronan, the signals that regulate its synthesis and deposition as well as its role in cellular communication. Hyaluronan is a polysaccharide containing thousands of disaccharide repeats of glucuronic acid and N-acetylglucosamine residues. It is abundantly found in free form or decorated by proteoglycans in the extracellular and pericellular matrices of mammals (Heldin and Pertoft, 1993; Laurent and Fraser, 1992), as well as in the surface coats of some bacteria (Weigel, 2004) and *Chlorella* virus infected algae (DeAngelis, 2001). Hyaluronan in the pericellular matrix interacts with the cell by sustained binding to its own membrane-associated synthase or to hyaluronan receptors and with other matrix molecules; these interactions influence intracellular signaling and thereby cellular functions such as cell migration, growth and differentiation (Heldin and Pertoft, 1993; Knudson and Knudson, 1993). Importantly, intracellular and nuclear hyaluronan has also been demonstrated in both normal and tumor cells (Evanko and Wight, 2001; Li et al., 2007b). Because of its remarkable physicochemical properties and hygroscopic nature, hyaluronan has important physiological properties, including tissue organization and tissue hydration. Thus, an accumulation of hyaluronan is a common feature of remodeling tissues, for example during embryonic development, followed by its clearance. However, an aberrant increase in the amount of hyaluronan of a more polydisperse character, with a preponderance of lower molecular mass forms, is seen during inflammation and tumor progression.

Both high and low molecular mass hyaluronan can function as signaling molecules through their interactions with cell surface receptors, e.g. CD44 and extracellular matrix proteins, e.g., versican (Toole, 1990; Turley et al., 2002; Wu et al., 2005). CD44 is an adhesion receptor that is found in different splice variants on immune cells and stromal cells in a low hyaluronan binding state (Aruffo et al., 1990). However, external stimuli by cytokines can induce the transition of CD44 to its high hyaluronan binding state. Active CD44 with high hyaluronan binding capacity is found on activated leukocytes and tumor cells (Cichy and Pure, 2003; Ponta et al., 2003). The ability of CD44 to bind hyaluronan is tightly controlled. High Mw hyaluronan facilitates CD44 oligomerization whereas hyaluronan fragments bind to monomeric CD44 molecules. West and colleagues were first to demonstrate that hyaluronan oligomers are angiogenic (West et al., 1985). Subsequently, a number of laboratories, including ours, revealed that hyaluronan fragments is an important initiation factor in fibrotic tissue remodeling by the induction of collagen genes (Li et al., 2000), chemokine genes (McKee et al., 1996; Teder et al., 2002), but also an angiogenic factor, by the induction of distinct and/or common sets of genes with the known angiogenic factor fibroblast growth factor-2 (FGF-2) (Takahashi et al., 2005).

Both FGF-2 and hyaluronan oligosaccharides promote tubulogenesis in a process dependent on the co-ordinated induction of ornithine decarboxylase (Odc) and ornithine decarboxylase antizyme inhibitor (Oazi) genes. Among the genes induced selectively by hyaluronan oligosaccharides was the chemokine *CXCL1/Gro1* gene (the human homolog is IL-8); the endothelial cell differentiation was CD44-mediated leading to activation of chemokine receptor 2 which is involved in endothelial cell retraction, a common phenomenon observed during angiogenesis (Takahashi et al., 2005). Thus, hyaluronan oligomers have an important function during the inflammatory and angiogenic responses in injuries and malignancies through a sustained production of chemokines.

EXPRESSION OF HYALURONAN SYNTHASES AND HYALURONIDASES

Vertebrate, bacterial and plant hyaluronan molecules have identical chemical structure. There exist three related yet distinct hyaluronan synthase (*HAS*) genes encoding the mammalian HAS-1, HAS-2, and HAS-3 isoforms (Weigel and DeAngelis, 2007). Notably, HAS-2 is required for embryonic development, but not HAS-1 and HAS-3 (Camenisch et al., 2000); moreover the *HAS-2* gene is under tighter regulatory control than the other *HAS* genes (Nairn et al., 2007). The expression patterns of the *HAS* genes was found to vary between normal mesenchymal cells, and between normal and their transformed counterparts. In general, the expression was higher in sub-confluent than in confluent cultures (Jacobson et al., 2000; Li et al., 2007b). Each one of the *HAS* genes encodes plasma membrane proteins that are independently active, with multiple transmembrane and membrane-associated domains; the majority of the protein is inside the cell and possesses consensus sequences for phosphorylation by protein kinases (Shyjan et al., 1996; Spicer et al., 1996; Spicer et al., 1997). Recently, studies have demonstrated that HAS activities can be regulated through extracellular signal-regulated kinase (ERK) (Bourguignon et al., 2007). Each of the three HAS proteins synthesizes hyaluronan chains of high molecular mass, *in situ* ($\geq 4 \times 10^6$ Da). However, *in vitro*, the HAS-2 isoform synthesizes hyaluronan chains of high molecular mass ($\geq 4 \times 10^6$ Da), whereas HAS-3 produce polydisperse hyaluronan (average molecular mass of 0.8×10^6 Da), and HAS-1 even smaller hyaluronan chains (average molecular mass of 0.1×10^6 Da). Furthermore, the HAS-3 protein was catalytically more active than HAS-2 which in turn was more active than HAS-1. It is possible that different cytoplasmic proteins specifically interact with each HAS protein and may have accessory or regulatory roles in hyaluronan biosynthesis (Brinck and Heldin, 1999). The nature of these proteins have not yet been identified.

The newly synthesized and growing hyaluronan chain is extruded through the plasma membrane while the synthesis is in progress, contributing to the assembly of pericellular matrices by remaining attached to its own membrane-associated synthase before being released into the extracellular matrix (Heldin and Pertoft, 1993). The transfer process of newly synthesized hyaluronan is not yet known. It has been proposed that hyaluronan is transported through a pore-like passage and/or uses the multidrug resistance system (Prehm and Schumacher, 2004; Tlapak-Simmons et al., 1999). Hyaluronan overexpression amplifies MDR1 multidrug transporter expression and increases doxorubicin resistance in breast cancer cells MCF-7 (Misra et al., 2005; Misra et al., 2003); further studies are necessary to elucidate the inter-relationship between hyaluronan synthesis/export and multidrug transporters.

The turnover rate of hyaluronan in mammals is high; its intravenous $t_{(1/2)}$ is about 5 min and in epidermis it is less than 24 h (Fraser et al., 1981; Tammi and Tammi, 1998). Hyaluronidases, the enzymes involved in hyaluronan degradation also exist in several isoforms (*HYAL-1*, *HYAL-2*, *HYAL-3*, *HYAL-4*, and *PH-20*) and are localized in lysozomes or are glycosylphosphatidyl-inositol linked to the plasma membrane. HYAL-1 and HYAL-2 proteins are widely expressed in tissues and act in a concerted manner to degrade hyaluronan chains (Csoka et al., 2001; Lepperdinger et al., 2001).

HYALURONAN SIGNALING PROMOTES THE MALIGNANT PHENOTYPE OF TUMOR CELLS

Studies on human cancers from different origins and various malignancy grades have demonstrated a positive correlation between tumor aggressiveness and stromal hyaluronan expression (Auvinen et al., 2000; Boregowda et al., 2006). The aberrant amounts of hyaluronan in the desmoplastic stroma can be produced by the tumor cells themselves or by the stromal cells commandeered by the tumor cells (Asplund et al., 1993; Toole, 2004). Notably, a differential expression of *HAS* genes is seen during tumor progression. For example, aggressive breast cancer cells and ovarian cancer express higher levels of *HAS-2* than *HAS-3* compared to non-aggressive ones (Bourguignon et al., 2007; Li et al., 2007b), whereas metastatic prostate and colon cancer express higher levels of *HAS-3* than *HAS-2* (Bullard et al., 2003; Simpson et al., 2001). *HAS-1* was expressed only at low levels in these tumors. An important question awaiting an answer is whether there are functional differences between tumor cell-produced and stromal fibroblast- and/or mesothelial cell-synthesized hyaluronan for tumor progression. It is also important to elucidate which regulatory factors modulate the expressions and activities of HAS and HYAL proteins.

Earlier studies in our laboratory demonstrated marked differences in hyaluronan synthesis and expression of CD44 between non-aggressive and aggressive breast cancer cells. Metastatic breast carcinoma cells were found to express high levels of CD44 with high hyaluronan binding capacity and to synthesize hyaluronan. In contrast, breast cancer cell lines which have a non-invasive character synthesized much lower amounts of hyaluronan and did not express CD44 (Heldin et al., 1996). Importantly, CD44 exhibiting high hyaluronan binding capacity was expressed on malignant mesotheliomas but not on normal mesothelial cells, suggesting an up-regulation of hyaluronan–CD44 interaction upon transformation (Asplund and Heldin, 1994). In view of these observations it is possible that tumor cell invasiveness could be related to tumor cell surface CD44–matrix hyaluronan interaction or tumor cell-presented hyaluronan interaction with soluble CD44 or CD44 expressed by endothelial cells (Hill et al., 2005). However, blocking hyaluronan–CD44 interaction does not lead to complete inhibition of tumor cell migration/invasion, suggesting that also other mechanisms are involved (Fig. 3.1). Using two-photon fluorescence correlation microscopy, hyaluronan molecules were demonstrated to form continuous cage-like structures partitioning the space of melanoma tumor matrix into aqueous and viscous compartments (Alexandrakis et al., 2004), thereby facilitating cell migration.

Several approaches have been used to elucidate the importance of hyaluronan for tumor progression; manipulation of tumor cell-produced

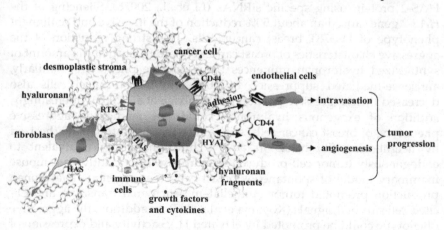

FIGURE 3.1   Tumor–host cross-talk in tumor progression. Growth factors and cytokines released by tumor cells, immune cells, and "activated" stromal cells trigger signaling events that increase the deposition of extracellular matrix macromolecules and activate tumor cell-expressed CD44 and RTK. Hyaluronan molecules form biological networks that bridges CD44 expressed by tumor and endothelial cells facilitating intravasation. Hyaluronan fragments, produced for example by the action of HYAL, bind to CD44 on endothelial cells and promote angiogenesis.

hyaluronan by overexpression of *HAS* or *HYAL* transcripts, overexpression of soluble CD44, administration of hyaluronan fragments, and treatment with antibodies that prevent hyaluronan–CD44 interactions (Toole, 2002; Toole, 2004). The impact of hyaluronan for the malignant phenotype of colon carcinomas was studied by overexpressing *HAS-2* and *HYAL-1* both *in vitro* and *in vivo*. The analysis revealed that *HAS-2* gene overexpression leads to a faster development of transplantable tumors in syngeneic rats, compared to mock-transfectants. In contrast, *HYAL-1* overexpression suppressed the growth rate of tumor cells (Jacobson et al., 2002). Similarly, inhibition of hyaluronan synthesis in prostate cancer cells impaired their growth (Liu et al., 2001; Simpson et al., 2001). Importantly, administration of hyaluronidase to mice bearing human breast cancer xenografts reduced tumor volume, hyaluronan content, and CD44 isoforms in the cancerous growth, supporting the hypothesis that loss of tumor cell-produced hyaluronan interactions is crucial for the maintenance of the malignant phenotype (Shuster et al., 2002).

Hyaluronan is abundant in highly aggressive breast cancer cells and has been shown to be a prognostic factor for patient survival of clinical breast carcinomas (Auvinen et al., 2000). *HAS-2* overexpression correlates with the promotion of the malignant phenotype of colon cancer and mesotheliomas (Jacobson et al., 2002; Li and Heldin, 2001). We therefore investigated the importance of HAS-2-synthesized hyaluronan for the malignant properties of breast cancer cells, by investigating the consequences of suppressing HAS-2 protein using specific siRNAs (Li et al., 2007b). Silencing of the *HAS-2* gene caused an about 50% reduction of the invasive and malignant phenotype of Hs578T breast cancer cells. This strong reduction of the aggressive characteristics of breast cancer cells suggests that the amount of synthesized hyaluronan influences their invasive phenotype. Similarly, antisense-mediated suppression of HAS-2 in breast cancer cells also decreased their aggressive phenotype (Udabage et al., 2005). Interestingly, addition of exogenous hyaluronan could not rescue the aggressive phenotype of breast cancer cells, suggesting that hyaluronan synthesized by neighboring fibroblasts *in vivo* is functionally not equivalent to endogenously tumor cell-produced hyaluronan. Importantly, in a mouse mammary model of spontaneous breast cancer, endogenous hyaluronan production promoted tumor epithelial–mesenchymal transition and elicited cell survival signals (Koyama et al., 2007). In addition, the aggressive phenotype could be promoted by elevated HAS activity and expression of CD44 receptors resulting in retention of hyaluronan on the surface of the neoplastic cells, facilitating their binding to the bone marrow endothelium (Draffin et al., 2004; Simpson et al., 2001). However, the involvement of hyaluronan in tumor progression is complex. Tumor cells often exhibit elevated levels of HYALs (Li et al., 2007b; Lokeshwar et al., 1999), leading to the production of angiogenic hyaluronan fragments. These observations demonstrate a cooperativity between HAS and HYAL activities, as well as CD44 hyaluronan receptors in the maintenance of the aggressive character of breast cancer cells (Fig. 3.1).

Interestingly, hyaluronan synthesized by mammary and colon carcinomas can, through interactions with tumor cell-expressed CD44, promote activation of several receptor tyrosine kinases (RTK), including ErbB2, and thereby promote cell survival and drug resistance. Perturbating the interaction between endogenously synthesized hyaluronan and CD44, Bryan Toole and his colleagues demonstrated the necessity of hyaluronan for the malignant properties of some cancer cells; addition of hyaluronan oligosaccharides, suppressed the tyrosine kinase activities of RTKs resulting in suppression of the phosphoinositol-3-kinase/Akt survival pathway (Ghatak et al., 2002; Ghatak et al., 2005; Misra et al., 2006).

The general concept emerging from these studies is that increased hyaluronan synthesis promotes tumorigenesis and plays an important role in the local aggressive spread of tumor cells. Thus, suppression of hyaluronan synthesis and/or prevention of its binding to cell surface receptors may provide a therapeutic opportunity to suppress tumor invasion.

REGULATION OF HYALURONAN LEVELS PRODUCED BY TUMOR CELLS

The levels of hyaluronan and differences in the size of hyaluronan molecules seen in rapidly remodeling tissues, e.g. tumor tissues, are due to the concerted action of HAS and HYAL enzymes that most likely are targets of local environmental cell specific factors. During tumor development and progression, besides the epithelial malignant cells and stromal cells (fibroblasts, mesothelial cells, and endothelial cells), a large number of inflammatory cells are also present; chronic inflammation goads premalignant cells to become malignant through the influx of innate immune cells that release cytokines and chemokines, promoting the growth and invasion of tumors (Mantovani, 2005). Tumor cells themselves release a variety of growth factors, for example TGF-β and PDGF, that may function in autocrine stimulation of tumor cells or in paracrine mechanisms involving stromal cells. Importantly, these growth signals can "activate" stromal cells to produce and release growth factors and cytokines, resulting in modulation of the matrix macromolecular structure (desmoplasia). Additionally, these signals can activate adhesive receptors (CD44, integrins) so that they can transmit signals for cell survival and metastasis (Hanahan and Weinberg, 2000; Hill et al., 2005). During inflammation in malignancies a possible co-existence of reactive oxygen species and overexpression of HYALs results in the accumulation of hyaluronan fragments in tissues.

Very little is known about the hyaluronan-stimulatory activities in various carcinomas. Mesothelioma cells produce PDGF-BB- and bFGF-like factors that most likely stimulate hyaluronan synthesis by the neighboring mesothelial cells and fibroblasts, creating a matrix that supports the colonization of malignant cells (Teder et al., 1996). However, it is possible that these factors also in an autocrine manner stimulate mesotheliomas to synthesize hyaluronan and thereby acquire a more malignant phenotype than the non-hyaluronan producing mesotheliomas (Li and Heldin, 2001). Furthermore, the phosphoglycoprotein osteopontin is implicated in breast cancer progression and metastasis probably through its interaction with CD44 resulting in the induction of HAS-2 expression and hyaluronan synthesis (Cook et al., 2006). Additionally, heregulin (HRG) activates members of the epidermal growth factor receptor family of tyrosine kinases leading to ERK activation and subsequent phosphorylation and activation of HAS-1, HAS-2 and HAS-3, affecting ovarian cancer progression (Bourguignon et al., 2007). Notably, the HRG-ErbB2-ERK signaling caused a decrease in the size of hyaluronan from about 400 kDa to about 80 kDa. Whether this reduction in the size of hyaluronan is due to HRG-mediated activation of HYAL in ovarian cancer cells, remains to be elucidated. Notably, HYAL activity was upregulated in TGFβ-stimulated human dermal fibroblasts cultures (Li et al., 2007a). Furthermore, oncostatin M, TGFβ and phorbol 12-myristate 13-acetate (PMA) induce the hyaluronan binding capacity of CD44 in lung tumor cells (Cichy and Pure, 2000; Teder et al., 1995). Because the molecular mechanisms underlying malignancy-induced hyaluronan production are not well understood, further studies on the regulatory mechanisms modulating the activities of HASs, HYALs, and CD44 hyaluronan binding capacity in tumors are necessary.

REGULATION OF HYALURONAN SYNTHESIS BY PERITUMORAL STROMA CELLS

The emphasis in this section is on the regulation of hyaluronan synthesis by stromal mesothelial cells and fibroblasts, particularly in response to PDGF-BB and TGFβ released by cancer cells, endothelial cells, and immune cells during the malignant progression. A large body of studies revealed that HAS-2 is the most abundantly expressed among the three HAS isoforms in mesothelial cells (Jacobson et al., 2000), corneal keratocytes (Guo et al., 2007), chondrocytes (Recklies et al., 2001), synovial cells (Stuhlmeier and Pollaschek, 2004), as well as in dermal, oral and lung fibroblasts (Li et al., 2007a; Li et al., 2000; Meran et al., 2007). HAS-3 is also found in appreciable amounts in these cells, whereas HAS-1 is hardly detected. However, the importance of each HAS isoform in the overall hyaluronan synthesis and assembly of the matrix surrounding cells is not known.

PDGF-BB had a potent stimulatory effect on hyaluronan synthesis through the induction of *HAS-2* in mesothelial, foreskin or dermal fibroblasts, and smooth muscle cell cultures (Heldin et al., 1992; Jacobson et al., 2000; Li et al., 2007a; Suzuki et al., 1995; van den Boom et al., 2006). Interestingly, PDGF-BB-mediated proliferation of human dermal fibroblasts and smooth muscle cells is promoted by the binding of hyaluronan to CD44 (Li et al., 2007a; van den Boom et al., 2006). Importantly, TGFβ stimulation suppresses *HAS-2* in mesothelial cells (Jacobson et al., 2000) and orbital fibroblasts from patients with Graves' ophthalmopathy (Wang et al., 2005), but potently activates *HAS-1* in synovial fibroblasts (Oguchi and Ishiguro, 2004; Recklies et al., 2001). The stimulatory effects of PDGF-BB and TGFβ were partly dependent on protein synthesis since the stimulations were partly inhibited by cycloheximide. Similarly, the protein kinase C stimulator PMA, powerfully induced hyaluronan synthesis probably via regulatory phosphorylation of a HAS isoform (Suzuki et al., 1995). Regulatory phosphorylation of each one of the three HAS isoforms has been demonstrated in a recently published study (Bourguignon et al., 2007). Combinations of PDGF-BB and TGFβ additively stimulated hyaluronan production in foreskin cultures (Suzuki et al., 1995). In contrast, in dermal fibroblast cultures TGFβ reduced the PDGF-BB-mediated hyaluronan production, probably because of activation of HYALs activity (Li et al., 2007a). These observations demonstrate that different cell types respond differentially to hyaluronan-modulating factors. More recently the downstream signaling pathways, through which PDGF-BB stimulates hyaluronan synthesis in human dermal fibroblasts were investigated. Using specific inhibitors for the major PDGF-BB-induced intracellular signaling pathways revealed that ERK MAPK and PI3K pathways are crucial for PDGF-BB-dependent *HAS-2* transcriptional activity and hyaluronan synthesis. Similarly, inhibition of NF-kB action completely suppressed hyaluronan production. The fact that the *HAS-2* promoter has putative transcription factor binding sites for CREB, NF-kB, and STAT (Monslow et al., 2003; Saavalainen et al., 2005), which are downstream of PDGF β-receptor signaling, is consistent with an important role of these signaling pathways in hyaluronan production (Fig. 3.2A).

The importance of high amounts of hyaluronan for PDGF-BB-mediated stromal fibroblast growth and migration were recently studied. The analysis revealed that hyaluronan-stimulated CD44 suppresses the activation state of the PDGF β-receptor, in PDGF-BB-stimulated human dermal fibroblasts, by the activation of a CD44-associated tyrosine phosphatase to the receptor, decreasing PDGF-BB-mediated fibroblast migration. Additionally, hyaluronan binding to CD44 is important for the mitogenic PDGF-BB response (Li et al., 2007a; Li et al., 2006) (Fig. 3.2B). Thus, dermal fibroblast CD44 binding to exogenous hyaluronan negatively regulates PDGF β-receptor-mediated migration, but positively

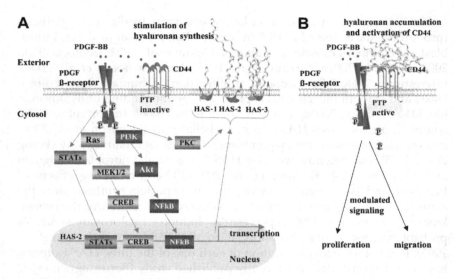

FIGURE 3.2 PDGF β-receptor-mediated hyaluronan production and its interaction with hyaluronan-activated CD44. (A) MEK1/2 and PI3K signaling pathways are important in mediating PDGF-BB-induced hyaluronan synthesis. Furthermore, activation of PKC is involved in the enhancement of HAS isoforms activities. (B) The interaction between hyaluronan-activated CD44 and PDGF β-receptor activates a CD44-associated PTP and modulates PDGF signaling, leading to cell migration and growth.

regulates its mitogenic response. Further studies are needed in order to elucidate the physiological importance of these observations during normal and abnormal tissue remodeling.

FUTURE PERSPECTIVES

Several observations support the notion that hyaluronan has an important role in tumorigenesis. Future studies should aim at unraveling the molecular mechanisms responsible for the synthesis and degradation of hyaluronan. Moreover, the mechanisms behind the expression and induction of CD44 from its low to high hyaluronan binding state as well as the functional importance of the interaction between RTK and CD44, remains to be elucidated.

ACKNOWLEDGMENTS

We thank Dr. Carl-Henrik Heldin for constructive criticism with the preparation of this article. The work was supported by grants from The Swedish Cancer Foundation, Wenner-Gren Foundation, and Mizoutani Foundation for Glycoscience.

References

Alexandrakis, G., Brown, E. B., Tong, R. T., et al. (2004). Two-photon fluorescence correlation microscopy reveals the two-phase nature of transport in tumors. *Nat Med* **10**, 203–207.

Aruffo, A., Stamenkovic, I., Melnick, M., Underhill, C. B., and Seed, B. (1990). CD44 is the principal cell surface receptor for hyaluronate. *Cell* **61**, 1303–1313.

Asplund, T. and Heldin, P. (1994). Hyaluronan receptors are expressed on human malignant mesothelioma cells but not on normal mesothelial cells. *Cancer Res* **54**, 4516–4523.

Asplund, T., Versnel, M. A., Laurent, T. C., and Heldin, P. (1993). Human mesothelioma cells produce factors that stimulate the production of hyaluronan by mesothelial cells and fibroblasts. *Cancer Res* **53**, 388–392.

Auvinen, P., Tammi, R., Parkkinen, J., et al. (2000). Hyaluronan in peritumoral stroma and malignant cells associates with breast cancer spreading and predicts survival. *Am J Pathol* **156**, 529–536.

Boregowda, R. K., Appaiah, H. N., Siddaiah, M., et al. (2006). Expression of hyaluronan in human tumor progression. *J Carcin* **5**, 2.

Bourguignon, L. Y., Gilad, E., and Peyrollier, K. (2007). Heregulin-mediated ErbB2-ERK signaling activates hyaluronan synthases leading to CD44-dependent ovarian tumor cell growth and migration. *J Biol Chem* **282**, 19426–19441.

Brinck, J. and Heldin, P. (1999). Expression of recombinant hyaluronan synthase (HAS) isoforms in CHO cells reduces cell migration and cell surface CD44. *Exp Cell Res* **252**, 342–351.

Bullard, K. M., Kim, H. R., Wheeler, M. A., et al. (2003). Hyaluronan synthase-3 is upregulated in metastatic colon carcinoma cells and manipulation of expression alters matrix retention and cellular growth. *Int J Cancer* **107**, 739–746.

Camenisch, T. D., Spicer, A. P., Brehm-Gibson, T., et al. (2000). Disruption of hyaluronan synthase-2 abrogates normal cardiac morphogenesis and hyaluronan-mediated transformation of epithelium to mesenchyme. *J Clin Invest* **106**, 349–360.

Cichy, J. and Pure, E. (2000). Oncostatin M and transforming growth factor-beta 1 induce post-translational modification and hyaluronan binding to CD44 in lung-derived epithelial tumor cells. *J Biol Chem* **275**, 18061–18069.

Cichy, J. and Pure, E. (2003). The liberation of CD44. *J Cell Biol* **161**, 839–843.

Cook, A. C., Chambers, A. F., Turley, E. A., and Tuck, A. B. (2006). Osteopontin induction of hyaluronan synthase 2 expression promotes breast cancer malignancy. *J Biol Chem* **281**, 24381–24389.

Csoka, A. B., Frost, G. I., and Stern, R. (2001). The six hyaluronidase-like genes in the human and mouse genomes. *Matrix Biol* **20**, 499–508.

DeAngelis, P.L. (2001). Novel hyaluronan synthases from chlorella viruses and pasteurella bacteria. http://www.glycoforum.gr.jp/science/hyaluronan/HA19/HA19E.html.

Draffin, J. E., McFarlane, S., Hill, A., Johnston, P. G., and Waugh, D. J. (2004). CD44 potentiates the adherence of metastatic prostate and breast cancer cells to bone marrow endothelial cells. *Cancer Res* **64**, 5702–5711.

Evanko, S.P. and Wight, T. (2001). Intracellular hyaluronan. http://www.glycoforum.gr.jp/science/hyaluronan/HA20/HA20E.html.

Fraser, J. R. E., Laurent, T. C., Pertoft, H., and Baxter, E. (1981). Plasma clearance, tissue distribution and metabolism of hyaluronic acid injected intravenously in the rabbit. *Biochem J* **200**, 415–424.

Ghatak, S., Misra, S., and Toole, B. P. (2002). Hyaluronan oligosaccharides inhibit anchorage-independent growth of tumor cells by suppressing the phosphoinositide 3-kinase/Akt cell survival pathway. *J Biol Chem* **277**, 38013–38020.

Ghatak, S., Misra, S., and Toole, B. P. (2005). Hyaluronan constitutively regulates ErbB2 phosphorylation and signaling complex formation in carcinoma cells. *J Biol Chem* **280**, 8875–8883.

Guo, N., Kanter, D., Funderburgh, M. L., Mann, M. M., Du, Y., and Funderburgh, J. L. (2007). A rapid transient increase in hyaluronan synthase-2 mRNA initiates secretion of hyaluronan by corneal keratocytes in response to transforming growth factor beta. *J Biol Chem* **282**, 12475–12483.

Hanahan, D. and Weinberg, R. A. (2000). The hallmarks of cancer. *Cell* **100**, 57–70.

Heldin, P., Asplund, T., Ytterberg, D., Thelin, S., and Laurent, T. C. (1992). Characterization of the molecular mechanism involved in the activation of hyaluronan synthetase by platelet-derived growth factor in human mesothelial cells. *Biochem J* **283**, 165–170.

Heldin, P., de la Torre, M., Ytterberg, D., and Bergh, J. (1996). Differential synthesis and binding of hyaluronan by human breast cancer cell lines: Relationship to hormone receptor status. *Oncol Rep* **3**, 1011–1016.

Heldin, P. and Pertoft, H. (1993). Synthesis and assembly of the hyaluronan-containing coats around normal human mesothelial cells. *Exp Cell Res* **208**, 422–429.

Hill, A., McFarlane, S., Johnston, P. G., and Waugh, D. J. (2005). The emerging role of CD44 in regulating skeletal micrometastasis. *Cancer Lett* **237**, 1–9.

Jacobson, A., Brinck, J., Briskin, M. J., Spicer, A. P., and Heldin, P. (2000). Expression of human hyaluronan synthases in response to external stimuli. *Biochem J* **348**, 29–35.

Jacobson, A., Rahmanian, M., Rubin, K., and Heldin, P. (2002). Expression of hyaluronan synthase 2 or hyaluronidase 1 differentially affect the growth rate of transplantable colon carcinoma cell tumors. *Int J Cancer* **102**, 212–219.

Knudson, C. B. and Knudson, W. (1993). Hyaluronan-binding proteins in development, tissue homeostasis and disease. *FASEB J* **7**, 1233–1241.

Koyama, H., Hibi, T., Isogai, Z., et al. (2007). Hyperproduction of hyaluronan in neu-induced mammary tumor accelerates angiogenesis through stromal cell recruitment: possible involvement of versican/PG-M. *Am J Pathol* **170**, 1086–1099.

Laurent, T. C. and Fraser, J. R. E. (1992). Hyaluronan. *FASEB J* **6**, 2397–2404.

Lepperdinger, G., Mullegger, J., and Kreil, G. (2001). HYAL-2 – less active, but more versatile? *Matrix Biol* **20**, 509–514.

Li, L., Asteriou, T., Bernert, B., Heldin, C.-H. and Heldin, P. (2007a). Growth factor regulation of hyaluronan synthesis and degradation in human dermal fibroblasts: Importance of hyaluronan for the mitogenic response of PDGF-BB. *Biochem J* 327–336.

Li, L., Heldin, C.-H., and Heldin, P. (2006). Inhibition of PDGF-BB-induced receptor activation and fibroblast migration by hyaluronan activation of CD44. *J Biol Chem* **281**, 26512–26519.

Li, Y. and Heldin, P. (2001). Hyaluronan production increases the malignant properties of mesothelioma cells. *Br J Cancer* **85**, 600–607.

Li, Y., Li, L., Brown, T.J. and Heldin, P. (2007b). Silencing of hyaluronan synthase 2 suppresses the malignant phenotype of invasive breast cancer cells. *Int J Cancer* **120**, 2557–2567.

Li, Y., Rahmanian, M., Widström, C., Lepperdinger, G., Frost, G. I., and Heldin, P. (2000). Irradiation-induced expression of hyaluronan (HA) synthase 2 and hyaluronidase 2 genes in rat lung tissue accompanies active turnover of HA and induction of types I and III collagen gene expression. *Am J Respir Cell Mol Biol* **23**, 411–418.

Liu, N., Gao, F., Han, Z., Xu, X., Underhill, C. B., and Zhang, L. (2001). Hyaluronan synthase 3 overexpression promotes the growth of TSU prostate cancer cells. *Cancer Res* **61**, 5207–5214.

Lokeshwar, V. B., Young, M. J., Goudarzi, G., Iida, N., Yudin, A. I., Cherr, G. N., and Selzer, M. G. (1999). Identification of bladder tumor-derived hyaluronidase: its similarity to HYAL-1. *Cancer Res* **59**, 4464–4470.

Mantovani, A. (2005). Cancer: inflammation by remote control. *Nature* **435**, 752–753.

McKee, C. M., Penno, M. B., Cowman, M., Burdick, M. D., Strieter, R. M., Bao, C., and Noble, P. W. (1996). Hyaluronan (HA) fragments induce chemokine gene expression in alveolar macrophages. The role of HA size and CD44. *J Clin Invest* **98**, 2403–2413.

Meran, S., Thomas, D., Stephens, P., Martin, J., Bowen, T., Phillips, A., and Steadman, R. (2007). Involvement of hyaluronan in regulation of fibroblast phenotype. *J Biol Chem* **282**, 25687–25697.

Misra, S., Ghatak, S., and Toole, B. P. (2005). Regulation of MDR1 expression and drug resistance by a positive feedback loop involving hyaluronan, phosphoinositide 3-kinase and ErbB2. *J Biol Chem* **280**, 20310–20315.

Misra, S., Ghatak, S., Zoltan-Jones, A., and Toole, B. P. (2003). Regulation of multidrug resistance in cancer cells by hyaluronan. *J Biol Chem* **278**, 25285–25288.

Misra, S., Toole, B. P., and Ghatak, S. (2006). Hyaluronan constitutively regulates activation of multiple receptor tyrosine kinases in epithelial and carcinoma cells. *J Biol Chem* **281**, 34936–34941.

Monslow, J., Williams, J. D., Norton, N., et al. (2003). The human hyaluronan synthase genes: genomic structures, proximal promoters and polymorphic microsatellite markers. *Int J Biochem Cell Biol* **35**, 1272–1283.

Nairn, A. V., Kinoshita-Toyoda, A., Toyoda, H., et al. (2007). Glycomics of proteoglycan biosynthesis in murine embryonic stem cell differentiation. *J Proteome Res* **6**, 4374–4387.

Oguchi, T. and Ishiguro, N. (2004). Differential stimulation of three forms of hyaluronan synthase by TGF-beta, IL-1beta and TNF-alpha. *Connect Tissue Res* **45**, 197–205.

Paget, S. (1889). The distribution of secondary growth in cancer of the breast. *Lancet* **1**, 571–573.

Ponta, H., Sherman, L., and Herrlich, P. A. (2003). CD44: from adhesion molecules to signalling regulators. *Nat Rev Mol Cell Biol* **4**, 33–45.

Prehm, P. and Schumacher, U. (2004). Inhibition of hyaluronan export from human fibroblasts by inhibitors of multidrug resistance transporters. *Biochem Pharmacol* **68**, 1401–1410.

Recklies, A. D., White, C., Melching, L., and Roughley, P. J. (2001). Differential regulation and expression of hyaluronan synthases in human articular chondrocytes, synovial cells and osteosarcoma cells. *Biochem J* **354**, 17–24.

Saavalainen, K., Pasonen-Seppanen, S., Dunlop, T. W., Tammi, R., Tammi, M. I., and Carlberg, C. (2005). The human hyaluronan synthase 2 gene is a primary retinoic acid and epidermal growth factor responding gene. *J Biol Chem* **280**, 14636–14644.

Schor, S. L. and Schor, A. M. (2001). Tumor–stroma interactions. Phenotypic and genetic alterations in mammary stroma: implications for tumour progression. *Breast Cancer Res* **3**, 373–379.

Shuster, S., Frost, G. I., Csoka, A. B., Formby, B., and Stern, R. (2002). Hyaluronidase reduces human breast cancer xenografts in SCID mice. *Int J Cancer* **102**, 192–197.

Shyjan, A. M., Heldin, P., Butcher, E. C., Yoshino, T., and Briskin, M. J. (1996). Functional cloning of the cDNA for a human hyaluronan synthase. *J Biol Chem* **271**, 23395–23399.

Simpson, M. A., Reiland, J., Burger, S. R., et al. (2001). Hyaluronan synthase elevation in metastatic prostate carcinoma cells correlates with hyaluronan surface retention, a prerequisite for rapid adhesion to bone marrow endothelial cells. *J Biol Chem* **276**, 17949–17957.

Spicer, A. P., Augustine, M. L., and McDonald, J. A. (1996). Molecular cloning and characterization of a putative mouse hyaluronan synthase. *J Biol Chem* **271**, 23400–23406.

Spicer, A. P., Olson, J. S., and McDonald, J. A. (1997). Molecular cloning and characterization of a cDNA encoding the third putative mammalian hyaluronan synthase. *J Biol Chem* **272**, 8957–8961.

Stuhlmeier, K. M. and Pollaschek, C. (2004). Differential effect of transforming growth factor beta (TGF-beta) on the genes encoding hyaluronan synthases and utilization of the p38 MAPK pathway in TGF-beta-induced hyaluronan synthase 1 activation. *J Biol Chem* **279**, 8753–8760.

Suzuki, M., Asplund, T., Yamashita, H., Heldin, C.-H., and Heldin, P. (1995). Stimulation of hyaluronan biosynthesis by platelet-derived growth factor-BB and transforming growth factor-β1 involves activation of protein kinase C. *Biochem J* **307**, 817–821.

Takahashi, Y., Li, L., Kamiryo, M., Asteriou, T., Moustakas, A., Yamashita, H., and Heldin, P. (2005). Hyaluronan fragments induce endothelial cell differentiation in a CD44- and CXCL1/GRO1-dependent manner. *J Biol Chem* **280**, 24195–24204.
Tammi, M.I. and Tammi, R. (1998). Hyaluronan in the epidermis. In http://www.glycoforum. gr.jp/science/hyaluronan/hyaluronanE.html.
Teder, P., Bergh, J., and Heldin, P. (1995). Functional hyaluronan receptors are expressed on squamous cell lung carcinoma cell line but not on other lung carcinoma cell lines. *Cancer Res* **55**, 3908–3914.
Teder, P., Vandivier, R. W., Jiang, D., et al. (2002). Resolution of lung inflammation by CD44. *Science* **296**, 155–158.
Teder, P., Versnel, M. A., and Heldin, P. (1996). Stimulatory effects of pleura fluids from mesothelioma patients on CD44 expression, hyaluronan production and cell proliferation in primary cultures of normal mesothelial and transformed cells. *Int J Cancer* **67**, 393–398.
Tlapak-Simmons, V. L., Baggenstoss, B. A., Clyne, T., and Weigel, P. H. (1999). Purification and lipid dependence of the recombinant hyaluronan synthases from *Streptococcus pyogenes* and *Streptococcus equisimilis*. *J Biol Chem* **274**, 4239–4245.
Toole, B. P. (1990). Hyaluronan and its binding proteins, the hyaladherins. *Curr Opin Cell Biol* **2**, 839–844.
Toole, B. P. (2002). Hyaluronan promotes the malignant phenotype. *Glycobiology* **12**, 37R–42R.
Toole, B. P. (2004). Hyaluronan: from extracellular glue to pericellular cue. *Nat Rev Cancer* **4**, 528–539.
Turley, E. A., Noble, P. W., and Bourguignon, L. Y. (2002). Signaling properties of hyaluronan receptors. *J Biol Chem* **277**, 4589–4592.
Udabage, L., Brownlee, G. R., Waltham, M., et al. (2005). Antisense-mediated suppression of hyaluronan synthase 2 inhibits the tumorigenesis and progression of breast cancer. *Cancer Res* **65**, 6139–6150.
van den Boom, M., Sarbia, M., von Wnuck Lipinski, K., et al. (2006). Differential regulation of hyaluronic acid synthase isoforms in human saphenous vein smooth muscle cells: possible implications for vein graft stenosis. *Circ Res* **98**, 36–44.
Wang, H. S., Tung, W. H., Tang, K. T., et al. (2005). TGF-beta induced hyaluronan synthesis in orbital fibroblasts involves protein kinase C beta II activation in vitro. *J Cell Biochem* **95**, 256–267.
Weigel, P.H. (2004). Bacterial hyaluronan synthases. http://www.glycoforum.gr.jp/science/ hyaluronan/HA06/HA06E.html.
Weigel, P. H. and DeAngelis, P. L. (2007). Hyaluronan synthases: a decade-plus of novel glycosyltransferases. *J Biol Chem* **282**, 36777–36781.
West, D. C., Hampson, I. N., Arnold, F., and Kumar, S. (1985). Angiogenesis induced by degradation products of hyaluronic acid. *Science* **228**, 1324–1326.
Wu, Y. J., La Pierre, D. P., Wu, J., Yee, A. J., and Yang, B. B. (2005). The interaction of versican with its binding partners. *Cell Res* **15**, 483–494.

CHAPTER

4

Hyaluronan Binding Protein 1 (HABP1/p32/gC1qR): A New Perspective in Tumor Development

Anindya Roy Chowdhury, Anupama Kamal, Ilora Ghosh, and Kasturi Datta

Tumor development is a multistage disease that can be divided into tumor initiation, promotion, malignant conversion and progression. The role of the extracellular matrix (ECM) in tumor development is reflected in several reports that document interactions with cell surface receptors regulating the process of inflammation, wound repair, and tissue organization (Wall et al., 2003). The ECM not only acts as a barrier for growth, invasion and metastasis of tumors, it also profoundly affects the behavior of malignant cells in later stages of progression. One of the major ECM components, hyaluronan (HA), is a negatively charged, high molecular weight component that has unique capability for retaining large volumes of water (Lee et al., 2000). This molecule is extruded from the cells through the plasma membrane into extracellular space, facilitating cell motility, thereby decreasing cell–cell contacts and impeding intracellular communication (Toole and Hascall, 2002). Recent developments show that HA is critical for anchorage-independent growth in culture, and can be considered as an indicator of tumorigenicity *in vivo*. The level of HA becomes crucial in tissue organization as its elevation in malignant tumor promotes anchorage-independent cell proliferation (Laurent et al., 1996).

These diverse functions of HA are regulated by interactions with a family of proteins having specific affinity towards HA, termed hyaladherins (Toole, 1990). Overexpression of HAS in fibrosarcoma cells stimulates both tumor growth *in vivo* and anchorage-independent growth in soft agar (Kosaki et al., 1999). Likewise, perturbation of endogenous HA interactions either by overexpression of soluble CD44 (Peterson et al., 2000) or by addition of HA oligomers (Auvinen et al., 2000) inhibit anchorage-independent growth.

HA interactions directly influence various intracellular signaling pathways important for cell behavior, through its binding to CD44, leading to internalization and degradation of HA. In this respect, it is significant that some tumor cells exhibit elevated levels of hyaluronidase and the ability to internalize and degrade HA. Thus, penetration of HA rich stroma or production of angiogenic breakdown products of HA may also promote tumor progression (Liu et al., 1996; Yu and Stamenkovic, 1999).

HA-receptor for HA-mediated motility (RHAMM) interactions has also been implicated in tumor cell behavior *in vitro* and *in vivo*. RHAMM is involved in the Ras and extracellular signal-regulated kinase signaling pathways and associates with the cytoskeleton (Turley et al., 2002).

HA–RHAMM interactions induce transient phosphorylation of p^{125}_{FAK} in concert with turnover of focal adhesions in Ras-transformed cells, thus leading to initiation of locomotion. Suppression of this interaction inhibits both cellular locomotion and proliferation *in vitro* and leads to inhibition of tumor growth *in vivo*, whereas overexpression leads to enhanced tumor growth and metastasis. A detailed discussion on the role of both CD44 and RHAMM has been made in this book. However, in addition to CD44 and RHAMM, there is at least another member of hyaladherin family, HABP1 that needs to be discussed in respect to cellular processes.

HYALURONAN BINDING PROTEIN 1 (HABP1)

Purification, Cloning and Characterization

HA binding protein 1 (HABP1) was first isolated from rat kidney by HA sepharose affinity chromatography and was shown to bind to HA with high specificity with Kd of 2.38×10^{-6}/Mol-disaccharide-equivalents but not with any other GAGs (Gupta et al., 1991). The antibodies raised against the purified protein were used to immunoscreen λgt11 cDNA expression library of human skin fibroblast, and a cDNA clone encoding HABP1 was isolated and characterized (Deb and Datta, 1996). The internal polypeptide sequence (83 residues) of the purified HABP1 was found to be identical to the predicted protein sequence derived from HABP1 cDNA, thus confirming the authenticity of the clone. Interestingly, the cDNA sequence of HABP1 shows complete identity with p32, a protein co-purified with the human pre-mRNA splicing factor SF2 (Krainer et al., 1991) and gC1qR, receptor for the globular head of complement subcomponent 1q (Ghebrehiwet et al., 1994).

The open reading frame of HABP1/p32/gC1qR encodes a pro-protein of 282 amino acid residues (Das et al., 1997), which after posttranslational cleavage of the first 73 amino acid residues, form the mature protein of 209 amino acid residues. The mature protein corresponds to 23.7 kDa, but migrates at 34 kDa on SDS-PAGE due to a high ratio of polar to hydrophobic amino acid residues. The mature protein is preceded by a 60 residue-long hydrophobic stretch containing five cysteines, which in turn is preceded by a 13 residues long leader peptide, which contains the mitochondrial targeting signal sequence. The precise role of these 60 residues immediately preceding the mature protein has not yet been determined. However, it has been predicted to play a role in cellular translocation. The mature protein has a calculated pI of 4.15 suggesting its acidic nature. There are 28 glutamic acid, 20 aspartic acid, 16 lysine, 5 histidine, and 4 arginine residues present in the 209 residue long mature protein. In contrast, the first 73 residue long stretch of the pro-protein does not have any glutamic acid, aspartic acid, lysine or histidine residues, but does have 11 arginine residues. The mature protein has only one cysteine at residue 186.

Specific Features of HABP1 Primary Structure

The computer analysis of the open reading frame (ORF) of HABP1 predicts it to contain three potential N-glycosylation sites spread all over the polypeptide chain. The residues involved are $^{109}WELELN^*GTEA^{118}$, $^{131}VTFNIN^*NSIPPTFD^{144}$, and $^{218}EWKDTN^*YTLNT^{228}$. In addition it is found to have one potential putative non-classical dibasic HA binding motif spanning from $^{119}KLVRKVAGEK^{128}$, one potential tyrosine sulfation site ($^{185}DCHY^*PEDEV^{193}$), one protein kinase C phosphorylation site ($^{202}DIFS^*IREVS^{210}$), and five casein kinase II phosphorylation sites involving the residues $^{74}LHT^*DGDKAFVD^{84}$, $^{200}ESDIFS^*IREV^{209}$, $^{208}EVSFQS^*TGESEWKD^{221}$, $^{246}RGVDNT^*FADELVEL^{259}$, and $^{256}LVEL$-$ST^*ALEHQEYI^{269}$ in its sequence. HABP1 shows a potential proline directed $^{160}PELTSTP^{166}$ sequence, which acts as the substrate phosphorylation site for protein kinases like ERK (Majumdar et al., 2002a). The presence of basic amino acid rich nuclear localization sequence $^{94}RKIQKHK^{100}$ and $^{118}AKLVRK^{123}$ also raises the possibility of the protein to be a nuclear protein. The amino acid sequence alignment studies confirm that it does not have any homology with other HA binding proteins (Deb and Datta, 1996).

HA Binding Motif in HABP1/p32/gC1qR

HABP1/p32 is the first HA binding protein belonging to a non-link module superfamily, whose three dimensional structure was solved by X-ray crystallography (Yang et al., 1994). The HA binding motif ($^{119}KLVRKVAGEK^{128}$) (Deb and Datta, 1996) is found to be in the loop region between the β_2 and β_3 sheet, solvent exposed and independent in topological arrangement. Structure-based alignment of p32/HABP1 sequence from human, mouse, C. elegans and S. cerevisiae shows identical HA binding motifs in human and mouse, whereas basic residues like Arg-122, Lys-123 and acidic residue Glu-127 are conserved in all four species. In contrast to the canonical definition of HA binding motif (Yang et al., 1994), HABP1 shows conservation of an acidic amino acid residue (E^{127}) in its HA binding motif. It has been shown that E^{127} is involved in the salt bridge formation with R^{246} allowing the clustering of positive charge along the putative HA-binding motif of required spacing critical for HA binding capacity for most of the non-link module hyaladherins. In addition, it has been shown that the whole three-dimensional structure is important for the HA–HABP1 interactions (Jiang et al., 1999). HABP1 is suggested to exist in a molten globule-like state in the absence of salt or at very low salt conditions, which seem to be responsible for its molecular chaperone-like activities. Presence of counter ions in the molecular environment reduces the electrostatic repulsion by screening the charges, and hence affecting the three dimensional structure and thermodynamic stability of HABP1, due to the existence of more compact structure under acidic pH or at high ionic strength at neutral pH (Jha et al., 2003). Among all the tested ligands, only HA–HABP1 interaction is found to be tightly regulated with change in pH and ion concentration suggesting this interaction to be highly significant in terms of the biological functions. However, it is also observed that the N- and C-termini truncated variants of HABP1 with truncated α-helices do not undergo oligomerization, but their binding affinities for hyaluronan remains comparable to that of HABP1 (Sengupta et al., 2004).

Three-Dimensional Structure of HABP1/p32/gC1qR

The crystal structure of HABP1 determined by Jiang et al. (1999) at 2.25 Å resolution reveals that HABP1 adopts a novel fold with seven consecutive anti-parallel β-strands flanked by one N-terminal and two C-terminal α-helices. These monomers form a doughnut-shaped quaternary structure with an unusually asymmetric charge distribution on the surface. The doughnut-shaped trimer has an outer diameter of approximately 75 Å, an average inner diameter of approximately 20 Å, and a thickness of approximately 30 Å. All three subunits have similar conformations. Monomeric HABP1 does not have any obviously distinct domains. Each monomer consists of seven consecutive β-strands, designated from β1 through 07, which form a highly twisted anti-parallel β-sheet with β1 nearly perpendicular to β7. The β-strands are flanked by one N-terminal and one C-terminal α-helices. All the helices are located on the same side of the β-sheet. Helix αB lies parallel to the β-sheet, with helix axis perpendicular to the direction of individual strands.

Helix αB and the N-terminal portion of the helix αC make extensive hydrophobic contacts with the β-sheet, which appears to be essential for the stability of the structure. The N-terminal helix αA does not contact the β-sheet within the same monomer, but forms an anti-parallel coiled coil with the C-terminal portion of αC. This coiled coil region is important for homo-oligomerization.

Three HABP1 monomers form a doughnut-shaped quaternary structure. The channel has a diameter of about 20 Å. But the loops connecting β6 and β7 partially cover the channel, reducing the diameter of the opening to about 10 Å. The channel wall is formed by β-sheets from all the three subunits. But this β wall is distinct from all other β structures such as β-sheet and β-barrels. Because of high degree of twisting of β-strands, adjacent monomers do not form contiguous sheets. The coiled coil region of αA and αB forms extensive intermolecular contacts: αA packs with anti-parallel αB of an adjacent monomer and C-terminal region of αC packs against the back of the β-sheet. Most of these interactions are hydrophobic in nature, except for the intermolecular electrostatic pairing of Arg-246 and Asp-79. In summary, the overall structure of the trimer can be visualized as if the β-sheets form a hyperboloid-shaped spool with the α-helices wrapping around it.

HABP1 is a very acidic protein with a pI of 4.15. However, the charge distribution on the protein surface is very asymmetric. Both the sides of the doughnut and the inside of the channel have a high concentration of negatively charged residues. In contrast, the opposite side of the doughnut is much less negatively charged. This polarity in charge distribution clearly suggests asymmetric functional roles for the two sides of this protein.

Chromosomal Localization and Genomic Organization of HABP1

By using fluorescence in situ hybridization (FISH), chromosomal localization of gC1qR/HABP1 is assigned to human chromosome 17p13.3, showing 99.5% similarity, from base 928 to base 1163, with STS WI-9243, a STS flanking marker of human chromosome 17p12-p13 (Majumdar and Datta, 2003). The cDNA sequence of HABP1 shows complete identity with p32 (Genbank accession no. M69039), a protein co-purified with the human pre-mRNA splicing factor SF2 (Krainer et al., 1991) and gC1qR (Genbank accession no. X75913), receptor for the globular head of complement subcomponent 1q (Ghebrehiwet et al., 1994). The gene has been named as HABP1 by HUGO nomenclature Committee of GDB (Accession no. 9786126) and referred as the synonym C1QBP/p32/HABP1 in human genome.

The genomic organization of HABP1/gC1qR gene including its 5'- and 3'-flanking regions spans about 7.8-kilobase pair (kb). From the first codon of the initiation methionine to the stop codon, the gene spans 6055 bp. There are six exons and five introns in the C1qBP gene. The size of the exons range from 94 bp (exon 3) to 232 bp (exon 1), and that of the introns range from 128 bp (intron 5) to 3156 bp (intron 2). A poly (A) signal is located at 369 bp from the stop codon (Majumdar et al., 2002b; Lim et al., 1998).

The first pseudogene for HABP1 was reported in the gene poor region of chromosome 21 and was flanked by six (three on each side) pseudogenes (Hattori et al., 2000). The complete sequencing of the human genome paved way for the search for more such elements in the human genome. Genomic sequence search with the HABP1 cDNA further revealed the existence of more HABP1 pseudogenes, extending to chromosomes 15, 11, and 4, each varying in length and similarity to the parental cDNA sequence. All these pseudogene sequences lack a 5' promoter sequence and possess multiple mutations, with the insertion of premature stop codons in all three reading frames, confirming their identity as processed pseudogenes (Majumdar et al., 2002b). The gene encoding hyaluronan binding protein 1 (HABP1) and its homologs have been reported across eukaryotes, from yeast to human. Further, we reported not only the presence of HABP1 pseudogene in other animal species, but also the presence of a homologous sequence in *Methanosarcina barkeri*, an ancient life form. This sequence has 44.8% homology to the human HABP1 cDNA and 45.3% homology with the HABP1 pseudogene in human chromosome 21. The presence of this HABP1 cDNA-like fragment in *M. barkeri* might enable us to shed light on the evolution of the HABP1 gene and whether it was present in a common ancestral organism before the lineages separated (Sengupta et al., 2004).

Subcellular Localization

The complete cDNA sequence of HABP1, corresponding to the proprotein, shows the presence of only a mitochondrial localization signal (Dedio et al., 1998). However, immunofluorescence and subcellular fractionation studies have detected this protein or its homologues not only in the mitochondrial matrix in yeast (Seytter et al., 1998) and mammals (Muta et al., 1997) but also on the cell surface (Gupta et al., 1991), cytoplasm (Dedio et al., 1998), and nucleus (Simos and Georgatos, 1994) of cultured cell lines.

Primary sequence of HABP1 has demonstrated a distinct mitochondrial localization signal at the N-terminal part of the pro-protein. This signal comprises of an amphipathic helix with basic residues (Muta et al., 1997). Transient expression system in PLC cells has demonstrated that the cDNA encoding full length protein localizes to the mitochondria, while the cDNA encoding the mature form of HABP1 targets it to the cytosol (Muta et al., 1997). This clearly shows that the mitochondrial localization signal exists only in the first 73 amino acids. Furthermore, the yeast homologue of HABP1 (Yeast Mam33p), is located in the mitochondrial matrix and plays an important role in the oxidative phosphorylation. The acidic surface of HABP1 is rich in aspartic and glutamic acid residues. This enables it to serve as a high capacity divalent cation storage protein and modulate the mitochondrial matrix cation concentration. Another interesting feature of HABP1 is its high negative charge density and asymmetric charge distribution, which suggests its possible mode of association with the inner mitochondrial membrane. The negatively charged side of HABP1 may potentially bind to the inner mitochondrial membrane in the presence of divalent metal ions.

Immunolocalization studies using tagged HABP1 have shown that the protein is localized mainly in the nucleus and cytosol (Van Leeuwen and O'Hare, 2001). Interaction of HABP1 with viral proteins also targets it to the nucleus (Matthews and Russell, 1998). These diverse subcellular localizations of HABP1, coupled with its various interacting proteins, led to the proposition that it could be a component of the trafficking pathway connecting the nucleus, mitochondria, and cytoplasm and the export pathway to the cell surface (Van Leeuwen and O'Hare, 2001). Though this evolutionary conserved protein is said to be ubiquitous in nature, it is not expressed equally all over the cell, as it shows differential localization in different cell lines under various physiological conditions. It belongs to a class of protein that is initially targeted to mitochondria, but found subsequently at highly specific locations (Soltys et al., 2000).

Though enough evidence exists for mitochondrial matrix localization of HABP1, the explicit function of the protein in mitochondria and other subcellular sites is still unclear.

EVIDENCE FOR HABP1 TO BE INVOLVED IN TUMOR DEVELOPMENT

It is evident that HABP1 is present in the human fibroblast and binds specifically to HA, so it would be interesting to study the physiological significance of HABP1.

HABP1 as an Adhesive Protein

Hyaluronan binding protein, initially referred to as hyaluronectin, the term first coined by Delpech et al. (1993), was reported on the cell surface in normal fibroblasts, tumor cell lines, and sperm using indirect immunofluorescence (Gupta and Datta, 1991), while its secretory nature was established on its detection in the serum free medium of macrophage tumor cell line. A differential expression of HABP1 on the cell surface of the sub-populations of AK-5 cells, a transplantable histiocytic tumor cell line, was observed. Cell fractions responsible for developing both ascites and solid tumors were found to contain higher amounts of HABP1 than fractions which are capable of producing only ascites, suggesting its involvement in solid tumor formation. HABP1 coating on the plates allowed more cells to attach, which could be specifically blocked by antibody against HABP1, indicating its possible role in cell attachment (Gupta and Datta, 1991). The pre-treatment of AK-5 cells with HABP1 antibody abolished their capacity to grow as solid tumors, but the cells retained their capacity to grow as ascites tumor, further confirming the adhesive property of HABP1 (Gupta and Datta, 1991). However, in our consecutive papers, we made an observation that according to amino acid compositions and other properties (D'Souza and Datta, 1986), this protein is different from other HA binding proteins described by Delpech et al. (1993), such as hyaluronectin. Therefore, in later reports we refrained from using the term hyaluronectin and referred to it as hyaluronic acid or hyaluronan binding protein. Finally, after its molecular cloning from human fibroblast (Deb and Datta, 1996) and reporting its localization on human chromosome 17p12-p13 (Majumdar and Datta, 1998), we referred it as to hyaluronan binding protein 1 as accepted by HUGO and we followed this nomenclature subsequently.

Earlier the above observations led us to propose that the adhesive interaction of this protein may be crucial for HABP1 in cellular signaling, anchoring the tumor cells to the endothelium during the initial phases of tumor formation. The presence of ^{247}RGVD^{250} instead of RGD alone by sequence analysis also supported these observations.

HABP1 and Its Proposed Roles in Signal Transduction

Among the proposed ligands interacting with HABP1, HA is the only ligand which falls under the category of non-protein in nature and can interact at physiological conditions (Jha et al., 2003). A few physiological roles had been proposed for HABP1 in macrophage cell adhesion: signal transduction, mammalian reproduction, and pathological infection on the basis of *in vitro* and *in vivo* experiments.

Signal Transduction

Cell surface receptors are known to recognize various extracellular matrix components as their ligands and play a crucial role as signal transducers. HABP1 was reported to strongly bind to HA & was present on the cell surface too.

This protein was observed to be highly phosphorylated in transformed fibroblasts compared with normal fibroblasts. Phosphorylation was enhanced in the presence of its ligand, i.e. hyaluronan, but not in the presence of other glycosaminoglycans. The regulation of the cellular and cell surface phosphorylation of HA binding protein by HA, PMA, and calyculin-A was demonstrated in different cell lines. Hyaluronan enhanced the phosphorylation of PLC-γ in association with increased formation of IP$_3$, both of which are specifically blocked by pretreatment of the cells with purified anti-hyaluronan binding protein antibodies (Gupta and Datta, 1991).

The role of HABP1 in hyaluronic acid-induced cellular signaling in lymphocytes has been examined. The binding of ^{125}I-HA to lymphocytes *in vivo* was found to be inhibited by pre-incubation of the cells with anti-34 kDa HA binding protein antibodies, thus confirming 34 kDa HA binding protein as the specific HA-receptor in lymphocytes. The HA-induced cell aggregation, tyrosine phosphorylation and cytoskeletal protein phosphorylation demonstrates early cellular signaling events in lymphocytes. Further, during mitogen-induced lymphocyte signaling, *in vivo* phosphorylation by HA and the inhibition of cellular aggregation and

IP$_3$ formation by anti-HA binding protein antibodies revealed that 34 kDa HA binding protein is one of the potential mediators in HA-induced signal transduction (Gupta and Datta, 1991). The role of hyaluronan binding protein 1 (HABP1) in cell signaling was further investigated and *in vitro* kinase assay demonstrated that it is a substrate for MAP kinase (Majumdar et al., 2002a). Phosphorylation of endogenous HABP1 was also observed following treatment of J774 cells with PMA. HABP1 was co-immunoprecipitated with activated ERK, confirming their physical interaction in the cellular context. Upon PMA stimulation of normal rat fibroblast (F111) and transformed (HeLa) cells, the HABP1 level in the cytoplasm gradually decreased with a parallel increase in the nucleus. In HeLa cells, within 6 h of PMA treatment, HABP1 was completely translocated to the nucleus, which was prevented by PD98059, a selective inhibitor of ERK. We also observed that the nuclear translocation of HABP1 is concurrent with that of ERK, suggesting that ERK activation is a requirement for the translocation of HABP1. It is thus established for the first time that HABP1 is a substrate for ERK and is an integral part of the MAP kinase cascade implying its universal role in signaling.

Upregulation of HABP1 in Apoptosis Induction

The MAP kinase cascade and cell cycle regulation pathway are closely related since the activation of the former and its downstream signaling determines whether the cell will follow normal growth and division or undergo apoptosis. Thus in continuation, the upregulation of HABP1 expression was reported in HeLa cells at 18 and 24 h of cisplatin treatment that induces apoptosis in HeLa cell (Fig. 4.1). Quantification of HABP1 expression by flow cytometry confirmed a two-fold increase in total intracellular HABP1 expression at 24 h of cisplatin treatment. Under the same conditions the HABP1 transcript level measured by semi-quantitative RT PCR showed 2.5-fold increase ascertaining transcriptional regulation of HABP1 during cisplatin-induced apoptosis. Further, in normal HeLa cells, though a small amount of HABP1 can be detected in nucleus, but with apoptosis induction, the protein is mainly concentrated around the nuclear periphery at 18 h of cisplatin treatment. It is present both in the nucleus as well as in the cytosol at 24 h of treatment, suggesting its nuclear translocation during apoptosis. To substantiate our findings, prior to the cisplatin treatment, the expression of HABP1 was reduced by short interference of RNA (siRNA) mediated knockdown. Significant reduction in apoptotic cell population was detected in cisplatin-treated HeLa cells with disrupted HABP1 conferring resistance to cisplatin-induced apoptosis (Guo et al., 1999). Like cisplatin, TNF-α, an inducer of apoptosis upregulates HABP1/gC1qR in bone marrow cells (Guo et al., 1999).

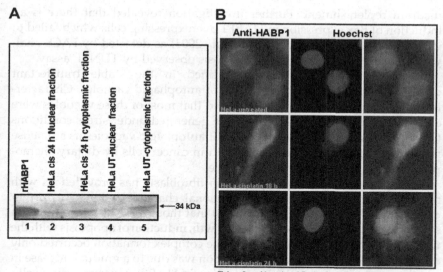

Taken from Kamal and Datta, Apoptosis, 2006, 11, 861-874

FIGURE 4.1 Translocation of HABP1 from cytosol to nucleus during cisplatin-induced apoptosis. (A) Immunodetection of isolated nuclear and cytoplasmic fractions of HeLa untreated and cisplatin-treated cells with anti-HABP1 antibody. Densitometry analysis shows a two-fold increase in nuclear levels of HABP1 in cisplatin treated HeLa cells at 24 h of treatment (Lane 2) in comparison to nuclear levels in untreated control (Lane 4). However, the cytoplasmic levels do not show any considerable change in cisplatin treated (Lane 5) and untreated HeLa cells (Lane 3). (B) Immunofluorescence analysis of HeLa untreated and cisplatin treated cells at 100× magnification. HABP1 is present predominantly in the cytoplasm with very little amounts in the nucleus of untreated HeLa cells. At 18 h of cisplatin treatment, HABP1 is concentrated around the nuclear periphery and by 24 h of cisplatin treatment, i.e. which marks apoptosis induction, HABP1 is present in the whole nucleus as well as in the cytoplasm. (See Page 1 in Color Section at the back of the book).

ECTOPIC EXPRESSION OF HABP1 INDUCES APOPTOSIS, AUTOPHAGIC VACUOLES, AND MITOCHONDRIAL DYSFUNCTION

Being an endogenous MAP kinase substrate, HABP1 may have wider implication in maintaining the delicate balance between cell cycle progression and apoptosis as the other proteins linked with MAP kinase pathway are associated with such functions.

We demonstrate that HABP1, when overexpressed in normal rat skin fibroblasts, remained in the cytosol, primarily concentrated around the nuclear periphery. However, HABP1 overexpressing cells showed extensive vacuolation and reduced growth rate, which was restored by frequent medium replenishment. Further investigation revealed that there is an induction of Bax expression in HABP1 overexpressing cells which failed to enter into the S-phase (Meenakshi et al., 2003) as detected by FACS analysis, and this finally undergoes apoptosis observed by TUNEL assay.

Another interesting feature identified in the stable transfectant expressing HABP1 is the appearance of autophagic vacuoles. Characterization of cytoplasmic vacuoles revealed that most of these vacuoles were autophagic in nature, resembling those generated under stress conditions (Sengupta et al., 2004). The induction of autophagic vacuoles as a defense mechanism against apoptotic cell death in cancer cells by dietary chemopreventive agents is a common practice.

The ectopic expression of HABP1 in fibroblasts has provided us with a model system to examine its functional characterization. We further demonstrated (Chowdhury et al., 2008) that though HABP1 accumulation started in mitochondria from 48 h of growth, induction of apoptosis with the release of cytochrome c and apoptosome complex formation occurred only after 60 h. This mitochondrial dysfunction was due to a gradual increase in reactive oxygen species (ROS) generation in HABP1 overexpressing cells. Along with ROS generation, increased Ca^{2+} influx in mitochondria leading to a drop in membrane potential was evident. Interestingly, upon expression of HABP1, the respiratory chain complex I was shown to be significantly inhibited, followed by defective mitochondrial ultrastructure. The reduction in oxidant generation and drop in apoptotic cell population accomplished by disruption of HABP1 expression, corroborated the fact that excess ROS generation in HABP1 overexpressing cells which lead to apoptosis, was due to mitochondrial HABP1 accumulation.

Differential Expression of Hyaluronic Acid Binding Protein 1 (HABP1)/P32/C1QBP During Progression of Epidermal Carcinoma

Keeping in mind that HA level is critical for development of carcinoma, and its significant role in cellular signaling, the expression profile of HABP1 was investigated from initiation to progression of epidermal carcinoma in mice, induced by benzo[α] pyrene (B[α]P) exposure (Ghosh et al., 2004). During tumor initiation, HABP1 accumulated in inflammatory subsquamous tissue and with progression, the protein was also seen to overexpress in papillomatic and acanthotic tissue (Fig. 4.2). With the onset of metastasis, HABP1 overexpression was confined to metastatic islands, while it disappeared gradually from the surrounding mass. Such expression profiles in metastasized tissue were supported by decreased levels of HABP1, both at protein and transcript levels. These observations taken together suggest that the changes in HABP1 level coincide with specific

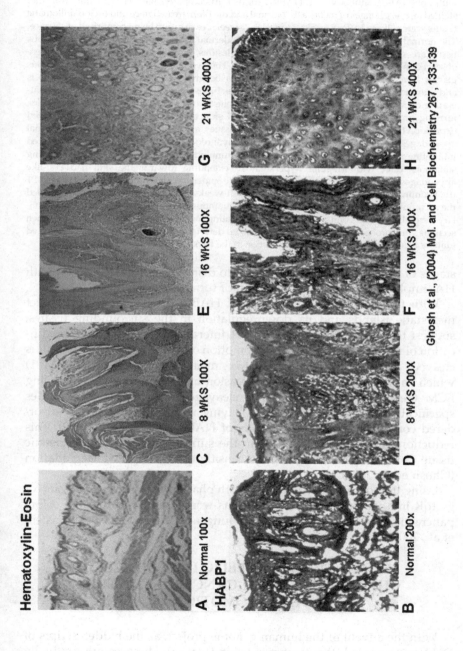

Hematoxylin-Eosin

A Normal 100x C 8 WKS 100X E 16 WKS 100X G 21 WKS 400X

rHABP1

B Normal 200x D 8 WKS 200X F 16 WKS 100X H 21 WKS 400X

Ghosh et al., (2004) Mol. and Cell. Biochemistry 267, 133-139

◀ **FIGURE 4.2**   Immunohistochemical analysis. Immunohistochemistry of papilloma sections with anti-HABP1 antibody (anti-HABP1) [b, d, f, h] compared with hematoxilin eosin (he) stained [a, c, e, g] stages of metastasis. Normal and the B(a)P treated mice and their papillomatic tissues were collected after 8 weeks, 16 weeks, 21 weeks, each group being comprised of six mice. The normal skin tissue displaying reticulopapillodermal appendages into fatty subcutaneous tissue are shown in (A). With the initiation of lesions, varying degrees of dysplasia were observed in squamous cells at about 8 weeks (C). The hypoplastic hair follicles and sebaceous glands of early lesions gradually became hyperplastic, giving rise to advanced hyperkeratosis and acanthosis in 16-week lesions (E). The 21-week biopsy plates (G) demonstrated the proliferating clumped cells deep down, with pleomorphism and atypia in rapidly dividing nuclei, forming proliferating metastatic islands. Normal skin tissue was examined which showed expression throughout the epidermis squamous tissues, as well as in the reticulopapillodermal appendages (B). At 8 weeks of carcinogenesis, advanced chronic inflammation was observed infiltrating into the submucosa with HABP1 accumulation (D). The infiltrating proliferating squamous cells, with irregular hyperkeratosis extending towards the skin had cyst-like appendages of acanthosis and papillomatosis of the epidermis, and the accumulation of HABP1 (F). Examination of the epidermal carcinoma at 21 weeks with anti-HABP1 antibody revealed the accumulation of HABP1 in all proliferating squamous epidermal tissues, embedded papillary dermal cells, and compact islands of proliferating metastasis cells (H). As a control when sections were probed with pre-immune serum, no detectable DAB precipitation was observed (data not shown). (See Page 2 in Color Section at the back of the book).

stages of tumor progression, that lead to disruption of its interaction with HA, implying a role in the regulation of tumor metastasis.

With this in view, accumulation of HABP1 in proliferating compact metastatic islands and its downregulation in the surrounding tissues, suggest the disruption of HA–HABP1 interactions. This leads to modification of tissue organization and promotion of invasion. HABP1/gC1qR is cleaved by membrane-type1 matrix metalloproteinase (MT1-MMP1) which has a crucial role in tumor invasion by degrading and remodeling ECM (Seiki and Yana, 2003). The cleavage site of MTI-MMP resides specifically in HABP1 at Gly^{79} Gln^{80} lying within the structurally disordered connecting $\beta3$ and $\beta4$ strands of HABP1 (Jiang et al., 1999). This reduction in the level of HABP1 in the surrounding mass of metastatic tissue can be attributed to its high sensitivity to MTI-MMP degradation (Ghosh et al., 2004).

Using the combinatory Ig library with phage display, overexpression of gC1qR in a variety of adenocarcinomas was observed in thyroid, colon, pancreatic, gastric, esophageal, and lung adenocarcinomas (Rubinstein et al., 2004).

ASSOCIATION OF ALLELIC LOSS AT 17P13.3 CHROMOSOME WITH CANCER INITIATION AND PROMOTION

With the advent of the human genome project, as the hidden scripts of DNA are revealed, the mysteries behind various diseases are gradually becoming clear. Population genetics aims to discover genetic variants (or polymorphisms) that influence traits by assessing correlations between genetic and phenotypic variation. The most abundant source of genetic variation in the human genome comprises single nucleotide polymorphisms (SNPs). Associating SNPs with human disease phenotypes has great potential for direct clinical application by providing new and more accurate genetic markers for diagnostic and prognostic purposes and, possibly, novel therapeutic targets.

The gene encoding HABP1/p32/gC1qR is reported to be localized on chromosome 17p13.3. Literature reviews showed that multiple chromosome 17 loci may be involved in cancer initiation and promotion. Loss of heterozygosity (LOH) studies have established that chromosome 17 may be a hotspot for chromosomal aberrations in various cancers like ovarian cancer, colorectal cancer, breast cancer, esophageal squamous cell carcinoma, and others. 17p13.3 is one such allele that gets lost in these carcinomas. Phillips et al. (1996) examined sporadic epithelial tumors for LOH at different loci on chromosome 17p. They found 80% of informative tumors had allelic loss in 17p13.3. Huang et al. (2000) showed that loss of this locus is associated with esophageal squamous cell carcinoma while Haga et al. (2001) reported its possible involvement in breast cancer. Therefore, it is hypothesized that HABP1 may be involved in the cancer initiation and promotion. Thus, it is now important to screen the SNPs in HABP1 to find out its association with various carcinomas as we already know that HABP1 is expressed differentially during the various stages of tumor induction and it is shown to be involved in several cellular processes that are linked with tumor initiation and promotion.

Here, it is pertinent to mention the recent report (Fogal et al., 2008), that not only supports our hypothesis but elucidates the specific role of this protein as a molecular target in tumor cells and tumor stroma. p32/gC1qR/HABP1 was shown as a cellular receptor for LYP-1, a tumor homing peptide that binds to tumor associated lymphatic vessels having antitumor activity. Corroborating our findings, they too consider the significant elevation of HABP1 cell surface level as a new marker of tumor cells.

CONCLUSION

HABP1 is a ubiquitously present glycoprotein having specific affinity towards HA, originally purified from rat tissue. Having immunological cross-reactivity with human tissue, it was used to identify the human homologue and shown to be located at human chromosome 17p13.3. Genomic organization and sequence analysis confirm its multifunctionality; it is identical with SF2/p32, the protein co-purified with splicing factor on the receptor of globular C1q and shown as a synonym in the human genome.

Experimental evidence suggests the probable involvement of HABP1 in tumor development. It is shown to be an endogenous substrate of MAP kinase, and is involved in cell cycle regulation and apoptosis induction. Upon ectopic expression of HABP1 there is generation of excess ROS in the fibroblasts, creating intrinsic mitochondrial dysfunction along with induction of autophagic vacuoles and ultimately apoptosis. Finally, differential expression of HABP1 and loss of HABP1 in skin papilloma suggest its involvement in tumor initiation and promotion. Linkage analysis with different cancer patient samples also suggests the loss of heterozygosity of human chromosome 17p13.3, where HABP1 is located.

ACKNOWLEDGMENTS

Financial assistance for this work is duly acknowledged to Department of Biotechnology, Department of science and Technology and Council of Scientific and Industrial Research (Govt. of India).

References

Auvinen, P., Tammi, R., Parkkinen, J., et al. (2000). Hyaluronan in peritumoral stroma and malignant cells associates with breast cancer spreading and predicts survival. *Am J Pathol* **156**, 529–536.

Chowdhury, A. R., Ghosh, I., and Datta, K. (2008). Excessive reactive oxygen species induces apoptosis in fibroblasts, role of mitochondrially accumulated hyaluronic acid binding protein 1 (HABP1/p32/gC1qR). *Exp Cell Res* **314** (3), 651–667.

Das, S., Deb, T. B., Kumar, R., and Datta, K. (1997). Multifunctional activities of human fibroblast 34-kDa hyaluronic acid-binding protein. *Gene* **190**, 223–225.

Deb, T. B. and Datta, K. (1996). Molecular cloning of human fibroblast hyaluronic acid binding protein confirms its identity with P-32, a protein co-purified with splicing factor SF2. *J Biol Chem* **269**, 2206–2212.

Dedio, J., Jahnen-Dechent, W., Bachmann, M., and Muller-Esterl, W. (1998). The multiligand-binding protein gC1qR, putative C1q receptor, is a mitochondrial protein. *J Immunol* **160** (7), 3534–3542.

Delpech, B., Maingonnat, C., Girard, N., et al. (1993). Hyaluronan and hyaluronectin in the extracellular matrix of human brain tumor stroma. *Eur J Cancer* **29**, 1012–1017.

D'Souza, M. and Datta, K. (1986). A novel glycoprotein that binds to hyaluronic acid. *Biochem Int* **13**, 79–88.

Fogal, V., Zhang, L., Krajewski, S., and Ruoslahti, E. (2008). Mitochondrial/Cell-surface protein p32/gC1qR as a molecular target in tumor cells and tumor stroma. *Cancer Res* **68** (17), 7210–7218.

Ghebrehiwet, B., Lim, B. L., Peerschke, E. I., Willis, C. A., and Reid, K. B. (1994). Isolation, cDNA cloning, and overexpression of a 33-kDa cell surface glycoprotein that binds to the globular "Heads" of C1q. *J Exp Med* **179**, 1809–1821.

Ghosh, I., Roy Chowdhury, A., Rajeswari, M. R., and Datta, K. (2004). Differential expression of hyaluronic acid binding protein 1 during progression of epidermal carcinoma. *Mol Cell Biochem* **267**, 133–139.

Guo, W. X., Ghebrehiwet, B., Weksler, B., Schweitzer, K., and Peerschke, E. I. (1999). Up-regulation of endothelial cell binding proteins/receptors for complement component C1q by inflammatory cytokines. *J Lab Clin Med* **133**, 541–550.

Gupta, S., Babu, B. R., and Datta, K. (1991). Purification, partial characterization of rat kidney hyaluronic acid binding protein and its localization on cell surface. *Eur. J. Cell Biol.* **56**, 58–67.

Gupta, S. and Datta, K. (1991). Possible role of hyaluronectin on cell adhesion in rat histiocytoma. *Exp Cell Res.* **195** (2), 386–394.

Haga, S., Emi, M., Hirano, A., et al. (2001). Association of allelic losses at 3p25.1, 13q12, or 17p13.3 with poor prognosis in breast cancers with lymph node metastasis. *Jpn J Cancer Res.* **92**, 1199–1206.

Hattori, M., Fujiyama, A., Taylor, T. D., et al. (2000). Chromosome 21 mapping and sequencing consortium. The DNA sequence of human chromosome 21. *Nature* **405** (6784), 311–319.

Honore, B., Madsen, P., Rasmussen, H. H., et al. (1993). Cloning and expression of a cDNA covering the complete coding region of the P32 subunit of human pre-mRNA splicing factor SF2. *Gene* **134**, 283–287.

Huang, J., Hu, N., Goldstein, A. M., et al. (2000). High frequency allelic loss on chromosome 17p13.3-p11.1 in esophageal squamous cell carcinomas from a high incidence area in northern China. *Carcinogenesis* **21**, 2019–2026.

Jha, B. K., Mitra, N., Rana, R., et al. (2004). pH and cation-induced thermodynamic stability of human hyaluronan binding protein 1 regulates its hyaluronan affinity. *J Biol Chem* **279**, 23061–23072.

Jha, B. K., Salunke, D. M., and Datta, K. (2003). Structural flexibility of multifunctional HABP1 may be important for regulating its binding to different ligands. *J Biol Chem* **278** (30), 27464–27472.

Jiang, J., Zhang, Y., Krainer, A. R., and Xu, R. M. (1999). Crystal structure of human p32, a doughnut-shaped acidic mitochondrial matrix protein. *Proc Natl Acad Sci USA* **96**, 3572–3577.

Kamal, A. and Datta, K. (2005). Upregulation of hyaluronan binding protein 1 (HABP1/p32/gC1qR) is associated with cisplatin induced apoptosis. *Apoptosis* **11**, 861–874.

Kosaki, R., Watanabe, K., and Yamaguchi, Y. (1999). Overproduction of hyaluronan by expression of the hyaluronan synthase HAS-2 enhances anchorage-independent growth and tumorigenicity. *Cancer Res* **59** (5), 1141–1151.

Krainer, A. R., Mayeda, A., Kozak, D., and Binns, G. (1991). Functional expression of cloned human splicing factor SF2, homology to RNA-binding proteins, U1 70K, and Drosophila splicing regulators. *Cell* **66**, 383–394.

Laurent, T. C., Laurent, U. B., and Fraser, J. R. (1996). Serum hyaluronan as a disease marker. *Ann Med* **28** (3), 241–253.

Lee, J. Y. and Spicer, A. P. (2000). Hyaluronan, a multifunctional, megadalton, stealth molecule. *Curr Opin Cell Biol* **12**, 581–586.

Lim, B. L., White, R. A., Hummel, G. S., et al. (1998). Characterization of the murine gene of gC1qBP, a novel cell protein that binds the globular heads of C1q, vitronectin, high molecular weight kininogen and factor XII. *Gene* **209**, 229–237.

Liu, D., Pearlman, E., and Diaconu, E. (1996). Expression of hyaluronidase in tumor cells induces angiogenesis *in vivo*. *Proc Natl Acad Sci USA* **93**, 7832–7837.

Majumdar, M., Bharadwaj, A., Ghosh, I., Ramachandran, S., and Datta, K. (2002a). Evidence for the presence of HABP1 pseudogene in multiple locations of mammalian genome. *DNA Cell Biol* **21**, 727–735.

Majumdar, M. and Datta, K. (1998). Assignment of cDNA encoding hyaluronic acid binding protein 1 to human chromosome 17 p12-13. *Genomics* **51**, 476–477.

Majumdar, M., Meenakshi, J., Goswami, S.K. and Datta K. (2002b). Hyaluronan binding protein 1 (HABP1)/C1QBP/p32 is an endogenous substrate for MAP kinase and is translocated to the nucleus upon mitogenic stimulation. *Biochem Biophys Res Commun* **291**, 829–837.

Matthews, D. A. and Russell, W. C. (1998). Adenovirus core protein V interacts with p32 – a protein which is associated with both the mitochondria and the nucleus. *J Gen Virol* **79** (7), 1677–1685.

Meenakshi, J., Anupama, Goswami, S. K., and Datta, K. (2003). Constitutive expression of hyaluronan binding protein 1 (HABP1/p32/gC1qR) in normal fibroblast cells perturbs its growth characteristics and induces apoptosis. *Biochem Biophys Res Commun* **300**, 686–693.

Muta, T., Kang, D., Kitajima, S., Fujiwara, T., and Hamasaki, N. (1997). p32 protein, a splicing factor 2-associated protein, is localized in mitochondrial matrix and is functionally important in maintaining oxidative phosphorylation. *J Biol Chem* **272** (39), 24363–24370.

Peterson, R. M., Yu, Q., Stamenkovic, I., and Toole, B. P. (2000). Perturbation of hyaluronan interactions by soluble CD44 inhibits growth of murine mammary carcinoma cells in ascites. *Am J Pathol* **156** (6), 2159–2167.

Phillips, N. J., Ziegler, M. R., Radford, D. M., et al. (1996). Allelic deletion on chromosome 17p13.3 in early ovarian cancer. *Cancer Res* **56** (3), 606–611.

Rubinstein, D. B., Stortchevoi, A., Boosalis, M., et al. (2004). Receptor for the globular heads of C1q (gC1q-R, p33, hyaluronan-binding protein) is preferentially expressed by adeno-carcinoma cells. *Int J Cancer* **110**, 741–750.

Seiki, M. and Yana, I. (2003). Roles of pericellular proteolysis by membrane type-1 matrix metalloproteinase in cancer invasion and angiogenesis. *Cancer Res* **94** (7), 569–574.

Sengupta, A., Ghosh, I., Mallick, J., Thakur, A. R., and Datta, K. (2004). Presence of a human hyaluronan binding protein 1 (HABP1) pseudogene-like sequence in *Methanosarcina barkeri* suggests its linkage in evolution. *DNA Cell Biol* **23**, 301–310.

Sengupta, A., Tyagi, R. K., and Datta, K. (2004). Truncated variants of hyaluronan-binding protein 1 bind hyaluronan and induce identical morphological aberrations in COS-1 cells. *Biochem J* **380**, 837–844.

Seytter, T., Lottspeich, F., Neupert, W., and Schwarz, E. (1998). Mam33p, an oligomeric, acidic protein in the mitochondrial matrix of *Saccharomyces cerevisiae* is related to the human complement receptor gC1q-R. *Yeast* **14** (4), 303–310.

Simos, G. and Georgatos, S. D. (1994). The lamin B receptor-associated protein p34 shares sequence homology and antigenic determinants with the splicing factor 2-associated protein p32. *FEBS Lett* **346** (2–3), 225–228.

Soltys, B. J., Kang, D., and Guptam, R. S. (2000). Localization of P32 protein (gC1q-R) in mitochondria and at specific extramitochondrial locations in normal tissues. *Histochem Cell Biol* **114**, 245–255.

Toole, B. P. and Hascall, V. C. (2002). Hyaluronan and tumor growth. *Am J Pathol* **161** (3), 745–747.

Toole, B. P. (1990). Hyaluronan and its binding proteins, the hyaladherins. *Curr Opin Cell Biol* **2** (5), 839–844.

Turley, E. A., Noble, P. W., and Bourguignon, L. Y. W. (2002). Signaling properties of hyaluronan receptors. *J Biol Chem* **277**, 4589–4592.

Van Leeuwen, H. C. and O'Hare, P. (2001). Retargeting of the mitochondrial protein p32/gC1Qr to a cytoplasmic compartment and the cell surface. *J Cell Sci* **114** (11), 2115–2123.

Wall, S. J., Jiang, Y., Muschel, R. J., and DeClerck, Y. A. (2003). Meeting report. Proteases, extracellular matrix, and cancer. An AACR Special Conference in Cancer Research. *Cancer Res* **63** (15), 4750–4755.

Yang, B., Yang, B. L., Savani, R. C., and Turley, E. A. (1994). Identification of a common hyaluronan binding motif in the hyaluronan binding proteins RHAMM, CD44 and link protein. *EMBO J* **13**, 286–296.

Yu, Q. and Stamenkovic, I. (1999). Localisation of matrix metalloproteinase 9 to the cell surface provides a mechanism for CD44 mediated tumor invasion. *Genes Dev* **13**, 35–48.

HYALURONAN RECEPTORS AND SIGNAL TRANSDUCTION PATHWAYS

CD44 Meets Merlin and Ezrin: Their Interplay Mediates the Pro-Tumor Activity of CD44 and Tumor-Suppressing Effect of Merlin

Ivan Stamenkovic and Qin Yu

OUTLINE

CD44 PROMOTES TUMOR INVASION AND METASTASIS

Long before the CD44 gene was identified, the hyaluronan (HA)-enriched extracellular matrix had been shown to support and promote cell migration during embryogenesis, inflammation, and tumorigenesis (Lesley et al., 1993; Sherman et al., 1994; Kincade et al., 1997; Stamenkovic, 2000). Moreover, HA production is often increased at the sites of tumor invasion and metastasis, and the binding affinity of CD44 for HA is often increased in proliferating and migrating cells (for a review, see Toole, 2004; Fig. 5.1A,B). As the primary cell-surface receptor for HA (Aruffo et al., 1990), CD44 binds to HA through its link protein module (Stamenkovic et al., 1989; Banerji et al., 2007). Many of CD44's functions are mediated through its binding to HA, including its ability to promote tumor invasion and angiogenesis by localizing matrix metalloproteinase-9 (MMP-9) on the tumor cell surface and by activating latent TGF-beta through its associated MMP-9 (Yu and Stamenkovic, 1999; 2000).

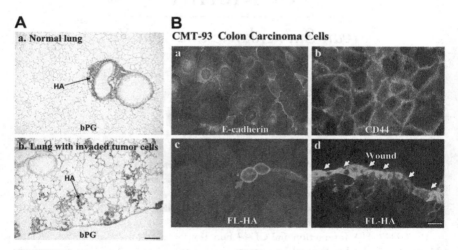

FIGURE 5.1 HA is upregulated at the sites of tumor invasion and metastasis, and the binding affinity of CD44 to HA is tightly regulated. **A**, HA production in normal mouse lung (A-a) and at the sites of infiltration by metastatic TA3 mammary carcinoma cells (A-b). Mouse lung sections, with or without tail-vein injection of TA3 mammary carcinoma cells 72 hours earlier, were used to detect HA production *in situ* with a biotinylated HA-binding proteoglycan (bPG, Yu et al., 1997) probe. **B**, Binding of fluorescein-labeled HA (FL-HA) to CMT-93 colon carcinoma cells (Xu and Yu, 2003) is enhanced in proliferating (B-c) and migrating (B-d) cells. CMT-93 cells express a high level of E-cadherin (B-a) and CD44 (B-b). A arrow in B-c indicates the proliferating CMT-93 cells and the white arrows in B-d show the migrating tumor cells at the edge of a newly formed wound, which was formed by a single cross of the monolayer of CMT-93 cells with a sterile yellow pipette tip. (See Page 3 in Color Section at the back of the book).

A recent study has further confirmed the functional relationship between CD44 and activation of TGF-beta signaling. This study demonstrated that CD44+ breast carcinoma cells are enriched in stem-cell markers and display a poor clinical outcome and an activated TGF-beta signaling pathway (Shipitsin et al., 2007). An interaction between CD44 and HA has also been found to promote chemoresistance of non-small cell lung carcinoma cells (Ohashi et al., 2007). Furthermore, studies have shown that CD44 expression is higher in tumors than in the corresponding normal tissues (Masumura and Tarin, 1992; Notterman et al., 2001; Sun et al., 2006 – data derived from www.ncbi.nlm.nih.gov/geo/query/acc.cgi?acc= GSE4290) and that highlevels of CD44 expression correlate with poor prognosis in non-Hodgkin's lymphoma (Pals et al., 1989; Jalkanen et al., 1991).

Several lines of evidence support the notion that CD44 promotes tumor progression: for example, overexpression of CD44 promotes tumor growth and metastasis in several experimental tumor models (Sy et al., 1991; 1992; Bartolazzi et al., 1994); and the v6-containing CD44 isoform confers metastatic behavior on rat pancreatic carcinoma cells (Gunthert et al., 1991). In addition, there is a good correlation between increased expression of CD44 and enhanced tumor invasiveness by some tumor types. For example, a CD44 proteoglycan can promote the motility and invasiveness of a melanoma cell line (Faassen et al., 1992), the expression of CD44 isoforms is correlated with the invasive capacity of various human breast carcinoma cell lines (Culty et al., 1994), increased CD44 expression is required for AP-1-mediated tumor invasion (Lamb et al., 1997), and CD44 promotes breast cancer invasion and metastasis to the liver (Ouhtit et al., 2007). Moreover, CD44 plays an essential role in promoting invasiveness and survival of metastatic breast carcinoma cells (Yu and Stamenokovic, 1999; 2000; Yu et al., 1997) and disruption of the CD44 gene has been reported to prevent the metastasis of osteosarcomas (Weber et al., 2002).

In contrast to these results, several other reports have suggested a very different role for CD44 in influencing the growth and metastasis of certain types of cancer. In particular, they have indicated that the expression of CD44 is repressed in neuroblastoma cells (Shtivelman and Bishop, 1991) and it is not required for the growth and metastasis of MDAY-D2 lymphosarcoma cells (Driessens et al., 1995). In addition, CD44 has been implicated in tumor suppression in prostate cancers (Kallakury et al., 1996; Gao et al., 1997; Lou et al., 1999); however, a recent study has shown that highly purified CD44+ prostate carcinoma cells are enriched in prostate cancer stem/initiating cells that display a high level of tumorigenicity and metastatic capacity (Patrawala et al., 2006). Furthermore, CD44+CD24+ESA+ pancreatic cancer cells (Li et al., 2007) and the CD44+ population of human head and neck squamous cell carcinoma cells (HNSCCs) (Prince et al., 2007) have been described as displaying the stem cell properties and CD44+CD24−/low breast cancer-initiating cells are more resistant to radiotherapy (Phillips et al., 2006) and

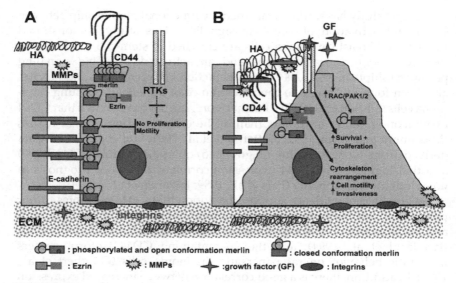

FIGURE 5.2 A working model of CD44 signaling through merlin and the ERM proteins. A detailed description of this model is given in the text.

display higher tumorigenicity than do other carcinoma cell subtypes (Liu et al., 2007). Moreover, an essential role has been suggested for CD44 in engraftment of leukemia stem/initiating cell in the bone marrow, an event that is essential to leukemia development (Williams and Cancelas, 2006). Together, these results point to an essential role for CD44 in the formation and/or maintenance of cancer stem cells.

The apparent discrepancies in the results of these studies most likely reflect the fact that CD44 plays different roles in different types of tumors or at different stages of tumor development, in part, because of differences in the availability of downstream CD44 effectors in the cellular microenvironment. In particular, recent studies have shown that the interaction of CD44 with the ERM proteins and/or merlin can lead to different functional outcomes (Figs. 5.2 and 5.3), and these differences may help explain the apparent inconsistencies in the observed effects of CD44.

CD44-MEDIATED SIGNALS AND THE DOWNSTREAM EFFECTORS OF CD44

CD44 has been linked to numerous signaling pathways: The CD44–HA interaction activates c-Src and focal adhesion kinase (FAK, Turley et al., 2002); CD44 serves as a co-receptor for c-Met (Matzke et al., 2007), and it modulates signals by members of the ErbB family of receptor tyrosine

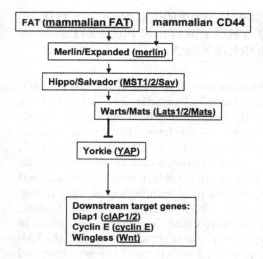

FIGURE 5.3   Merlin-Hpo–Wts–Yki signaling pathway in *Drosophila*. The mammalian homologs/orthologs of the components of the *Drosophila* mer-Hippo (Hpo)–Warts (Wts)–Yorkie (Yki) pathway are underlined.

kinases (RTKs) and of the receptors of transforming growth factor (TGF)-beta (Turley et al., 2002; Bourguignon et al., 2002; 2007); it signals through its own COOH-terminal cytoplasmic proteolytic cleavage fragment after shedding its extracellular domain (Itoh and Seiki, 2006); it promotes cell migration by activating Rac1 (Murai et al., 2004); and finally, it signals through its downstream effectors, the best characterized of which are the ERM proteins and merlin (Okada et al., 2007; Edgar, 2007; McClatchey and Giovanni, 2005; Herrlich et al., 2000), which will be the focus of this discussion.

The ERM proteins and merlin serve as cross-linkers between cortical actin filaments and the plasma membrane and regulate actin-cytoskeleton organization and cell motility (Okada et al., 2007; Edgar, 2007; McClatchey and Giovanni, 2005; Hughes and Fehon, 2007). CD44 is among several transmembrane proteins that bind to the ERM proteins and merlin (Tsukita et al., 1994; 1997; Hirao et al., 1996; Sainio et al., 1997; Morrison et al., 2001). Several basic residues in the cytoplasmic tail of CD44 are essential for the interaction between CD44 and the ERM proteins (Yonemura et al., 1998; Legg and Isacke, 1998). Merlin is the product of the neurofibromatosis type 2 (NF2) gene. Mutations in merlin or a loss of its expression are responsible for the NF2 disease, which is characterized by the development of schwannomas, meningiomas, and ependymomas (Baser et al., 2003; McClatchey and Giovanni, 2005; Edgar, 2007; Okada et al., 2007). Two major merlin isoforms, I and II, are products of alternative splicing near the 3′ end of the NF2 gene. Only merlin isoform I has been shown to exhibit tumor-suppressing activity (Sherman et al., 1997).

MERLIN ACTS AS A TUMOR SUPPRESSOR, AND THE MUTANT FORMS OF THIS PROTEIN PROMOTE TUMORIGENESIS

Like the ERM proteins, merlin regulates cytoskeletal remodeling and cell motility (McClatchey and Giovanni, 2005; Edgar, 2007; Okada et al., 2007). Merlin and the ERM proteins display a similar subcellular localization that includes areas of dynamic cytoskeleton remodeling, such as microspikes and membrane ruffles (Arpin et al., 1994; Bretscher et al., 2002). It has been well established that mutations and deletions of merlin cause NF2 and that loss of heterozygosity (LOH) of merlin is associated with sporadic schwannomas, ependymomas, meningiomas, and malignant mesotheliomas (Gutmann et al., 1997; Baser et al., 2003). Genetic analysis of NF2 patients has demonstrated that the deletion mutations in the NH_2-terminal the band four-point-one/ezrin/radixin/moesin (FERM) domain occur frequently and are associated with the early tumor onset and poor prognosis (Ruttledge et al., 1996; Koga et al., 1998).

Overexpression of several merlin mutants causes overproliferation of *Drosophila* wing epithelial cells by interfering with the activity of endogenous wild-type (wt) merlin (LaJeunesse, et al., 1998). Transgenic mice expressing a naturally occurring merlin mutant or having a conditional homozygous merlin knockout in Schwann cells develop schwannomas that resemble those associated with NF2 (Giovanni et al., 1999; 2000). Loss of merlin is embryonically lethal in both the mouse and the fruit fly, suggesting broader and essential roles for this protein during embryonic development (Fehon et al., 1997; McClatchey et al., 1997). In addition, mice with a heterozygous merlin knockout (NF2+/−) genotype develop metastatic osteosarcomas, fibrosarcomas, and hepatocellular carcinomas and nearly all of these tumors have lost the wt NF2 allele (McClatchey et al., 1998). Together, these results suggest that merlin serves as a tumor suppressor in a wide array of cells and that loss of merlin may play an important role in tumor progression.

THE FUNCTION OF MERLIN IS REGULATED BY ITS POSTTRANSLATIONAL MODIFICATIONS

The domain organization of merlin is similar to that of the ERM proteins. The highest sequence homology between merlin and the ERM proteins is in the conserved tri-lobe NH_2-terminal FERM domain. In addition to the NH_2-terminal domain (head), merlin has a central alpha-helical region and an extended COOH-terminal tail (Pearson et al., 2000; Shimizu et al., 2002). Like the ERM proteins, merlin can form head-to-tail intra- and intermolecular associations. The conserved residues in the COOH-terminus of merlin and the ERM proteins constitute the C-ERM association domains (C-ERMADs), which mediate the interaction with their corresponding NH_2-termini.

Head-to-tail self-association, which generates the closed conformation form of merlin, is required for the protein's tumor-suppressing activity (Sherman et al., 1997). Phosphorylation of merlin at the COOH-terminus abolishes this head-to-tail self-association, producing an open conformation and inactivating the tumor-suppressing activity of the molecule (Sherman et al., 1997; Shaw et al., 2001; Kissil et al., 2003). p21-activated kinase 1 (PAK1) and 2 (PAK2), as well as cAMP-dependent protein kinase A (PKA), phosphorylate merlin at Ser518 (Shaw et al., 2001; Kissil et al., 2003; Thaxton et al., 2007), whereas AKT phosphorylates merlin at Thr 230 and Ser 315 (Tang et al., 2007). Phosphorylation of merlin at S518 produces the open conformation (Shaw et al., 2001; Kissil et al., 2003; Thaxton et al., 2007), whereas AKT-mediated phosphorylation at Thr 230 and Ser 315 promotes merlin degradation through ubiquitination (Tang et al., 2007). Conversely, the myosin phosphatase MYPT1–PP1δ dephosphorylates Ser518, resulting in merlin activation (Jin et al., 2006). In addition, merlin is cleaved by calpain, a calcium-dependent cysteine protease, in schwannomas and meningiomas (Kimura et al., 2000), suggesting that merlin can be inactivated and downregulated by calpain-mediated proteolytic cleavage.

MERLIN COUNTERACTS THE EFFECT OF EZRIN AND INHIBITS THE CD44–HA INTERACTION AND CD44 FUNCTION

Studies have shown that merlin interacts with several transmembrane proteins, including CD44 (Tsukita et al., 1994; 1997; Hirao et al., 1996; Bai et al., 2007), NHE-RF/EBP50 (Murthy et al., 1998), β1 integrin (Obremski et al., 1998), and FAT, a large protocadherin (Willecke et al., 2006). Merlin also associates with spectrin (Scoles et al., 1998) and tubulin (Xu et al., 2004). Like the ERM proteins, merlin interacts with membrane-associated binding partners, including CD44, through its NH_2-terminal domain (Tsukita et al., 1994; 1997; Hirao et al., 1996; Bai et al., 2007). The ERM proteins play an important role in organizing the actin-cytoskeleton, which is essential for cell adhesion and migration. In particular, ezrin is involved in promoting tumor metastasis, most likely through its interaction with CD44 (Yu et al., 2004; Khanna et al., 2004). Unlike the ERM proteins, merlin lacks the conventional COOH-terminal actin-binding site, but it contains an unconventional NH_2-terminal actin-binding site (Xu et al., 1998) and interacts indirectly with the actin-cytoskeleton through an actin-binding protein, βII-spectrin (Scoles et al., 1998). The association of merlin with the actin-cytoskeleton likely underlies its effects on the actin-cytoskeletal organization and on cell spreading and motility.

III. HYALURONAN RECEPTORS AND SIGNAL TRANSDUCTION PATHWAYS

CD44 plays an important role in promoting tumor growth and metastasis, chiefly as a result of its ability to bind HA (for reviews, see Lesley et al., 1993; Sherman et al., 1994; Kincade et al., 1997; Stamenkovic, 2000; Toole, 2004), an interaction that is tightly regulated. Studies have shown that E-cadherin negatively regulates the interaction between CD44 and HA as well CD44 function (Xu and Yu, 2003), and overproduction of HA promotes anchorage-independent growth of CD44-positive breast epithelial cells by inducing an epithelial–mesenchymal transition (Zoltan-Jones et al., 2003), presumably as a result of downregulating E-cadherin-mediated cell–cell adhesion. Furthermore, merlin plays an important role in mediating contact inhibition of cell growth through its interaction with CD44 (Morrison et al., 2001), and loss of merlin destabilizes the cadherin-containing cell–cell junctions, leading to a loss of contact inhibition and elevation of extracellular signal-regulated kinase (ERK) and c-Jun-NH2-kinase (JNK) activity (Shaw et al., 2001; Lallemand et al., 2003; Okada et al., 2005). We have recently shown that increased expression of wt merlin in Tr6BC1 schwannoma cells inhibits the binding of HA to CD44 (Bai et al., 2007). Furthermore, we found that overexpression of merlin inhibits the subcutaneous growth of Tr6BC1 cells in immunocompromised Rag1 mice. In contrast, knocking down the expression of endogenous merlin with shRNAs or overexpression of a merlin deletion mutant that is incapable of inhibiting the CD44–HA interaction can promote subcutaneous growth of Tr6BC1 cells in Rag1 mice (Bai et al., 2007). We have also shown that increased expression of merlin not only inhibits the CD44–HA interaction but also reduces the interaction between CD44 and ezrin and that between CD44 and the actin-cytoskeleton. Together, these results indicate that inhibition of the CD44–HA interaction contributes to the tumor suppressor activity of merlin, and they suggest that the various downstream effectors of CD44, the ERM proteins or merlin, play different or even opposing roles in organizing cortical actin and in tumorigenesis, presumably by activating different signaling pathways (Bai et al., 2007; Figs. 5.2 and 5.3).

MERLIN IS INVOLVED IN SEVERAL IMPORTANT SIGNAL TRANSDUCTION PATHWAYS

Tight control of cell proliferation, death, and motility is critical for maintaining normal tissue homeostasis, and loss of this control is a hallmark of cancer. CD44 acts upstream of merlin, and inhibition of its interaction with HA is one mechanism that underlies the tumor-suppressing function of merlin (Bai et al., 2007). Merlin is uniquely localized for sensing and coordinating extracellular and intracellular signaling pathways. Studies have shown that increased expression of merlin not only inhibits cell proliferation and promotes apoptosis, but also impairs cell–cell and cell–matrix adhesion, cell spreading, and cell motility (Koga et al., 1998; Xu and Gutmann, 1998). The exact mechanism by which merlin exerts its tumor-suppressing function has not been fully established, but recent findings, including observations that place merlin in the Hippo pathway in *Drosophila* have shed some light on this issue.

Merlin has been shown to reverse the Ras-induced malignancy (Tikoo et al., 1994; Kim et al., 2002), suggesting that it negatively regulates the Ras-signaling pathway. The Rho family of small GTPases (Rho, CDC42, and Rac) are essential downstream effectors of the Ras signaling pathway: overexpression of activated Rac causes transformation of fibroblasts, and Rac activity is required for Ras-induced transformation (Qiu et al., 1995). Merlin has been shown to act downstream of Rac (Morrison et al., 2007), and its overexpression inhibits Rac-mediated anchorage-independent growth and transformation (Shaw et al., 2001). Merlin is thought to mediate contact inhibition of growth by promoting the formation and stabilization of adherens junctions (Lallemand et al., 2003) and to inhibit mitogenesis in confluent cells by blocking recruitment of Rac to the plasma membrane (Okada et al., 2005). In addition, expression of active Rac1 induces PAK-1-dependent phosphorylation of merlin at residue S518, causing merlin to adopt an inactive, open conformation (Kissil et al., 2003). PAKs (PAK1–4) are a group of serine/threonine kinases and the immediate downstream effectors of the Rac signaling pathway. Through a feedback loop, merlin inhibits PAK1 activity by binding to the Cdc42/Rac binding domain (PBD) of PAK1 and inhibiting recruitment of PAK1 to focal adhesions (Kissil et al., 2003). Therefore, loss of merlin results in increased activity of PAK1.

Genetic analyses in *Drosophila* have indicated a role for merlin in promoting endocytosis of a number of signaling receptors, including Notch, the epidermal growth factor receptor (EGFR), patched, and smoothened, and of adhesion receptors, E-cadherin and Fat (Maitra et al., 2006). Studies in mammalian cells have consistently indicated that merlin accumulates at and stabilizes the adherens junctions and that upon establishment of contact inhibition, merlin inhibits internalization of EGFR by sequestering it into an insoluble membrane compartment, thereby inhibiting EGFR signaling (Lallemand et al., 2003; Curto et al., 2007).

MERLIN ACTS UPSTREAM OF THE HIPPO SIGNALING PATHWAY IN *DROSOPHILA*

In *Drosophila*, merlin (*mer*) is one of two FERM domain-containing proteins that negatively regulate cell growth and proliferation, the other protein being, expanded (*ex*). *mer* is the functional homolog of human merlin, but whether a functional homolog of *ex* exists in mammals is not yet clear (Pellock et al., 2007). Inactivation mutants of *mer* or *ex* display reduced apoptosis and increased cell proliferation in *Drosophila* (McCartney et al., 2000; Hamaratoglu et al., 2006). Analysis of *mer* mutants has indicated that merlin signaling occurs upstream of the Hippo (Hpo)/ Salvador (Sav)/Warts (Wts)/Mob1 (Hippo) signaling pathway that regulates cell proliferation and survival (McCartney et al., 2000; Hamaratoglu et al., 2006; Curto et al., 2007; Pellock et al., 2007). Hpo and Wts are Thr/Ser kinases of the sterile 20 (Ste20) and the nuclear Dbf2-related (NDR) family, respectively. In *Drosophila*, Hpo phosphorylates and activates Wts, which in turn phosphorylates and inhibits Yorkie (Yki; Huang et al., 2005), leading to lose of its pro-proliferative and anti-apoptotic activity. Conversely, inactivation of the Hpo/Wts pathway results in upregulation and activation of Yki, which in turn upregulates *Drosophila* inhibitor of apoptosis protein 1 (DIAP1) and cyclin E, resulting in increased proliferation and cell survival (Huang et al., 2005; Pellock et al., 2007).

In *Drosophila*, studies have shown that FAT (Willecke et al., 2006; Hamaratoglu et al., 2006), merlin, and Hpo signals (McCartney et al., 2000; Huang et al., 2005; Hamaratoglu et al., 2006; Curto et al., 2007; Pellock et al., 2007) all converge in the activation of Wts and inactivation of Yki through which they influence the expression of a common set of downstream target genes that play essential roles in cell proliferation, survival, and motility, including cyclin E, DIAP1, and wingless (wg) (Cho et al., 2006; Fig. 5.3). The mammalian homologs/orthologs of Hpo, Wts, Yki, DIAP, and wg are mammalian sterile twenty-like (MST) kinase 1 and 2 (MST1/2) (Creasy and Chernoff, 1995), Lats1 and 2 (Chan et al., 2005; Takahashi et al., 2005; Aylon et al., 2006), yes-associated protein (YAP) (Overholtzer et al., 2006; Zender et al., 2006), cellular inhibitor of apoptosis1/2 (cIAP1/2), and the Wnt family proteins (Fodde and Brabletz, 2007; Fig. 5.3), respectively. Like MST1 and MST2, the Lats1/2 genes encode serine/threonine kinases and display anti-tumor activity (St John et al., 1999; Li et al., 2003). Loss of Lats1 causes a predisposition to soft-tissue sarcoma and ovarian tumor development in mice (St John et al., 1999). In addition, mouse embryonic fibroblasts derived from the Lats2-null mice show a growth advantage and display a profound defect in contact inhibition of growth (McPherson et al., 2004). Furthermore, overexpression of Lats1 induces G2/M arrest and subsequent apoptosis (Xia et al., 2002), and Lats2 suppresses RasV12-induced transformation of NIH 3T3 cells (Li et al., 2003).

In contrast, YAP, which encodes the mammalian ortholog of *Yki*, is amplified in human cancers and displays oncogenic activity (Overholtzer et al., 2006; Zender et al., 2006). As a transcriptional co-activator, YAP promotes cell proliferation and inhibits apoptosis by upregulating the expression of cyclin E and cIAP1/2, respectively (Yagi et al., 1999; Vassilev et al., 2001). Like *Yki*, YAP rescues the pupal lethality caused by overexpression of Hpo or Wts in *Drosophila* (Huang et al., 2005). Similarly, human homologs of Hpo and Wts can rescue their corresponding *Drosophila* mutants (Tao et al., 1999; Wu et al., 2003; Lai et al., 2005). However, it is not yet know whether the entire Hpo–Wts–Yki signaling pathway is conserved in mammalian cells, and the potential relationships between merlin and MST1/2, Lats1/2, YAP, and cIAP1/2 have not been fully established in mammalian cells. A very recent study showed that merlin is a potent inhibitor of high-grade human glioma and that increased expression of merlin correlates with activation of MST1/2-Lats2 signaling pathway and inhibition of Wnt signals (Lau et al., 2008).

A MODEL FOR CD44 SIGNALING THROUGH MERLIN AND THE ERM PROTEINS

Observations in mammalian cells and *Drosophila* point toward a model of CD44 signaling through merlin and the ERM proteins (Fig. 5.2). When there is no contact inhibition, and signaling from the activated receptor tyrosine kinases (RTKs) has occurred (Fig. 5.2B), PAK1 and PAK2 are activated, and these molecules in turn phosphorylate and inactivate merlin. Under these circumstances, CD44 predominantly interacts with ezrin and displays a high binding affinity for HA. The HA–CD44 interaction allows the cells to recruit MMPs and to activate latent TGF-beta, which stimulates motility and invasiveness. The cell-surface bound MMPs may also contribute to the shedding or cleavage of E-cadherin, which leads to further loss of adherens junction integrity and a loss of cell–cell adhesion. In addition, the HA–CD44 interaction enhances RTK-derived mitogenic and survival signals, further promoting the activation of RAC and PAK1/2 and the deactivation of merlin (Fig. 5.2B).

In contrast, once the cells have established contact inhibition through the homophilic E-cadherin interactions and experienced a decrease in the RTK-derived activation signals (Fig. 5.2A), merlin adopts an unphosphorylated closed conformation and displays tumor suppressor activity. Merlin is recruited to the plasma membrane and interacts with CD44 and E-cadherin, stabilizing the homophilic E-cadherin interaction. At this point, the CD44 pool that predominantly interacts with merlin can no longer bind HA, and therefore cell–cell adhesion is enhanced further. Merlin thus stabilizes cell–cell adhesion complexes and renders the cells stationary. Under these conditions (Fig. 5.2A), CD44 cannot function to enhance RTK-derived signals to effectively promote cell survival and proliferation. On the other hand, CD44-associated merlin can activate MST1/2 and Lats1/2, which in turn inactivate YAP. Loss of YAP activity can lead to a reduction in the level of cycle E, cIAP1/2, and Wnt signals, which causes a reduction in cell proliferation and motility and enhanced sensitivity to apoptosis (Figs. 5.2 and 5.3). Thus, the data accumulated to date suggest that merlin serves as a key point of contact between the extracellular and intracellular signaling networks, and that its function is exerted through its interaction with cell-surface adhesion receptors including, and perhaps predominantly, CD44.

ACKNOWLEDGMENTS

This work was supported by the grants from the Swiss National Scientific Research Foundation and the Molecular Oncology NCCR (to IS) and from the DOD-U.S. Army Medical Research (DAMD17-02-1-0650 and W81XWH-05-1-0191 to QY). We thank Dr. Bryan Toole for his support and Dr. Deborah McClellan for her excellent editorial assistance.

References

Arpin, M., Algrain, M., and Louvard, D. (1994). Membrane-actin microfilament connections: an increasing diversity of players related to band 4.1. *Curr Opin Cell Biol* **6**, 136–141.

Aruffo, A., Stamenkovic, I., Melnick, M., Underhill, C. B., and Seed, B. (1990). CD44 is the principal cell surface receptor for hyaluronate. *Cell* **61**, 1303–1313.

Aylon, Y., Michael, D., Shmueli, A., et al. (2006). A positive feedback loop between the p53 and Lats2 tumor suppressors prevents tetraploidization. *Genes Dev* **20**, 2687–2700.

Bai, Y., Liu, Y. J., Wang, H., Xu, Y., Stamenkovic, I., and Yu, Q. (2007). Inhibition of the hyaluronan–CD44 interaction by merlin contributes to the tumor-suppressor activity of merlin. *Oncogene* **26** (6), 836–850.

Banerji, S., Wright, A. J., Noble, M., et al. (2007). Structures of the Cd44–hyaluronan complex provide insight into a fundamental carbohydrate protein interaction. *Nat Struct Mol Biol* **14** (3), 234–239.

Bartolazzi, A., Peach, R., Aruffo, A., and Stamenkovic, I. (1994). Interaction between CD44 and hyaluronan is directly implicated in the regulation of tumor development. *J Exp Med* **180**, 53–66.

Baser, M. E., Evans, D. G., and Gutmann, D. H. (2003). Neurofibromatosis 2. *Curr Opin Neurol* **16** (1), 27–33.

Bourguignon, L. Y., Gilad, E., and Peyrollier, K. (2007). Heregulin-mediated ErbB2-ERK signaling activates hyaluronan synthases leading to CD44-dependent ovarian tumor cell growth and migration. *J Biol Chem* **282** (27), 19426–19441.

Bourguignon, L. Y., Singleton, P. A., Zhu, H., and Zhou, B. (2002). Hyaluronan promotes signaling interaction between CD44 and the transforming growth factor beta receptor I in metastatic breast tumor cells. *J Biol Chem* **277** (42), 39703–39712.

Bretscher, A., Edwards, K., and Fehon, R. G. (2002). ERM proteins and merlin: integrators at the cell cortex. *Nat Rev Mol Cell Biol* **3**, 586–599.

Chan, E. H., Nousiainen, M., Chalamalasetty, R. B., et al. (2005). The Ste20-like kinase Mst2 activates the human large tumor suppressor kinase Lats1. *Oncogene* **24**, 2076–2086.

Cho, E., Feng, Y., Rauskolb, C., et al. (2006). Delineation of a Fat tumor suppressor pathway. *Nat Genet* **38**, 1142–1150.

Creasy, C. L. and Chernoff, J. (1995). Cloning and characterization of a member of the MST subfamily of Ste20-like kinases. *Gene* **167**, 303–306.

Culty, M., Shizari, M., Thompson, E. W., and Underhill, C. B. (1994). Binding and degradation of hyaluronan by human breast cancer cell lines expressing different forms of CD44: correlation with invasive potential. *J Cell Physiol* **160**, 275–286.

Curto, M., Cole, B. K., Lallemand, D., Liu, C. H., and McClatchey, A. I. (2007). Contact-dependent inhibition of EGFR signaling by Nf2/Merlin. *J Cell Biol* **177** (5), 893–903.

Driessens, M. H., Stroeken, P. J., Rodriguez Erena, N. F., et al. (1995). Targeted disruption of CD44 in MDAY-D2 lymphosarcoma cells has no effect on subcutaneous growth or metastatic capacity. *J Cell Biol* **131**, 1849–1855.

Edgar, B. A. (2007). From cell structure to transcription: Hippo forges a new path. *Cell* **124** (2), 267–273.

Faassen, A. E., Schrager, J. A., Klein, D. J., et al. (1992). A cell surface chondroitin sulfate proteoglycan, immunologically related to CD44, is involved in type I collagen-mediated melanoma cell motility and invasion. *J Cell Biol* **116**, 521–531.

Fehon, R. G., Oren, T., LaJeunesse, D. R., Melby, T. E., and McCartney, B. M. (1997). Isolation of mutations in the Drosophila homologues of the human neurofibromatosis 2 and yeast CDC42 genes using a simple and efficient reverse-genetic method. *Genetics* **146**, 245–252.

Fodde, R. and Brabletz, T. (2007). Wnt/beta-catenin signaling in cancer stemness and malignant behavior. *Curr Opin Cell Biol* **19** (2), 150–158.

Gao, A. C., Lou, W., Dong, J. T., and Isaacs, J. T. (1997). CD44 is a metastasis suppressor gene for prostatic cancer located on human chromosome 11p13. *Cancer Res* **57**, 846–849.

Giovannini, M., Robanus-Maandag, E., Niwa-Kawakita, M., et al. (1999). Schwann cell hyperplasia and tumors in transgenic mice expressing a naturally occurring mutant NF2 protein. *Genes Dev* **13** (8), 978–986.

Giovannini, M., Robanus-Maandag, E., van der Valk, M., et al. (2000). Conditional biallelic Nf2 mutation in the mouse promotes manifestations of human neurofibromatosis type 2. *Genes Dev* **14**, 1617–1630.

Gunthert, U., Hofmann, M., Rudy, W., et al. (1991). A new variant of glycoprotein CD44 confers metastatic potential to rat carcinoma cells. *Cell* **65**, 13–24.

Gutmann, D. H., Giordano, M. J., Fishback, A. S., and Guha, A. (1997). Loss of merlin expression in sporadic meningiomas, ependymomas and schwannomas. *Neurology* **49**, 267–270.

Hamaratoglu, F., Willecke, M., Kango-Singh, M., et al. (2006). The tumour-suppressor genes NF2/merlin and expanded act through Hippo signalling to regulate cell proliferation and apoptosis. *Nat Cell Biol* **8**, 27–36.

Herrlich, P., Morrison, H., Sleeman, J., et al. (2000). CD44 acts both as a growth- and invasiveness-promoting molecule and as a tumor-suppressing cofactor. *Ann NY Acad Sci* **910**, 106–118, discussion 118–120.

Hirao, M., Sato, N., Kondo, T., et al. (1996). Regulation mechanism of ERM protein/plasma membrane association: possible involvement of phosphatidylinositol turnover and rho-dependent signaling pathway. *J Cell Biol* **135**, 37–52.

Huang, J., Wu, S., Barrera, J., Matthews, K., and Pan, D. (2005). The Hippo signaling pathway coordinately regulates cell proliferation and apoptosis by inactivating Yorkie, the Drosophila Homolog of YAP. *Cell* **122**, 421–434.

Hughes, S. C. and Fehon, R. G. (2007). Understanding ERM proteins – the awesome power of genetics finally brought to bear. *Curr Opin Cell Biol* **19** (1), 51–56.

Itoh, Y. and Seiki, M. (2006). MT1-MMP, a potent modifier of pericellular microenvironment. *J Cell Physiol* **206** (1), 1–8.

Jalkanen, S., Joensuu, H., Soderstro, K. O., and Klemi, P. (1991). Lymphocyte homing and clinical behavior of non-Hodgkin's lymphoma. *J Clin Invest* **87**, 1835–1840.

Jin, H., Sperka, T., Herrlich, P., and Morrison, H. (2006). Tumorigenic transformation by CPI-17 through inhibition of a merlin phosphatase. *Nature* **442**, 576–579.

Kallakury, B. V., Yang, F., Figge, J., et al. (1996). Decreased levels of CD44 protein and mRNA in prostate carcinoma. Correlation with tumor grade and ploidy. *Cancer* **78**, 1461–1469.

Khanna, C., Wan, X., Bose, S., et al. (2004). The membrane-cytoskeleton linker ezrin is necessary for osteosarcoma metastasis. *Nat Med* **10** (2), 182–186.

Kim, H., Lim, J. Y., Kim, Y. H., et al. (2002). Inhibition of ras-mediated activator protein 1 activity and cell growth by merlin. *Mol Cells* **14**, 108–114.

Kimura, Y., Saya, H., and Nakao, M. (2000). Calpain-dependent proteolysis of NF2 protein, involvement in schwannomas and meningiomas. *Neuropathology* **20**, 153–160.

Kincade, P. W., Zheng, Z., Katoh, S., and Hanson, L. (1997). The importance of cellular environment to function of the CD44 matrix receptor. *Current Opin Cell Biol* **9**, 635–642.

Kissil, J. L., Wilker, E. W., Johnson, K. C., et al. (2003). Merlin, the product of the Nf2 tumor suppressor gene, is an inhibitor of the p21 activated kinase, Pak1. *Mol Cell* **12**, 841–849.

Koga, H., Araki, N., Takeshima, H., et al. (1998). Impairment of cell adhesion by expression of the mutant neurofibromatosis type 2 (NF2) genes which lack exons in the ERM-homology domain. *Oncogene* **17**, 801–810.

Lai, Z. C., Wei, X., Shimizu, T., et al. (2005). Control of cell proliferation and apoptosis by mob as tumor suppressor, mats. *Cell* **120**, 675–685.

LaJeunesse, D. R., McCartney, B. M., and Fehon, R. G. (1998). Structural analysis of Drosophila merlin reveals functional domains important for growth control and subcellular localization. *J Cell Biol* **141**, 1589–1599.

Lallemand, D., Curto, M., Saotome, I., Giovannini, M., and McClatchey, A. I. (2003). NF2 deficiency promotes tumorigenesis and metastasis by destabilizing adherens junctions. *Genes Dev* **17**, 1090–1100.

Lamb, R. F., Hennigan, R. F., Turnbull, K., et al. (1997). AP-1 mediated invasion requires increased expression of the hyaluronan receptor CD44. *Mol Cell Biol* **17**, 963–976.

Lau, Y. K., Murray, L. B., Houshmandi, S. S., et al. (2008). Merlin is a potent inhibitor of glioma growth. *Cancer Res* **68** (14), 5733–5742.

Legg, J. W. and Isacke, C. M. (1998). Identification and functional analysis of the ezrin-binding site in the hyaluronan receptor, CD44. *Curr Biol* **8**, 705–708.

Lesley, J., Hyman, R., and Kincade, P. (1993). CD44 and its interaction with extracellular matrix. *Adv Immunol* **54**, 271–335.

Li, C., Heidt, D. G., Dalerba, P., Burant, C. F., et al. (2007). Identification of pancreatic cancer stem cells. *Cancer Res* **67** (3), 1030–1037.

Li, Y., Pei, J., Xia, H., et al. (2003). Lats2, a putative tumor suppressor, inhibits G1/S transition. *Oncogene* **22**, 4398–4405.

Liu, R., Wang, X., Chen, G. Y., et al. (2007). The prognostic role of a gene signature from tumorigenic breast-cancer cells. *N Engl J Med* **356** (3), 217–226.

Lou, W., Krill, D., Dhir, R., et al. (1999). Methylation of the CD44 metastasis suppressor gene in human prostate cancer. *Cancer Res* **59**, 2329–2331.

Maitra, S., Kulikauskas, R. M., Gavilan, H., and Fehon, R. G. (2006). The tumor suppressors merlin and expanded function cooperatively to modulate receptor endocytosis and signaling. *Curr Biol* **16**, 702–709.

Masumura, Y. and Tarin, D. (1992). Significance of CD44 gene products for cancer diagnosis and disease evaluation. *Lancet* **340**, 1053–1058.

Matzke, A., Sargsyan, V., Holtmann, B., et al. (2007). Haploinsufficiency of c-Met in cd44−/− mice identifies a collaboration of CD44 and c-Met in vivo. *Mol Cell Biol* **27** (24), 8797–8806.

McCartney, B. M., Kulikauskas, R. M., LaJeunesse, D. R., and Fehon, R. G. (2000). The neurofibromatosis-2 homologue, merlin, and the tumor suppressor expanded function together in Drosophila to regulate cell proliferation and differentiation. *Development* **127**, 1315–1324.

McClatchey, A. I. and Giovannini, M. (2005). Membrane organization and tumorigenesis – the NF2 tumor suppressor. *Merlin. Genes Dev* **19** (19), 2265–2277.

McClatchey, A. I., Saotome, I., Mercer, K., et al. (1998). Mice heterozygous for a mutation at the Nf2 tumor suppressor locus develop a range of highly metastatic tumors. *Genes Dev* **12**, 1121–1133.

McClatchey, A. I., Saotome, I., Ramesh, V., Gusella, J. F., and Jacks, T. (1997). The Nf2 tumor suppressor gene product is essential for extraembryonic development immediately prior to gastrulation. *Genes Dev* **11**, 1253–1265.

McPherson, J. P., Tamblyn, L., Elia, A., et al. (2004). Lats2/Kpm is required for embryonic development, proliferation control and genomic integrity. *EMBO J* **23**, 3677–3688.

Morrison, H., Sherman, L. S., Legg, J., et al. (2001). The NF2 tumor suppressor gene product, merlin, mediates contact inhibition of growth through interactions with CD44. *Genes Dev* **15**, 968–980.

Morrison, H., Sperka, T., Manent, J., Giovannini, M., Ponta, H., and Herrlich, P. (2007). Merlin/neurofibromatosis type 2 suppresses growth by inhibiting the activation of Ras and Rac. *Cancer Res* **67**, 520–527.

Murai, T., Miyazaki, Y., Nishinakamura, H., et al. (2004). Engagement of CD44 promotes Rac activation and CD44 cleavage during tumor cell migration. *J Biol Chem* **279** (6), 4541–4550.

Murthy, A., Gonzalez-Agosti, C., Cordero, E., et al. (1998). NHE-RF, a regulatory cofactor for Na(+)-H+ exchange, is a common interactor for merlin and ERM (MERM) proteins. *J Biol Chem* **273**, 1273–1276.

Notterman, D. A., Alon, U., Sierk, A. J., and Levine, A. J. (2001). Transcriptional gene expression profiles of colorectal adenoma, adenocarcinoma, and normal tissue examined by oligonucleotide arrays. *Cancer Res* **61** (7), 3124–3130.

Obremski, V. J., Hall, A. M., and Fernandez-Valle, C. (1998). Merlin, the neurofibromatosis type 2 gene product, and beta1 integrin associate in isolated and differentiating Schwann cells. *J Neurobiol* **37** (4), 487–501.

Ohashi, R., Takahashi, F., Cui, R., et al. (2007). Interaction between CD44 and hyaluronate induces chemoresistance in non-small cell lung cancer cell. *Cancer Lett* **252** (2), 225–234.

Okada, T., Lopez-Lago, M., and Giancotti, F. G. (2005). Merlin/NF-2 mediates contact inhibition of growth by suppressing recruitment of Rac to the plasma membrane. *J Cell Biol* **171**, 361–371.

Okada, T., You, L., and Giancotti, F. G. (2007). Shedding light on Merlin's wizardry. *Trends Cell Biol* **17** (5), 222–229.

Ouhtit, A., Abd Elmageed, Z. Y., Abdraboh, M. E., Lioe, T. F., and Raj, M. H. (2007). In vivo evidence for the role of CD44s in promoting breast cancer metastasis to the liver. *Am J Pathol* **171** (6), 2033–2039.

Overholtzer, M., Zhang, J., Smolen, G. A., et al. (2006). Transforming properties of YAP, a candidate oncogene on the chromosome 11q22 amplicon. *Proc Natl Acad Sci USA* **103**, 12405–12410.

Pals, S., Horst, E., Ossekoppele, G., et al. (1989). Expression of lymphocyte homing receptor as a mechanism of dissemination in non-Hodgkin's lymphoma. *Blood* **37**, 885–888.

Patrawala, L., Calhoun, T., Schneider-Broussard, R., et al. (2006). Highly purified CD44+ prostate cancer cells from xenograft human tumors are enriched in tumorigenic and metastatic progenitor cells. *Oncogene* **25** (12), 1696–1708.

Pearson, M. A., Reczek, D., Bretscher, A., and Karplus, P. A. (2000). Structure of the ERM protein moesin reveals the FERM domain fold masked by an extended actin binding tail domain. *Cell* **101**, 259–270.

Pellock, B. J., Buff, E., White, K., and Hariharan, I. K. (2007). The Drosophila tumor suppressors expanded and merlin differentially regulate cell cycle exit, apoptosis, and Wingless signaling. *Dev Biol* **304**, 102–115.

Phillips, T. M., McBride, W. H., and Pajonk, F. (2006). The response of CD24(−/low)/CD44+ breast cancer-initiating cells to radiation. *J Natl Cancer Inst* **98** (24), 1777–1785.

Prince, M. E., Sivanandan, R., Kaczorowski, A., et al. (2007). Identification of a subpopulation of cells with cancer stem cell properties in head and neck squamous cell carcinoma. *Proc Natl Acad Sci USA* **104** (3), 973–978.

Qiu, R. G., Chen, J., Kirn, D., McCormick, F., and Symons, M. (1995). An essential role for Rac in Ras transformation. *Nature* **374**, 457–459.

Ruttledge, M. H., Andermann, A. A., Phelan, C. M., et al. (1996). Type of mutation in the neurofibromatosis type 2 gene (NF2) frequently determines severity of disease. *Am J Hum Genet* **59**, 331–342.

Sainio, M., Zhao, F., Heiska, L., et al. (1997). Neurofibromatosis 2 tumor suppressor protein colocalizes with ezrin and CD44 and associates with actin-containing cytoskeleton. *J Cell Sci* **110**, 2249–2260.

Scoles, D. R., Huynh, D. P., Morcos, P. A., et al. (1998). Neurofibromatosis 2 tumour suppressor schwannomin interacts with betaII-spectrin. *Nat Genet* **18**, 354–359.

Shaw, R. J., Paez, J. G., Curto, M., et al. (2001). The Nf2 tumor suppressor, merlin, functions in Rac-dependent signaling. *Dev Cell* **1**, 63–72.

Sherman, L., Sleeman, J., Herrlich, P., and Ponta, H. (1994). Hyaluronate receptors: key players in growth, differentiation, migration and tumor progression. *Current Opin Cell Biol* **6**, 726–733.

Sherman, L., Xu, H.-M., Geist, R. T., et al. (1997). Interdomain binding mediates tumor growth suppression by the NF2 gene product. *Oncogene* **15**, 2505–2509.

Shimizu, T., Seto, A., Maita, N., et al. (2002). Structural basis for neurofibromatosis type 2. Crystal structure of the merlin FERM domain. *J Biol Chem* **277**, 10332–10336.

Shipitsin, M., Campbell, L. L., Argani, P., et al. (2007). Molecular definition of breast tumor heterogeneity. *Cancer Cell* **11** (3), 259–273.

Shtivelman, E. and Bishop, J. M. (1991). Expression of CD44 is repressed in neuroblastoma cells. *Mol Cell Biol* **11**, 5446–5453.

St John, M. A., Tao, W., Fei, X., et al. (1999). Mice deficient of Lats1 develop soft-tissue sarcomas, ovarian tumours and pituitary dysfunction. *Nat Genet* **21**, 182–186.

Stamenkovic, I. (2000). Matrix metalloproteinases in tumor invasion and metastasis. *Cancer Biol* **10**, 415–433.

Stamenkovic, I., Amiot, M., Pesando, J. M., and Seed, B. (1989). A lymphocyte molecule implicated in lymph node homing is a member of the cartilage link protein family. *Cell* **56**, 1056–1062.

Sun, L., Hui, A. M., Su, Q., et al. (2006). Neuronal and glioma-derived stem cell factor induces angiogenesis within the brain. *Cancer Cell* **9** (4), 287–300.

Sy, M. S., Guo, Y.-J., and Stamenkovic, I. (1991). Distinct effects of two CD44 isoforms on tumor growth in vivo. *J Exp Med* **174**, 859–866.

Sy, M. S., Guo, Y.-J., and Stamenkovic, I. (1992). Inhibition of tumor growth in vivo with a soluble CD44-immunoglobulin fusion protein. *J Exp Med* **176**, 623–627.

Takahashi, Y., Miyoshi, Y., Takahata, C., et al. (2005). Down-regulation of LATS1 and LATS2 mRNA expression by promoter hypermethylation and its association with biologically aggressive phenotype in human breast cancers. *Clin Cancer Res* **11**, 1380–1385.

Tang, X., Jang, S. W., Wang, X., et al. (2007). Akt phosphorylation regulates the tumour-suppressor merlin through ubiquitination and degradation. *Nat Cell Biol* **9**, 1199–1207.

Tao, W., Zhang, S., Turenchalk, G. S., et al. (1999). Human homologue of the Drosophila melanogaster lats tumour suppressor modulates CDC2 activity. *Nat Genet* **21**, 177–181.

Thaxton, C., Lopera, J., Bott, M. and Fernandez-Valle, C. (2008). Neuregulin and laminin stimulate phosphorylation of the NF2 tumor suppressor in Schwann cells by distinct protein kinase A and p21-activated kinase-dependent pathways. *Oncogene* **27**, 2705-2715.

Tikoo, A., Varga, M., Ramesh, V., Gusella, J., and Maruta, H. (1994). An anti-Ras function of neurofibromatosis type 2 gene product (NF2/merlin). *J Biol Chem* **269**, 23387–23390.

Toole, B. P. (2004). Hyaluronan, from extracellular glue to pericellular cue. *Nat Rev Cancer* **4** (7), 528–539.

Tsukita, S., Oishi, K., Sato, N., et al. (1994). ERM family members as molecular linkers between the cell surface glycoprotein CD44 and actin-based cytoskeletons. *J Cell Biol* **126**, 391–401.

Tsukita, S., Yonemura, S., and Tsukita, S. (1997). ERM (ezrin/radixin/moesin) family, from cytoskeleton to signal transduction. *Current Opin Cell Biol* **9**, 70–75.

Turley, E. A., Noble, P. W., and Bourguignon, L. Y. (2002). Signaling properties of hyaluronan receptors. *J Biol Chem* **277** (7), 4589–4592.

Vassilev, A., Kaneko, K. J., Shu, H., Zhao, Y., and DePamphilis, M. L. (2001). TEAD/TEF transcription factors utilize the activation domain of YAP65, a Src/Yes-associated protein localized in the cytoplasm. *Genes Dev* **15**, 1229–1241.

Weber, G. F., Bronson, R. T., Ilagan, J., et al. (2002). Absence of the CD44 gene prevents sarcoma metastasis. *Cancer Res* **62** (8), 2281–2286.

Willecke, M., Hamaratoglu, F., Kango-Singh, M., et al. (2006). The fat cadherin acts through the hippo tumor-suppressor pathway to regulate tissue size. *Curr Biol* **16**, 2090–2100.

Williams, D. A. and Cancelas, J. A. (2006). Leukaemia: niche retreats for stem cells. *Nature* **444** (7121), 827–828.

Wu, S., Huang, J., Dong, J., and Pan, D. (2003). Hippo encodes a Ste-20 family protein kinase that restricts cell proliferation and promotes apoptosis in conjunction with salvador and warts. *Cell* **114**, 445–456.

Xia, H., Qi, H., Li, Y., et al. (2002). LATS1 tumor suppressor regulates G2/M transition and apoptosis. *Oncogene* **21**, 1233–1241.

Xu, H.-M. and Gutmann, H. (1998). Merlin differentially associates with the microtubule and actin cytoskeleton. *J Neurosci Res* **51**, 403–415.

Xu, Y. and Yu, Q. (2003). E-cadherin negatively regulates CD44–hyaluronan interaction and CD44-mediated tumor invasion and branching morphogenesis. *J Biol Chem* **278** (10), 8661–8668.

Yagi, R., Chen, L. F., Shigesada, K., Murakami, Y., and Ito, Y. (1999). A WW domain-containing yes-associated protein (YAP) is a novel transcriptional co-activator. *EMBO J* **18**, 2551–2562.

Yonemura, S., Hirao, M., Doi, Y., et al. (1998). Ezrin/radixin/moesin (ERM) proteins bind to a positively charged amino acid cluster in the juxta-membrane cytoplasmic domain of CD44, CD43, and ICAM-2. *J Cell Biol* **140**, 885–895.

Yu, Q. and Stamenkovic, I. (1999). Localization of matrix metalloproteinase 9 (MMP-9) to the cell surface provides a mechanism for CD44-mediated tumor invasion. *Genes Dev* **13**, 35–48.

Yu, Q. and Stamenkovic, I. (2000). Cell surface-localized matrix metalloproteinase-9 proteolytically activates TGF-beta and promotes tumor invasion and angiogenesis. *Genes Dev* **14**, 163–176.

Yu, Q., Toole, B. P., and Stamenkovic, I. (1997). Induction of apoptosis of metastatic mammary carcinoma cells in vivo by disruption of tumor cell surface CD44 function. *J Exp Med* **186**, 1985–1996.

Yu, Y., Khan, J., Khanna, C., Helman, L., Meltzer, P. S., and Merlino, G. (2004). Expression profiling identifies the cytoskeletal organizer ezrin and the developmental homeoprotein Six-1 as key metastatic regulators. *Nat Med* **10** (2), 175–181.

Zender, L., Spector, M. S., Xue, W., et al. (2006). Identification and validation of oncogenes in liver cancer using an integrative oncogenomic approach. *Cell* **125**, 1253–1267.

Zoltan-Jones, A., Huang, L., Ghatak, S., and Toole, B. P. (2003). Elevated hyaluronan production induces mesenchymal and transformed properties in epithelial cells. *J Biol Chem* **278** (46), 45801–45810.

CHAPTER

6

Hyaluronan-Mediated CD44 Interaction with Receptor and Non-Receptor Kinases Promotes Oncogenic Signaling, Cytoskeleton Activation and Tumor Progression

Lilly Y.W. Bourguignon

HYALURONAN (HA) METABOLISM IN CANCER PROGRESSION

Hyaluronan (HA) is a non-sulfated, unbranched glycosaminoglycan consisting of the repeating disaccharide units, D-glucuronic acid and N-acetyl-D-glucosamine and found in the extracellular matrix (ECM) of most mammalian tissues (Toole, 1991; Laurent and Fraser, 1992). HA is often overexpressed at tumor cell attachment sites and appears to play an important role in promoting tumor cell-specific behaviors (Lee and Spicer, 2000; Toole, 2001; Turley et al., 2002). The biosynthesis of HA is regulated by three mammalian HA synthase isozymes, HA synthase1 (HAS-1), HA synthase 2 (HAS-2), and HA synthase 3 (HAS-3) (Weigel et al., 1997; Itano and Kimata, 1996; Itano and Kimata, 1998; Spicer and Nguyen, 1999). Although the three HAS genes are located on different chromosomes, they share a great deal of sequence homology (Spicer and McDonald, 1998; Shyjan et al., 1996; Watanabe and Yamaguchi, 1996). HAS-1 synthesizes a low level of a large-size HA ($\sim 1 \times 10^6 - 1 \times 10^7$ Da) (Spicer and McDonald, 1998; Shyjan et al., 1996). HAS-2 produces relatively high amounts of large-size HA ($\sim 1 \times 10^6 - 1 \times 10^7$ Da) and is definitely involved in cell proliferation, angiogenesis as well as embryonic and cardiac development (Watanabe and Yamaguchi, 1996). HAS-3, which synthesizes a small-size HA ($\sim 1 \times 10^5$ Da), is one of the most biologically active HAS molecules and is known to contribute to the malignant phenotype in many different cell types (Spicer and McDonald, 1998; Spicer and Nguyen, 1999). Recent studies indicate that the expression of HAS-1 restores the metastatic potential of mouse mammary carcinoma mutant cells that have low HA production and low metastatic capability (Itano et al., 1999). Furthermore, overexpression of HAS-2 and HAS-3 stimulates both tumorigenicity and tumor progression (Kosaki et al., 1999; Liu et al., 2001). In ovarian cancer patients, all three HAS molecules are detected in ovarian carcinoma tissues (Yabushita et al., 2004). In particular, HAS-1 expression in ovarian cancer patients appears to be closely associated with tumor angiogenesis and disease progression; and is also an important predictor of patient survival (Yabushita et al., 2004). Thus, dysregulation of HAS expression and activities often results in abnormal production of HA and directly contributes to aberrant cellular processes such as transformation and metastasis (Zhang et al., 1995).

HA can also be digested into various smaller-sized molecules by hyaluronidases (Stern and Jedrzejas, 2006). Somatic hyaluronidases (acid-active enzymes) have been found to be significantly increased in breast tumor metastases as compared to primary tumors (Bertrand et al., 1997). Currently, at least six hyaluronidase-like gene sequence products [e.g. HYAL-1, HYAL-2, HYAL-3, HYAL-4, PHYAL-1, PH-20 (or Spam1)] have been reported. HYAL-1 is considered to be a candidate of tumor suppressor gene product (Csoka et al., 1998; Frost et al., 2000; Lerman and Minna, 2000), whereas HYAL-2 appears to function primarily (but not exclusively) as an oncogene (Lepperdinger et al., 1998; Strobl et al., 1998; Lepperdinger et al., 2001). Both HYAL-1 and HYAL-2 genes are found tightly clustered on chromosome 3p21.3 and are the major hyaluronidases in human somatic tissues (Linebaugh et al., 1999; Nicoll et al., 2002).

HYAL-3 has been detected in chondrocytes (Flannery et al., 1998). The level of HYAL-3 expression increases when fibroblasts undergo chondrocyte differentiation (Nicoll et al., 2002). HYAL-4 based on preliminary evidence appears to be a chondroitinase and PHYAL-1 (a pseudogene) is not commonly translated in humans (Csoka et al., 1999). PH-20 (or Spam1), the neutral-active hyaluronidase is detected primarily in testes and in breast tumors (Cherr et al., 2001; Beech et al., 2002). Low pH environment promotes tumor cell-specific behaviors including activation of ECM-degrading enzymes such as hylauronidases for tumor cell invasion (Chambers et al., 1997; Duffy, 1992; Mignatti and Rifkin, 1993). Among all mammalian hylauronidases, HYAL-2 appears to be present in virtually all tissues (Lepperdinger et al., 1998; Strobl et al., 1998). HYAL-2 belongs to a family of endo-N-acetylhexosaminidases that hydrolyze HA and to a lesser extent chondroitin sulfates (Lepperdinger et al., 1998; 2001; Strobl et al., 1998). It is a cell surface enzyme via glycosylphosphatidyl-inositol (GPI) linkage to the plasma membrane (Lepperdinger et al., 2001; Rai et al., 2001; Bourguignon et al., 2004). The enzymatic activity of HYAL-2 is upregulated in acidic compartments derived from invaginated plasma membrane microdomains such as lipid rafts (Lepperdinger et al., 2001; Rai et al., 2001). Activated HYAL-2 hydrolyzes high molecular mass HA into intermediate sized fragments of approximately 20 kDa corresponding to about 50 disaccharide units (Lepperdinger et al., 1998). Interestingly, fragments of such size stimulate inflammatory cytokine synthesis (Nobel, 2002). Overexpression of HYAL-2 in murine astrocytoma cells accelerates intracerebral but not subcutaneous tumor formation and invasion (Novak et al., 1999). HYAL-2 also functions as a receptor for the jaagsiekte sheep retroviruses (Rai et al., 2001). In this system, HYAL-2 negatively regulates receptor tyrosine kinase (Danilkovitch-Miagkova et al., 2003) and mediates tumor cell transformation by jaagsiekte sheep retroviruses in lung cancers (Rai et al., 2001). Clearly, activation of ECM-degrading enzymes such as HYAL-2 appears to be closely associated with tumor progression.

CD44 (AN HA RECEPTOR) IN TUMOR PROGRESSION

CD44 denotes a family of cell-surface glycoproteins which are expressed in a variety of cells and tissues including tumor cells and carcinoma tissues (Gunther et al., 1991; Iida and Bourguignon, 1995; Iida and Bourguignon, 1997; Zhu and Bourguignon, 1998; Auvinen et al., 2005; Bourguignon, 2001; Bourguignon et al., 1997; 1998; 1999; 2000; 2001a,b; 2002; 2003; 2004; 2006; 2007; Turley et al., 2002). Nucleotide sequence analyses reveal that many CD44 isoforms (derived by alternative splicing mechanisms) are variants of the standard form, CD44s (Screaton et al., 1992; 1993). The presence of high levels of CD44 variant (CD44v) isoforms is emerging as an important metastatic tumor marker in a number of cancers (Gunther et al., 1991; Iida and Bourguignon, 1995; Iida and Bourguignon, 1997; Zhu and Bourguignon, 1998; Auvinen et al., 2005; Bourguignon et al., 1997; 1998a,b; 1999; 2000; 2001a,b,c; 2002; 2003; Turley et al., 2002). Recent studies have shown that CD44 is also expressed in tumor stem cells which have the unique ability to initiate tumor cell-specific properties (Al-Hajj et al., 2003). In fact, CD44 is proposed to be one of the important surface markers for cancer stem cells (Al-Hajj et al., 2003). All CD44 isoforms contain an HA-binding site in their extracellular domain and thereby serve as a major cell surface receptor for HA (Underhill, 1992). Both CD44v isoforms and HA are overexpressed at sites of tumor attachment (Toole et al., 2002). It has been determined that HA binding to CD44v isoforms is involved in the stimulation of both receptor kinases (e.g. ErbB2, EGFR, and TGFβ receptors) (Bourguignon et al., 1997; 2001b; 2002; 2006; 2007; Wang and Bourguignon, 2006a,b) and non-receptor kinases (e.g. c-Src and ROK) (Bourguignon et al., 1999; 2001b; 2003) required for a variety of tumor cell-specific functions leading to tumor progression. Some of the cellular and molecular mechanisms controlling the ability of HA to activate CD44v isoform-associated kinases in tumor cells to growth, migrate, and invade other tissues will be described below.

HA-Mediated CD44 Interaction with Receptor Kinases in Tumor Cell Activation

HA-Mediated CD44 Interaction with ErbB2 (p185^{HER2}) Kinase

The HER2 oncogene (also called ErbB2 or neu) encodes a 185 kDa membrane protein (p185^{HER2}) which contains a single transmembrane spanning region, and a tyrosine kinase-associated cytoplasmic domain (Bargmann et al., 1986). This protein, initially discovered as an activated oncogenic variant (neu), is overexpressed in many epithelial tumors including mammary and ovarian carcinomas (Press et al., 1990; Reese and Slamon, 1997; Meden and Kuhn, 1997). Overexpression or amplification of the HER2 oncogene (ErbB2) in both breast and ovarian cancers is generally associated with a poor prognosis (Raspollini et al., 2006; Malamou-Mitsi et al., 2007; Sasaki et al., 2007). In recent years, much excitement has surrounded the remarkable therapeutic effects of HER-2/neu-blocking antibodies, such as Herceptin™ (Bookman et al., 2003; Longva et al., 2005). Nevertheless, the molecular mechanisms by which HER-2/neu enhances the growth and survival of cancer cells, and induces tumor progression are not completely understood.

Hyaluronan (HA) is known to constitutively regulate ErbB2 tyrosine kinase activity and to influence ErbB2 interaction with CD44 signaling in tumor cells (Bourguignon et al., 1997; 2001b; 2007). Previously, we determined that CD44 and ErbB2 are physically linked to each other via interchain disulfide bonds in human ovarian tumor cells (Bourguignon et al., 1997). Most importantly, HA binding to a CD44-ErbB2 complex activates the ErbB2 tyrosine kinase activity and promotes tumor cell growth (Bourguignon et al., 1997). HA-mediated CD44-ErbB2 signaling complexes in ovarian tumor cells are also associated with molecular scaffolds and adaptors such as Vav2 (a Rac-specific GEF) and Grb2 (Bourguignon et al., 2001b). HA treatment induces the recruitment of both Vav2 and Grb2 into CD44v3-ErbB2 (p185^{HER2}) containing multi-molecular complexes leading to the co-activation of Rac1 and Ras signaling and ovarian tumor cell growth and migration (Bourguignon et al., 2001b). Most recently, we have found that HA/CD44–ErbB2 interaction [mediated by N-WASP (Neural Wiskott-Aldrich Syndrome Protein–an actin-cytoskeleton activator)] promotes β-catenin signaling and actin-cytoskeleton functions leading to certain tumor-specific behaviors (e.g. transcriptional activation and tumor cell migration) and ovarian cancer progression (Bourguignon et al., 2007). Clearly, there are important direct signaling interactions among HA, CD44, and ErbB2 during tumor cell activation (summarized in Fig. 6.1).

HA-Mediated CD44 Interaction with EGFR

Epidermal growth factor receptor (EGFR) contains a single transmembrane spanning region, and a tyrosine kinase-associated cytoplasmic domain (Cohen et al., 1982). Previous studies have found that guanidine nucleotide (GDP/GTP) exchange on Ras is significantly stimulated by tyrosine phosphorylation of EGFR (Cohen et al., 1982). Thus, it appears that EGFR activation mediates Ras-mediated stimulation of a downstream kinase cascade which includes the Raf-1/MEK/MAPK pathway leading to tumor cell growth (Cohen et al., 1982). EGFR is overexpressed in many tumors including head and neck squamous cell carcinomas (HNSCC) (Bonner et al., 2002; Choong and Cohen, 2006; Hoyek-Gebeily et al., 2007). Overexpression or amplification of EGFR in head and neck cancer is generally associated with a poor prognosis (Bonner et al., 2002; Choong and Cohen, 2006; Hoyek-Gebeily et al., 2007). Recently, much excitement has surrounded the remarkable therapeutic effects of FDA-approved EGFR-blocking antibodies, such as cetuximab (Gebbia et al., 2007). Nevertheless, the molecular mechanisms by which EGFR enhances the growth and survival of HNSCC, and induces resistance to chemotherapy are not completely understood.

Hyaluronan (HA) is known to constitutively regulate EGFR tyrosine kinase activity and to influence EGFR interaction with CD44 signaling in HNSCC (Bourguignon et al., 2006; Wang and Bourguignon, 2006a,b). Previously, we determined that CD44 and EGFR are physically linked to each other in human HNSCC (Bourguignon et al., 2006). Most importantly,

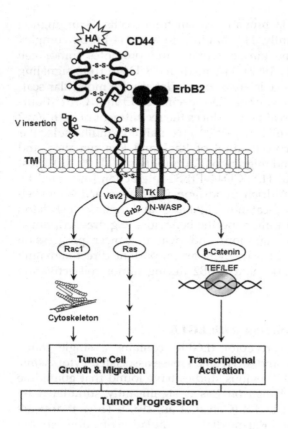

FIGURE 6.1 HA-mediated CD44 interaction with ErbB2 signaling in tumor cells. The binding of HA to CD44 promotes ErbB2 kinase activation leading to Vav2-mediated Rac1 signaling and Grb2-associated Ras activation. The co-activation of Rac1 and Ras induces tumor cell growth and migration. HA binding to CD44 also stimulates N-WASP-mediated activation and β-catenin signaling leading to transcriptional upregulation. All these events (e.g. transcriptional activation, tumor cell growth, and migration) are required for tumor progression.

HA binding to a CD44–EGFR complex activates the EGFR tyrosine kinase activity and promotes HNSCC functions (Bourguignon et al., 2006; Wang and Bourguignon, 2006a,b). HA-mediated CD44–EGFR signaling complexes in HNSCC are also associated with molecular scaffolds and adaptors such as LARG (Leukemia-Associated RhoGEF, a RhoA-specific GEF) (Bourguignon et al., 2006). HA treatment induces the recruitment of LARG into CD44–EGFR containing multimolecular complexes leading to the co-activation of RhoA and Ras signaling and HNSCC cell growth and migration (Bourguignon et al., 2006). Most recently, we have found that HA/CD44–EGFR interaction also promotes Ca^{2+} signaling (Bourguignon et al., 2006; Wang and Bourguignon, 2006a) and cytoskeleton functions (Bourguignon et al., 2006) as well topoisomerase II activation (Wang et al., 2007) leading, HNSCC migration, growth, and multidrug resistance (Wang and Bourguignon, 2006a, b; Wang et al., 2007). Clearly, there are important direct signaling interactions among HA, CD44, and EGFR during HNSCC activation (summarized in Fig. 6.2).

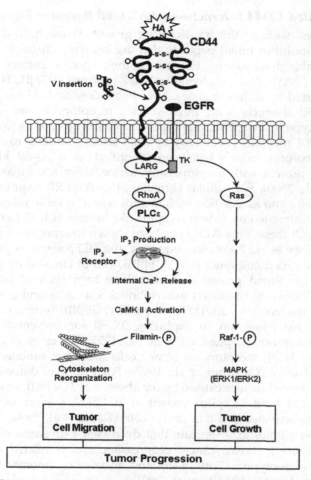

FIGURE 6.2 HA-mediated CD44 interaction with EGFR signaling in tumor cells. HA–CD44 binding is tightly coupled with LARG and EGFR in a complex which can induce LARG-mediated RhoA activation and EGFR signaling-regulated Ras activation. HA/CD44 and LARG-activated RhoA then stimulates PLCε-mediated IP3 production and IP3 receptor-triggered intracellular Ca2+ mobilization resulting in CaMKII activation. CaMKII then phosphorylates the cytoskeletal protein, filamin leading to cytoskeleton reorganization and tumor cell migration. In the meantime, Ras activation by HA-mediated CD44–LARG–EGFR complex formation also promotes Raf-1 phosphorylation, MAPK (in particular, ERK1 and ERK2) activation, and tumor cell growth. Together, we believe that CD44–LARG interaction with EGFR plays an important role in promoting HA-dependent RhoA and Ras co-activation leading to the concomitant stimulation of HNSCC growth and migration required for tumor progression.

HA-Mediated CD44 Interaction with TGFβ Receptor Signaling

Cytokines, such as the transforming growth factor β (TGFβ) super-family are multifunctional peptides that are known to regulate a diverse set of cellular processes by binding to their specific surface receptors (Blobe et al., 2000). Three mammalian TGFβ isoforms (TGFβ1, TGFβ2, and TGFβ3), coded by different genes, have been identified (Messague et al., 1992). TGFβ interacts with three surface receptors known as type I (TGFβRI), type II (TGFβRII), and type III (TGFβIII) receptors (Blobe et al., 2000). TGFβ1 mediates its activity by high affinity binding to the type II (TGFβII) receptor, which has been identified as a 70–80 kDa trans-membrane protein with a cytoplasmic serine/threonine kinase domain (Blobe et al., 2000). For cellular signaling, the TGFβRII requires both its kinase activity and association with members of a series of related 55 kDa TGFβRIs [designated as activin receptor-like kinase-ALK (1 to 6 different subtypes)]. Of these, only ALK5 has been shown to represent a functional TGFβRI (Blobe et al., 2000). Subsequently, the TGFβ signal is propagated from the plasma membranes (via TGFβRII/TGFβI kinases) by phosphor-ylation of the Smad proteins which belong to a class of intracellular mediators known to regulate transcriptional responses and gene expres-sion in the nucleus (Shi, 2001). The type III (TGFβIII) receptor also binds TGFβ and may function in capturing TGFβ for presentation to the signaling receptors (Lopez-Casilla et al., 1991; Wang et al., 1991). In cancers, the TGFβ receptors on tumor cells are often mutated or func-tionally defective (Carcamo et al., 1995). For example, defective ligand binding to the cell surface caused by the absence of TGFβRII, or expression of a truncated form or splice variant of TGFβRII, may account for the resistance to activated TGFβ in cancer cells (Kadin et al., 1994; Park et al., 1994). Some studies also indicate that decreased expression of TGFβRII may contribute to breast cancer progression, and is related to a more aggressive phenotype in both in-situ and invasive carcinomas (Oft et al., 1998; Yin et al., 1999; Yoneda et al., 2001).

In order to gain a better understanding of TGFβ receptor signaling in breast cancer cells, we have examined the interaction between CD44 and the transforming growth factor β (TGFβ) receptors (a family of serine/threonine kinase membrane receptors) in human metastatic breast tumor cells (MDA-MB-231 cell line). Immunological data indicate that both CD44 and TGFβ receptors are expressed in MDA-MB-231 cells; and that CD44 is physically linked to the TGFβ receptor I (TGFβRI) [and to a lesser extent to the TGFβ receptor II (TGFβRII)] as a complex in vivo. Scatchard plot analyses and in vitro binding experiments show that the cytoplasmic domain of CD44 binds to TGFβRI at a single site with high affinity [an apparent dissociation constant (K_d) of ~1.78 nM]. These findings indicate that TGFβRI contains a CD44 binding site (Bourguignon et al., 2002).

Furthermore, we have found that the binding of HA to CD44 in MDA-MB-231 cells stimulates TGFβRI serine/threonine kinase activity which, in turn, increases Smad2/Smad3 phosphorylation and parathyroid hormone-related protein (PTHrP) production (well-known downstream effector functions of TGFβ signaling). Most importantly, TGFβRI kinase activated by HA phosphorylates CD44 which enhances its binding interaction with the cytoskeletal protein, ankyrin, leading to HA-mediated breast tumor cell migration. Overexpression of TGFβRI by transfection of MDA-MB-231 cells with TGFβRIcDNA stimulates formation of the CD44-TGFβRI complex, the association of ankyrin with membranes, and HA-dependent/CD44-specific breast tumor migration. (Bourguignon et al., 2002). Taken together, these findings strongly suggest that CD44 interaction with the TGFβRI kinase promotes concomitant activation of both HA-dependent and TGFβ-specific signaling pathways required for ankyrin–membrane interaction, tumor cell migration, and important oncogenic events (e.g. Smad2/Smad3 phosphorylation and PTHrP production) during metastatic breast tumor progression (summarized in Fig. 6.3).

HA-Mediated CD44 Interaction with Non-Receptor Kinases in Tumor Cell Activation

HA-Mediated CD44 Interaction with c-Src Signaling

The Src family kinases are classified as oncogenic proteins due to their ability to activate cell proliferation (Barone and Courtneidge, 1995; Broome and Hunter, 1996), spreading (Kaplan et al., 1995; Rodier et al., 1995), and migration (Hansen et al., 1996; Hall et al., 1996; Rahimi et al., 1998) in many cell types including epithelial tumor cells (Tanaka et al., 1996; Summy and Gallick, 2003). The amino terminus of Src contains a myristoylation (or palmiotylation) site which is important for membrane association (Thomas and Brugge, 1997; Schlessinger, 2000). Src also contains several functional domains including SH3 and SH2 (Src homology) domains, the catalytic protein tyrosine kinase (PTK) core, and a conserved regulatory tyrosine phosphorylation site (Thomas and Brugge, 1997; Schlessinger, 2000). Certain amino acid residues in the c-Src molecule play an important role in modulating its kinase activity. Mutations of specific key amino acids result in either upregulation or downregulation of c-Src kinase activity. For example, replacement of tyrosine 527 with phenylalanine (e.g. Y527F, the dominant-active form of c-Src kinase) strongly activates c-Src kinase transforming capability and enzyme activities (Kmiecik and Shalloway, 1987). Mutation of Lysine 295 to Arginine (e.g. K295R, the dominant-negative form of c-Src kinase) renders c-Src kinase defective and reduces c-Src kinase-mediated biological activities (Kmiecik and Shalloway, 1987; Bagrodia et al., 1991).

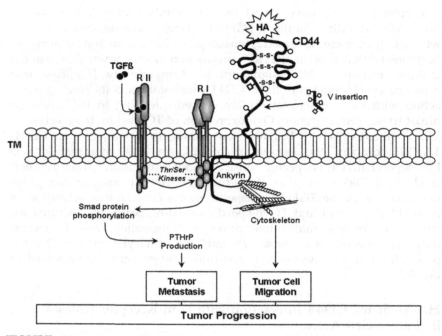

FIGURE 6.3 HA-mediated TGFβ receptor signaling in tumor cells. CD44 is tightly complexed with TGFβRI. This CD44-associated TGFβRI kinase can be activated by HA and/or TGFβ leading to phosphorylation of Smad proteins (Smad2 and Smad 3) and PTHrP production which is known to cause metastasis. Moreover, HA and/or TGFβ-activated CD44-TGF-βRI kinase is also capable of phosphorylating CD44. Most importantly, CD44 phosphorylation enhances its binding to the cytoskeletal protein, ankyrin, which, in turn, interacts with the cytoskeleton and induces tumor cell migration. Both oncogenic signaling events (e.g. Smad2/Smad3 phosphorylation and PTHrP production) and ankyrin-cytoskeleton function contribute to tumor progression.

In addition, it has been observed that the interaction between Src kinase and membrane-linked molecules regulates receptor signaling and various cellular functions (Thomas and Brugge, 1997; Schlessinger, 2000). In fact, CD44-mediated cellular signaling has been suggested to involve Src kinase family members (Taher et al., 1996; Ilangumaran et al., 1998). For example, Lck, one of the Src kinase family members is found to be closely complexed with CD44 during T-cell activation (Taher et al., 1996). CD44 also selectively associates with active Src family tyrosine kinases (e.g. Lck and Fyn) in glycosphingolipid-rich plasma membrane domains (lipid rafts) of human peripheral blood lymphocytes (Ilangumaran et al., 1998). Moreover, the cytoplasmic domain of CD44s has been shown to be involved in the recruitment of the Src family (e.g., Src, Yes, and Fyn) in prostate tumor cells during anchorage-independent colony growth (Zhu and Bourguignon,

1998). Collectively, all these observations support the notion that c-Src kinases participate in CD44-mediated cellular signaling.

Overexpression or increased activity of c-Src is frequently detected in human ovarian cancers, implicating the involvement of c-Src in the etiology of ovarian carcinomas (Summy and Gallick, 2003; Tanaka et al., 2004). CD44 is known to be tightly coupled with c-Src kinase in ovarian tumor cells (SK-OV-3.ipl cells) (Bourguignon et al., 2001c). Our previous work determined that the cytoplasmic domain of CD44 binds to c-Src kinase at a single site with high affinity (Bourguignon et al., 2001a). Most importantly, HA interaction with CD44 stimulates c-Src kinase activity which, in turn, increases tyrosine phosphorylation of the cytoskeletal protein, cortactin. Subsequently, tyrosine phosphorylation of cortactin alters its interaction with actin-cytoskeleton leading to tumor cell migration (Bourguignon et al., 2001c). Therefore, HA/CD44-mediated cellular signaling clearly involves Src kinase family members during tumor cell activation (summarized in Fig. 6.4).

HA-Mediated CD44 Interaction with RhoA-ROK Signaling and Cytoskeleton Function

Members of the Rho subclass of the Ras superfamily [small molecular weight GTPases, (e.g. RhoA, Rac1, and Cdc42)] are known to transduce a variety of signals regulating many different cellular processes (Hall, 1998). Overexpression of certain RhoGTPases in human tumors often correlates with a poor prognosis (Fritz et al., 1999; Li and Lim, 2003). In particular, coordinated RhoGTPase signaling is considered to be a possible mechanism underlying cell proliferation and motility, an obvious prerequisite for metastasis (Fritz et al., 1999; Li and Lim, 2003).

Several enzymes have been identified as possible downstream targets for RhoGTPases (e.g., RhoA) in regulating cytoskeleton-mediated cell motility (Kimura et al., 1996; Amano et al., 1997; Fukata et al., 1999). One such enzyme is Rho-Kinase (ROK, also called Rho-binding kinase) which is a serine-threonine kinase (Kimura et al., 1996; Amano et al., 1997; Fukata et al., 1999). ROK interacts with RhoA in a GTP-dependent manner and phosphorylates a number of cytoskeletal proteins such as adducin (Fukata et al., 1999) and myosin phosphatase (Kimura et al., 1996). Structurally, ROK is composed of catalytic (CAT), coiled-coil, Rho-binding (RB), and pleckstrin-homology (PH) domains (Kimura et al., 1996; Amano et al., 1997; Fukata et al., 1999). Overexpression of the Rho-binding domain (a dominant-negative form) of ROK by transfecting breast tumor cells with RB cDNA induces reversal of tumor cell-specific phenotypes (Bourguignon et al., 1999; 2003). A previous study showed that ROK is responsible for the phosphorylation of CD44-associated cytoskeletal proteins during actin filament and plasma membrane interaction. When ROK is overexpressed

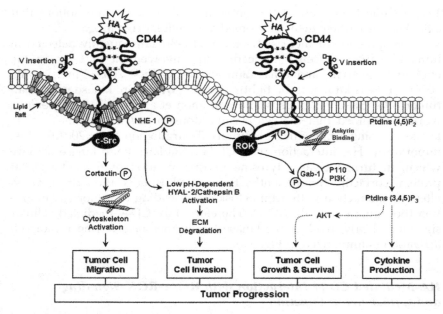

FIGURE 6.4 HA-mediated c-Src signaling and ROK activation in tumor cells. The binding of HA to CD44 stimulates Src signaling in lipid rafts leading to cortactin phosphorylation and cytoskeleton-mediated tumor cell migration. HA binding to CD44 also promotes RhoA-mediated ROK activation. Activated ROK then phosphorylates NHE1 (in lipid rafts). Most importantly, the phosphorylation of NHE1 by ROK promotes Na^+-H^+ exchange activity, intra-endosomal/lysosomal pH changes, and extracellular acidification leading to a concomitant activation of at least two low pH-dependent enzymes-HYAL-2 (located at lipid rafts) and cathepsin B (secreted in the medium) required for extracellular matrix (ECM) degradation, HA modification and tumor cell invasion. HA/CD44-activated ROK is also capable of phosphorylating the cytoplasmic domain of CD44. CD44 phosphorylation by ROK enhances its binding interaction with ankyrin. Finally, HA/CD44-activated ROK phosphorylates the linker molecule, Gab-1. Most importantly, phosphorylation of Gab-1 by ROK promotes the membrane localization of Gab-1 and PI3 kinase to CD44 and activates certain isoforms of PI3 kinase to convert PtdIns $(4,5)P_2$ to PtdIns $(3,4,5)P_3$ leading to AKT activation, cytokine production, and tumor cell behaviors (e.g. tumor cell growth, survival, and invasion) required for tumor progression.

or constitutively activated, changes in actin-cytoskeleton organization occur which are similar to those observed during normal Rho-activated conditions (Bourguignon et al., 1999; 2003). ROK is overexpressed in breast tumor cells and is capable of phosphorylating the cytoplasmic domain of CD44 (Bourguignon et al., 1999). Moreover, phosphorylation of the cytoplasmic domain of CD44 by ROK enhances its binding interaction with ankyrin (Bourguignon et al., 1999). Overexpression of the Rho-binding domain (a dominant-negative form) of ROK by transfecting breast tumor cells with RB cDNA induces reversal of tumor cell-specific phenotypes

(Bourguignon et al., 1999). These findings support the notion that ROK plays a pivotal role in CD44–ankyrin interaction and RhoA-mediated oncogenic signaling required for membrane-cytoskeleton function and metastatic tumor cell migration.

HA/CD44-activated RhoA also promotes ROK phosphorylation of the linker molecule, Grb2-associated binder-1 (Gab-1) (Bourguignon et al., 2003). Most importantly, phosphorylation of Gab-1 by ROK promotes the membrane localization of Gab-1 and phosphatidylinositol 3-kinase (PI3 kinase) to CD44 and activates certain isoforms of PI3 kinase to convert PtdIns $(4,5)P_2$ to PtdIns $(3,4,5)P_3$ leading to AKT activation, cytokine production, and tumor cell behaviors (e.g. tumor cell growth, survival, and invasion) required for breast tumor progression (Bourguignon et al., 2003). These findings suggest that there is a strong connection between RhoA-ROK activation and Gab-1-associated PI3 kinase activation during the stimulation of breast tumor cells by HA–CD44 interaction.

Previous studies indicate that Na^+-H^+ exchanger-1 (NHE1) is one of the principal intracellular pH (pHi) regulatory molecules in breast tumor cells (Boyer and Tannock, 1992). NHE1 appears to be involved in the aberrant regulation of both extracellular pH (pH_e) and intracellular pH (pH_i) in human breast tumor cells under nutrient-depleted conditions (Reshkin et al., 2000). Certain observations on the localization and regulation of certain NHE isoforms suggest that there are connections between NHE activity and cholesterol-enriched lipid rafts (Poli et al., 1991). Interestingly, NHE1 also serves as one of the cellular substrates for RhoA-activated Rho-kinase (ROK) (Tominaga et al., 1998). NHE1 phosphorylation by RhoA-activated ROK induces actin stress fiber assembly (Tominaga et al., 1998). Recent evidence indicates that the binding of HA to CD44 activates RhoA-ROK activity which, in turn, promotes NHE1 phosphorylation and Na^+-H^+ exchange activity leading to intracellular acidification and creates an acidic extracellular matrix environment. These events result in hyaluronidase (HYAL-2)-mediated HA catabolism, HA modification, cysteine proteinase (cathepsin B)-mediated matrix degradation, and cancer progression (Bourguignon et al., 2004) (summarized in Fig. 6.4).

In conclusion, we believe that the binding of HA to CD44 is capable of stimulating the activation of receptor kinases (e.g., ErbB2, EGFR, and TGFβ receptor) and non-receptor kinases (e.g., c-Src kinases and ROK) in a variety of tumor cells. These events result in a coordinated "cross-talk" among multiple signaling pathways (see Figs 6.1–6.4) leading to onco-genesis and subsequent cancer progression.

ACKNOWLEDGMENTS

I gratefully acknowledge Dr. Gerard J. Bourguignon's assistance in the preparation of this paper. I am also grateful for Ms. Christine Camacho and Christine Earle for their help in preparing graphs, illustrations, and manuscript. This work was supported by United States Public Health grants (R01 CA66163, R01 CA78633, and P01 AR39448), a VA Merit Review grant, and a DOD grant. L.Y.W.B. is a VA Research Career Scientist.

References

Al-Hajj, M., Wicha, M. S., Benito-Hernandez, A., Morrison, S. J., and Clarke, M. F. (2003). Prospective identification of tumorigenic breast cancer cells. *Proc Natl Acad Sci USA* **100**, 3983–3988.

Amano, M., Chihara, K., Kimura, K., Fukata, Y., Nakamura, N., Matsuura, Y., and Kaibuchi, K. (1997). Formation of actin stress fibers and focal adhesions enhanced by Rho-kinase. *Science* **275**, 1308–1311.

Auvinen, P., Tammi, R., Tammi, M., Johansson, R., and Kosma, V. M. (2005). Expression of CD 44 s, CD 44 v 3 and CD 44 v 6 in benign and malignant breast lesions: correlation and colocalization with hyaluronan. *Histopathology* **47**, 420–428.

Bagrodia, S., Chackalaparampil, I., Kmiecik, T. E., and Shalloway, D. (1991). Altered tyrosine 527 phosphorylation and mitotic activation of p60c-src. *Nature* **349**, 172–175.

Bargmann, C. I., Hung, M. C., and Weinberg, R. A. (1986). Multiple independent activations of the neu oncogene by a point mutation altering the transmembrane domain of p185. *Cell* **45**, 649–657.

Barone, M. V. and Courtneidge, S. A. (1995). Myc but not Fos rescue of PDGF signalling block caused by kinase-inactive Src. *Nature* **378**, 509–512.

Beech, D. J., Madan, A. K., and Deng, N. (2002). Expression of PH-20 in normal and neoplastic breast tissue. *J Surg Res* **103**, 203–207.

Bertrand, P., Girard, N., Duval, C., et al. (1997). Increased hyaluronidase levels in breast tumor metastases. *Int J Cancer* **73**, 327–331.

Blobe, G. C., Schiemann, W. P., and Lodish, H. F. (2000). Role of transforming growth factor beta in human disease. *N Engl J Med* **342**, 1350–1358.

Bonner, J. A., De Los Santos, J., Waksal, H. W., et al. (2002). Epidermal growth factor receptor as a therapeutic target in head and neck cancer. *Semin Radiat Oncol* **12**, 11–20.

Bookman, M. A., Darcy, K. M., Clarke-Pearson, D., et al. (2003). Evaluation of monoclonal humanized anti-HER2 antibody, trastuzumab, in patients with recurrent or refractory ovarian or primary peritoneal carcinoma with overexpression of HER2: a phase II trial of the Gynecologic Oncology Group. *J Clin Oncol* **21**, 283–290.

Bourguignon, L. Y. (2001). CD44-mediated oncogenic signaling and cytoskeleton activation during mammary tumor progression. *J Mammary Gland Biol Neoplasia* **6**, 287–297.

Bourguignon, L. Y., Gilad, E., Brightman, A., Diedrich, F., and Singleton, P. (2006). Hyaluronan–CD44 interaction with leukemia-associated RhoGEF and epidermal growth factor receptor promotes Rho/Ras co-activation, phospholipase C epsilon-Ca^{2+} signaling and cytoskeleton modification in head and neck squamous cell carcinoma cells. *J Biol Chem* **281**, 14026–14040.

Bourguignon, L. Y., Gunja-Smith, Z., Iida, N., et al. (1998). CD44v(3,8-10) is involved in cytoskeleton-mediated tumor cell migration and matrix metalloproteinase (MMP-9) association in metastatic breast cancer cells. *J Cell Physiol* **176**, 206–215.

Bourguignon, L. Y., Peyrollier, K., Gilad, E., and Brightman, A. (2007). Hyaluronan–CD44 interaction with neural Wiskott–Aldrich syndrome protein (N-WASP) promotes actin polymerization and ErbB2 activation leading to beta-catenin nuclear translocation, transcriptional up-regulation and cell migration in ovarian tumor cells. *J Biol Chem* **282**, 1265–1280.

Bourguignon, L. Y., Singleton, P. A., Diedrich, F., Stern, R., and Gilad, E. (2004). CD44 interaction with Na^+-H^+ exchanger (NHE1) creates acidic microenvironments leading to hyaluronidase-2 and cathepsin B activation and breast tumor cell invasion. *J Biol Chem* **279**, 26991–27007.

Bourguignon, L. Y., Singleton, P. A., Zhu, H., and Diedrich, F. (2003). Hyaluronan-mediated CD44 interaction with RhoGEF and Rho kinase promotes Grb2-associated binder-1 phosphorylation and phosphatidylinositol 3-kinase signaling leading to cytokine (macrophage-colony stimulating factor) production and breast tumor progression. *J Biol Chem* **278**, 29420–29434.

Bourguignon, L. Y., Singleton, P. A., Zhu, H., and Zhou, B. (2002). Hyaluronan promotes signaling interaction between CD44 and the transforming growth factor beta receptor I in metastatic breast tumor cells. *J Biol Chem* **277**, 39703–39712.

Bourguignon, L. Y., Zhu, H., Chu, A., et al. (1997). Interaction between the adhesion receptor, CD44 and the oncogene product, p185HER2, promotes human ovarian tumor cell activation. *J Biol Chem* **272**, 27913–27918.

Bourguignon, L. Y., Zhu, H., Shao, L., and Chen, Y. W. (2000). CD44 interaction with tiam1 promotes Rac1 signaling and hyaluronic acid-mediated breast tumor cell migration. *J Biol Chem* **275**, 1829–1838.

Bourguignon, L.Y., Zhu, H., Shao, L. and Chen, Y.W. (2001a). CD44 interaction with c-Src kinase promotes cortactin-mediated cytoskeleton function and hyaluronic acid-dependent ovarian tumor cell migration. *J Biol Chem* **276**, 7327–7336.

Bourguignon, L. Y., Zhu, H., Shao, L., Zhu, D., and Chen, Y. W. (1999). Rho-kinase (ROK) promotes CD44v(3,8-10)–ankyrin interaction and tumor cell migration in metastatic breast cancer cells. *Cell Motil Cytoskeleton* **43**, 269–287.

Bourguignon, L.Y., Zhu, H., Zhou, B., Diedrich, F., Singleton, P.A. and Hung, M.C. (2001b). Hyaluronan promotes CD44v3–Vav2 interaction with Grb2-p185(HER2) and induces Rac1 and Ras signaling during ovarian tumor cell migration and growth. *J Biol Chem* **276**, 48679–48692.

Boyer, M. J. and Tannock, I. F. (1992). Regulation of intracellular pH in tumor cell lines: influence of microenvironmental conditions. *Cancer Res* **52**, 4441–4447.

Broome, M. A. and Hunter, T. (1996). Requirement for c-Src catalytic activity and the SH3 domain in platelet-derived growth factor BB and epidermal growth factor mitogenic signaling. *J Biol Chem.* **271**, 16798–16806.

Carcamo, J., Zentella, A., and Massague, J. (1995). Disruption of transforming growth factor beta signaling by a mutation that prevents transphosphorylation within the receptor complex. *Mol Cell Biol* **15**, 1573–1581.

Chambers, A. F. and Matrisian, L. M. (1997). Changing views of the role of matrix metalloproteinases in metastasis. *J Natl Cancer Inst* **89**, 1260–1270.

Cherr, G. N., Yudin, A. I., and Overstreet, J. W. (2001). The dual functions of GPI-anchored PH-20: hyaluronidase and intracellular signaling. *Matrix Biol* **20**, 515–525.

Choong, N. W. and Cohen, E. E. (2006). Epidermal growth factor receptor directed therapy in head and neck cancer. *Crit Rev Oncol Hematol* **57**, 25–43.

Cohen, S., Ushiro, H., Stoscheck, C., and Chinkers, M. (1982). A native 170,000 epidermal growth factor receptor–kinase complex from shed plasma membrane vesicles. *J Biol Chem* **257**, 1523–1531.

Csoka, A. B., Frost, G. I., Heng, H. H., et al. (1998). The hyaluronidase gene HYAL-1 maps to chromosome 3p21.2-p21.3 in human and 9F1–F2 in mouse, a conserved candidate tumor suppressor locus. *Genomics* **48**, 63–70.

Csoka, A. B., Scherer, S. W., and Stern, R. (1999). Expression analysis of six paralogous human hyaluronidase genes clustered on chromosomes 3p21 and 7q31. *Genomics* **60**, 356–361.

Danilkovitch-Miagkova, A., Duh, F. M., Kuzmin, I., et al. (2003). Hyaluronidase 2 negatively regulates RON receptor tyrosine kinase and mediates transformation of epithelial cells by jaagsiekte sheep retrovirus. *Proc Natl Acad Sci USA* **100**, 4580–4585.

Duffy, M. J. (1992). The role of proteolytic enzymes in cancer invasion and metastasis. *Clin Exp Metastasis* **10**, 145–155.

Flannery, C. R., Little, C. B., Hughes, C. E., and Caterson, B. (1998). Expression and activity of articular cartilage hyaluronidases. *Biochem Biophys Res Commun* **251**, 824–829.

III. HYALURONAN RECEPTORS AND SIGNAL TRANSDUCTION PATHWAYS

Fritz, G., Just, I., and Kaina, B. (1999). Rho GTPases are over-expressed in human tumors. *Int J Cancer* **81**, 682–687.

Frost, G. I., Mohapatra, G., Wong, T. M., Csoka, A. B., Gray, J. W., and Stern, R. (2000). HYAL-1 LUCA-1, a candidate tumor suppressor gene on chromosome 3p21.3, is inactivated in head and neck squamous cell carcinomas by aberrant splicing of pre-mRNA. *Oncogene* **19**, 870–877.

Fukata, Y., Oshiro, N., Kinoshita, N., et al. (1999). Phosphorylation of adducin by Rho-kinase plays a crucial role in cell motility. *J Cell Biol* **145**, 347–361.

Gebbia, V., Giuliani, F., Valori, V. M., et al. (2007). Cetuximab in squamous cell head and neck carcinomas. *Ann Oncol* **18**, Suppl 6, vi5–7.

Gunthert, U., Hofmann, M., Rudy, W., et al. (1991). A new variant of glycoprotein CD44 confers metastatic potential to rat carcinoma cells. *Cell* **65**, 13–24.

Hall, A. (1998). Rho GTPases and the actin cytoskeleton. *Science* **279**, 509–514.

Hall, C. L., Lange, L. A., Prober, D. A., Zhang, S., and Turley, E. A. (1996). pp60(c-src) is required for cell locomotion regulated by the hyaluronanreceptor RHAMM. *Oncogene* **13**, 2213–2224.

Hansen, K., Johnell, M., Siegbahn, A., et al. (1996). Mutation of a Src phosphorylation site in the PDGF beta-receptor leads to increased PDGF-stimulated chemotaxis but decreased mitogenesis. *EMBO J* **15**, 5299–5313.

Hoyek-Gebeily, J., Nehme, E., Aftimos, G., Sader-Ghorra, C., Sargi, Z., and Haddad, A. (2007). Prognostic significance of EGFR, p53 and E-cadherin in mucoepidermoid cancer of the salivary glands: a retrospective case series. *J Med Liban* **55**, 83–88.

Iida, N. and Bourguignon, L. Y. (1995). New CD44 splice variants associated with human breast cancers. *J Cell Physiol* **162**, 127–133.

Iida, N. and Bourguignon, L. Y. (1997). Coexpression of CD44 variant (v10/ex14) and CD44S in human mammary epithelial cells promotes tumorigenesis. *J Cell Physiol* **171**, 152–160.

Ilangumaran, S., Briol, A., and Hoessli, D. C. (1998). CD44 selectively associates with active Src family protein tyrosine kinases Lck and Fyn in glycosphingolipid-rich plasma membrane domains of human peripheral blood lymphocytes. *Blood* **91**, 3901–3908.

Itano, N. and Kimata, K. (1996). Expression cloning and molecular characterization of HAS protein, a eukaryotic hyaluronan synthase. *J Biol Chem* **271**, 9875–9878.

Itano, N. and Kimata, K. (1998). Hyaluronan synthase: new directions for hyaluronan research. *Trends Glycosci Glycotech* **10**, 23–38.

Itano, N., Sawai, T., Miyaishi, O., and Kimata, K. (1999). Relationship between hyaluronan production and metastatic potential of mouse mammary carcinoma cells. *Cancer Res* **59**, 2499–2504.

Kadin, M. E., Cavaille-Coll, M. W., Gertz, R., et al. (1994). Loss of receptors for transforming growth factor beta in human T–cell malignancies. *Proc Natl Acad Sci USA* **91**, 6002–6006.

Kaplan, K. B., Swedlow, J. R., Morgan, D. O., and Varmus, H. E. (1995). C-Src enhances the spreading of src–/– fibroblasts on fibronectin by a kinase-independent mechanism. *Genes Dev* **9**, 1505–1517.

Kimura, K., Ito, M., Amano, M., et al. (1996). Regulation of myosin phosphatase by Rho and Rho-associated kinase (Rho-kinase). *Science* **273**, 245–248.

Kmiecik, T. E. and Shalloway, D. (1987). Activation and suppression of pp60c–src transforming ability by mutation of its primary sites of tyrosine phosphorylation. *Cell* **49**, 65–73.

Kosaki, R., Watanabe, K., and Yamaguchi, Y. (1999). Overproduction of hyaluronan by expression of the hyaluronan synthase HAS-2 enhances anchorage-independent growth and tumorigenicity. *Cancer Res* **59**, 1141–1145.

Laurent, T. C. and Fraser, J. R. (1992). Hyaluronan. *FASEB J* **6**, 2397–2404.

Lee, J. Y. and Spicer, A. P. (2000). Hyaluronan: a multifunctional, megaDalton, stealth molecule. *Curr Opin Cell Biol* **12**, 581–586.

Lepperdinger, G., Mullegger, J., and Kreil, G. (2001). HYAL-2 – less active, but more versatile? *Matrix Biol* **20**, 509–514.

Lepperdinger, G., Strobl, B., and Kreil, G. (1998). HYAL-2, a human gene expressed in many cells, encodes a lysosomal hyaluronidase with a novel type of specificity. *J Biol Chem* **273**, 22466–22470.

Lerman, M. I. and Minna, J. D. (2000). The 630-kb lung cancer homozygous deletion region on human chromosome 3p21.3: identification and evaluation of the resident candidate tumor suppressor genes. The International Lung Cancer Chromosome 3p21.3 Tumor Suppressor Gene Consortium. *Cancer Res* **60**, 6116–6133.

Li, X. and Lim, B. (2003). RhoGTPases and their role in cancer. *Oncol Res* **13**, 323–331.

Linebaugh, B. E., Sameni, M., Day, N. A., Sloane, B. F., and Keppler, D. (1999). Exocytosis of active cathepsin B enzyme activity at pH 7.0, inhibition and molecular mass. *Eur J Biochem* **264**, 100–109.

Liu, N., Gao, F., Han, Z., Xu, X., Underhill, C. B., Creswell, K., and Zhang, L. (2001). Hyaluronan synthase 3 overexpression promotes the growth of TSU prostate cancer cells. *Cancer Res* **61**, 5702–5714.

Longva, K. E., Pedersen, N. M., Haslekas, C., Stang, E., and Madshus, I. H. (2005). Herceptin-induced inhibition of ErbB2 signaling involves reduced phosphorylation of Akt but not endocytic down-regulation of ErbB2. *Int J Cancer* **116**, 359–367.

Lopez-Casillas, F., Cheifetz, S., Doody, J., et al. (1991). Structure and expression of the membrane proteoglycan betaglycan, a component of the TGF-beta receptor system. *Cell* **67**, 785–795.

Malamou-Mitsi, V., Crikoni, O., Timotheadou, E., et al. (2007). Prognostic significance of HER-2, p53 and Bcl-2 in patients with epithelial ovarian cancer. *Anticancer Res* **27**, 1157–1165.

Massague, J., Cheifetz, S., Laiho, M., et al. (1992). Transforming growth factor-beta. *Cancer Surv* **12**, 81–103.

Meden, H. and Kuhn, W. (1997). Overexpression of the oncogene c-erbB-2 (HER2/neu) in ovarian cancer: a new prognostic factor. *Eur J Obstet Gynecol Reprod Biol* **71**, 173–179.

Mignatti, P. and Rifkin, D. B. (1993). Biology and biochemistry of proteinases in tumor invasion. *Physiol Rev* **73**, 161–195.

Nicoll, S. B., Barak, O., Csoka, A. B., Bhatnagar, R. S., and Stern, R. (2002). Hyaluronidases and CD44 undergo differential modulation during chondrogenesis. *Biochem Biophys Res Commun* **292**, 819–825.

Nobel, P. W. (2002). Hyaluronan and its catabolic products in tissue injury and repair. *Matrix Biol* **21**, 25–29.

Novak, U., Stylli, S. S., Kaye, A. H., and Lepperdinger, G. (1999). Hyaluronidase-2 over-expression accelerates intracerebral but not subcutaneous tumor formation of murine astrocytoma cells. *Cancer Res* **59**, 6246–6250.

Oft, M., Heider, K. H., and Beug, H. (1998). TGFbeta signaling is necessary for carcinoma cell invasiveness and metastasis. *Curr Biol* **8**, 1243–1252.

Park, K., Kim, S. J., Bang, Y. J., et al. (1994). Genetic changes in the transforming growth factor beta (TGF-beta) type II receptor gene in human gastric cancer cells: correlation with sensitivity to growth inhibition by TGF-beta. *Proc Natl Acad Sci USA* **91**, 8772–8776.

Poli de Figueiredo, C. E., Ng, L. L., Davis, J. E., et al. (1991). Modulation of Na-H antiporter activity in human lymphoblasts by altered membrane cholesterol. *Am J Physiol* **261**, C1138–1142.

Press, M. F., Jones, L. A., Godolphin, W., Edwards, C. L., and Slamon, D. J. (1990). HER-2/neu oncogene amplification and expression in breast and ovarian cancers. *Prog Clin Biol Res* **354A**, 209–221.

Rahimi, N., Hung, W., Tremblay, E., Saulnier, R., and Elliott, B. (1998). C-Src kinase activity is required for hepatocyte growth factor-induced motility and anchorage-independent growth of mammary carcinoma cells. *J Biol Chem* **273**, 33714–33721.

Rai, S. K., Duh, F. M., Vigdorovich, V., et al. (2001). Candidate tumor suppressor HYAL-2 is a glycosylphosphatidylinositol (GPI)-anchored cell-surface receptor for jaagsiekte sheep retrovirus, the envelope protein of which mediates oncogenic transformation. *Proc Natl Acad Sci USA* **98**, 4443–4448.

Raspollini, M. R., Amunni, G., Villanucci, A., et al. (2006). HER-2/neu and bcl-2 in ovarian carcinoma: clinicopathologic, immunohistochemical and molecular study in patients with shorter and longer survival. *Appl Immunohistochem Mol Morphol* **14**, 181–186.

Reese, D. M. and Slamon, D. J. (1997). HER-2/neu signal transduction in human breast and ovarian cancer. *Stem Cells* **15**, 1–8.

Reshkin, S. J., Bellizzi, A., Albarani, V., et al. (2000). Phosphoinositide 3-kinase is involved in the tumor-specific activation of human breast cancer cell Na(+)/H(+) exchange, motility and invasion induced by serum deprivation. *J Biol Chem* **275**, 5361–5369.

Rodier, J. M., Valles, A. M., Denoyelle, M., Thiery, J. P., and Boyer, B. (1995). pp60c-src is a positive regulator of growth factor-induced cell scattering in a rat bladder carcinoma cell line. *J Cell Biol* **131**, 761–773.

Sasaki, N., Kudoh, K., Kita, T., et al. (2007). Effect of HER-2/neu overexpression on chemoresistance and prognosis in ovarian carcinoma. *J Obstet Gynaecol Res* **33**, 17–23.

Schlessinger, J. (2000). New roles for Src kinases in control of cell survival and angiogenesis. *Cell.* **100**, 293–296.

Screaton, G. R., Bell, M. V., Bell, J. I., and Jackson, D. G. (1993). The identification of a new alternative exon with highly restricted tissue expression in transcripts encoding the mouse Pgp-1 (CD44) homing receptor. Comparison of all 10 variable exons between mouse, human and rat. *J Biol Chem* **268**, 12235–12238.

Screaton, G. R., Bell, M. V., Jackson, D. G., et al. (1992). Genomic structure of DNA encoding the lymphocyte homing receptor CD44 reveals at least 12 alternatively spliced exons. *Proc Natl Acad Sci USA* **89**, 12160–12164.

Shi, Y. (2001). Structural insights on Smad function in TGFbeta signaling. *Bioessays* **23**, 223–232.

Shyjan, A. M., Heldin, P., Butcher, E. C., Yoshino, T., and Briskin, M. J. (1996). Functional cloning of the cDNA for a human hyaluronan synthase. *J Biol Chem* **271**, 23395–23399.

Spicer, A. P. and McDonald, J. A. (1998). Characterization and molecular evolution of a vertebrate hyaluronan synthase gene family. *J Biol Chem* **273**, 1923–1932.

Spicer, A. P. and Nguyen, T. K. (1999). Mammalian hyaluronan synthases: investigation of functional relationships in vivo. *Biochem Soc Trans* **27**, 109–115.

Stern, R. and Jedrzejas, M. J. (2006). Hyaluronidases: their genomics, structures and mechanisms of action. *Chem Rev* **106**, 818–839.

Strobl, B., Wechselberger, C., Beier, D. R., and Lepperdinger, G. (1998). Structural organization and chromosomal localization of HYAL-2, a gene encoding a lysosomal hyaluronidase. *Genomics* **53**, 214–219.

Summy, J. M. and Gallick, G. E. (2003). Src family kinases in tumor progression and metastasis. *Cancer Metastasis Rev* **22**, 337–358.

Taher, T. E., Smit, L., Griffioen, A. W., et al. (1996). Signaling through CD44 is mediated by tyrosine kinases. Association with p56lck in T lymphocytes. *J Biol Chem* **271**, 2863–2867.

Tanaka, Y., Kobayashi, H., Suzuki, M., Kanayama, N., and Terao, T. (2004). Transforming growth factor-beta1-dependent urokinase up-regulation and promotion of invasion are involved in Src-MAPK-dependent signaling in human ovarian cancer cells. *J Biol Chem* **279**, 8567–8576.

Thomas, S. M. and Brugge, J. S. (1997). Cellular functions regulated by Src family kinases. *Annu Rev Cell Dev Biol* **13**, 513–609.

Tominaga, T., Ishizaki, T., Narumiya, S., and Barber, D. L. (1998). p160ROCK mediates RhoA activation of Na-H exchange. *EMBO J* **17**, 4712–4722.

Toole, B. P. (1991). Proteoglycans and hyaluronan in morphogenesis and differentiation. Cell Biology of Extracellular Matrix. *Vol. 19* Hey, E. D. (ed.) Plenum Press, New York. pp.305–314.

Toole, B. P. (2001). Hyaluronan in morphogenesis. *Semin Cell Dev Biol* **12**, 79–87.

Toole, B. P., Wight, T. N., and Tammi, M. I. (2002). Hyaluronan–cell interactions in cancer and vascular disease. *J Biol Chem* **277**, 4593–4596.

Turley, E. A., Noble, P. W., and Bourguignon, L. Y. (2002). Signaling properties of hyaluronan receptors. *J Biol Chem* **277**, 4589–4592.

Wang, S.J. and Bourguignon, L.Y. (2006a). Hyaluronan and the interaction between CD44 and epidermal growth factor receptor in oncogenic signaling and chemotherapy resistance in head and neck cancer. *Arch Otolaryngol Head Neck Surg* **132**, 771–778.

Wang, S.J. and Bourguignon, L.Y. (2006b). Hyaluronan–CD44 promotes phospholipase C-mediated Ca2+ signaling and cisplatin resistance in head and neck cancer. *Arch Otolaryngol Head Neck Surg* **132**, 19–24.

Wang, S. J., Peyrollier, K., and Bourguignon, L. Y. (2007). The influence of hyaluronan–CD44 interaction on topoisomerase II activity and etoposide cytotoxicity in head and neck cancer. *Arch Otolaryngol Head Neck Surg* **133**, 281–288.

Wang, X. F., Lin, H. Y., Ng-Eaton, E., et al. (1991). Expression cloning and characterization of the TGF-beta type III receptor. *Cell* **67**, 797–805.

Watanabe, K. and Yamaguchi, Y. (1996). Molecular identification of a putative human hyaluronan synthase. *J Biol Chem* **271**, 22945–22948.

Weigel, P. H., Hascall, V. C., and Tammi, M. (1997). Hyaluronan synthases. *J Biol Chem* **272**, 13997–14000.

Yabushita, H., Noguchi, M., Kishida, T., et al. (2004). Hyaluronan synthase expression in ovarian cancer. *Oncol Rep* **12**, 739–743.

Yin, J. J., Selander, K., Chirgwin, J. M., et al. (1999). TGF-beta signaling blockade inhibits PTHrP secretion by breast cancer cells and bone metastases development. *J Clin Invest* **103**, 197–206.

Yoneda, T., Williams, P. J., Hiraga, T., Niewolna, M., and Nishimura, R. (2001). A bone-seeking clone exhibits different biological properties from the MDA-MB-231 parental human breast cancer cells and a brain-seeking clone in vivo and in vitro. *J Bone Miner Res* **16**, 1486–1495.

Zhang, L., Underhill, C. B., and Chen, L. (1995). Hyaluronan on the surface of tumor cells is correlated with metastatic behavior. *Cancer Res* **55**, 428–433.

Zhu, D. and Bourguignon, L. Y. (1998). The ankyrin-binding domain of CD44s is involved in regulating hyaluronic acid-mediated functions and prostate tumor cell transformation. *Cell Motil Cytoskeleton* **39**, 209–222.

CHAPTER

7

Adhesion and Penetration: Two Sides of CD44 Signal Transduction Cascades in the Context of Cancer Cell Metastasis

David J.J. Waugh, Ashleigh McClatchey, Nicola Montgomery, and Suzanne McFarlane

INTRODUCTION: HA, CD44, THE TUMOR MICROENVIRONMENT, AND METASTASIS

Hyaluronan (HA) is a major glycosaminoglycan (GAG) of the extracellular matrix, constituted by a variable number of repeating glucuronic acid and *N*-acetylglucosamine disaccharide subunits. Synthesized on the inner face of the plasma membrane by three distinct isoforms of hyaluronan synthase (HAS), this GAG is then extruded into the surrounding extracellular space (Toole, 2004). Increased HA deposition in the stroma has been observed in many cancers and is believed to result from increased HA synthesis by both malignant cancer cells and tumor stromal cells. Elevations in stromal HA levels have been correlated with poor prognosis and reduced survival. For example, elevated HA expression in the tumor stroma in breast cancer has been shown to correlate with poorly differentiated tumors, auxiliary lymph node status, and short overall survival (Auvinen et al., 2000; Karihtala et al., 2007). It is proposed that the tumor-progressing effect of HA is mediated through a combination of its biophysical, hydrodynamic properties on the matrix and its capacity to activate a number of different cell-surface receptors that are expressed on tumor and stromal cells. This chapter will focus on how one of the principal HA receptors, CD44, is associated with conferring the malignant progression of tumor cells, by contributing to the modulation of cancer cell adhesion and cancer cell invasion within the tumor microenvironment.

CD44 is a cell-surface glycoprotein receptor with a well-defined HA binding domain within its amino-terminal ectodomain. As a consequence of the insertion of differential exon products or alternatively, through extensive addition of carbohydrate moieties, the extracellular domain of CD44 has also been shown to bind other ligands besides HA, including osteopontin and fibroblast growth factor. The intracellular carboxy-terminal tail of the receptor has been shown to interact with a number of adaptor proteins including ankyrin, ezrin, GAB1, Src kinases, and Rho-family GTPases (Ponta et al., 2003; Thorne et al., 2004). Consequently, through the association with these intracellular proteins, CD44 is coupled to a range of distinct signaling pathways. As a result of the diversity of signaling pathways induced downstream of CD44, HA has been shown to modulate cell adhesion, cell motility, cell proliferation, cell survival, and cancer cell invasion (see Toole, 2004). In this regard, these experimental observations in cell-based studies provide evidence to implicate HA-induced CD44 signaling in conferring many of the properties essential for tumor cells to successfully negotiate each of the defined stages of the metastatic cascade.

In addition to the cancer cell, CD44 is also expressed on numerous other cell types within the tumor microenvironment, including endothelial cells and specialist cells such as osteoclasts. Therefore, given that cancer cells synthesize and secrete HA and OPN into the extracellular space, these signaling competent molecules may bind to CD44 and induce responses in other cell types present within the tumor stroma. For example, HA is a chemotactic factor for fibroblasts and breast cancer cells (Tzircotis et al., 2005; Tzircotis et al., 2006), thus cancer cell-derived HA may enhance fibroblast recruitment at the tumor site. Other studies show that activation of CD44 is critical in driving osteoclast motility within the bone (Chellaiah et al., 2003a,b). This suggests a potential role for breast cancer cell-derived HA or OPN in stimulating osteoclasts to resorb bone, a common clinical feature of advanced stages of this disease. Therefore, activation of CD44 receptors in multiple cell types present within the tumor microenvironment may have added significance to cancer metastasis, besides that of simply modulating cancer cell function. Indeed, the potential importance of host-cell CD44 expression to cancer metastasis is inferred from studies conducted exploiting the CD44 knock-out mouse in which CD44 deficiency was shown to lead to an inability of osteosarcoma cells to develop spontaneous metastasis (Weber et al., 2002).

Interest in the significance of CD44 with regard to tumor progression has also been fueled by the characterization of this receptor as a cell surface marker of cancer stem cells (Al-Hajj et al., 2003; Dontu et al, 2005). In respect of metastasis, the capacity of stem cells to disseminate and colonize distant organs during the early stage of disease has been suggested to underpin a later re-emergence of metastatic disease during patient relapse. Breast cancer cells that disseminate to the bone marrow cavity during the early stage of disease have been shown to express the characteristic breast cancer stem cell signature, i.e., high CD44+/low CD24− expression (Balic et al., 2006). Further clinical studies have reported that the prevalence of CD44+/CD24− cells also favors distant metastasis in breast cancer patients (Abraham et al., 2005). Therefore, experimental and clinical studies both suggest a key role for HA and CD44 in promoting metastasis of cancer. While HA and CD44 signaling has been reported to increase epithelial cell proliferation, the predominant pathological function of CD44 is associated with the role of this receptor in orchestrating and co-ordinating cell–cell and cell–matrix interactions within the tumor microenvironment. Consequently, this has implications in facilitating the passage of malignant cells through the matrix at the primary tumor site, in addition to promoting the initial arrest, penetration, and outgrowth of metastatic cancer cells at secondary tissues.

PENETRATION: CD44 PROMOTED INVASION AND INTRAVASATION

The capacity of tumor cells to invade their surrounding space is regarded as one of the ascribed "hallmarks of cancer" (Hanahan and Weinberg, 2000) and is a key factor in the acquisition of malignancy. CD44 signal transduction has been associated in regulating cell migration, cell invasion, and cell adhesion; properties that facilitate the cell–cell and cell–matrix interactions that underlie tumor-promoted angiogenesis, localized cellular invasion, the entry (intravasation) and exit (extravasation) of cancer cells from the bloodstream, and ultimately, penetration and colonization of a secondary organ or tissue. Loss of E-cadherin expression has been reported in the early stages of localized epithelial tumor invasion. Decreased E-cadherin expression not only facilitates adherens junction disassembly, but is reported to lead to aberrant activation of the β-catenin transcription factor resulting in increased transcription of genes driving cell proliferation, survival, and cell invasion. Studies conducted in colorectal epithelium revealed that mutation of the tumor suppressor protein APC, a negative regulator of β-catenin activation, resulted in enhanced expression of CD44 in these epithelial cells, suggesting that this cell surface receptor is a downstream transcriptional target of β-catenin (Wielenga et al., 1999). Therefore, deregulation of β-catenin transcription resulting from either the inactivation of tumor suppressor genes or alternatively, as suggested in colorectal cancer, from the disruption of adherens junctions may contribute to the emergence of increased CD44 expression in epithelial cells during the early stages of tumor development (Kim et al., 1994). More recently, CD44 signaling cascades have also been shown to induce activation of the β-catenin signaling pathway suggesting that disruption of β-catenin signaling may serve as an important initiating and potentiating event in facilitating the localized tumor invasion promoted by CD44 (Bourguignon et al., 2007).

HA-induced activation of CD44 and/or RHAMM has been shown to induce the invasion of multiple cancer cell lines. In addition to facilitating a cytoskeletal-dependent promotion of cell migration, our increasing understanding of how CD44 signaling regulates the activity of an expanding list of extracellular proteases is fundamental to the ability of cancer cells to degrade the physical barriers provided by the extracellular matrix and basement membrane. Enzymes of the matrix metalloproteinase (MMP), serine protease, and cysteine protease families have each been reported to underpin cancer cell invasion and metastasis (Deryugina and Quigley, 2006; Mohamed and Sloane, 2006; Gocheva and Joyce, 2007; Lopez-Ortin and Matrisian, 2007). It is increasingly acknowledged that there is significant interplay between these different protease families in promoting tumor invasion. For example, cysteine proteases have been shown to regulate the activity of serine proteases through the promotion of enzymatic cleavage and removal of pro-peptide sequences. In addition, the differential patterns of enzymatic expression, their complex mode of activation, the breadth of substrate-specificity and indeed the redundancy amongst the enzymes have all been contributing factors in the sub-optimal targeting of these species in the clinic. Consequently, targeting of individual proteases may not necessarily provide a favorable therapeutic intervention. Instead, understanding how the expression and/or activity of these species is co-ordinately regulated may identify alternative means of controlling the spectrum of proteolytic activity acquired by malignant cancer cells and may show enhanced clinical benefit in treating locally advanced or metastatic cancer. As discussed below, CD44 has been shown to regulate proteolytic activity in cancer cells as a consequence of the signal transduction properties of the receptor and its role in serving as a physical scaffold for proteases on the cell surface (Fig. 7.1).

MMP-9 or Gelatinase B has high avidity to degrade type IV collagen, an extracellular matrix protein that is enriched in the basement membrane. MMP-9 has been shown to associate with the large ectodomain of CD44 in murine mammary carcinoma cells (Yu and Stamenkovic, 1999).

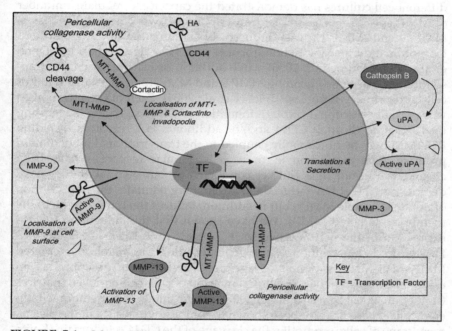

FIGURE 7.1 Schematic depiction of the role played by CD44 in regulating and co-ordinating proteolytic-dependent invasion of cancer cells through extracellular matrix. CD44 signaling regulates the transcription of multiple proteolytic species, whose activation in the extracellular space is promoted through autocatalysis or following protease-mediated cleavage of the inactive precursor, e.g., cathepsin B-mediated activation of urokinase plasminogen activator (uPA). In addition, CD44 serves to recruit the membrane-tethered matrix metalloproteinase MT1-MMP and increase the expression of the cytoskeletal protein cortactin to the invasive front of cells to promote the formation of invadopodia, membrane protrusions that underpin cell invasion. The association of CD44 with MT1-MMP and the secreted MMP-9 concentrates the extracellular enzymatic activity at the invasive front of the cell, enabling cells to exhibit the pericellular collagenolytic activity that is essential for degrading collagen-enriched matrices. In addition, the capacity of MT1-MMP to associate and cleave CD44 results in the shedding of the ectodomain of CD44 from the cell surface, decreasing the adhesion of the cell to substrates in the extracellular matrix. The resultant loss of adhesion at the invasive edge of the cell facilitates further propagation of cell motility.

Consequently, this interaction localizes MMP-9 proteolytic activity to the invasive surface of the cells, promoting physical degradation of the matrix in the immediate vicinity of the cell. In concentrating MMP-9 activity at the invasive front of the cell, the emergence or increased expression of CD44 in epithelial cells may thus enhance their capacity to disrupt the integral structure of the basement membrane and facilitate the access of these tumor cells to the underlying connective tissue.

More recent studies have identified a different mode through which HA-induced CD44 signaling may regulate cellular proteolytic activity; at the level of transcription. Addition of HA-oligosaccharides to several different cell cultures has demonstrated the capacity to regulate a number of MMP species. For example, transcription of two potent collagenases, MMP-3 and MMP-13 was induced by stimulation of articular and temporomandibular joint chondrocytes with HA-oligosaccharides (Ohno et al., 2005; Ohno et al., 2006) while addition of HA increased the transcription of the genes encoding MMP-9 and MMP-13 in murine 3LL tumor cells and primary fibroblasts (Fieber et al., 2004). Each of these responses was partially disrupted through inhibition of CD44 signaling, suggesting that signaling through CD44 and/or additional HA receptors including RHAMM may be functionally coupled to the transcriptional regulation of protease gene expression in these cells. Similar responses have been observed in tumor cells. Studies conducted by our own laboratory have demonstrated that HA-induced CD44 signaling regulates the transcription of not only soluble MMPs, but in addition, induces the transcription of the membrane-tethered MMPs. We have shown that the induction of CD44 expression in the non-invasive MCF-7 breast cancer cell line underpins HA-induced transcription of the gene encoding the membrane-tethered MMP, MT1-MMP, a potent collagenase that has been described as an essential component of pericellular invasion of collagen-enriched matrices (Itoh et al., 2006; Szabova et al., 2007). Interestingly, this membrane-tethered collagenase has also been shown to complex with CD44 in the plasma membrane of cells, promoting the cleavage of the CD44 ectodomain (Kajita et al., 2001; Mori et al., 2002; Ueda et al., 2003). The cleavage of the ectodomain of the receptor is proposed to attenuate the CD44 promoted adhesion of the cell with substrates that are enriched within the extracellular matrix. Accordingly, this reduces the avidity or strength of the interaction between the cell cytoskeleton and the extracellular matrix, thus increasing the potential of cell motility.

Furthermore, CD44-regulated protease activity is not solely restricted to the MMP family but may also drive transcription of serine protease and cysteine cathepsin genes in invasive breast cancer cells. For example CD44 has been shown to induce expression and activity of the serine protease urokinase plasminogen activator (uPA) in chondrosarcoma cells (Kobayashi et al., 2002). Using a tetracycline-regulated CD44 expression system in the weakly invasive MCF-7F breast cancer cell line, in addition to several other metastatic breast cancer cell lines, we have shown that HA-CD44 signaling induces the transcription of cysteine proteases including cathepsins B and K (Hill et al., 2008) and increases expression of uPA. Of particular interest, cysteine cathepsins are known to induce cleavage-promoted activation of uPA in the extracellular milieu, suggesting that CD44 signaling may regulate the transcription and/or secretion of several intermediate species within this pathway, thus defining a mechanism by which this cell adhesion receptor can regulate a co-ordinated activation of this proteolytic cascade. Similarly, in the context of regulating MMP activity, MT1-MMP has been shown to induce proteolytic cleavage of MMP-13 and MMP-2, processing the inactive zymogen to the mature activated species (Atkinson et al., 1995; Knauper et al., 2002). Accordingly, CD44 signaling may function as a key upstream regulator through which a number of proteolytic cascades may be induced and co-ordinately regulated in invasive cancer cells.

In addition to degrading the basement membrane that separates epithelial cells from the underlying connective tissue, proteolytic enzymes are essential in enabling cancer cells to transmigrate across the vessel wall of capillaries within the extracellular matrix. In localizing MMP-9 activity to the invasive front (Yu and Stamenkovic, 1999), one can rationalize an argument for CD44 contributing to the invasion of cancer cells across the type IV collagen-rich basement membrane of blood vessels. However, somewhat contrary to expectation, the inhibition of MMP-9 activity has been reported to potentiate rather than inhibit the intravasation of fibrosarcoma cells in an *in vivo* model. Instead, activation of uPA activity was shown to differentiate those fibrosarcoma cells with high intravasation potential from low intravasation potential (Madsen et al., 2006). Furthermore, pharmacological inhibition of uPA activity attenuated the rate of fibrosarcoma cell intravasation. Therefore, although the linkage of CD44 to the regulation of uPA activity remains to be validated in a range of cancer cell-based models, the emerging experimental data from *in vitro* experiments suggest that CD44 signaling may increase the spectrum of proteolytic activity that enhances the efficiency with which malignant tumor cells gain access to circulatory systems. This is also supported by recent observations from *in vivo* experiments that have reported a marked enrichment of CD44 in endocrine-resistant breast cancer cells that have actively disseminated to the lymphatic vessels or formed metastases in the regional lymph nodes (Harrell et al., 2006).

However, there are several caveats to the proposition that CD44 actively contributes to the intravasation and extravasation of cancer cells from circulatory systems. Firstly, the paradigm that uPA is a universal determinant of a cancer cell's ability to successfully complete the process of intravasation remains to be established in different experimental models.

III. HYALURONAN RECEPTORS AND SIGNAL TRANSDUCTION PATHWAYS

Similarly, it remains to be established whether the lack of involvement of MMP-9 activity in the case of fibrosarcoma cells is universal or whether this gelatinase may contribute to the intravasation of other cancer cell lines. In addition, the role of uPA, MMP-9, and other proteases in promoting the degradation of the basement membrane at the endothelial boundary at the secondary site of metastasis, in order to permit the escape of the cancer cell from the circulation, remains to be established. Finally, the involvement of CD44 in contributing to and/or regulating the rate of intravasation and extravasation of cancer cells remains to be directly investigated in relevant experimental systems.

Functionally, CD44 expression and activation are clearly linked with regulating cancer cell invasion of experimental matrices. Neutralizing antibodies that block the CD44 receptor attenuate the invasion of numerous cancer cell lines including representative models of gastric colorectal, ovarian, breast, and prostate cancer. We have observed that RNAi-mediated depletion of CD44 expression also reduces the efficiency with which metastatic breast cancer cells invade through a collagen I-enriched matrix, reflecting the emerging functional importance of this cell-surface receptor in regulating collagenase expression, activation, and distribution (Hill et al., 2008). In the context of metastasis, collagen I is a significant constituent of the extracellular matrix in numerous tissues, including the specialized matrices of bone and skin, and soft tissues including the lungs and liver, all tissues to which breast cancer preferentially disseminates to. Further research is likely to expand the number of proteolytic enzymes whose expression is regulated by CD44 signaling and which contribute to the invasion promoted by HA-induced signaling through this receptor. In extending the characterization of such activity, a more detailed appreciation of the functional significance of CD44 to each of the proteolytic-dependent processes of the metastatic cascade will also emerge.

CD44 signaling may also contribute to tumor invasion through additional mechanisms. Several key effector signaling proteins such as ankyrin, the ezrin-radixin-moiesin (ERM) family of proteins, and Neural–Wiskott–Aldrich syndrome protein (N-WASP) are reported to couple to the intracellular domain of CD44 (Ponta et al., 2003; Thorne et al., 2004; Bourguignon et al., 2007), facilitating the ability of this receptor to modulate the actin-cytoskeleton (discussed in more detail in a further chapter of this volume). The formation of elongated actin-cytoskeletal projections at the leading edge of a motile cell has been shown to promote the migration and invasion of cells through tissue matrix. Recent studies demonstrate a strong interplay between the cytoskeletal protein cortactin and the matrix-metalloproteinase MTI-MMP within the invadopodia (membranous projections) of invasive breast cancer cells (Artym et al., 2006). These invadipodia are key sites of proteolytic enzyme activity, where the enzymes are concentrated to degrade the matrix components at the leading edge of the cell. Therefore, CD44 may have a dual function in invadopodia, promoting the phosphorylation of cortactin to promote actin-cytoskeletal projections (Bourguignon et al., 2001; McFarlane et al., 2008) while acting as a scaffold protein to localize MT1-MMP activity to these sites on the cell membrane (Mori et al., 2002). In addition, since CD44 signaling regulates the transcription of the genes encoding cortactin and MT1-MMP (Hill et al., 2006; 2008), the resulting increase in cortactin and MT1-MMP expression identifies a potential mechanism that potentiates the timeframe and magnitude of the invasive response that is initiated by the activation of CD44.

CELL–CELL ADHESION: ROLE OF CD44 IN PROMOTING EXTRAVASATION

The ability of CD44 to regulate the expression, activation, and distribution of proteolytic enzymes on the cell surface of cancer cells indicates a role for this receptor in facilitating localized tumor invasion, and ultimately, a potential role in promoting the entry (intravasation) of invasive cancer cells into the bloodstream. However, work from a number of laboratories suggests that CD44 may further contribute to metastasis at later stages of the cascade. For example, the regulation of proteolytic activity is equally important in enabling the penetration of invasive cancer cells into and through the extracellular matrix at the secondary site of tumor growth. In addition, CD44 has been proposed to contribute in part to the establishment of pericellular HA coats around chondrocytes, endothelial cells, and invasive cancer cells (Knudson et al., 1996; Nandi et al., 2000; Draffin et al., 2004). The maintenance of a HA coat around metastatic cancer cells has been proposed to enable them to withstand the vascular shear stresses encountered during passage through capillaries, promote their survival during their transportation in the circulation, enable cancer cells to escape immune surveillance, and finally, as discussed in detail below, contribute to the promotion of cell–cell adhesion at vascular endothelial cell boundaries (Simpson et al., 2002). In addition, pericellular HA has been shown to promote adhesion-independent growth of cancer cell lines, suggesting an important role in enabling cancer cells to new environment and promote outgrowth of a new tumor (Itano et al., 2004; Zoltan-Jones et al., 2003).

The role of CD44 in contributing to cell–cell adhesion has been confirmed by studies from several laboratories examining both its physiological and pathophysiological significance. Studies conducted by Siegelman and colleagues were the first to confirm that CD44 was functionally important in initiating the rolling and adhesion of T-lymphocytes on endothelial cells *in vitro*. The physiological significance of their findings was also demonstrated from observations that the inhibition of CD44 abrogated the infiltration of lymphocytes into sites of peritoneal infection *in vivo* (DeGrendele et al., 1996; 1997). Subsequently, CD44 was shown to promote integrin receptor activation, in order to mediate the arrest and firm adhesion of these cells on endothelium (Siegelman et al., 2000; Nandi et al., 2004). Like lymphocytes, cancer cells are proposed to exploit a similar "docking and locking" mechanism in order to exit the bloodstream at capillary beds and infiltrate into underlying tissues (Fig. 7.2). Consequently, we and others have demonstrated the role of CD44 in promoting the adhesion and transmigration of myeloma, prostate cancer and breast cancer cells to bone marrow endothelial cell monolayers (Okada et al., 1999; Draffin et al., 2004). Specifically, we have confirmed that CD44 expression on breast and prostate cancer cell lines correlates with their ability to adhere to bone marrow endothelial cells. Blockade of CD44 function using neutralizing antibodies or inhibition of CD44 expression using siRNA attenuated the adhesion of the PC3 prostate cancer cell line or the MDA-MB-231 and MDA-MB-157 breast cancer cell lines. Furthermore, overexpression of CD44 in the DU145 prostate cancer cell line or the T47D breast cancer cell line (each shown to exhibit negligible endogenous CD44 expression), increased the capacity for these cells to adhere to bone marrow endothelial cells. Therefore, the capacity with which metastatic cancer cell lines adhere to this endothelial boundary can be dictated by manipulating the expression of CD44 in these cancer cells.

Pericellular HA has also been shown to contribute to the promotion of cell–cell adhesion. Enzyme (hyaluronidase)-mediated inhibition of a pericellular HA matrix around PC3 cells was shown by McCarthy and colleagues to attenuate the ability of these cells to adhere to bone marrow endothelial cell monolayers (Simpson et al., 2001; 2002). Furthermore, inhibition of HAS-2 and HAS-3 expression in PC3 cells also attenuated their adhesion suggesting that the HA promoting the formation of the pericellular HA coat and driving adhesion was sourced from the cancer cell. This was confirmed by further experiments in which hyaluronidase treatment of PC3 cells but not the bone marrow endothelial cells failed to affect the adhesion of prostate cancer cells to this monolayer (Simpson et al., 2001; Draffin et al., 2004). In addition, we demonstrated that CD44 expression on the surface of endothelial cells facilitated the adhesion of the HA-presenting cancer cell (Draffin et al., 2004). In contrast, at sites of inflammation, CD44-positive T-cells adhere to a HA layer deposited on the surface of the endothelial boundary. The action of inflammatory cytokines and chemokines have been shown to stimulate HAS expression and activity in the endothelial cells, ultimately generating this HA coating (Estess et al., 1999). Thus, the experimental data indicate that the

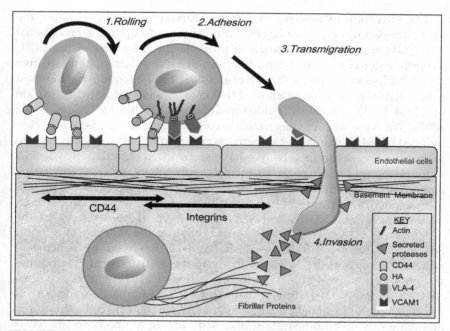

FIGURE 7.2   Schematic depiction of CD44-promoted adhesion to distal endothelium and penetration of the underlying matrix. CD44 initiates the rolling and arrest of metastatic breast cancer cells upon the distal endothelium. CD44 expressed on the cancer cell contributes to the retention of a pericellular HA coat that is presented to and interacts with CD44 receptors expressed on the surface of the receptive endothelium. The engagement of CD44 receptors initiates the activation of α4β1-integrin (VLA-4) receptors on the cancer cell that then engage with their counter-receptor on the endothelium (VCAM). The activation of CD44 and VLA-4 results in the activation of signaling cascades that promote cytoskeletal reorganization promoting the assembly of focal adhesions to underpin the arrest of the cell on the endothelium. The transmigration of the arrested cancer cell across the endothelium may be enhanced by CD44 signaling events regulating the contraction of the endothelium and secondly, through the action of CD44-regulated gelatinase activity (MMP-2/MMP-9) facilitating the degradation of the basement membrane. CD44 signaling may also contribute to the subsequent invasion of the cancer cell into the underlying tissue while HA/CD44 signaling may assist in promoting the survival and outgrowth of the cancer cell within the new tissue environment.

orientation of HA presentation in mediating cancer cell arrest is markedly different from that engaged in promoting the tethering of lymphocytes at sites of inflamed endothelial boundaries. This also suggests that cancer cells that acquire elevated HAS activity and express CD44 receptors, perhaps in direct response to increased autocrine/paracrine cytokine and chemokine signaling, may define a sub-population of cancer cells that have increased potential to adhere to vascular endothelial boundaries and thus have increased metastatic potential.

The activation of integrin receptors is fundamental in promoting the firm adhesion of both lymphocytes and cancer cells upon the endothelial cell. CD44 was originally shown to promote the activation of the $\alpha4\beta1$-integrin heterodimer in T-lymphocytes, driving a latrunculin-sensitive cell–cell adhesion response (Nandi et al., 2004). Studies conducted in breast cancer cell lines have also identified that cross-linking mediated activation of CD44 facilitates an $\alpha4\beta1$-integrin-dependent adhesion (Wang et al., 2005). We have also characterized a complex interaction between HA-induced CD44 signaling and integrin receptor expression and activation in metastatic breast cancer cells. We have shown that stimulation with HA induces a rapid change in $\beta1$-integrin subunit expression and using conformation-specific antibodies confirmed that HA increases the pool of activated $\beta1$-integrin receptors in these adhesive breast cancer cells (McFarlane et al., 2008). Furthermore, we have demonstrated that the adhesion of MDA-MB-231 breast cancer cells and the PC3 prostate cancer cell line to bone marrow endothelial cells is inhibited by concurrent administration of neutralizing antibody to the $\alpha4\beta1$-integrin heterodimer but not by an antibody blocking the activation of $\alpha2\beta1$-integrin receptors. In addition, we have shown that a HA/CD44-induced activation of a cortactin-paxillin signaling pathway that underpins cell–cell adhesion, is attenuated by co-administration of a neutralizing antibody to the $\alpha4\beta1$-integrin heterodimer. Consequently, the results from several laboratories suggest a physical interaction of CD44 with the $\alpha4\beta1$-integrin heterodimer, to regulate the cytoskeletal reorganization that underpins cell–cell adhesion. The precise nature of the intermolecular association of these two cell-surface receptors remains unknown but may result from a direct physical interaction of the ectodomains of the receptors or alternatively, by the action of scaffolding/adaptor proteins interacting with the cytoplasmic tails of these receptors.

Further studies also indicate that CD44 may contribute to cell–cell adhesion through interactions with P-selectins. This ability of CD44 to interact with selectins is associated with the extensive incorporation of O-linked sialofucosylated glycans within the variant exons of CD44. In a series of recent studies primarily conducted on colorectal cancer cell lines, the expression of variant CD44 was shown to selectively slow the rolling and promote adhesion of these cancer cells on P-selectin with diminished effects reported in facilitating rolling and/or adhesion to either E- or L-selectin (Hanley et al., 2006; Napier et al., 2007).

As such, the expression of standard and variant isoforms of CD44 on cancer cells may contribute to the promotion of cell–cell adhesion through their ability to engage in (i) a HA-dependent reciprocal interaction with CD44 expressed on endothelial cells; (ii) through regulating inside-out activation of integrin receptors; and finally (iii) through serving as a ligand for selectins expressed on the surface of endothelial cells.

THE ROLE OF CD44 IN PROMOTING TISSUE-SPECIFIC METASTASIS

Tumors show a high degree of selectivity with regard to the organs in which secondary tumors develop. The development of metastases within the liver is a prevalent clinical feature of advanced and aggressive colorectal cancers. CD44 expression, particularly that of large molecular weight variant isoforms of CD44, has been shown to correlate with increased metastasis of colorectal cancer cells to the liver (Yue et al., 2003; Ohji et al., 2007; Kuhn et al., 2007). However, other diseases such as breast cancer metastasize to a diffuse number of tissues including the lungs, liver, brain, skin, and bone. Therefore, does CD44 expression on cancer cells promote tissue-specific homing of cancers? Tetracycline-regulated induction of CD44 expression in MCF-7 breast cancer cells has recently been shown to potentiate their dissemination from the mammary fat pad to the liver, without incidence of histologically detectable metastasis in the lungs or other soft tissues (Ouhtit et al., 2007). In contrast, consistent with CD44 promoting an efficient adhesion of breast cancer cells to bone marrow endothelium cells in *in vitro* models, clinical reports have confirmed an enrichment of CD44 expression (studied in the context of the breast cancer stem-cell markers) in breast cancer cells resident within the bone marrow of patients diagnosed with early stage breast cancer (Balic et al., 2006). Enrichment of CD44 expression has also been detected in endocrine-resistant breast cancer cells that have disseminated to the lymphatic vessels or regional lymph nodes in athymic mice (Harrell et al., 2006), experimental findings that are again supported by a prior correlation of CD44 expression with nodal metastasis in breast cancer patients (Berner and Nesland, 2001). The apparent disparity of the results observed from these range of studies with breast cancer cells may reflect the fact that the resultant site of metastasis observed can be heavily influenced by the cell line exploited, the stresses that they are exposed to, and by the route of tumor cell administration to the animal. The use of *in vivo*-selected, tissue-homing derivatives of breast cancer cells in experimental models of metastasis allied with an extensive pathological examination of metastatic lesions from patients will be useful in elucidating the role of CD44 in promoting the dissemination of breast cancer to particular target organs. The current evidence from clinical and laboratory-based studies suggests that CD44 expression correlates with the escape of breast cancer cells to the lymph nodes, liver, and bone marrow cavity. Although some studies suggest a differential role for standard and variant isoforms, expression of CD44 in multiple tumor types typically correlates with an increased metastatic capability as opposed to directing tissue-specific metastasis.

III. HYALURONAN RECEPTORS AND SIGNAL TRANSDUCTION PATHWAYS

CONCLUSIONS

Expression of CD44 on the surface of cancer cells facilitates their adhesion to other cells and the avidity with which they may adhere to substituents of the extracellular matrix. In addition to regulating adhesion, CD44 facilitates the penetration of cancer cells through tissue by acting as a scaffold to concentrate proteases on the invasive surface of cells and by regulating protease expression in cancer cells. Further understanding of the downstream signaling proteins and mediators of CD44-promoted adhesion and penetration will provide novel therapeutic opportunities to attenuate the metastasis-promoting action of HA and CD44 within the tumor microenvironment. Alternatively, an increased biophysical knowledge of CD44 and its interacting proteins may identify unique strategies to uncouple CD44 from proteases and other cell-surface adhesion receptors, providing more selective strategies to target the adhesion and penetration of cancer cells.

References

Abraham, B. K., Fritz, P., McClellan, M., et al. (2005). Prevalence of CD44+/CD24−/low cells in breast cancer may not be associated with clinical outcome but may favour distant metastasis. *Clinical Cancer Res* **11**, 1154–1159.

Al-Hajj, M., Wicha, M. S., Benito-Hernandez, A., Morrison, S. J., and Clarke, M. F. (2003). Prospective identification of tumorigenic BrCa cells. *Proc Natl Acad Sci USA* **100**, 3983–3988.

Artym, V. V., Zhang, Y., Seillier-Moiseiwitch, F., Yamada, K. M., and Mueller, S. C. (2006). Dynamic interactions of cortactin and membrane type-1 matrix metalloproteinase at invadopodia: defining the stages of invadopodia formation and function. *Cancer Res* **66**, 3034–3043.

Atkinson, S. J., Crabbe, T., Cowell, S., et al. (1995). Intermolecular autolytic cleavage can contribute to the activation of pro-gelatinase A by cell membranes. *J Biol Chem* **270**, 30479–30485.

Auvinen, P., Tammi, R., Parkkinen, J., et al. (2000). Hyaluronan in peritumoral stroma and malignant cells associates with breast cancer spreading and predicts survival. *Am J Pathol* **156**, 529–536.

Balic, M., Lin, H., Young, L., et al. (2006). Most early disseminated cancer cells detected in bone marrow of BrCa patients have a putative BrCa stem cell phenotype. *Clin Cancer Res* **12**, 5615–5621.

Berner, H. S. and Nesland, J. M. (2001). Expression of CD44 isoforms in infiltrating lobular carcinoma of the breast. *Breast Cancer Res Treat* **65**, 23–29.

Bourguignon, L. Y., Peyrollier, K., Gilad, E., and Brightman, A. (2007). Hyaluronan–CD44 interaction with neural Wiskott–Aldrich syndrome protein (N-WASP) promotes actin polymerization and ErbB2 activation leading to beta-catenin nuclear translocation, transcriptional up-regulation and cell migration in ovarian tumor cells. *J Biol Chem* **282**, 1265–1280.

Bourguignon, L. Y., Zhu, H., Shao, L., and Chen, Y. W. (2001). CD44 interaction with c-Src kinase promotes cortactin-mediated cytoskeleton function and hyaluronic-acid dependent ovarian tumor cell migration. *J Biol Chem* **271**, 7327–7336.

Chellaiah, M. A., Biswas, R. S., Rittling, S. R., Denhardt, D. T., and Hruska, K. A. (2003). Rho-dependent Rho-kinase activation increases CD44 surface expression and bone resorption in osteoclasts. *J Biol Chem* **278**, 29086–29097.

Chellaiah, M. A., Kizer, N., Biswas, R., et al. (2003). Osteopontin deficiency produces osteoclast dysfunction due to reduced CD44 surface expression. *Mol Biol Cell* **14**, 173–189.

DeGrendele, H. C., Estess, P., Picker, L. J., and Siegelman, M. H. (1996). CD44 and its ligand hyaluronate mediate rolling under physiologic flow: a novel lymphocyte–endothelial cell primary adhesion pathway. *J Exp Med* **183**, 1119–1130.

DeGrendele, H. C., Estess, P., and Siegelman, M. H. (1997). Requirement for CD44 in activated T cell extravasation into an inflammatory site. *Science* **278**, 672–675.

Deryugina, E. I. and Quigley, J. P. (2006). Matrix metalloproteinases and tumor metastasis. *Cancer Metast Rev* **25**, 9–34.

Dontu, G., Liu, S., and Wicha, M. S. (2005). Stem cells in mammary development and carcinogenesis: implications for prevention and treatment. *Stem Cell Rev* **1**, 207–213.

Draffin, J. E., McFarlane, S., Hill, A., Johnston, P. G., and Waugh, D. J. J. (2004). CD44 potentiates the adhesion of metastatic prostate and breast cancer cells to bone marrow endothelium. *Cancer Res* **64**, 5702–5711.

Estess, P., Nandi, A., Mohamadzadeh, M., and Siegelman, M. H. (1999). Interleukin-15 induces hyaluronan expression in vitro and promotes activated T cell extravasation through a CD44-dependent pathway in vivo. *J Exp Med* **190**, 9–19.

Fieber, C., Baumann, P., Vallon, R., et al. (2004). Hyaluronan-oligosaccharide-induced transcription of metalloproteases. *J Cell Sci* **117**, 359–367.

Gocheva, V. and Joyce, J. A. (2007). Cysteine cathepsins and the cutting edge of cancer invasion. *Cell Cycle* **6**, 60–64.

Hanahan, D. and Weinberg, R. A. (2000). The hallmarks of cancer. *Cell* **100**, 57–70.

Hanley, W. D., Napier, S. L., Burdick, M. M., et al. (2006). Variant isoforms of CD44 are P- and L-selectin ligands on colon carcinoma cells. *FASEB J* **20**, 337–339.

Harrell, J. C., Dye, W. W., Allred, D. C., et al. (2006). Estrogen receptor positive breast cancer metastasis: altered hormonal sensitivity and tumor aggressiveness in lymphatic vessels and lymph nodes. *Cancer Res* **66**, 9308–9315.

Hill, A., McFarlane, S., Johnston, P. G., and Waugh, D. J. J. (2008). CD44 signaling increases cathepsin K and MT1-MMP expression to potentiate breast cancer cell invasion through collagen I. *Breast Cancer Res* **10** Suppl 2, P38.

Hill, A., McFarlane, S., Mulligan, K., et al. (2006). Cortactin underpins CD44-promoted invasion and adhesion of breast cancer cells to bone marrow endothelial cells. *Oncogene* **25**, 6079–6091.

Itano, N., Sawai, T., Atsumi, F., et al. (2004). Selective expression and functional characteristics of three mammalian hyaluronan synthases in oncogenic malignant transformation. *J Biol Chem* **279**, 18679–18687.

Itoh, Y. and Seiki, M. (2006). MT1-MMP: a potent modifier of pericellular microenvironment. *J Cell Physiol* **206**, 1–8.

Kajita, M., Itoh, Y., Chiba, T., Mori, H., Okada, A., Kinoh, H., and Seiki, M. (2001). Membrane-type 1 matrix metalloproteinase cleaves CD44 and promotes cell migration. *J Cell Biol* **153**, 893–904.

Karihtala, P., Soini, Y., Auvinen, P., Tammi, R., Tammi, M., and Kosma, V. M. (2007). Hyaluronan in breast cancer: correlations with nitric oxide synthases and tyrosine nitrosylation. *J Histochem Cytochem* **55**, 1191–1198.

Kim, H. R., Wheeler, M. A., Wilson, C. M., et al. (2004). Hyaluronan facilitates invasion of colon carcinoma cells in vitro via interaction with CD44. *Cancer Res* **64**, 4569–4576.

Kim, H., Yang, X. L., Rosada, C., Hamilton, S. R., and August, J. T. (1994). CD44 expression in colorectal adenomas is an early event occurring prior to K-ras and p53 gene mutation. *Arch Biochem Biophys* **310**, 504–507.

Knauper, V., Bailey, L., Worley, J. R., et al. (2002). Cellular activation of pro-MMP-13 by MT1-MMP depends on the C-terminal domain of MMP-13. *FEBS Lett* **532**, 127–130.

Knudson, W., Aguiar, D. J., Qua, Q., and Knudson, C. B. (1996). CD44-anchored hyaluronan-rich pericellular matrices: an ultrastructural and biochemical analysis. *Exp Cell Res* **228**, 216–228.

Kobayashi, H., Suzuki, M., Kanayama, N., et al. (2002). CD44 stimulation by fragmented hyaluronic acid induces upregulation of urkinase-type plasminogen activator and its receptor and subsequently facilitates invasion of human chondrosarcoma cells. *Int J Cancer* **102**, 379–389.

Kuhn, S., Koch, M., Nubel, T., et al. (2007). A complex of EpCAM, claudin-7, CD44 variant isofroms and tetraspanins promotes colorectal cancer progression. *Mol Cancer Res* **5**, 553–567.

Lopez-Ortin, C. and Matrisian, L. M. (2007). Emerging roles of proteases in tumour suppression. *Nat Rev Cancer* **7**, 800–808.

Madsen, M. A., Deryugina, E. I., Niessen, S., Cravatt, B. F., and Quigley, J. P. (2006). Activity based protein profiling implicates urokinase activation as a key step in human fibrosarcoma intravasation. *J Biol Chem* **281**, 15997–16005.

Mohamed, M. M. and Sloane, B. F. (2006). Cysteine cathepsins: multifunctional enzymes in cancer. *Nat Rev Cancer* **6**, 764–775.

Mori, H., Tomari, T., Koshikawa, N., et al. (2002). CD44 directs membrane type-1 matrix metalloproteinase to lamellipodia by associating with its hemopexin-like domain. *EMBO J* **21**, 3949–3959.

McFarlane, S., Hill, A., Johnston, P. G., and Waugh, D. J. J. (2008). Characterization of a cytoskeletal signaling pathway underpinning CD44-initiated, integrin-mediated adhesion of breast cancer cells to bone marrow endothelium. *Breast Cancer Res* **10** Suppl 2, 36.

Nandi, A., Estess, P., and Siegelman, M. H. (2000). Hyaluronan anchoring and regulation on the surface of vascular endothelial cells is mediated through the functionally active form of CD44. *J Biol Chem* **275**, 14939–14948.

Nandi, A., Estess, P., and Siegelman, M. (2004). Bimolecular complex between rolling and firm adhesion receptors required for cell arrest; CD44 association with VLA-4 in T cell extravasation. *Immunity* **20**, 455–465.

Napier, S. L., Healy, Z. R., Schnaar, R. L., and Konstantopoulos, K. (2007). Selectin ligand expression regulates the initial vascular interactions of colon carcinoma cells: the roles of CD44v and alternative sialofucosylated selectin ligands. *J Biol Chem* **282**, 3433–3441.

Ohji, Y., Yao, T., Eguchi, T., et al. (2007). Evaluation of risk of liver metastasis in colorectal adenocarcinoma based on the combination of risk factors including CD10 expression: multivariate analysis of clinicopathological and immunohistochemical factors. *Oncol Rep* **17**, 525–530.

Ohno, S., Im, H. J., Knudson, C. B., and Knudson, W. (2006). Hyaluronan oligosaccharides induce matrix metalloproteinase 13 via transcriptional activation of NFkappaB and p38 MAP kinase in articular chondrocytes. *J Biol Chem* **281**, 17952–17960.

Ohno, S., Ohno-Nakahara, M., Knudson, C. B., and Knudson, W. (2005). Induction of MMP-3 by hyaluronan oligosaccharides in temporomandibular joint chondrocytes. *J Dent Res* **84**, 1005–1009.

Okada, T., Hawley, R. G., Kodaka, M., and Okuna, H. (1999). Significance of VLA-4-VCAM-1 interaction and CD44 for transendothelial invasion in a bone marrow metastatic myeloma model. *Clin Exp Metastasis* **17**, 623–629.

Ouhtit, A., Abd Elmegeed, Z. Y., Abdraboh, M. E., Lioe, T. F., and Raj, M. H. (2007). In vivo evidence for the role of CD44s in promoting breast cancer metastasis to the liver. *Am J Pathol* **171**, 2033–2039.

Ponta, H., Sherman, L., and Herrlich, P. A. (2003). CD44: from adhesion molecules to signaling regulators. *Nat Rev Mol Cell Biol* **4**, 33–45.

Siegelman, M. H., Stanescu, D., and Estess, P. (2000). The CD44-initiated pathway of T-cell extravasation uses VLA-4, but not LFA-1 for firm adhesion. *J Clin Invest* **105**, 683–691.

Simpson, M. A., Reiland, J., Burger, S. R., et al. (2001). Hyaluronan synthase inhibition in metastatic prostate carcinoma cells correlates with hyaluronan surface retention, a prerequisite for rapid adhesion to bone marrow endothelial cells. *J Biol Chem* **276**, 17949–17957.

Simpson, M. A., Wilson, C. M., Furcht, L. T., et al. (2002). Manipulation of hyaluronan–synthase expression in prostate adenocarcinoma cells alters pericellular matrix retention and adhesion to bone marrow endothelial cells. *J Biol Chem* **277**, 10050–10057.

Szabova, L., Chrysovergis, K., Yamada, S.S. and Holmbeck, K. (2007). MT1-MMP is required for efficient tumor dissemination in experimental metastatic disease. *Oncogene* Dec 10, epub ahead of print.

Thorne, R. F., Legg, J. W., and Isacke, C. M. (2004). The role of the CD44 transmembrane domain and cytoplasmic domains in co-ordinating adhesive and signaling events. *J Cell Sci* **117**, 373–380.

Toole, B. P. (2004). Hyaluronan: from extracellular glue to pericellular cue. *Nat Rev Cancer* **4**, 528–539.

Tzircotis, G., Thorne, R. F., and Isacke, C. M. (2005). Chemotaxis towards hyaluronan is dependent on CD44 expression and modulated by cell type variation in CD44–hyaluronan binding. *J Cell Sci* **118**, 5119–5128.

Tzircotis, G., Thorne, R. F., and Isacke, C. M. (2006). Directional sensing of a phorbol ester gradient requires CD44 and is regulated by CD44 phosphorylation. *Oncogene* **25**, 7401–7410.

Ueda, J., Kajita, M., Suenaga, N., Fujii, K., and Seiki, M. (2003). Sequence-specific silencing of MT1-MMP expression suppresses tumour cell migration and invasion: importance of MT1-MMP as a therapeutic target for invasive tumours. *Oncogene* **22**, 8716–8722.

Wang, H. S., Hung, Y., Su, C. H., et al. (2005). CD44 cross-linking induces integrin-mediated adhesion and transendothelial migration in breast cancer cell line by up-regulation of LFA-1 (alphaL beta2) and VLA-4 (alpha4 beta1). *Exp Cell Res* **304**, 116–126.

Weber, G. F., Bronson, R. T., Ilagan, J., Cantor, H., Schmits, R., and Mak, T. W. (2002). Absence of the CD44 gene prevents sarcoma metastasis. *Cancer Res* **62**, 2281–2286.

Wielenga, V. J., Smits, R., Korinek, V., et al. (1999). Expression of CD44 in Apc and Tcf mutant mice implies regulation by the Wnt pathway. *Am J Pathol* **154**, 515–523.

Yu, Q. and Stamenkovic, I. (1999). Localization of matrix metalloproteinase-9 to the cell surface provides a mechanism for CD44-mediated tumor invasion. *Genes Dev* **13**, 35–48.

Yue, S. Q., Yang, Y. L., Dou, K. F., and Li, K. Z. (2003). Expression of PCNA and CD44 mRNA in colorectal cancer with venous invasion and its relationship to liver metastasis. *World J Gastroenterol* **9**, 2863–2865.

Zoltan-Jones, A., Huang, L., Ghatak, S., and Toole, B. P. (2003). Elevated hyaluronan production induces mesenchymal and transformed properties in epithelial cells. *J Biol Chem* **278**, 45801–45810.

III. HYALURONAN RECEPTORS AND SIGNAL TRANSDUCTION PATHWAYS

CHAPTER

8

Involvement of CD44, a Molecule with a Thousand Faces, in Cancer Dissemination

David Naor, Shulamit B. Wallach-Dayan,
Muayad A. Zahalka, and Ronit Vogt Sionov

PROLOGUE

Cancer progression and metastasis are complex processes associated with multiple and highly organized sequential steps that are both well coordinated and organ selective. The dissemination of malignant cells is dependent on growth and local invasion at the primary site. First, the cells must lose local adhesiveness, e.g., in epithelial cells by breaking homophilic interactions mediated by cadherins. The cells, which are detached from the local growth (now designated as metastatic cells) penetrate into the lymphatic system and the blood circulation. The metastatic cells migrate in the lymph vessels and blood vasculature until they become trapped at remote sites through adhesion to endothelial cells. Then they transverse the endothelium (in a process known as trans-endothelial migration) and establish new colonies of malignant cells at the extravascular site (Yeatman and Nicolson, 1993). Well orchestrated interplay of adhesion molecules and chemoattractants is indispensable for the cell's locomotion in the blood-lymph circulation, transmigration through the endothelium and colonization within the target organ (Stetler-Stevenson et al., 1993). Organ-derived chemokines, present in soluble form as well as confined to the cell membrane or extracellular matrix (ECM), are involved in the selective navigation of metastatic cells into specific organs (Yeatman and Nicolson, 1993). Adhesion molecules expressed on metastatic cells and their specific countermolecules (ligands) expressed on endothelial cells and on cells of the extravasculature are also involved in the selective lodgment of the disseminating cancer cells in the target tissues. In this context we have found (Zahalka et al., 1995) that spleen invasion by lymphoma cells is mediated by $\beta2$-integrin, whereas CD44, expressed on the same lymphoma cells, navigates these cells to the lymph nodes. Possibly, integrin ligand exclusively expressed in the spleen and CD44 ligand exclusively expressed in the lymph node differentially arrest the lymphoma cells in the respective organs. The docking of the metastatic cell in the target tissue is another essential step in the malignant process, as their adherence to substrate blocks delivery of apoptotic signals and thereby rescues the cancer cell from apoptotic death (Meredith and Schwartz, 1997; Tian et al., 2000).

The three steps model of Dr. Springer (Springer, 1994) describes the mechanism of cell locomotion in the vasculature and the cell trans-endothelial migration. In the first step, cells flowing in the blood vessel form lose attachment with the endothelial cells, resulting in cell rolling interactions mediated by adhesion de-adhesion processes (i.e. the cells lose and gain adhesiveness). In the second step, chemoattractants derived from the extravascular tissue or from the endothelium stimulate G-protein-like receptors on the rolling cells. The stimulated G-protein receptors activate cell surface integrins. In the third step, the activated integrins generate firm attachments with the endothelial cells resulting in extravasation and cell accumulation in the tissue. Acquisition of rolling capacity by a cell of multicellular organism is highly dependent on the ability of the cell to gain and lose adhesiveness. In order to move, the affinity of cell surface molecules to the substratum countermolecules must be intermediate. Too strong affinity may arrest the cell on the endothelium, whereas too weak affinity may allow the blood shear stress to release the cell from the endothelium. The process of cell motility is essential for embryonic development, and reoccurs in the adult organism at sites of wound healing or tissue remodeling and also during metastasis, where malignant cells probably recapitulate processes of embryonic behavior.

The interplay between chemokines, cytokines, growth factors, cell-bound or tissue-bound enzymes, adhesion and homing molecules, as well as apoptosis-promoting and survival-inducing signals, determines the tumor fate for life or death. Therefore, comprehensive understanding of the sophisticated biological networks is a prerequisite for developing novel therapeutic strategies. Realizing that the cell migration machinery is an important factor for tumor spread and knowing that adhesion and homing molecules are critical elements for this function, we have focused our research on these molecules. The malignant lymphoma LB cell line was used as an experimental model in our study, since we were familiar with its structural and functional phenotype, that has been extensively explored in our laboratory (e.g., see Ish-Shalom et al., 1995; Pillemer et al., 1992; Sharon et al., 1993). Of the two dominant families of adhesion/homing molecules that could be involved in the support of LB cell migration, i.e. integrins or selectins, we focused our attention on the former, simply because the latter is not expressed on these cells (Zahalka et al., 1993). Using the protocol described by Zahalka et al. (1993), we found that injection of the $\beta2$-integrin-specific anti-CD18 monoclonal antibodies (mAb) into BALB/c mice almost completely blocked the spleen colonization by LB lymphoma cells, as indicated by measuring the tumor cell proliferation in the organ, using the H^3-thymidine incorporation assay. Surprisingly, the invasion of proliferating LB cells into the lymph nodes was not affected by targeting the $\beta2$-integrin with the same antibody. Furthermore, spleen-derived LB cells (i.e., LB cells that were recovered from the spleen) colonized upon their subcutaneous (s.c.) injection both in the spleen and the lymph nodes, whereas lymph node-derived LB cells colonized in the lymph nodes only (Zahalka et al., 1993). These findings suggest that LB cells trapped in the lymph nodes lose their spleen homing receptor, whereas LB cells trapped in the spleen retain both their spleen and lymph node homing receptors. We subsequently asked, if not $\beta2$ integrin, which adhesion/homing receptor guides LB cells into the lymph node? At that time a report by Mackay et al. (1990) attracted our attention. They showed that CD44

III. HYALURONAN RECEPTORS AND SIGNAL TRANSDUCTION PATHWAYS

expression on activated lymphocytes correlated with their localization in the peripheral lymph nodes. Ultimately, we decided to explore the involvement of CD44 in lymph node colonization by LB cells.

CD44, A POLYGAMIC MOLECULE, INTERACTING WITH MULTIPLE LIGANDS

CD44 Structure

CD44 is a single chain glycoprotein encoded by a single copy of a gene located on the short arm of chromosome 11 in humans and on chromosome 2 in mice, spanning ~50 kb of genomic DNA. The CD44 glycoprotein is an acidic molecule whose charge is largely determined by sialic acid. The $t_{1/2}$ of CD44 turnover is estimated to 8 h. Mouse CD44 contains 20 exons. In the mouse, the first 5 exons (1 to 5) are constant. The next 10 exons (6 to 15, also numbered v1 to v10) are variant, i.e., they are differentially incorporated into the CD44 molecule by alternative splicing. Exons 16, 17, and 18 are again constant whereas exons 19 and 20 are variant. Exons 16 and 17 encode the C' terminus part of the extracellular domain, exon 18 encodes the transmembrane spanning domain, and exon 19 or 20 differentially encodes either a short (three amino acids) cytoplasmic tail or the more abundant long one (72 amino acids). The human CD44 does not contain exon v1 (the first variant exon of human CD44 is numbered v2). The differential utilization of the 10 variant exons generates multiple CD44 variants (CD44v) with different combinations of variant exon products (reviewed in Naor et al., 2002; Naor et al., 1997; Ponta et al., 2003). Theoretically, over 800 membrane-bound CD44 isoforms can be generated, although apparently not all combinations are expressed (Naor et al., 2002; Naor et al., 1997). To date, tens of different CD44 isoforms have been discovered; the most common one is standard CD44 (CD44s), in which exon 5 is spliced directly to exon 16, skipping the entire variant exon sequence.

The amino terminus of the CD44 molecule contains six cysteine residues, which are possibly utilized to form a globular domain or three globular subdomains. A stretch of 92 amino acids (from position 12 to position 103) at the N-terminus of the human CD44 receptor displays ~35% homology with a sequence from a family of proteins that bind hyaluronic acid (HA). This group of proteins, known as hyaladherins or the link protein superfamily, includes cartilage link protein, aggrecan, versican, tumor necrosis factor-inducible protein-6 (TSG-6) and lymphatic vessel endothelial HA receptor-1 (LYVE-1), but not the receptor for hyaluronate-mediated motility (RHAMM) or its variant: intracellular HA binding protein (IHABP), both of which are structurally unrelated to the hyaladherins (Naor et al., 2002). A structural model of the CD44 link-homologous region was constructed based on the TSG-6 link module determined by NMR (Bajorath, 2000). Residues Arg-21, Tyr-22, Arg-58, and Tyr-59 of human CD44 form a cluster critical for HA binding.

The structural variability of the CD44 molecule is further enriched by posttranslational modifications. Massive N-linked and O-linked glyco-sylations, as well as glycosaminoglycan (GAG) attachments double the molecular mass of the CD44s core protein from 37 kDa to 80–95 kDa. The molecular mass of the different CD44 isoforms ranges from $\sim$80 kDa to $\sim$250 kDa. Both alternative splicing and posttranslational modifica-tions contribute to this diversity. Most of the potential N-linked glycosyl-ation sites (Asn-residues) are found in the extreme N-terminal of the extracellular domain, which includes the link module, and in the alterna-tively spliced variable region. The potential O-linked glycosylation sites (Ser/Thr residues) and GAG attachments (Ser-Gly motifs) are distributed in the more carboxyl terminal regions of the extracellular domain, including the membrane proximal region, and the variable region (Naor et al., 2002). The human CD44 cytoplasmic tail contains 6 potential serine phosphorylation sites. Schematic maps depicting the CD44 molecule are illustrated in our review articles (Naor et al., 2002; Naor et al., 1997) and in a more recent review by Ponta et al. (2003).

CD44 Ligands and the Mechanism of Ligand Binding

The principal ligand of the CD44 receptor is hyaluronic acid (HA, hyaluronate, hyaluronan), a linear polymer of repeating disaccharide units [D-glucuronic acid (1-β-3) N-acetyl-D-glucosamine (1-β-4)]$_n$. CD44 can, however, interact with several additional molecules such as galectin-8 (Eshkar Sebban et al., 2007), collagen, fibronectin, fibrinogen, laminin, chondroitin sulfate, mucosal vascular addressin, serglycin/gp600, osteopontin (OPN) and the major histocompatibility complex class II invariant chain (Ii), as well as L-selectin and E-selectin (reviewed in Naor et al., 2002, and Naor et al., 1997; and see also Hanley et al., 2005; Napier et al., 2007). In many cases CD44 does not bind to its ligand unless activated by external stimuli. As both CD44 and its ligand are ubiquitous molecules, this mechanism should avoid unnecessary engagement of the receptor. In fact, three states of CD44 activation have been identified in cell lines and normal cell populations (Lesley et al., 1995): active CD44, which constitutively binds HA; inducible CD44, which does not bind HA or binds it only weakly, unless activated by inducing factors (e.g., mAbs, cytokines, or phorbol ester); and inactive CD44, which does not bind HA, even in the presence of inducing factors. In many cases the N-glycosylation pattern of the CD44 receptor dictates the status of its HA binding capacity, where active CD44 is the least glycosylated, inactive CD44 the most glycosylated, and inducible CD44 holds the intermediate position (Lesley et al., 1995). Complete or partial removal of N-glycans from the CD44 molecule prevents HA from binding to the receptor of some types of cells, whereas in others partial removal enables HA binding. Trimming of the exterior sugar unit, alpha 2,3-linked sialic acid, confers HA binding upon some CD44 molecules that normally do not engage this ligand. Acquisition or enhancement of HA binding can be achieved through desialylation of cell surface CD44 or by treating cells expressing this molecule with TGF-β1 (Cichy and Pure, 2000) or phorbol 12-myristate 13-acetate (PMA) (Rochman et al., 2000). Desialylation reduces the negative charge of CD44 thereby seemingly enabling the interaction with the negatively charged HA. Exposure of HA binding sites, either directly or due to a conformational change and/or increase in the net positive charge of the receptor, could account for the acquisition of HA binding by the CD44 modified molecule. O-glycosylation may negatively influence HA binding in several, but by no means all, cell lines. Additionally, GAG modifications may negatively or positively influence HA binding (Naor et al., 2002; Naor et al., 1997). For example, decoration of CD44 by heparan sulfate (HS) or chondroitin sulfate (CS) is required for binding of several growth factors, fibronectin or collagen (Ehnis et al., 1996; Jalkanen and Jalkanen, 1992), and for acquisition of migration capacity on collagen and fibrinogen substrates by melanoma and endothelial cells, respectively (Faassen et al., 1992; Henke et al., 1996; Knutson et al., 1996). In some cell lines HA binds exclusively to CD44s, whereas in others only CD44 variants bind HA (reviewed in Naor et al., 1997). Expression of variant epitopes may enforce aggregation of the CD44 receptors, leading to acquisition of HA binding (Sleeman et al., 1996), or it might simply expose the distal HA binding site for interaction with the ligand. Note that HA aggregation also enhances its binding to the CD44 receptor (Zhuo et al., 2006).

THE INTERACTION BETWEEN THE CD44 OF LB CELLS AND HYALURONATE – IN VITRO STUDIES

Flow cytometry analysis revealed that LB cells co-express CD18 and CD44 (Zahalka et al., 1995). Before analyzing how the CD44 molecule supports the invasion of LB cells into mouse lymph nodes, we will discuss first our in vitro findings, which elucidate the mechanism of interaction between the CD44 of LB cells and hyaluronate. Having this information in mind, it might be easier to evaluate the data derived from animal studies. LB cells directly obtained from culture or from local tumor growth (following s.c. injection), as well as from spleen or axillary lymph nodes (again, after their s.c. injection into BALB/c mice), expressed pan-CD44 (detected by flow cytometry, using anti-CD44 mAb recognizing a constant epitope shared by all CD44 isoforms) and v6-containing CD44 isoform (detected with anti-CD44v6 mAb recognizing epitope included in the v6 variant exon product). These cells expressed to a lesser extent v4-containing epitope detected with anti-CD44v4 mAb (Sionov and Naor, 1997). Analyzing the CD44 transcript repertoire with RT-PCR provides additional information related to the CD44 expression at the nucleotide level: lymphoma dissemination, especially to the lymph node, is associated with enhanced expression of cell surface CD44 variants not expressed on parental LB cells (Wallach et al., 2000). However, all these types of lymphoma cells neither bound soluble fluoresceinamine-labeled HA (Fl-HA), as indicated by flow cytometry, nor attached to HA immobilized on plastic. Pretreatment of the culture or animal-derived cells with hyaluronidase did not restore the HA binding capacity, implying that the lack of binding cannot be attributed to external-bound HA, derived from culture or animal, that blocks the CD44 receptor. In independent experiments we showed that pretreatment of HA9 cells (a subline of LB cells that constitutively bind HA) with hyaluronidase does not affect the subsequent binding of Fl-HA to HA9 cells (see below and Sionov and Naor, 1997). Acquisition of HA binding to all types of LB cells was detected by both methods after their incubation with PMA (Rochman et al., 2000; Sionov and Naor, 1997; 1998). Increasing the dose of PMA gradually enhanced the upregulation of CD44 on LB cells, finally reaching a threshold of cell surface CD44 expression, which allows HA binding (Sionov and Naor, 1998). A constitutive HA-binder HA9 cell line was derived from low HA-binder LB2.3 subline (S. Elbaz from our laboratory) by repeatedly selecting for cells binding to immobilized HA. HA9 cells expressed CD44 and CD44 variants more intensively than the parental LB cells, as indicated by both flow cytometry (Sionov and Naor, 1997) and RT-PCR (Wallach et al., 2000). Flow cytometry revealed that this difference was maintained even after PMA activation, which enhanced the CD44 expression on both cell types. Linear relationships were registered for CD44 upregulation and acquisition of HA binding by HA9 cells and PMA-activated LB cells, the latter after reaching an expression threshold level (Sionov and Naor, 1998). Interestingly, presence of IM7.8.1 anti-CD44 mAb (but not anti-CD44v6 or anti-CD44v4 mAbs) reduced the binding of Fl-HA to HA9 cells, yet, excess of soluble HA, while blocking the binding of Fl-HA to HA9 cells, did not interfere with the binding of the anti-CD44 mAb to the same cells (Sionov and Naor, 1997). This finding suggests that HA and the anti-pan CD44 mAb interact with distinct sites on the CD44 molecule, but the antibody allosterically reduced the binding of the CD44 molecule for HA. Immobilized HA treated with hyaluronidase lost its ability to bind PMA-activated LB cells (Zahalka et al., 1995) or HA9 cells (Sionov and Naor, 1997).

To acquire HA binding potential, LB cells must be activated either *in vitro* (e.g. by PMA) or *in vivo* by a still unknown factor. We suggest that the *in vivo*-activated LB cells are a rare cell population, represented by HA9 cells isolated in our study by serial repeating selections of LB cells attached to immobilized HA. These cells are characterized by their ability to form CD44-dependent aggregates (Wallach et al., 2000). Hence, the activation of LB cells by PMA and HA9 cells, respectively, represent two phases in HA binding process: the induction phase mediated by phorbol ester stimulation and the effector phase of constitutively HA-binder cells. This discrepancy enables to distinguish between intracellular signals involved in the induction phase and those involved in the effector phase of HA binding. The *de novo* synthesis of cell surface CD44 of LB cells activated with PMA as well as the acquisition of HA binding potential by these cells were inhibited by cycloheximide. The basic expression of cell surface CD44 on non-activated LB and HA9 cells as well as the spontaneous HA binding capacity of HA9 cells were not affected by addition of cycloheximide, suggesting a relative long half life of CD44. In contrast, after PMA activation, the newly synthesized cell surface CD44 and the accompanied acquisition of HA binding were inhibited by this reagent (Sionov and Naor, 1998). PMA is a well-known protein kinase C (PKC) activator; e.g. such a stimulated PKC is involved in activation of CD44, leading to its ability to mediate chemotaxis (Tzircotis et al., 2006). However, staurosporine, sphingosine, polymyxin B, and quercetin that inhibit PKC activity or genestein that inhibits tyrosine protein kinase, did not affect PMA-induced HA binding by LB cells (Sionov and Naor, 1998), indicating that these signaling pathways are not involved in the acquisition of HA binding capacity. In contrast, the PMA-induced HA binding by LB cells (and the corresponding enhanced expression of CD44), but not the constitutive binding of HA9 cells, was strongly suppressed by calmodulin antagonists (chlorpromazine, trifluoperazine, W-7), the calcium channel blocker verapamil, and the calcium ionophore ionomycin (possibly owing to overdose paralyzing effect) (Sionov and Naor, 1998). Hence, in LB cells, PMA-induced calcium mobilization leads to cell surface CD44 activation and acquisition of HA binding potential. We propose that these effects are mediated by calmodulin, that following calcium binding, upregulates Ca^{2+}/calmodulin-dependent kinase II (CaMKII). Mishra et al. (2005) demonstrated that calmodulin and CaMKII are involved in upregulation of monocyte CD44 after stimulation with TNFα, but not lipopolysaccharide. Lipopolysaccharide, however, induced CD44 transcription through JNK and Egr-1. The mechanisms by which calmodulin affect CD44 expression and HA binding capacity, are still unknown. There is one report showing that CaMKII phosphorylates CD44 at Ser325 (Lewis et al., 2001).

This modification was essential for CD44-dependent migration on HA. It remains to be determined whether calmodulin is involved in CD44 deglycosylation (or desialylation), or alternatively, in induction of differential splicing, both functions are associated with acquisition of HA binding (see below).

The PMA-induced HA binding acquisition described above is an example of inside-out activation process which generates oncogenic phenotype, as interaction of cell surface CD44 with HA supports cell migration and resistance to apoptosis (Naor et al., 2002). Dr. Bourguignon and colleagues described a reversed outside-in model, which also leads to oncogenic phenotype. Using a head and neck squamous cell carcinoma (HNSCC) cell line, it was demonstrated that interaction of HA with cell surface CD44 leads to calcium mobilization, CaMKII activation, and filamin (cytoskeletal protein) phosphorylation, resulting in reorganization of the cytoskeleton and tumor cell migration (Bourguignon et al., 2006). In these cells calcium mobilization, induced by HA–CD44 interaction, generates signals that enhance also cell survival and growth as well as the subsequent resistance to the anti-cancer cytotoxic drug cisplatin (Wang and Bourguignon, 2006). Hence, calcium signaling elements such as calmodulin and CaMKII as well as upstream factors such as phospholipase C and inositol 1,4,5-triphosphate are potential therapeutic targets in malignant diseases, using, for example, the relevant small interfering RNA (siRNA) approach. The question whether interaction between the cytoplasmic domain of CD44 and cytoskeletal proteins is important for HA binding is controversial and possibly cell type-related (Naor et al., 2002). However, integrated cytoskeleton is not required for PMA-induced acquisition of HA binding by LB cells or the constitutive HA binding by HA9 cells. PMA-activated LB cells and HA9 cells lost their ability to bind HA following treatment with microfilament-disrupting dose of cytochalasin D (which prevents actin polymerization) or with microtubule-disrupting dose of colchicine (which blocks microtubule formation) (Rochman et al., 2000; Sionov and Naor, 1998). Both functions are essential for cytoskeleton activity. Furthermore, we found that the acquisition of HA binding is not related to changes in the macroaggregation state of cell surface CD44 or to changes in membrane microviscosity (Rochman et al., 2000; Sionov and Naor, 1998). It should be emphasized that the question how PMA confers HA binding upon LB cells is only partially resolved.

ELUCIDATION OF HA BINDING MECHANISM IN LB CELLS

PMA activation enhanced the expression of standard CD44 and the expression and number of CD44 variants on LB cells as shown by western blot, flow cytometry, and RT-PCR (Wallach et al., 2000; Rochman et al., 2000; Sionov and Naor, 1997). Although we cannot role out the significance of quantitative changes in CD44 expression (i.e. upregulation of this receptor), which allow HA binding to LB cells, our data rather point to the significance of quantitative alterations in the context of this effect. Parental LB cells, that do not bind HA, acquired such binding potential after treatment with neuraminidase (Rochman et al., 2000). This enzyme cleaves sialic acid from cell surface CD44 glycoform, thereby increasing the net positive charge of this molecule. This in turn enables its interaction with the negatively charged HA (Naor et al., 1997). Neuraminidase treatment did not enhance HA binding to PMA-activated LB cells or HA9 cells, either because such cells are deficient in CD44 sialic acid and/or they display enhanced expression of CD44 variants. Both phenotypes may be involved, either separately or in combination, in enhancing the positive net charge of this molecule. Indeed, we presented evidence for the existence of these two parameters in PMA-activated LB cells and HA9 cells (Rochman et al., 2000; Sionov and Naor, 1997), which may explain their ability to bind HA. Glycosaminoglycans (GAGs; e.g. HA, heparin, heparan sulfate, chondroitin sulfate, or keratan sulfate) can be co-precipitated with their glycoproteins by the cationic detergent cetylpyridinium chloride (CPC). CD44 western blot analysis of the precipitated proteins revealed that HA, but not the other GAGs, binds exclusively to CD44 variants of PMA-activated LB cell extracts, but not to parental LB cell extracts. These data show that it is possible to confer HA binding upon LB cells by either removing sialic acid with neuraminidase, or by enhancing CD44 variant expression by phorbol ester (Rochman et al., 2000). Our results further suggest that CD44 variants can bind HA even if their sialic acid (if exists) was not removed (Rochman et al., 2000), since some of the alternative splicing products (but not standard CD44) may carry sufficient net positive charge to allow their interaction with HA. Low concentrations of neuraminidase and PMA, which by themselves were insufficient to confer HA binding upon LB cells, displayed a significant synergistic binding effect when combined. Similarly, tunicamycin, which inhibits N-glycosylation of cell surface glycoproteins, does not confer HA binding upon LB cells unless combined with sub-optimal doses of PMA. Interestingly, CD44 expression on LB cells under these experimental conditions (tunicamycin plus low doses of PMA) was similar to the CD44 expression on parental LB cells, yet HA binding could be detected on the double treated cells only (Rochman et al., 2000). This finding clearly shows that enhanced expression of CD44 is not obligatory for HA binding to LB cells, but qualitative changes are rather essential. The simultaneous effect of increasing the CD44 positive charge by desialylation with neuraminidase (or by deglycosylation with tunicamycin) and augmenting the CD44 variant repertoire by treatment with PMA may explain this synergistic effect. It should be stressed that the desialylation or deglycosylation procedures can affect, in addition to CD44, other cell surface proteins, whose modified version may indirectly influence HA binding to CD44 of LB cells. However, it has been observed by other investigators that desialylation of CD44-coated beads (Katoh et al., 1995) as well as desialylation or deglycosylation of soluble CD44 (Skelton et al., 1998) generates HA binding sites in the absence of other proteins, implying that the modified CD44 can independently interact with hyaluronan.

We have seen that standard CD44 of LB cells can interact with HA only after desialylation with neuraminidase, whereas CD44 variants of PMA-activated LB cells interact with HA, but not with other GAGs, even without chopping their sialic acid (Rochman et al., 2000). The last observation was supported by transfection experiments. LB cells were transfected with CD44v4-v10 (CD44v) cDNA or with standard CD44 (CD44s) to overexpress these isoforms. The CD44v-transfected LB cells bound HA and chondroitin sulfate, whereas the CD44s did not. Anti-CD44 mAb and excess of soluble HA blocked the binding of Fl-HA to CD44v-transfected LB cells, implying that CD44 receptor is involved in this interaction. The CPC precipitation assay revealed that CD44 variants (but not standard CD44) of cell extracts derived from CD44v-transfected LB cells can bind HA and other GAGs, but standard CD44 of cell extracts derived from CD44s-transfected LB cells cannot (Wallach-Dayan et al., 2001). Yet, CD44 variants from PMA-activated LB cells bind HA only, as indicated by the CPC assay (Rochman et al., 2000). This is not surprising since the CD44 variant repertoire of CD44v-transfected LB cells differs from the variant repertoire of PMA-activated LB cells.

Measuring LB cell binding under dynamic conditions, rather than static conditions, reflects more precisely the physiological *in vivo* state when the cells are bound to the endothelium of the blood vessels under the flow pressure of the blood (known as a wall shear stress; the physiological shear stress is around 2 dyn/cm^2). The LB transfectants were settled on substrates coated with different GAGs, assembled in flow chamber apparatus, and then subjected to increasing shear flow. The number of cells accumulating and rolling under the shear flow was counted. Three to four times more CD44v-transfected LB cell accumulated and rolled on HA substrate (but not on other GAGs) under shear stress of 2 dyn/cm^2 than did CD44s-transfected LB cells. The accumulation and rolling of the CD44v-transfected LB cells was blocked with anti-CD44 mAb or excess of soluble HA. In conclusion, LB cells transfected with a CD44 variant resist shear stress and remained attached to the HA substrate at much higher rates than parental LB cells or LB cells transfected with standard CD44 (Wallach-Dayan et al., 2001). Extrapolating these *in vitro* data into the animal milieu may predict that the *in vivo* migration of CD44v-transfected LB cells should be much more efficient than that of parental LB cells or LB cells transfected with standard CD44. The HA binding site of CD44v4-v10 was destroyed by point mutation, replacing arginine with alanine at position 43. This construct was transfected into LB cells. LB cells expressing the CD44 mutant, although normally presenting their CD44 isoforms, lost their ability to bind soluble Fl-HA from the solution, and to display rolling interactions on HA substrate under shear stress (Wallach-Dayan et al., 2001). Since only CD44 variants bind HA, we must assume that the variable region, generated by alternative splicing, influences the configuration of the constant domain, hence, creating an HA binding site.

THERAPY OF THE MALIGNANT LYMPHOMA: BLOCKING THE INTERACTION BETWEEN CD44 AND ITS LIGAND BY TARGETING THE CD44 VARIANT WITH RELEVANT ANTIBODY

If the *in vitro* migratory advantage of CD44v-transfected LB cells is also reflected *in vivo*, we may find that these cells disseminate faster and more efficiently than CD44s-transfected LB cells or parental LB cells. To allow detection of the disseminated lymphoma, all cell lines were labeled by transfected green fluorescence protein (GFP). Indeed, the skin local tumor growth and the lymph node invasion of CD44v-transfected LB cells were much faster than that of the corresponding CD44s-LB cells and parental LB cells. Furthermore, the mutated CD44v-LB cells, that lost their ability to bind HA, lost also their ability to develop local tumors and to invade the lymph nodes following s.c. injection (Wallach-Dayan et al., 2001). Hence, we may suggest a parallel between the *in vitro* and the *in vivo* findings, showing that the interaction between cell surface CD44v and HA is essential for both cell accumulation/rolling *in vitro* and tumor cell progression *in vivo*. However, LB cells invade the spleen and the lymph nodes substantially faster than HA9 cells, which display constitutive HA binding potential (Sionov and Naor, 1997). This finding suggests that the high affinity of CD44 isoforms expressed on HA9 cells to tissue HA slows their migration to the lymphoid organs. Hence, the following hierarchy is suggested in the context of dissemination capacity (faster > slower) of the lymphoma cells: CD44v-transfected LB cells > CD44s-transfected LB cells = LB cells > HA9 cells. To explain this hierarchy we should assume that intermediate affinity (represented by CD44v-transfected LB cells) between cell surface CD44 and tissue HA is required to allow a rapid dissemination. Too high affinity (represented by HA9 cells) slows the migration capacity and too weak affinity allows release of the cells from the tissue substrate, e.g. by the shear stress flow. The remaining cell lines adjust their positions between these two extremes.

As indicated earlier, s.c. inoculated LB cells invade the spleen and the peripheral lymph nodes. Intravenous injection of anti-CD18 mAb blocked spleen, but not lymph node invasion (Zahalka et al., 1995). It was predicted, therefore, that LB cells exploit a different molecule for lymph node infiltration. At that time it was shown (Mackay et al., 1990) that CD44 expression on activated lymphocytes correlated with their localization in the peripheral lymph nodes. We hypothesized, therefore, that the lymph node invasion by LB lymphoma cells is CD44-depended and subsequently challenged this hypothesis by analyzing the ability of anti-CD44 mAb to block the tumor infiltration into this organ. Subcutaneously injected LB cells enter the spleen via the blood and the lymph nodes - via the afferent lymphatics. LB cells cannot enter the lymph node via the high endothelial venule (HEV) even when injected intravenously (Zahalka et al., 1995). This is hardly surprising, because LB cells do not express MEL-14, a molecule essential for HEV binding. Therefore, to efficiently avoid LB cell invasion into the lymph node the relevant antibody should be s.c. injected adjacent to this organ, whereas to avoid spleen invasion the relevant antibody should be injected intravenously (i.v.). Indeed i.v. injection of anti-CD18 mAb (directed against the $\beta2$ integrin) suppressed the invasion of LB cell into the spleen, but not into the lymph nodes, whereas s.c. injection of anti-CD44 mAb or its Fab' fragments suppressed the invasion of this lymphoma into the axillary and brachial lymph nodes, but not into the spleen (LB cells co-express CD18 and CD44), as indicated by measuring the proliferation of the organ-infiltrating cells using ^{3}H-thymidine incorporation assay (Zahalka and Naor, 1994; Zahalka et al., 1995). This anti-CD44 mAb, designated I.M.7.81, recognizes a constant epitope, shared by standard CD44 and all CD44 variants. The antibody was injected (150 µg mAb per injection, per mouse) two hours after the tumor inoculation and then every other day until the termination of the experiment (day 12). We further found that at the end of the antibody treatment, the number of the lymphoma cells invading the lymph nodes decreased by at least two orders of magnitude, from hundreds of thousands to several hundreds dpm (Zahalka et al., 1995). Using fluorescent-labeled lymphoma cells (transfected with GFP) and tracing the labeled cells in the local tumor growth as well as in the peripheral lymph node, we found that the I.M.7.81 anti-CD44 antibody suppressed both the local growth and remote lymph node invasions by CD44v-transfected LB cells even when injected 6 days after the tumor inoculation (Wallach-Dayan et al., 2001). Our *in vitro* studies demonstrated that the rolling attachments of CD44-transfected LB cells were restricted to HA substrate. Moreover, CD44v-transfected LB cells mutated at the HA binding site could not invade the local site and remote peripheral lymph nodes. Hence, digestion of tissue HA by the enzyme hyaluronidase may also disrupt the tumor progression. Using the protocol of antibody treatment, s.c. injection of hyaluronidase either two

III. HYALURONAN RECEPTORS AND SIGNAL TRANSDUCTION PATHWAYS

hours (Zahalka et al., 1995) or 6 days (Wallach-Dayan et al., 2001) after tumor inoculation suppressed the local lymphoma growth and lymph node invasion, in most but not all experiments (Sionov and Naor, 1997). These apparent contradictable findings can be reconciled when considering the double-edged sword role of hyaluronidase (Kovar et al., 2006; Lokeshwar et al., 2005). Delicate balance determines whether hyaluronidase enhances the lymph node metastases by destroying the matrix resistance to cell migration or retards them by tuning and targeting the enzymatic activity to the HA-depended tumor migration phase. When the two vectors display an equal power the enzyme effect cannot be detected. Further apparent contradiction was raised by the hyaluronidase experiment. We reported here earlier that lymph node-infiltrating LB cells do not bind HA, yet in some experiments we showed that hyaluronidase injection reduces the lymph node invasion by the lymphoma. If lymph node-infiltrating LB cells do not recognize HA, how could they be sensitive to enzyme treatment? This paradox can be reconciled by assuming that the metastatic LB cells acquire HA binding capacity before their invasion into the lymph node, e.g. to support their migration, in the afferent lymphatics, but they lose it upon penetration into the lymphoid organ. Alternatively, we may hypothesize that only a very small fraction of lymph node-infiltrating LB cells (which is not detected by flow cytometry) binds HA. This cell fraction, which includes perhaps cancer stem cells (Lobo et al., 2007), is responsible for the HA-depended lymph node colonization and therefore displays hyaluronidase sensitivity.

The enhanced accumulation of CD44v-transfected LB cells over CD44s-transfected LB cells or parental LB cells in the lymph nodes may be attributed to more efficient cell migration to these organs or to more extensive cell proliferation inside them. We found that, after their lymph node invasion, all the above mentioned cell lines (including CD44v- and CD44s-transfected LB cells) display similar proportions of cells in S-phase, i.e. they show almost the same proliferation rate inside the lymphoid organ. This finding implies that the enhanced accumulation of CD44v-transfected LB cells in the lymph nodes must be attributable to the efficient migration into the lymphoid organ rather than accelerated cell division inside it (Wallach-Dayan et al., 2001).

GENE VACCINATION WITH CD44 VARIANT CONSTRUCT, RATHER THAN WITH STANDARD CD44 CONSTRUCT, GENERATES RESISTANCE TO TUMOR GROWTH

The lesson that we have learned from the LB cell lymphoma model in particular and from reviewing the literature in general (Naor et al., 2002; Naor et al., 1997; Ponta et al., 2003) focused our attention not only on the substantial involvement of CD44 in tumor progression, but also to the role of the specific CD44 isoforms in the support of tumor spread and metastasis. We have seen that CD44 variant(s) of LB cells, rather than standard CD44, is engaged in binding the cells to HA and in supporting their resistance to detachment from HA substrate when subjected to physiological shear stress flow. Moreover, the same CD44 variant(s) is involved in the *in vivo* interaction of LB cells with HA, which is essential for the local tumor growth and its accumulation in remote lymph nodes. If this is the case, the therapeutic strategy should target the relevant CD44 variant on the tumor cells, rather than targeting constant epitopes that are shared by both CD44 variants and standard CD44. As CD44 variants associated with a certain tumor are less ubiquitous than the other CD44 isoforms, including standard CD44, the exclusive targeting of the tumor-expressing variant(s) should rescue all innocent cells that express the other CD44 isoforms, including a substantial number of cells that express standard CD44 only. In other words, limiting the therapeutic targeting to the CD44 variant(s) of the tumor cells should markedly reduce the potential of a side effect damage, which would ultimately be generated if standard CD44 is targeted. To cope with such a challenge we must first identify the CD44 variant(s) of the tumor, then show that this variant(s) is essential for the tumor survival and finally tailor a "therapeutic suit" that would selectively silence or kill the relevant tumor. The "therapeutic suit" could be either an antibody against the relevant tumor antigen (e.g., CD44v), or even better, a vaccine or a cDNA vaccine of such an antigen. Vaccination, in general, and DNA vaccinations, in particular, have some advantages over passive immunization by antibody. They provide prolonged antigen exposure that continuously stimulates the immune system and offer a unique opportunity to enhance the immunogenicity of the antigen in question by modifying the potential immunizing cDNA sequence. Using the above indicated principles, we generated a cDNA vaccine against CD44v and CD44s of DA3 mammary tumor of BALB/c mice. Vaccination with CD44v cDNA, but not with CD44s cDNA, generated resistance to the tumor growth (Wallach-Dayan et al., submitted).

EPILOGUE

CD44 isoforms support the local growth and the metastatic spread of LB cell lymphoma, DA3 mammary adenocarcinoma (this review article), and many other hematological and solid tumors described in our previous review articles (Naor et al., 2002; Naor et al., 1997). CD44 variants, but not necessarily standard CD44, are associated with the malignant phenotype (i.e., their migratory potency and the metastasis outcome) of the tumors described in this review as well as with many other tumors (Bourguignon et al., 2001; Carvalho et al., 2006; Hong et al., 2006; Wang et al., 2007; Naor et al., 2002; Naor et al., 1997). However this observation is not the rule. There are examples (Bartolazzi et al., 1994; Sy et al., 1991) that show that standard CD44, rather than its variants, enhances tumor progression. Even more, in some cases (e.g., prostate cancer and cervical neuroendocrine carcinoma) CD44 functions as a metastasis suppressor molecule (Gao et al., 1997; Kuo et al., 2006), which locks the tumor cells in the primary site, thus preventing the malignant spread (reviewed in Naor et al., 2002). Under such circumstances only CD44 downregulation allows release of the tumor from the primary site and the subsequent metastatic phase. However, later studies reported that CD44 rather supports prostate cancer progression (Desai et al., 2007; Patrawala et al., 2007). The internal and external environment of the tumor possibly influences the structure–function relationship between the tumor CD44 and its local substrate. A certain combination of conditions dictates very high affinity of CD44 to its tissue ligand (an example prostate cancer), resulting in tumor arrest in the local growth and a minimal metastatic phase. On the other hand, a different combination of conditions activates, in different tumors, either standard CD44 or CD44 variants, which their intermediate affinity to the substrate rather generates a migratory metastasis-supportive phenotype. Hence, the CD44 structure–function relationship must be separately analyzed in each tumor, and accordingly the therapeutic suit should be tailored. Note that the heterogeneity in CD44-dependency can be sometimes detected even in tumors derived from the same histological origin (e.g., prostate cancer).

The alternative splicing machinery when intensively activated, e.g., by inflammatory cascade (Nedvetzki et al., 2003) or perhaps malignant environment (Huang et al., 2007), may generate, especially at the splicing junction of cell surface CD44, modified sequences, which further enrich the CD44 repertoire and create new, sometimes disease-related structural specificities. For example, we have found that the transcript CD44v3-v10 of synovial fluid cells from joints of rheumatoid arthritis (RA) patients contains an extra tri-nucleotide CAG, which allows translation of an extra alanine between exon v4 and exon v5, without interfering with the reading frame. The extra CAG, was transcribed from the end of the intron, bridging exon v4 to exon v5, exactly at the splicing junction (Nedvetzki et al., 2003). This CD44 version of RA patients, that was designated CD44vRA, while being detected on synovial fluid inflammatory cells, was rarely found on peripheral blood leukocytes (PBLs) of the same patient and never found on synovial fluid cells of osteoarthritis patients or keratinocytes of normal donors, who do express the wild type variant CD44v3-v10 (Golan et al., 2007). The cell surface CD44vRA variant accumulates fibroblast growth factor from the environment and presents it at right orientation to cells with cognate receptor (e.g., fibroblasts, endothelial cells) thereby accelerating the RA inflammatory cascade, leading to bone and cartilage destruction (Nedvetzki et al., 2003). Furthermore, we suggest that soluble CD44vRA, that was enzymatically cleaved from the RA synovial fluid cell membrane, can interact with galectin-8 and reduce its ability to deliver apoptotic signals to the inflammatory cells, thereby rescuing their pathological activity (Eshkar Sebban et al., 2007). We recently generated anti-CD44vRA mAbs that induce apoptosis in synovial fluid cells of RA patients, but neither in PBLs of the same patients nor in synovial fluid cells of osteoarthritis patients. Furthermore, the anti-CD44vRA mAb induced resistance to joint inflammation in mice with collagen-induced arthritis (Golan et al., 2007). If the malignant process also creates specific modifications in the CD44 splice variants (Vela et al., 2006) as does the inflammatory cascade, mAbs recognizing this alteration may silent the malignant activity of the relevant tumor without causing damage to normal cells engaged in essential physiological functions. This prediction should be challenged, first in animal models, and if found true, in clinical settings.

ACKNOWLEDGMENTS

The research described in this article was supported by grants from the following institutions: The Society of Research Associates of the Lautenberg Center; The Concern Foundation of Los Angeles; Wakefern/Shoprite Endowment for Basic Research in Cancer Biology and Immunology; Deutsche Krebshilfe, Dr. Milderd Scheel Stiftung; Focused Giving Program, Johnson & Johnson; and the German-Israeli Foundation for Scientific Research and Development (GIF).

References

Bajorath, J. (2000). Molecular organization, structural features, and ligand binding characteristics of CD44, a highly variable cell surface glycoprotein with multiple functions. *Proteins* **39**, 103–111.

Bartolazzi, A., Peach, R., Aruffo, A., and Stamenkovic, I. (1994). Interaction between CD44 and hyaluronate is directly implicated in the regulation of tumor development. *J Exp Med* **180**, 53–66.

Bourguignon, L. Y., Gilad, E., Brightman, A., Diedrich, F., and Singleton, P. (2006). Hyaluronan–CD44 interaction with leukemia-associated RhoGEF and epidermal growth factor receptor promotes Rho/Ras co-activation, phospholipase C epsilon-Ca^{2+} signaling, and cytoskeleton modification in head and neck squamous cell carcinoma cells. *J Biol Chem* **281**, 14026–14040.

Bourguignon, L. Y., Zhu, H., Zhou, B., et al. (2001). Hyaluronan promotes CD44v3-Vav2 interaction with Grb2-p185(HER2) and induces Rac1 and Ras signaling during ovarian tumor cell migration and growth. *J Biol Chem* **276**, 48679–48692.

Carvalho, R., Milne, A. N., Polak, M., Offerhaus, G. J., and Weterman, M. A. (2006). A novel region of amplification at 11p12-13 in gastric cancer, revealed by representational difference analysis, is associated with overexpression of CD44v6, especially in early-onset gastric carcinomas. *Genes Chromosomes Cancer* **45**, 967–975.

Cichy, J. and Pure, E. (2000). Oncostatin M and transforming growth factor-beta 1 induce post-translational modification and hyaluronan binding to CD44 in lung-derived epithelial tumor cells. *J Biol Chem* 275, 18061–18069.

Desai, B., Rogers, M. J., and Chellaiah, M. A. (2007). Mechanisms of osteopontin and CD44 as metastatic principles in prostate cancer cells. *Mol Cancer* 6, 18.

Ehnis, T., Dieterich, W., Bauer, M., Lampe, B., and Schuppan, D. (1996). A chondroitin/dermatan sulfate form of CD44 is a receptor for collagen XIV (undulin). *Exp Cell Res* 229, 388–397.

Eshkar-Sebban, L., Ronen, D., Levartovsky, D., et al. (2007). The involvement of CD44 and its novel ligand galectin-8 in apoptotic regulation of autoimmune inflammation. *J Immunol* 179, 1225–1235.

Faassen, A. E., Schrager, J. A., Klein, D. J., et al. (1992). A cell surface chondroitin sulfate proteoglycan, immunologically related to CD44, is involved in type I collagen-mediated melanoma cell motility and invasion. *J Cell Biol* 116, 521–531.

Gao, A. C., Lou, W., Dong, J. T., and Isaacs, J. T. (1997). CD44 is a metastasis suppressor gene for prostatic cancer located on human chromosome 11p13. *Cancer Res* 57, 846–849.

Golan, I., Nedvetzki, S., Golan, I., et al. (2007). Expression of extra trinucleotide in CD44 variant of rheumatoid arthritis patients allows generation of disease-specific monoclonal antibody. *J Autoimmun* 28, 99–113.

Hanley, W. D., Burdick, M. M., Konstantopoulos, K., and Sackstein, R. (2005). CD44 on LS174T colon carcinoma cells possesses E-selectin ligand activity. *Cancer Res* 65, 5812–5817.

Henke, C. A., Roongta, U., Mickelson, D. J., Knutson, J. R., and McCarthy, J. B. (1996). CD44-related chondroitin sulfate proteoglycan, a cell surface receptor implicated with tumor cell invasion, mediates endothelial cell migration on fibrinogen and invasion into a fibrin matrix. *J Clin Invest* 97, 2541–2552.

Hong, S. C., Song, J. Y., Lee, J. K., et al. (2006). Significance of CD44v6 expression in gynecologic malignancies. *J Obstet Gynaecol Res* 32, 379–386.

Huang, C. S., Shen, C. Y., Wang, H. W., Wu, P. E., and Cheng, C. W. (2007). Increased expression of SRp40 affecting CD44 splicing is associated with the clinical outcome of lymph node metastasis in human breast cancer. *Clin Chim Acta* 384, 69–74.

Ish-Shalom, D., Tzivion, G., Christoffersen, C. T., et al. (1995). Mitogenic potential of insulin on lymphoma cells lacking IGF-1 receptor. *Ann. N.Y. Acad. Sci.* 766, 409–415.

Jalkanen, S. and Jalkanen, M. (1992). Lymphocyte CD44 binds the COOH-terminal heparin-binding domain of fibronectin. *J Cell Biol* 116, 817–825.

Katoh, S., Zheng, Z., Oritani, K., Shimozato, T., and Kincade, P. W. (1995). Glycosylation of CD44 negatively regulates its recognition of hyaluronan. *J Exp Med* 182, 419–429.

Knutson, J. R., Iida, J., Fields, G. B., and McCarthy, J. B. (1996). CD44/chondroitin sulfate proteoglycan and alpha 2 beta 1 integrin mediate human melanoma cell migration on type IV collagen and invasion of basement membranes. *Mol Biol Cell* 7, 383–396.

Kovar, J. L., Johnson, M. A., Volcheck, W. M., Chen, J., and Simpson, M. A. (2006). Hyaluronidase expression induces prostate tumor metastasis in an orthotopic mouse model. *Am J Pathol* 169, 1415–1426.

Kuo, K. T., Liang, C. W., Hsiao, C. H., et al. (2006). Downregulation of BRG-1 repressed expression of CD44s in cervical neuroendocrine carcinoma and adenocarcinoma. *Mod Pathol* 19, 1570–1577.

Lesley, J., English, N., Perschl, A., Gregoroff, J., and Hyman, R. (1995). Variant cell lines selected for alterations in the function of the hyaluronan receptor CD44 show differences in glycosylation. *J Exp Med* 182, 431–437.

Lewis, C. A., Townsend, P. A., and Isacke, C. M. (2001). Ca^{2+}/calmodulin-dependent protein kinase mediates the phosphorylation of CD44 required for cell migration on hyaluronan. *Biochem J* 357, 843–850.

Lobo, N. A., Shimono, Y., Qian, D., and Clarke, M. F. (2007). The biology of cancer stem cells. *Annu Rev Cell Dev Biol* **23**, 675–699.

Lokeshwar, V. B., Cerwinka, W. H., Isoyama, T., and Lokeshwar, B. L. (2005). HYAL-1 hyaluronidase in prostate cancer: a tumor promoter and suppressor. *Cancer Res* **65**, 7782–7789.

Mackay, C. R., Marston, W. L., and Dudler, L. (1990). Naive and memory T cells show distinct pathways of lymphocyte recirculation. *J Exp Med* **171**, 801–817.

Meredith Jr., J. E. and Schwartz, M. A. (1997). Integrins, adhesion and apoptosis. *Trends Cell Biol* **7**, 146–150.

Mishra, J. P., Mishra, S., Gee, K., and Kumar, A. (2005). Differential involvement of calmodulin-dependent protein kinase II-activated AP-1 and c-Jun N-terminal kinase-activated EGR-1 signaling pathways in tumor necrosis factor-alpha and lipopolysaccharide-induced CD44 expression in human monocytic cells. *J Biol Chem* **280**, 26825–26837.

Naor, D., Nedvetzki, S., Golan, I., Melnik, L., and Faitelson, Y. (2002). CD44 in cancer. *Crit Rev Clin Lab Sci* **39**, 527–579.

Naor, D., Sionov, R. V., and Ish-Shalom, D. (1997). CD44: structure, function, and association with the malignant process. *Adv Cancer Res* **71**, 241–319.

Napier, S. L., Healy, Z. R., Schnaar, R. L., and Konstantopoulos, K. (2007). Selectin ligand expression regulates the initial vascular interactions of colon carcinoma cells: the roles of CD44v and alternative sialofucosylated selectin ligands. *J Biol Chem* **282**, 3433–3441.

Nedvetzki, S., Golan, I., Assayag, N., et al. (2003). A mutation in a CD44 variant of inflammatory cells enhances the mitogenic interaction of FGF with its receptor. *J Clin Invest* **111**, 1211–1220.

Patrawala, L., Calhoun-Davis, T., Schneider-Broussard, R., and Tang, D. G. (2007). Hierarchical organization of prostate cancer cells in xenograft tumors: the CD44$^+$alpha2beta1$^+$ cell population is enriched in tumor-initiating cells. *Cancer Res* **67**, 6796–6805.

Pillemer, G., Lugasi-Evgi, H., Scharovsky, G., and Naor, D. (1992). Insulin dependence of murine lymphoid T-cell leukemia. *Int J Cancer* **50**, 80–85.

Ponta, H., Sherman, L., and Herrlich, P. A. (2003). CD44: from adhesion molecules to signalling regulators. *Nat Rev Mol Cell Biol* **4**, 33–45.

Rochman, M., Moll, J., Herrlich, P., et al. (2000). The CD44 receptor of lymphoma cells: structure-function relationships and mechanism of activation. *Cell Adhes Commun* **7**, 331–347.

Sharon, R., Pillemer, G., Ish-Shalom, D., et al. (1993). Insulin dependence of murine T-cell lymphoma. II. Insulin-deficient diabetic mice and mice fed low-energy diet develop resistance to lymphoma growth. *Int J Cancer* **53**, 843–849.

Sionov, R. V. and Naor, D. (1997). Hyaluronan-independent lodgment of CD44+ lymphoma cells in lymphoid organs. *Int J Cancer* **71**, 462–469.

Sionov, R. V. and Naor, D. (1998). Calcium- and calmodulin-dependent PMA-activation of the CD44 adhesion molecule. *Cell Adhes Commun* **6**, 503–523.

Skelton, T. P., Zeng, C., Nocks, A., and Stamenkovic, I. (1998). Glycosylation provides both stimulatory and inhibitory effects on cell surface and soluble CD44 binding to hyaluronan. *J Cell Biol* **140**, 431–446.

Sleeman, J., Rudy, W., Hofmann, M., Moll, J., Herrlich, P., and Ponta, H. (1996). Regulated clustering of variant CD44 proteins increases their hyaluronate binding capacity. *J Cell Biol* **135**, 1139–1150.

Springer, T. A. (1994). Traffic signals for lymphocyte recirculation and leukocyte emigration: the multistep paradigm. *Cell* **76**, 301–314.

Stetler-Stevenson, W. G., Aznavoorian, S., and Liotta, L. A. (1993). Tumor cell interactions with the extracellular matrix during invasion and metastasis. *Annu Rev Cell Biol* **9**, 541–573.

Sy, M. S., Guo, Y. J., and Stamenkovic, I. (1991). Distinct effects of two CD44 isoforms on tumor growth in vivo. *J Exp Med* **174**, 859–866.

III. HYALURONAN RECEPTORS AND SIGNAL TRANSDUCTION PATHWAYS

Tian, B., Takasu, T., and Henke, C. (2000). Functional role of cyclin A on induction of fibroblast apoptosis due to ligation of CD44 matrix receptor by anti-CD44 antibody. *Exp Cell Res* **257**, 135–144.

Tzircotis, G., Thorne, R. F., and Isacke, C. M. (2006). Directional sensing of a phorbol ester gradient requires CD44 and is regulated by CD44 phosphorylation. *Oncogene* **25**, 7401–7410.

Vela, E., Roca, X., and Isamat, M. (2006). Identification of novel splice variants of the human CD44 gene. *Biochem Biophys Res Commun* **343**, 167–170.

Wallach, S. B., Friedmann, A., and Naor, D. (2000). The CD44 receptor of the mouse LB T-cell lymphoma: analysis of the isoform repertoire and ligand binding properties by reverse-transcriptase polymerase chain reaction and antisense oligonucleotides. *Cancer Detect Prev* **24**, 33–45.

Wallach-Dayan, S. B., Grabovsky, V., Moll, J., et al. (2001). CD44-dependent lymphoma cell dissemination: a cell surface CD44 variant, rather than standard CD44, supports in vitro lymphoma cell rolling on hyaluronic acid substrate and its in vivo accumulation in the peripheral lymph nodes. *J Cell Sci* **114**, 3463–3477.

Wallach-Dayan, S. B., Rubinstein, A. M., Hand, C., Breuer, R., and Naor, D. (2008). DNA vaccination with CD44 variant isoform reduces mammary tumor local growth and lung metastasis. *Mol Cancer Ther* **7**, 1615–1623.

Wang, S. J. and Bourguignon, L. Y. (2006). Hyaluronan-CD44 promotes phospholipase C-mediated Ca2+ signaling and cisplatin resistance in head and neck cancer. *Arch Otolaryngol Head Neck Surg* **132**, 19–24.

Wang, S. J., Wreesmann, V. B., and Bourguignon, L. Y. (2007). Association of CD44 V3-containing isoforms with tumor cell growth, migration, matrix metalloproteinase expression, and lymph node metastasis in head and neck cancer. *Head Neck* **29**, 550–558.

Yeatman, T. J. and Nicolson, G. L. (1993). Molecular basis of tumor progression: mechanisms of organ-specific tumor metastasis. *Semin Surg Oncol* **9**, 256–263.

Zahalka, M. A. and Naor, D. (1994). Beta 2-integrin dependent aggregate formation between LB T cell lymphoma and spleen cells: assessment of correlation with spleen invasiveness. *Int Immunol* **6**, 917–924.

Zahalka, M. A., Okon, E., Gosslar, U., Holzmann, B., and Naor, D. (1995). Lymph node (but not spleen) invasion by murine lymphoma is both CD44- and hyaluronate-dependent. *J Immunol* **154**, 5345–5355.

Zahalka, M. A., Okon, E., and Naor, D. (1993). Blocking lymphoma invasiveness with a monoclonal antibody directed against the beta-chain of the leukocyte adhesion molecule (CD18). *J Immunol* **150**, 4466–4477.

Zhuo, L., Kanamori, A., Kannagi, R., et al. (2006). SHAP potentiates the CD44-mediated leukocyte adhesion to the hyaluronan substratum. *J Biol Chem* **281**, 20303–20314.

CHAPTER

9

RHAMM/HMMR: An Itinerant and Multifunctional Hyaluronan Binding Protein That Modifies CD44 Signaling and Mitotic Spindle Formation

James B. McCarthy and Eva A. Turley

INTRODUCTION

RHAMM was originally isolated as a secreted fibroblast protein in a model of wounding *in culture* (Fig. 9.1) (Turley, 1982). In this study, embryonic heart tissue was explanted and the supernatant medium was collected for further analysis between 18 and 24 hours after explantation, when fibroblasts were rapidly migrating and producing high levels of HA. Sampling at this time excluded analysis of dividing cells and interphase cells that had formed actin stress fibers/focal contacts, since these events do not occur until approximately 36 hours after explantation (Fig. 9.1). Hyaluronan binding proteins were isolated using hyaluronan sepharose chromatography, analyzed on SDS-PAGE, and purified proteins were extracted and used for preparation of polyclonal and monoclonal antibodies (Turley et al., 1985; 1987). Expression cloning using these reagents resulted in the isolation of RHAMM (Hardwick et al., 1992). RHAMM is a coiled coil protein that first appeared in vertebrates and is not found in lower organisms or in insects (Fig. 9.2).

Given its roles in such important cellular processes as motility and cell division *in culture* (Turley et al., 2002; Slevin et al., 2007), it is surprising that germline genetic deletion of RHAMM does not result in a detectable phenotype during embryogenesis or adult homeostasis (Tolg et al., 2003; 2006). However, loss of RHAMM does affect the tissue response to injuries

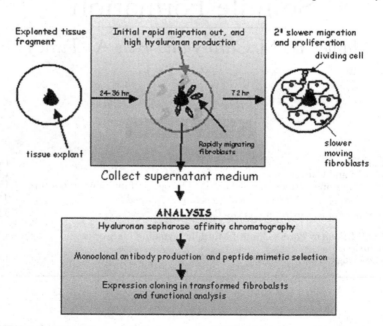

FIGURE 9.1   Schemata used to isolate secreted RHAMM from rapidly migrating fibroblasts *in culture*. Embryonic chick hearts were explanted and fibroblasts allowed to rapidly migrate out of the explants for 24–36 hours. Cell division and a secondary stage of slower cell migration does not occur until 72 hours. The supernatant medium was decanted and hyaluronan binding proteins were captured using hyaluronan-Sepharose affinity columns. Monoclonal antibodies were prepared against the captured proteins and these were used for expression cloning (Turley, 1982).

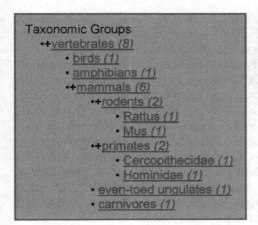

Taxonomic Groups
- **+**vertebrates *(8)*
  - birds *(1)*
  - amphibians *(1)*
  - **+**mammals *(6)*
    - **+**rodents *(2)*
      - Rattus *(1)*
      - Mus *(1)*
    - **+**primates *(2)*
      - Cercopithecidae *(1)*
      - Hominidae *(1)*
    - even-toed ungulates *(1)*
    - carnivores *(1)*

FIGURE 9.2 Taxonomic groups in which RHAMM has emerged. RHAMM first appears in vertebrates and has been identified in mammals, birds, and amphibians. No structural orthologues have yet been identified in lower organisms.

including excisional cutaneous wound repair and bleomyocin-induced lung damage (Tolg et al., 2006; Slevin et al., 2007). Excisional cutaneous wound, repair occurs via several well-regulated stages that include initial inflammation, proliferation, and remodeling (Dorsett-Martin, 2004; Gibran et al., 2007; Oberyszyn, 2007). The inflammatory and proliferative (fibrogenesis) stages are subtly altered by loss of RHAMM (Tolg et al., 2006). For example, neutrophil accumulation remains high throughout the inflammatory and post-inflammatory stages of wound repair in RHAMM$^{-/-}$ mice whereas these cells have undergone apoptosis within several days following injury in wild-type animals. Fibrogenesis of RHAMM$^{-/-}$ wounds aberrantly decreased due to reduced fibroblast infiltration compared to fibrogenesis/fibroblast infiltration in wild-type animals. Mesenchymal differentiation during fibrogenesis is also modified and/or inappropriately regulated. For example, granulation tissue of RHAMM$^{-/-}$ wounds is filled with adipocytes and Rhamm$^{-/-}$ fibroblasts explanted from wounds undergo adipogenesis *in culture* to a much greater extent than wild-type wound fibroblasts. Myofibroblast activity is reduced in Rhamm$^{-/-}$ wounds, as reflected by the relative lack of differentiation markers such as smooth muscle actin. Furthermore, muscle differentiation within the wound is impaired (Tolg et al., 2006). The many similarities between responses to tissue injury and cancer suggest that RHAMM may be important in the progression of malignant tumors.

RHAMM AND CANCER

Although RHAMM mRNA is rarely detected in most homeostatic adult tissues, it is commonly hyperexpressed in human cancers. Cell surface RHAMM has been detected as a tumor antigen in several malignancies, including leukemia, breast, melanoma, prostate, and ovarian cancers (Greiner et al., 2006; Schmitt et al., 2008). Intracellular RHAMM expression is also elevated above normal tissue levels in brain tumors (e.g., astrocytomas, gliomas, meningiomas) (Panagopoulos et al., 2008), gastric cancer (Li et al., 2000a), colorectal cancer (Zlobec et al., 2008), multiple myeloma (Maxwell et al., 2004), oral squamous cell carcinoma (Yamano et al., 2008), endometrial carcinomas (Rein et al., 2003), and bladder cancer (Kong et al., 2003). RHAMM is often most highly expressed in carcinomas at later stages of progression and the levels of RHAMM have been correlated to tumor stage and in some cases to prognosis. For example, RHAMM hyperexpression is an independent prognostic indicator of poor clinical outcome in breast cancer (Wang et al., 1998; Pujana et al., 2007), colorectal cancer (Zlobec et al., 2008), multiple myeloma (Maxwell et al., 2004), and oral squamous cell carcinoma (Yamano et al., 2008).

The prognostic value of RHAMM is linked to specific subtypes of organ specific malignancies, which have distinct anatomic and molecular pathologies that are correlated to different outcomes. For example, colorectal cancers can be grouped into DNA mismatch repair-proficient, MLH1 negative and presumed Lynch syndrome. Although RHAMM is expressed at all Dukes stages of ungrouped colorectal cancers (Yamada et al., 1999), hyperexpression of RHAMM mRNA is an independent prognostic factor for poor outcome and ranks higher than T stage, vascular invasion, tumor budding, and tumor grade in its association with increased risk of peripheral metastasis and with worse outcome in patients with metastatic disease (Zlobec et al., 2008a, b). RHAMM is a prognostic factor in DNA-mismatch repair-proficient (MMR-proficient) and presumed Lynch syndrome forms of colorectal cancer but not in MLH1 negative colorectal tumors (Duval et al., 2001; Lugli et al., 2006; Zlobec et al., 2008c). Furthermore, combining RHAMM hyperexpression with another prognostic factor, loss of p21, identifies a subgroup of MMR-proficient tumors with a high incidence of microsatellite instability that has a particularly poor prognosis (Zlobec et al., 2008c). These results predict that the association of elevated RHAMM expression with neoplastic disease may be selective in terms of tumor subtype. It may therefore be possible to use RHAMM levels as one marker to further classify specific tumors or to target RHAMM for therapy in such tumor subtypes.

In addition to increased levels of expression, distinct RHAMM isoforms that arise through multiple mechanisms are associated with tumors and some of these isoforms appear to behave as multifunctional oncogenes. Two major alternatively spliced mRNA species have been identified, which differ in the presence of exon 4, (which encodes a 15 amino acid sequence in the amino terminal region of RHAMM). For example, Rhamm^{-exon4} is expressed by aggressive multiple myeloma subsets (Maxwell et al., 2004).

Other tumor types (including breast carcinoma, astrocytoma, multiple myeloma, melanoma, gastric carcinoma, and astrocytomas), which have been reported to hyperexpress RHAMM mRNA also express truncated RHAMM protein forms (Li et al., 2000a, b; Ahrens et al., 2001; Zhou et al., 2002; Hamilton et al., 2007). Although the basis for the formation of these smaller proteins (which range from 40 to 70 kD compared to 86 kD full length Rhamm protein) is not known, it is possible that they arise as a result of partial proteolytic processing and/or the use of multiple start codons, which are present within the coding sequence of RHAMM. Human astrocytomas express a 70 kD tumor specific RHAMM isoform in addition to the full length protein (86 kDa) which is expressed in low amounts by normal astrocytes (Zhou et al., 2002). Gastric carcinomas, breast carcinomas, and melanomas also express multiple RHAMM isoforms (52–70 kDa) in addition to the full-length form (Li et al., 2000b; Ahrens et al., 2001; Hamilton et al., 2007). Highly invasive human breast cancer cell lines such as MDA-MB-231 similarly express a 70 and 52 kD Rhamm protein compared to less aggressive lines such as MCF7, which primarily express the full length RHAMM protein (Hamilton et al., 2007). It is also possible that some of these additional lower molecular weight isoforms are created from low abundance splice variants of RHAMM mRNA, however no such variants have been identified to date.

Several studies suggest that mutations in RHAMM may affect oncogenic properties. RHAMM contains mononucleotide repeat coding sequences that are mutated in human MMR type colorectal cancers with high microsatellite instability (Duval et al., 2001). These mutations result in conversion of intronic 3' non-coding to a coding sequence which creates additional protein sequence resembling the HA binding region of RHAMM. Malignant peripheral nerve sheath tumors arising from neurofibromatosis type 1 (risk is 10%) often exhibit an allelic (N = 1) deletion of a RHAMM gene (Levy et al., 2004; Mantripragada et al., 2008). Germ line polymorphisms upstream of the coding sequence in the RHAMM gene are also associated with a predisposition to breast cancer suggesting that mutated RHAMM may function as a novel breast cancer susceptibility gene (Pujana et al., 2007).

Finally, one of the more intriguing characteristics of RHAMM that is likely associated with oncogenesis is the change in both the subcellular distribution and functions of the protein (Maxwell et al., 2008). In examples such as gastric and highly malignant breast cancers, RHAMM is exported to the cell surface where it functions as a hyaluronan-responsive motogenic protein (Li et al., 2000a; Hamilton et al., 2007). Extracellular RHAMM is detected on a number of tumor cell lines cultured in vitro. In this case, cell surface RHAMM is more pronounced in subconfluent, rather than confluent cultures (Turley, 1982; Hardwick et al., 1992) and the addition of hyaluronan acts to further increase the amount of cell surface RHAMM

(Gao et al., 2008). As has been observed with many other non-conventionally exported proteins such as autocrine motility factor and galectins 1 and 3 (Radisky et al., 2003; Nickel, 2005; Prudovsky et al., 2008), extracellular RHAMM functions, at least in part, to enhance motility as well as invasion of tumor cells (Hamilton et al., 2007). However, in certain tumors such as mesenchymal desmoid tumors, alterations in RHAMM may function to promote both tumor initiation and as invasion (Tolg et al., 2003). RHAMM is highly expressed in human mesenchymal desmoid tumors (fibromatoses). In an animal model of desmoid tumor susceptibility, both the number (which equates to initiation) and size (which equates to growth and invasiveness) of tumors are decreased when RHAMM is germ line deleted as a result of homologous recombination (Tolg et al., 2003).

Although the molecular mechanisms for the unconventional export and other subcellular changes in the distribution of RHAMM during tumorigenesis are not yet known, similar changes occur during normal wound healing and tissue repair (Zaman et al., 2005; Tolg et al., 2006; Slevin et al., 2007), which are processes that require cellular infiltration and remodeling of the microenvironment. Thus, some of the functions of extracellular RHAMM in tumors may indeed mirror those observed in wound healing, which like tumor progression requires both cellular infiltration and tissue remodeling.

EXTRACELLULAR AND INTRACELLULAR RHAMM ONCOGENIC FUNCTIONS IN CULTURE

RHAMM was originally identified as a "secreted" hyaluronan binding protein that was isolated from the supernatant medium of rapidly locomoting, sparsely cultured chick embryonic fibroblast (Turley, 1982) (Fig. 9.1). The binding of purified, secreted RHAMM protein to fibroblast monolayers is both saturable and of high affinity (Turley et al., 1985). These and other data led to the prediction that secreted RHAMM acts as a co-receptor that associates with binding sites or other receptors located at the cell surface (Hardwick et al., 1992). More recently, cell surface RHAMM has been shown to function as a motility receptor in both dermal fibroblasts responding to injury and in aggressively invasive breast cancer cell lines that have undergone EMT by forming complexes with CD44 (Tolg et al., 2006; Hamilton et al., 2007). RHAMM coated beads were used to demonstrate that RHAMM interactions with the cell surface enhances the cell surface display of CD44 (Fig. 9.3). Additional studies using RHAMM$^{-/-}$ fibroblasts have shown that RHAMM-stimulated motility is sensitive to antibodies against CD44 and that RHAMM stimulates Erk1,2 activity through its interaction with CD44 (Tolg et al., 2006). These and other data (summarized below) indicate that cell surface RHAMM functions as a co-receptor for CD44 in wound fibroblasts and in breast cancer cells that have undergone an EMT (e.g., MDA-MB-231 cells) (Tolg et al., 2006; Hamilton et al., 2007). Furthermore, this interaction is necessary for sustained activation of motogenic signaling pathways through ERK1,2. Cell surface RHAMM also associates with RON (gene name, MST-1R), a tyrosine receptor kinase of the c-Met family, whose ligand is macrophage stimulating protein (MSP) (Manzanares et al., 2007). Thus extracellular RHAMM functions as a co-receptor with one or more adhesion/growth factor receptors on the cell surface, with the ability to alter the intensity and/or duration of key signal transduction pathways.

Extracellular export of RHAMM may be particularly relevant to tumor therapy. For example, patients with certain malignant tumors were shown to have increased levels of RHAMM in their circulation predicting that this protein might be used as a tumor specific marker (Greiner et al., 2006). Indeed, dendritic vaccination using RHAMM peptides has shown efficacy in phase I clinical trials of patients with acute myeloid leukemia (AML) and multiple myeloma (Schmitt et al., 2008). Furthermore, vaccination of mice using dendritic cells transfected with RHAMM mRNA induces an anti-tumor effect associated with increased activation of the immune system in a model of mouse glioma (Amano et al., 2007). Thus, tumors may be susceptible to these and other RHAMM-targeted therapies as a result of the increased levels of cell surface protein expressed on these tumors.

While initial reports on the function of RHAMM focused on its extracellular functions, the cloning and further characterization of RHAMM primary structure revealed the protein lacks a signal peptide for classical secretion through the Golgi/ER or a membrane spanning sequence typical of traditional transmembrane receptors (e.g., CD44) (Maxwell et al., 2008). Several subsequent studies therefore focused on identifying the functions of intracellular RHAMM protein forms, since its sequence was more in line with that of a cytoplasmic or nuclear protein. These studies demonstrated that intracellular RHAMM is associated with interphase microtubules, it can be localized to the nucleus (Hofmann et al., 1998a; Liska et al., 2004; Shakib et al., 2005), and it also co-distributes with centrosomes/mitotic spindles (Maxwell et al., 2003; Evanko et al., 2004). This complex subcellular distribution of RHAMM in tumor cells has created challenges for developing models to study the importance of cell surface verses intracellular RHAMM.

The addition of soluble, recombinant extracellular RHAMM to Ras-transformed fibroblasts was originally observed to inhibit TGF-β induced progression through the cell cycle by inhibiting cyclin B1 and cdc2 expression thus restricting passage through G2/M (Mohapatra et al., 1996). Additionally, it has been shown using an unbiased expression array screen in synchronized, proliferating cells that RHAMM mRNA levels peak at or near the G2/M boundary of the cell cycle (Cho et al., 2001;

Whitfield et al., 2002; Groen et al., 2004; Liska et al., 2004; Yang et al., 2005) and RHAMM is one of the numerous mitotic spindle proteins that are phosphorylated during spindle formation (Nousiainen et al., 2006). Furthermore, intracellular RHAMM is required for progression of certain cell types through G2/M (Maxwell et al., 2003). Collectively these studies imply that control of G2/M by RHAMM may be coordinated by both extracellular and intracellular forms of this protein. However, the precise relationship between the mitotic functions of these two pools of RHAMM has not yet been established.

Cells derived from RHAMM$^{-/-}$ animals provide an important tool to address this problem. Studies using recombinant RHAMM-coated beads to challenge RHAMM$^{-/-}$ fibroblasts have shown that these beads are sufficient to stimulate the motility of these cells over 4 hours indicating that cell surface RHAMM alone is sufficient to promote short spurts of motility (Tolg et al., 2006). On the other hand, recombinant RHAMM coated beads do not rescue the aberrant mitosis of RHAMM$^{-/-}$ fibroblasts (Tolg et al., 2006). Collectively, the evidence indicates that intracellular RHAMM protein levels (e.g. too much or too little) are necessary and sufficient to control mitotic spindle formation and stability. More recently, the spindle formation function of RHAMM has been shown to be part of the BRCA/BARD1 E3 ubiquitin ligase pathway (Pujana et al., 2007; Joukov et al., 2006). This pathway has the potential for promoting genomic instability in tumors in which BRCA1 is compromised by mutation. The mitotic spindle functions of RHAMM may also, at least in part, explain the importance of RHAMM as a novel breast tumor susceptibility gene.

These motogenic and mitogenic effects of RHAMM have been demonstrated *in culture*, and although they have the potential to contribute to cancer initiation and progression, is there any evidence that RHAMM performs similar motogenic and mitogenic functions *in vivo*? As noted above, RHAMM polymorphisms and hyperexpression have been linked to cancer susceptibility and poor clinical outcome in a number of human cancers. However, analyses of RHAMM gene dysregulation using animal models have not been extensively reported. To date, *in vivo* analyses of the mitogenic and motogenic effects of RHAMM have been reported in a mouse model of collagen-induced arthritis (Nedvetski et al., 2004), during lung repair of bleomycin induce injury (Zaman et al., 2003), during repair of full thickness, excisional skin wounds in RHAMM$^{-/-}$ mice (Tolg et al., 2006), and in desmoids/upper intestinal tract tumor formation in a transgenic mouse model of these tumors (Tolg et al., 2003).

Functions of RHAMM in Animal Models of Repair and Disease

Collagen-induced arthritis in both wild-type and CD44$^{-/-}$ mice is dependent upon the production of hyaluronan since destruction of this polysaccharide by hyaluronidase injected into the toe joint blocks its development (Naor et al., 2007). In wild-type mice, blocking CD44 function also stops the appearance of arthritis in joints but genetic loss of CD44 intensifies rather that abrogates arthritis development. In this model, extracellular RHAMM compensates for CD44/hyaluronan binding function since injection of anti-RHAMM antibodies or recombinant RHAMM reduces the inflammatory response in the absence of CD44 (Nedvetski et al., 2004). However, compensation is accompanied by an intensified disease course due to the upregulation of pro-inflammatory cytokines that are highly expressed in the absence of CD44. The molecular mechanisms by which RHAMM affects inflammation in this model have not yet been identified but the process is associated with increased accumulation of hyaluronan in the toe joints of $CD44^{-/-}$ animals, which acts to promote signaling through cell surface RHAMM. Although RHAMM partners with CD44 in some systems (see below), this is not possible in the $CD44^{-/-}$ mice, suggesting that cell surface RHAMM may partner with an as yet unidentified integral receptor on spleenocytes (Nedvetski et al., 2004).

Macrophage infiltration into injured lung tissues also requires cell surface expression of RHAMM (Zaman et al., 2005). In this model of lung injury, TGFβ1 production is stimulated by bleomycin and this cytokine promotes expression of cell surface RHAMM and motogenic influx of macrophages to the site of injury. Although macrophages express both cell surface RHAMM and CD44, blocking the motogenic function of cell surface RHAMM with anti-RHAMM antibodies alone reduced macrophage influx and the concomitant fibrotic response resulting from bleomycin. Collectively, these two studies using injury models demonstrate a role for cell surface RHAMM in controlling influx of inflammatory cells into wounded tissues thereby affecting the course of tissue repair.

Germline genetic deletion of RHAMM results in the wound mesenchymal defects noted above (Tolg et al., 2006). In $RHAMM^{-/-}$ mice, reduced fibroplasia is related to a motogenic defect associated with reduced responsiveness *in culture* to motility-promoting stimuli such as PDGFBB and hyaluronan oligosaccharides (Tolg et al., 2006). Surprisingly, neither increased apoptosis nor reduced proliferation is detected in the wound site although markers for motility are aberrantly expressed. A motogenic defect of $RHAMM^{-/-}$ fibroblasts is retained *in culture* and results from reduced display of cell surface CD44 and aberrant activation kinetics of ERK1,2. This study also demonstrated that CD44 and RHAMM co-associate to regulate the duration of ERK1,2 motogenic signaling during tissue repair. Furthermore, the motogenic defect of $RHAMM^{-/-}$ wound fibroblasts can be rescued either by the stable expression of constitutively active MEK1 or by addition of extracellular recombinant RHAMM, which was prevented from intracellular uptake by linkage to Sepharose beads (Fig. 9.3).

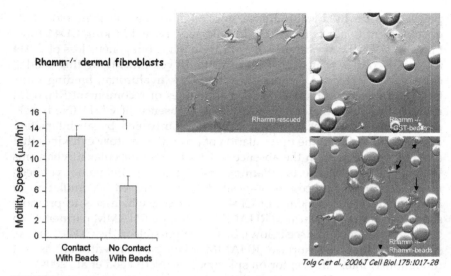

FIGURE 9.3   Cell surface but not intracellular RHAMM is required for acute migration in response to hyaluronan or growth factors. RHAMM$^{-/-}$ dermal fibroblasts were exposed to recombinant GST-RHAMM protein linked to Sepharose beads or to GST alone. GST-RHAMM-beads that touch RHAMM$^{-/-}$ fibroblasts promote migration while GST-beads or untouched beads (detected by real time analysis) do not. Since cells cannot internalize beads coated with recombinant RHAMM, showing that cell surface RHAMM is sufficient to promote migration, at least over the 4 h assay period. From Tolg et al. (2006).

Two studies have also investigated the roles of RHAMM in facilitating the growth of tumors *in vivo*. In a xenograft injection model using MDA-435 tumor cells, expression of the hyaluronan binding region of RHAMM (Liu et al., 2004) results in decreased growth rate compared to control tumor cells lacking this peptide. The slower growth produced by the expression of the hyaluronan binding domain of RHAMM results from the ability of this peptide to activate caspase 3 and 8 as well as poly-(ADP-ribose) polymerase, thereby promoting apoptosis of the transfected tumor cells. The effect of RHAMM expression in the initiation and growth of desmoid tumors in a syngeneic mouse environment has also been evaluated (Tolg et al., 2003). For these studies, RHAMM$^{-/-}$ mice were crossed to Apc/Apc1638N transgenic mice [which are predisposed to desmoids (fibromatoses) and upper intestinal tract tumors]. RHAMM deficiency significantly reduces both the number (equated to initiation) and size (equated to invasion) of desmoids but not upper intestinal tract tumors. These effects of RHAMM on tumorigenesis are apparently related to a desmoid tumor cell proliferation deficiency that is observed at low but not high cell density *in culture*. Currently therefore, these *in vivo* data support a role for RHAMM in both cell motility and proliferation and further suggest a potential role in apoptosis. Is there any clinical evidence that RHAMM performs similar functions in human cancers?

Functions of RHAMM in Human Tumors

The selectively high expression of RHAMM in tumor cell subsets (Wang et al., 1998; Pujana et al., 2007) and at the invading front of primary breast cancers (Assmann et al., 1998) is consistent with the ability of RHAMM to promote motogenic/invasive activities of tumors that have been identified *in culture*. Proteomics and integrative genomics of human cancers also point to a role for RHAMM in mitosis. For example, DNA microarray analyses have revealed that RHAMM expression is cell cycle G2/M regulated in hepatic carcinoma (Yang et al., 2005) and RHAMM expression is correlated with centrosomal structural abnormalities in multiple myeloma (Maxwell et al., 2005). These studies are consistent with evidence indicating that intracellular RHAMM has a mitotic spindle/centrosomal function *in culture*. But how do the multiple RHAMM protein forms control these disparate cellular functions?

RHAMM PROTEIN FORMS CONTROL MULTIPLE SIGNALING NETWORKS

Neither structural nor phylogenetic analyses of RHAMM (Fig. 9.2) have provided clues as to the molecular mechanisms by which the protein products of this gene influence cell motility and cell division. The secondary structure of RHAMM is largely a hydrophilic coiled coil (Turley et al., 2002; Groen et al., 2004), with the exception of an N-terminal 162 amino acid globular region (encoded by exons 1–4) and additional interspersed linker regions connecting the helical domains of the coiled coils (Fig. 9.4). These analyses suggest intracellular RHAMM performs mainly structural functions yet the majority of experimental data point to the importance of this protein in activation of signal transduction pathways. RHAMM primary sequence contains neither a signal peptide nor a membrane spanning sequence that would predict its ability to be exported from the cell or to activate signaling cascades once it is exported. Yet again, evidence *in vitro* shows a clear role for extracellular RHAMM in motogenic signaling (Hofmann et al., 1998b; Maxwell et al., 2008). Furthermore, although it contains putative nuclear localization and export sequences (and is found in the cell nucleus), it does not contain sequences that predict its localization to the mitotic spindle, mitochondria, interphase microtubules, or actin filaments. These results suggest that interaction of RHAMM with these subcellular

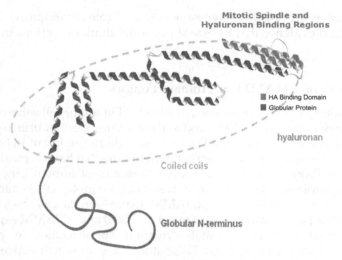

FIGURE 9.4 The predicted secondary structure of RHAMM. RHAMM is largely an acidic coiled coil protein. Approximately ¾ of its sequence occurs as a coiled coil similar to TACC, myosin, and other coiled coil proteins. The N-terminal 163 amino acids form a globular peptide. The mitotic spindle binding (a leucine zipper) and hyaluronan binding (highly basic region) are localized to the carboxyl terminus of RHAMM.

components is either indirect or involves unique structural domains that mediate binding.

Phylogenetic analysis of RHAMM sequence has also not clarified the molecular mechanisms by which RHAMM could affect such basic cell functions as motility and mitosis. RHAMM orthologs first appeared in vertebrates but there is no significant homology to proteins in lower organisms such as *Drosophila*, worms, or yeast that might provide hints as to the functions of this gene (Fig. 9.2). As a result, a conundrum has arisen surrounding not only the motogenic and mitogenic functions of RHAMM protein forms but also when and where these functions come into play. On the one hand experimental (but not proteomic) analysis of RHAMM shows it performs essential motogenic and cellular functions as a cell surface co-receptor and mitogenic functions as a mitotic spindle binding protein *in culture* and during tissue repair *in vivo*. The conundrum then is this: How can a cell surface co-receptor also function as a mitotic spindle protein?

Growing evidence has shown that this conundrum is not restricted to RHAMM. In recent years a number of proteins that are classified as cytoplasmic (due to lack of Golgi/ER export sequence and membrane spanning sequence) are secreted by unconventional means and associate with integral receptors to influence cell signaling (Fig. 9.5) (Radisky et al., 2003; Nickel, 2005; Maxwell et al., 2008; Prudovsky et al., 2008). The

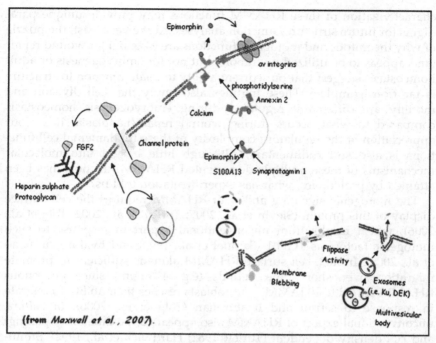

FIGURE 9.5   Diagram of mechanisms for unconventional export of cytoplasmic proteins. Several mechanisms have been identified that permit release of cytoplasmic proteins in the absence of cell death. These include export through channels, via an export protein complex, or as a result of flippase activity, membrane blebbing or release of multivesicular bodies. Adapted from Maxwell et al. (2008).

signaling functions of these multifunctional or "moonlighting" proteins at the cell surface so far appear to be unrelated to their intracellular functions. RHAMM is one example of this class of proteins and others include epimorphin/syntaxin 2; galectins 1, 3, autocrine motility receptor, and bFGF1,2. Export mechanisms are still being characterized but to date appear to involve one of five types: passage through membrane channels, flipping through the plasma membrane via protein export complexes, membrane blebbing, flippase activity, and exosomes (Fig. 9.5) (Maxwell et al., 2008).

Although the mechanisms for unconventional export of RHAMM have not yet been defined, it is secreted in response to specific stimuli as is the case with other unconventionally exported proteins. For example, TGFβ1 and hyaluronan both promote export of RHAMM to the extracellular compartment. The extent to which the extracellular functions of RHAMM are coordinated remain to be determined; however evidence suggests these apparently discrete pools of RHAMM are somehow linked. Further characterization of these RHAMM functions may provide unique paradigms for future structure/function analyses. At the very least, the puzzle of why the mitotic and motogenic functions are essential for wound repair (and appear to be utilized by tumors) but not for embryogenesis or adult homeostasis suggest that our current ability to relate function to structure is far from complete. These findings also imply that cell division and motility are differently regulated during embryogenesis/homeostasis compared to what occurs during wound repair/neoplasia. If so, our appreciation of the regulatory complexity of these fundamental cell functions is also still rudimentary. Although little insight into molecular mechanisms of these multiple documented RHAMM functions has been attained by proteomics, what has experimentation told us?

The motogenic signaling ability of RHAMM requires the cell surface display of this protein (Slevin et al., 2007; Turley et al., 2002; Tolg et al., 2006), which results from unconventional export in response to such motogenic factors as TGFβ1 (Samuel et al., 1993) and hyaluronan (Gao et al., 2008). In fact, cell surface RHAMM alone is sufficient to promote migration over short time periods (e.g. 4 hours) since exogenous RHAMM$^{-/-}$ added to RHAMM$^{-/-}$ fibroblasts rescues their ability to migrate in response to serum and hyaluronan (Tolg et al., 2006). *In culture,* unconventional export of RHAMM also appears to be time (after plating) and cell density dependent (Turley, 1982; Hardwick et al., 1992). Stimulated export is a characteristic of unconventionally secreted proteins that differs from classical export through the Golgi-ER, which is constitutive in nature.

Exported RHAMM promotes activation of signaling cascades such as protein tyrosine kinases (e.g., c-abl, src, MST-1R, FAK, Erk1,2, PI3 kinase and Ca^{++} fluxes in response to PDGF, hyaluronan, serum, bFGF, and MSP (Turley et al., 2002; Greiner et al., 2006; Evanko et al., 2007; Naor et al., 2007; Slevin et al., 2007). RHAMM, which lacks membrane-spanning sequence and is a likely peripheral surface protein, appears to associate with integral receptors in order to promote activation of signaling cascades. As described above, RHAMM physically associates with integral receptors (e.g., CD44 and MST-1R) (Tolg et al., 2006; Hamilton et al., 2007; Manzanares et al., 2007). Although a co-association between cell surface RHAMM and CD44 has been described in aggressive human breast cancer cell lines and in dermal wound fibroblasts, it clearly partners with other as of yet unidentified cell surface receptors in the absence of CD44. The RHAMM/CD44 pairing promotes cell surface retention of CD44 and sustained activation of ERK1,2 possibly through formation of RHAMM/CD44 dimers. However, this partnering may not always transduce signaling in response to hyaluronan. For example, hyaluronan-promoted invasion of glioma cells into brain slices requires RHAMM but not CD44 even though individual tumor cells express both proteins (Akiyama et al.,

2001). As another example, hyaluronan oligosaccharides promote protein tyrosine phosphorylation of focal adhesion kinase (FAK), paxillin, and ERK1,2 in endothelial cells (Lokeshwar and Selzer, 2000; Slevin et al., 2007; Gao et al., 2008). These signaling events require RHAMM although do not necessarily involve CD44. Furthermore, although both CD44 and RHAMM are required for endothelial tube formation, CD44 is uniquely required for endothelial cell adhesion while only RHAMM is required for migration of endothelial cells in 2D culture (Savani et al., 2001). Importantly, RHAMM but not CD44 is required for bFGF-induced neo-vascularization in mice (Slevin et al., 2007).

These results indicate that CD44 and cell surface RHAMM share some functions, which likely require their physical association, but they also regulate distinct processes possibly in part by partnering with distinct membrane and/or signaling components. For example, the results and evidence discussed earlier that a cell surface form of RHAMM compensates for the genetic loss of CD44 in a hyaluronan-dependent collagen-induced form of arthritis (Nedvetski et al., 2004) predict that RHAMM binds to additional hyaluronan receptors and/or to other integral membrane proteins. Indeed, the physical association of cell surface RHAMM with MST-1R in airway epithelia provides an example of a RHAMM partner that mediates growth factor (MSP1) regulated signaling (Manzanares et al., 2007).

To date, mitotic spindle/centrosome functions of intracellular RHAMM forms have been reported but the presence of RHAMM in multiple intracellular compartments (e.g. mitotic spindle, interphase microtubules, actin stress fibers, cell-substratum adhesions, nucleus, Fig. 9.6) and their association with disparate proteins such as BCL2 (Xu et al., 2003), calmodulin (Lynn et al., 2001), and glucose regulated proteins p78 and 75 (Kuwabara et al., 2006) predict additional intracellular functions. The mitotic spindle functions of RHAMM were first revealed when microinjected anti-RHAMM antibodies were observed to result in aberrant mitotic spindle formation (Maxwell et al., 2003). Consequently, several studies showed that RHAMM complexes with other mitotic spindle binding proteins including TPX2 and NuMa (Groen et al., 2004; Joukov et al., 2006; Pujana et al., 2007). Addition of recombinant RHAMM or use of anti-RHAMM antibodies to block endogenous RHAMM protein function results in aberrant mitotic spindle formation in extracts of Xenopus or HeLa cells (Joukov et al., 2006). Quantifying the amount of RHAMM in these cell extracts reveals that excess RHAMM protein results in formation of multiple and structurally aberrant spindles. The E3 ubiquitin ligase activity of BRCA1/BARD1 complexes attenuates the mitotic spindle functions of RHAMM by regulating the level of this protein at specific points of the cell cycle (Joukove et al., 2006). The proposed mechanisms by which cell surface RHAMM/CD44 activates motogenic signaling cascades and intracellular RHAMM/TPX2/BRCA1/

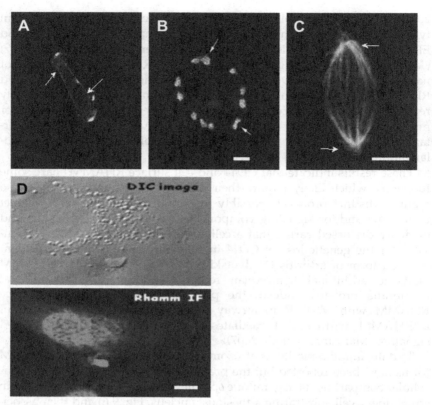

FIGURE 9.6 RHAMM proteins are localized to multiple subcellular compartments. Confocal analyses of fibroblasts show that compartmentalization of RHAMM proteins change over time after subculture. Thus RHAMM protein is (**A**) initially found at the cell membrane as cells flatten onto the growth surface [red is RHAMM (arrow), green is cortactin] then (**B**) becomes distributed in podosomes (red is RHAMM, green is cortactin, arrows show co-localization of RHAMM and cortactin). Bar = 20μm (**C**) RHAMM decorates the apex of mitotic spindles (arrow, Bar = 5μm) of dividing cells (red is RHAMM and green is phospho-Erk1,2) and is often present in the (**D**) nuclei of interphase fibroblasts. Bar = 25μm. (See Page 4 in Color Section at the back of the book).

BARD1 complexes regulate mitotic spindle integrity have recently been reviewed (Maxwell et al., 2008).

Subsequent studies using yeast two hybrid screens identified 28 additional protein partners for intracellular RHAMM of which several (BRCA1,2, CSPG6) were confirmed to associate with RHAMM in mammalian cells using immunoprecipitation assays (Pujana et al., 2007). From these and other published data a RHAMM "interactome" was constructed mainly of intracellular proteins that are components of various signaling pathways that impact upon cytoskeleton or centrosome integrity.

These analyses have cemented a dogma that the cell surface signaling and mitotic spindle protein of RHAMM are distinct. But are the known intracellular and extracellular functions of RHAMM somehow linked? Certain studies (discussed above) suggest that some functions of extracellular and intracellular forms of RHAMM might be linked in mitosis. To reiterate, extracellular and intracellular RHAMM forms both affect progression through G2/M, and RHAMM levels increase during transit through the cell cycle. One way to develop testable models to address this question of relatedness is to use a systems biology approach to construct potential functional interactomes, which predict networks of interacting or functionally linked proteins. For example, this approach was used to construct a RHAMM/centrosome interactome implicated in spindle stability (Pujana et al., 2007). We constructed a similar interactome *in silico* using the Ingenuity Pathway Analysis Software-Complete Pathways Database functional network program (http://www.ingenuity.com). With this program, we interrogated the functional interactions amongst reported RHAMM partner proteins, including both the intracellular and cell surface protein partners reviewed above. Four functional networks were identified but only one of these incorporated most of the well characterized RHAMM binding partners. This network (Fig. 9.7) includes RHAMM/HMMR, CD44, ERK1,2, BRCA1, 2, BARD1, RON, PDGFBB, PDGFR, MAD1L1, MAD2L1, HGS, XP06, and nuclear pore proteins. The functions of this network include cell cycle, cellular assembly, and organization, consistent with the dynamic nature of both cell motility and mitotic spindle formation.

The two signaling pathways incorporated with the highest level of significance within this functional network account for the role of BRCA1 in DNA damage response ($p < 0.000180$) and PDGF signaling ($p < 0.001$). However, these interactomes do not include the previously published BRCA1 pathway in mitotic spindle formation/stability, although many of the components of this pathway are included (Fig. 9.7). The role for RHAMM in PDGF and other growth factor signaling is well documented (Turley et al., 2002; Slevin et al., 2007), as is an involvement of ERK1,2 in RHAMM controlled signaling (Turley et al., 2002; Slevin et al., 2007). The direct interaction between RHAMM and CD44 is represented in Fig. 9.7 and CD44 is predicted to link RHAMM to PDGFR. The top physiological system associated with this signaling network is connective tissue function ($p < 0.000005$) and the top disease category is cancer ($p < 0.000005$), in particular tumor morphology. Collectively, these results support numerous experimental data which suggest that both cell surface and intracellular RHAMM forms contribute to a functionally interconnected network that controls processes essential to cell motility and mitosis important for both connective tissue function and cancer progression.

III. HYALURONAN RECEPTORS AND SIGNAL TRANSDUCTION PATHWAYS

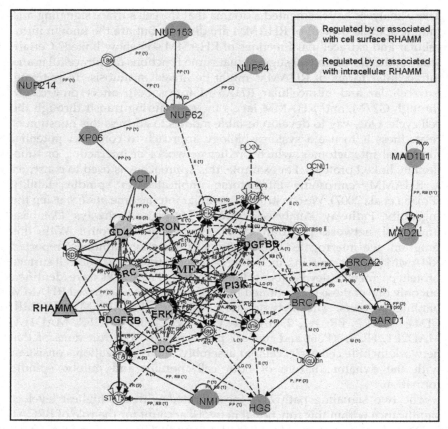

FIGURE 9.7   Functional signaling network regulated by cell surface and intracellular RHAMM. Using published RHAMM protein intracellular and extracellular partners, functional networks were constructed using Ingenuity pathway analysis. Several networks were derived from this analysis, but the network were shown here reveals that most of the proposed oncogenic effects of RHAMM can be functionally linked. For example, this network shows that CD44, RON, PDGF, MEK1/ERK1,2 are functionally linked to BRCA1,2, BARD 1 nuclear pore proteins and centrosomal proteins. The cell surface activities are shown in red and the intracellular activities are shown black. These results predict that at least some of the extracellular and intracellular functions of RHAMM may be coordinated. RHAMM (HMMR, receptor for hyaluronan mediated motility); RON (MSTR-1, macrophage stimulating 1 receptor); ACTN (alpha actinin); MEK1 (mitogen-activated protein kinase kinase); ERK1,2 (mitogen-activated protein kinase); PDGFRB (platelet derived growth factor receptor beta); PDGFBB (platelet derived growth factor beta dimer); SRC (Rous sarcoma oncogene); PI3K (phosphoinositide-3-kinase); CD44 [CD44 molecule (Indian blood group)]; BRCA1 (breast cancer 1); BRCA2 (breast cancer 2); BARD1 (BRCA1 associated RING domain 1); NMI [N-myc (and STAT) interactor]; HGS (hepatocyte growth factor-regulated tyrosine kinase substrate); MAD2L (MAD2 mitotic arrest deficient-like 1); MAD1L1 (MAD1 mitotic arrest deficient-like 1); XPO6 (Exportin 6); NUP62 (nucleoporin 62); NUP54 (nucleoporin 54); NUP153 (nucleoporin 153); and NUP214 (nucleoporin 214).

CONCLUSIONS

RHAMM is a gene that belongs to a multifunctional group of cytoplasmic proteins, which are unconventionally exported to the cell surface in response to specific stimuli. This process is highly active during response to injury and tumorigenesis and in certain situations/phenotypic backgrounds, RHAMM, or structural variants of RHAMM, can act as an oncogene. RHAMM is poorly expressed in normal tissues but highly expressed and required for response-to-injury processes, in particular affecting inflammation and the subsequent mesenchymal response for repair. It is also found in numerous human tumors and hyperexpression of RHAMM is associated with poor clinical outcome of many of these cancers. Cell surface RHAMM performs co-receptor functions by affecting signaling through hyaluronan and growth factor receptors while intracellular RHAMM is a mitotic spindle protein that interacts with BRCA1/ BARD1 complexes to control mitotic spindle stability. Further analyses of the molecular mechanisms by which RHAMM regulates these processes will undoubtedly identify new "inside–outside" paradigms that control motility and mitosis and may lead to the identification of novel key signaling/structural nodes that can be targeted in the treatment of chronic inflammation or cancer.

ACKNOWLEDGMENTS

Work described in this review was supported by the Canadian Institutes of Health Research [grant to E.A. Turley (MOP-57694)], the Translational Breast Cancer Research Unit at the London Regional Cancer Program (Breast Cancer Society of Canada salary support to E.A. Turley), US Army Medical Research and Material Command (J.B. McCarthy and E.A. Turley, W81XWH-06-1-0135).

References

Ahrens, T., et al. (2001). CD44 is the principal mediator of hyaluronic-acid-induced melanoma cell proliferation. *J Invest Dermatol* **116** (1), 93–101.

Akiyama, Y., et al. (2001). Hyaluronate receptors mediating glioma cell migration and proliferation. *J Neurooncol* **53** (2), 115–127.

Amano, T., et al. (2007). Antitumor effects of vaccination with dendritic cells transfected with modified receptor for hyaluronan-mediated motility mRNA in a mouse glioma model. *J Neurosurg* **106** (4), 638–645.

Assmann, V., et al. (1998). The human hyaluronan receptor RHAMM is expressed as an intracellular protein in breast cancer cells. *J Cell Sci* **111** (Pt 12), 1685–1694.

Cho, R. J., et al. (2001). Transcriptional regulation and function during the human cell cycle. *Nat Genet* **27** (1), 48–54.

Dorsett-Martin, W. A. (2004). Rat models of skin wound healing: a review. *Wound Repair Regen* **12** (6), 591–599.

Duval, A., et al. (2001). Evolution of instability at coding and non-coding repeat sequences in human MSI-H colorectal cancers. *Hum Mol Genet* **10** (5), 513–518.

Gibran, N. S., Boyce, S., and Greenhalgh, D. G. (2007). Cutaneous wound healing. *J Burn Care Res* **28** (4), 577–579.

Greiner, J., Dohner, H., and Schmitt, M. (2006). Cancer vaccines for patients with acute myeloid leukemia – definition of leukemia-associated antigens and current clinical protocols targeting these antigens. *Haematologica* **91** (12), 1653–1661.

Evanko, S. P., Parks, W. T., and Wight, T. N. (2004). Intracellular hyaluronan in arterial smooth muscle cells: association with microtubules, RHAMM, and the mitotic spindle. *J Histochem Cytochem* **52** (12), 1525–1535.

Evanko, S. P., et al. (2007). Hyaluronan-dependent pericellular matrix. *Adv Drug Deliv Rev* **59** (13), 1351–1365.

Gao, F., et al. (2008). Hyaluronan oligosaccharides are potential stimulators to angiogenesis via RHAMM mediated signal pathway in wound healing. *Clin Invest Med* **31** (3), E106–116.

Groen, A. C., et al. (2004). XRHAMM functions in ran-dependent microtubule nucleation and pole formation during anastral spindle assembly. *Curr Biol* **14** (20), 1801–1811.

Hamilton, S. R., et al. (2007). The hyaluronan receptors CD44 and Rhamm (CD168) form complexes with ERK1,2 that sustain high basal motility in breast cancer cells. *J Biol Chem* **282** (22), 16667–16680.

Hardwick, C., et al. (1992). Molecular cloning of a novel hyaluronan receptor that mediates tumor cell motility. *J Cell Biol* **117** (6), 1343–1350.

Hofmann, M., et al. (1998a). Identification of IHABP, a 95 kDa intracellular hyaluronate binding protein. *J Cell Sci* **111** (Pt 12), 1673–1684.

Hofmann, M., et al. (1998b). Problems with RHAMM, a new link between surface adhesion and oncogenesis? *Cell* **95** (5), 591–592, author reply 592–593.

Joukov, V., et al. (2006). The BRCA1/BARD1 heterodimer modulates ran-dependent mitotic spindle assembly. *Cell* **127** (3), 539–552.

Kong, Q. Y., et al. (2003). Differential expression patterns of hyaluronan receptors CD44 and RHAMM in transitional cell carcinomas of urinary bladder. *Oncol Rep* **10** (1), 51–55.

Kuwabara, H., et al. (2006). Glucose regulated proteins 78 and 75 bind to the receptor for hyaluronan mediated motility in interphase microtubules. *Biochem Biophys Res Commun* **339** (3), 971–976.

Li, H., et al. (2000a). Alternative splicing of RHAMM gene in Chinese gastric cancers and its *in vitro* regulation. *Zhonghua Yi Xue Yi Chuan Xue Za Zhi* **17** (5), 343–347.

Li, H., et al. (2000b). Expression of hyaluronan receptors CD44 and RHAMM in stomach cancers, relevance with tumor progression. *Int J Oncol* **17** (5), 927–932.

Levy, P., et al. (2004). Molecular profiling of malignant peripheral nerve sheath tumors associated with neurofibromatosis type 1, based on large-scale real-time RT-PCR. *Mol Cancer* **3**, 20.

Liska, A. J., et al. (2004). Homology-based functional proteomics by mass spectrometry: application to the Xenopus microtubule-associated proteome. *Proteomics* **4** (9), 2707–2721.

Liu, N., et al. (2004). Hyaluronan-binding peptide can inhibit tumor growth by interacting with Bcl-2. *Int J Cancer* **109** (1), 49–57.

Lokeshwar, V. B. and Selzer, M. G. (2000). Differences in hyaluronic acid-mediated functions and signaling in arterial, microvessel, and vein-derived human endothelial cells. *J Biol Chem* **275** (36), 27641–27649.

Lugli, A., et al. (2006). Overexpression of the receptor for hyaluronic acid mediated motility is an independent adverse prognostic factor in colorectal cancer. *Mod Pathol* **19** (10), 1302–1309.

Lynn, B. D., et al. (2001). Identification of sequence, protein isoforms, and distribution of the hyaluronan-binding protein RHAMM in adult and developing rat brain. *J Comp Neurol* **439** (3), 315–330.

Mantripragada, K. K., et al. (2008). High-resolution DNA copy number profiling of malignant peripheral nerve sheath tumors using targeted microarray-based comparative genomic hybridization. *Clin Cancer Res* **14** (4), 1015–1024.

Manzanares, D., et al. (2007). Apical oxidative hyaluronan degradation stimulates airway ciliary beating via RHAMM and RON. *Am J Respir Cell Mol Biol* **37** (2), 160–168.

Maxwell, C. A., et al. (2003). RHAMM is a centrosomal protein that interacts with dynein and maintains spindle pole stability. *Mol Biol Cell* **14** (6), 2262–2276.

Maxwell, C. A., et al. (2004). RHAMM expression and isoform balance predict aggressive disease and poor survival in multiple myeloma. *Blood* **104** (4), 1151–1158.

Maxwell, C. A., et al. (2005). Receptor for hyaluronan-mediated motility correlates with centrosome abnormalities in multiple myeloma and maintains mitotic integrity. *Cancer Res* **65** (3), 850–860.

Maxwell, C. A., McCarthy, J., and Turley, E. (2008). Cell-surface and mitotic-spindle RHAMM: moonlighting or dual oncogenic functions? *J Cell Sci* **121** (Pt 7), 925–932.

Mohapatra, S., et al. (1996). Soluble hyaluronan receptor RHAMM induces mitotic arrest by suppressing Cdc2 and cyclin B1 expression. *J Exp Med* **183** (4), 1663–1668.

Naor, D., et al. (2007). CD44 involvement in autoimmune inflammations, the lesson to be learned from CD44-targeting by antibody or from knockout mice. *Ann NY Acad Sci* **1110**, 233–247.

Nedvetzki, S., et al. (2004). RHAMM, a receptor for hyaluronan-mediated motility, compensates for CD44 in inflamed CD44-knockout mice, a different interpretation of redundancy. *Proc Natl Acad Sci USA* **101** (52), 18081–18086.

Nickel, W. (2005). Unconventional secretory routes, direct protein export across the plasma membrane of mammalian cells. *Traffic* **6** (8), 607–614.

Nousiainen, M., et al. (2006). Phosphoproteome analysis of the human mitotic spindle. *Proc Natl Acad Sci USA* **103** (14), 5391–5396.

Oberyszyn, T. M. (2007). Inflammation and wound healing. *Front Biosci* **12**, 2993–2999.

Panagopoulos, A. T., et al. (2008). Expression of cell adhesion proteins and proteins related to angiogenesis and fatty acid metabolism in benign, atypical, and anaplastic meningiomas. *J Neurooncol* **89** (1), 73–87.

Prudovsky, I., et al. (2008). Secretion without Golgi. *J Cell Biochem* **103** (5), 1327–1343.

Pujana, M. A., et al. (2007). Network modeling links breast cancer susceptibility and centrosome dysfunction. *Nat Genet* **39** (11), 1338–1349.

Radisky, D. C., Hirai, Y., and Bissell, M. J. (2003). Delivering the message, epimorphin and mammary epithelial morphogenesis. *Trends Cell Biol* **13** (8), 426–434.

Rein, D. T., et al. (2003). Expression of the hyaluronan receptor RHAMM in endometrial carcinomas suggests a role in tumour progression and metastasis. *J Cancer Res Clin Oncol* **129** (3), 161–164.

Samuel, S. K., et al. (1993). TGF-beta 1 stimulation of cell locomotion utilizes the hyaluronan receptor RHAMM and hyaluronan. *J Cell Biol* **123** (3), 749–758.

Savani, R. C., et al. (2001). Differential involvement of the hyaluronan (HA) receptors CD44 and receptor for HA-mediated motility in endothelial cell function and angiogenesis. *J Biol Chem* **276** (39), 36770–36778.

Schmitt, M., et al. (2008). RHAMM-R3 peptide vaccination in patients with acute myeloid leukemia, myelodysplastic syndrome, and multiple myeloma elicits immunologic and clinical responses. *Blood* **111** (3), 1357–1365.

Shakib, K., et al. (2005). Proteomics profiling of nuclear proteins for kidney fibroblasts suggests hypoxia, meiosis, and cancer may meet in the nucleus. *Proteomics* **5** (11), 2819–2838.

Slevin, M., et al. (2007). Hyaluronan-mediated angiogenesis in vascular disease: uncovering RHAMM and CD44 receptor signaling pathways. *Matrix Biol* **26** (1), 58–68.

Still, I. H., et al. (2004). Structure-function evolution of the transforming acidic coiled coil genes revealed by analysis of phylogenetically diverse organisms. *BMC Evol Biol* **4**, 16.

III. HYALURONAN RECEPTORS AND SIGNAL TRANSDUCTION PATHWAYS

Tolg, C., et al. (2006). Rhamm–/– fibroblasts are defective in CD44-mediated ERK1,2 motogenic signaling, leading to defective skin wound repair. *J Cell Biol* **175** (6), 1017–1028.

Tolg, C., et al. (2003). Genetic deletion of receptor for hyaluronan-mediated motility (Rhamm) attenuates the formation of aggressive fibromatosis (desmoid tumor). *Oncogene* **22** (44), 6873–6882.

Turley, E. A. (1982). Purification of a hyaluronate-binding protein fraction that modifies cell social behavior. *Biochem Biophys Res Commun* **108** (3), 1016–1024.

Turley, E. A., Bowman, P., and Kytryk, M. A. (1985). Effects of hyaluronate and hyaluronate binding proteins on cell motile and contact behaviour. *J Cell Sci* **78**, 133–145.

Turley, E. A., Moore, D., and Hayden, L. J. (1987). Characterization of hyaluronate binding proteins isolated from 3T3 and murine sarcoma virus transformed 3T3 cells. *Biochemistry* **26** (11), 2997–3005.

Turley, E. A., Noble, P. W., and Bourguignon, L. Y. (2002). Signaling properties of hyaluronan receptors. *J Biol Chem* **277** (7), 4589–4592.

Wang, C., et al. (1998). The overexpression of RHAMM, a hyaluronan-binding protein that regulates ras signaling, correlates with overexpression of mitogen-activated protein kinase and is a significant parameter in breast cancer progression. *Clin Cancer Res* **4** (3), 567–576.

Whitfield, M. L., et al. (2002). Identification of genes periodically expressed in the human cell cycle and their expression in tumors. *Mol Biol Cell* **13** (6), 1977–2000.

Xu, X. M., et al. (2003). A peptide with three hyaluronan binding motifs inhibits tumor growth and induces apoptosis. *Cancer Res* **63** (18), 5685–5690.

Yamada, Y., et al. (1999). Receptor for hyaluronan-mediated motility and CD44 expressions in colon cancer assessed by quantitative analysis using real-time reverse transcriptase-polymerase chain reaction. *Jpn J Cancer Res* **90** (9), 987–992.

Yamano, Y., et al. (2008). Hyaluronan-mediated motility, a target in oral squamous cell carcinoma. *Int J Oncol* **32** (5), 1001–1109.

Yang, C. W., et al. (2005). Integrative genomics based identification of potential human hepatocarcinogenesis-associated cell cycle regulators: RHAMM as an example. *Biochem Biophys Res Commun* **330** (2), 489–497.

Zaman, A., et al. (2005). Expression and role of the hyaluronan receptor RHAMM in inflammation after bleomycin injury. *Am J Respir Cell Mol Biol* **33** (5), 447–454.

Zhou, R., Wu, X., and Skalli, O. (2002). The hyaluronan receptor RHAMM/IHABP in astrocytoma cells: expression of a tumor-specific variant and association with microtubules. *J Neurooncol* **59** (1), 15–26.

Zlobec, I., et-al. (2008a). Role of RHAMM within the hierarchy of well-established prognostic factors in colorectal cancer. *Gut* **57** (10): 1413–1419.

Zlobec, I., et al. (2008b). Node-negative colorectal cancer at high risk of distant metastasis identified by combined analysis of lymph node status, vascular invasion, and Raf-1 kinase inhibitor protein expression. *Clin Cancer Res* **14** (1), 143–148.

Zlobec, I., et-al. (2008c). RHAMM, p21 combined phenotype identifies microsatellite instability-high colorectal cancers with a highly adverse prognosis. *Clin Cancer Res* **14** (12), 3798–3806.

HYALURONAN SYNTHESIS

CHAPTER

10

Altered Hyaluronan Biosynthesis in Cancer Progression

Naoki Itano and Koji Kimata

INTRODUCTION

Hyaluronan (HA) is a sugar-chain macromolecule in which N-acetylglucosamine and glucuronic acid are linked together by alternating β-1,3 and β-1,4 linkages (Fig. 10.1) (Laurent and Fraser, 1992). Despite its apparently simple structure, HA exhibits multiple properties depending on its molecular size and its binding molecules (Fraser and Laurent, 1989). For instance, high molecular weight HA forms part of the extracellular matrix (ECM) by linking HA-binding molecules into macromolecular aggregates and regulating a variety of cell behaviors, such as cell adhesion, motility, growth, and differentiation. HA oligosaccharides also

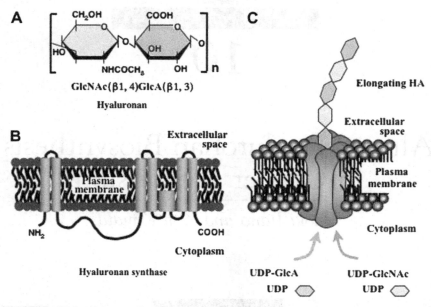

FIGURE 10.1   Scheme illustrating the HA structure (**A**), a predicted structure of mammalian HAS (**B**), and a proposed secretion process of HA (**C**).

regulate such cell behaviors in different ways by acting on intracellular signaling pathways through interaction with cell surface receptors (Toole et al., 1989). Accumulating evidence has demonstrated that the production of HA is excessive in cancer malignancies (Knudson, 1996; Toole, 2004); increased serum levels and deposition in tumor tissue are often associated with malignant progression in many cancers, including breast cancer and colorectal cancer (Ponting et al., 1992; Ropponen et al., 1998).

Although the close association of HA production in the progression of cancer cells is now being established, the entire picture of diverse and complex HA functions in cancer progression remains to be elucidated. Fortunately, animal models with genetically manipulated HA Synthase (HAS) expression provide powerful tools for understanding the *in vivo* function of HA, particularly in connection with cancer cell behavior. Thus the central aim of the present review is to highlight the role of HA in cancer progression from the viewpoint of abnormal HA biosynthesis.

HA BIOSYNTHESIS

The discovery of three members of the HAS gene family (HAS-1, HAS-2, and HAS-3) has enabled great strides in understanding the unique process of HA biosynthesis and mode of chain elongation (Weigel et al., 1997; Itano and Kimata, 2002). Structurally, all HAS proteins are composed of multiple membrane-spanning regions and large cytoplasmic loops (Fig. 10.1). Unlike typical glycosyltransferases, the cytoplasmic loop in HAS molecules possesses two active sites which participate in the transfer of UDP-GlcNAc and UDP-GlcA substrates. Characterization of the three HAS isoforms has revealed differences in enzymatic properties, particularly in their ability to form HA matrices and determine product size (Itano et al., 1999a). The expression profiles of HAS genes are temporally and spatially regulated during embryogenesis and pathogenesis (Sugiyama et al., 1998; Kennedy et al., 2000; Recklies et al., 2001; Pienimaki et al., 2001), and divergence in the transcriptional regulation of HAS genes during these processes can be explained to some extent by upstream signaling pathways that are triggered by various growth factors, cytokines, cellular stress, and so on. The dynamic turnover of HA is therefore tightly regulated by altering the expression profiles of HAS isoforms to have different enzymatic properties (Weigel et al., 1997; Itano and Kimata, 2002).

ALTERED HA SYNTHESIS IN CANCER

The malignant transformation of cells frequently impairs regulation of HA synthesis and induces excessive HA production (Hamerman et al., 1965; Hopwood and Dorfman, 1977; Leonard et al., 1978). During this process, multiple transcriptional regulation of HAS genes allows cells to optimize the extracellular environment for tumor growth and malignant progression, and a transcriptional switch in HAS isoforms has been demonstrated in cells undergoing oncogenic transformation (Itano et al., 2004). Of the three HAS isoforms, only HAS-2 gene expression was increased in *v-Ha-ras* transformed cells, which showed lowered malignancy. Conversely, both HAS-1 and HAS-2 expression were elevated in highly malignant cells transformed with *v-src*. This implies that HAS isoforms may be involved in different stages of malignant tumor progression. From this point of view, it is increasingly necessary to confirm the relationship between HAS expression and prognosis by statistical analysis using clinical samples. Thus far, these clinicopathological studies have indicated that elevated HAS-1 expression and/or intronic gene splicing correlate with poor prognosis in human colon cancer, ovarian cancer, and multiple myelomas (Yamada et al., 2004; Yabushita et al., 2004; Adamia et al., 2005).

Emerging evidence is providing new insight into the functional aspect of this polysaccharide, particularly in respect to the involvement of HA in tumor malignancy; forced expression of HAS-2 and HAS-3 genes results in excess HA production and enhanced tumorigenic ability of fibrosarcoma and melanoma cells (Kosaki et al., 1999; Liu et al., 2001; Li and Heldin, 2001). Moreover, induced expression of HAS-1 restores the metastatic potential of mouse mammary carcinoma mutants, previously having low levels of HA synthesis and metastatic ability (Itano et al., 1999b). Inversely, suppression of HAS-2 or HAS-3 decreases HA production and reduces the tumorigenic potential of various cell lines (Simpson et al., 2002a, b; Nishida et al., 2005; Udabage et al., 2005). Although the above clearly demonstrates the important role of HA in tumorigenesis, the tumor promoting ability of excess HA is still somewhat controversial since HAS-2 overexpression also suppresses the tumorigenesis of glioma cells (Enegd et al., 2002). Furthermore, the *in vitro* growth of human prostate cancer cells decreased dramatically when transfected with an HAS-2-expression plasmid, but co-expression of HAS-2 and hyaluronidase HYAL-1 restores the growth of these cancer cells (Bharadwaj et al., 2007). Here, since HA accumulation is the result of a balance between the activities of HAS and hyaluronidases, the presence of hyaluronidase may have promoted HA turnover in the cancer cells and overcome the tumor suppression by excess amounts of HA. Alternatively, the biphasic effects of HA on tumorigenesis can be explained by considering its dose-dependent properties (Itano et al., 2004). To assess this idea, we generated stable transfectants expressing various levels of HAS genes and examined their tumorigenicity in nude mice. Although significant growth promotion was observed within a narrow range of HAS-2 expression, this growth was inhibited at high expression levels. The dose-dependency of HA may help us consider statements regarding the physiological significance of changes in HA concentration with tumor grade or stage, since HA accumulation in clinical samples varies and occasionally shows little statistical changes with tumor grade.

The involvement of HAS in tumor progression has also been evaluated using genetically manipulated animal models. In one study, a transgenic (Tg) mouse model allowing overexpression of murine HAS-2 in mammary glands was generated in order to simulate hyperproduction of HA found in human breast cancer (Koyama et al., 2007). In this model, the expression of exogenous HAS-2 was conditionally controlled by the expression of Cre-recombinase driven by a mammary epithelial cell-specific MMTV promoter. By intercrossing the Tg mice with a mouse mammary tumor model expressing rat c-*neu* protooncogenes in mammary epithelial cells, mammary tumors with aggressive growth rates were developed. Histo-logically, these tumors were classified as poorly differentiated adenocar-cinomas with numerous intratumoral stroma (Fig. 10.2).

TUMOR–STROMAL INTERACTION

Most aggressive tumors are composed not only of cancer cells, but also of many host stromal cells (Kalluri and Ziesberg, 2006), and the importance of interactions between cancer cells and their surrounding stroma in facilitating tumor progression has been demonstrated both by clinical and

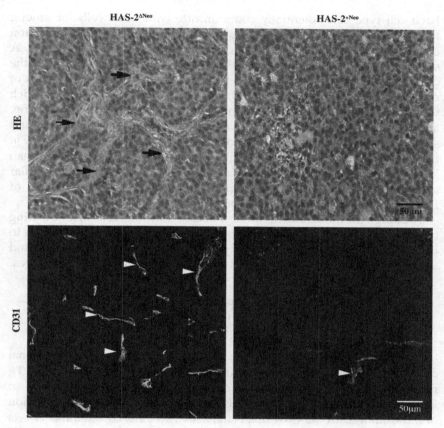

FIGURE 10.2 HA overproduction promotes the formation of intratumoral stroma. Tumor sections from HAS-2-overexpressing transgenic (HAS-2$^{\Delta Neo}$) and control (HAS-2^{+Neo}) mice were stained with hematoxylin and eosin (upper panels). The most prominent histological feature of the HAS-2$^{\Delta Neo}$ tumors was increased formation of intratumoral stroma (arrows). In contrast, control tumors had the characteristics of ductal carcinoma with much less stroma. Tissue sections from HAS-2$^{\Delta Neo}$ and HAS-2^{+Neo} tumors were stained with an antibody against murine CD31 (lower panels). Tumor microvessels (arrowheads) of smaller size were more numerous in the HAS-2$^{\Delta Neo}$ tumors compared with control tumors. (See Page 5 in Color Section at the back of the book).

experimental evidence (Bhowmick and Moses, 2005). Carcinoma cells actively recruit several distinct stromal cells, such as inflammatory cells, vascular cells, and fibroblasts within the tumor, by secreting chemoattractant factors (Desmouliere et al., 2004; Mantovani et al., 2006). Furthermore, crosstalk between carcinoma cells and adjacent stromal cells influences the composition and arrangement of the tumor microenvironment to support tumor progression by allowing angiogenesis and facilitating the invasion and metastasis of tumor cells (Bhowmick et al., 2004).

Each cell type can potentially communicate with other cells, or among themselves, through the release of auto-/paracrine factors and formation of a complex ECM network (Orimo et al., 2005; Shekhar et al., 2003). A strong correlation has been drawn between cancer progression and the degree of HA accumulation within it, especially in the invading edges of carcinomas (Setala et al., 1999; Auvinen et al., 2000). As such, HA-rich tumor microenvironments, which may be favorable for cancer invasion, are likely generated from complex interactions between tumor cells and stromal cells infiltrating from adjacent host connective tissue. In fact, *in vitro* HA synthesis was synergistically increased in co-cultures of human lung tumor cells with fibroblasts (Knudson et al., 1984; 1989), and similar synergistic effects have been demonstrated with the combination of fibroblasts and other tumor cell types (Merrilees and Finlay, 1985).

Although the main roles of tumor-associated stromal cells in modulating tumor cell behavior is well-established, critical questions still remain as to the molecular mechanisms underlying communication carcinoma and stromal cells and regulating stromal cell recruitment within tumor tissues.

NOVEL FUNCTION OF HA IN STROMAL CELL RECRUITMENT

Genetic evidence supporting the role of tumor-derived HA in stromal cell recruitment has recently come from our experiments using HAS-2 Tg mice; ectopic expression of HAS-2 in mammary tumor cells leads to a marked recruitment of stromal cells inside tumors followed by formation of intratumoral stroma (Koyama et al., 2007). To date, considerable efforts have been made to purify the stromal cell chemotactic factors produced by tumor cells, but only a few factors, such as PDGF, have been identified (Dong et al., 2004). The above notion therefore strongly suggests the function of an HA-rich ECM as a stromal cell chemotactic factor.

Several complex and multifaceted mechanisms can be considered for understanding how tumor-derived HA intratumorally recruits stromal cells during tumor formation. Extracellular accumulation of highly hydrated HA provides microenvironments amenable to easy fibroblast-penetration by increasing turgidity (Laurent and Fraser, 1992), and the fact that forced expression of HAS-2 impairs the intercellular adhesion machinery of tumor cells. This also explains how HA-rich matrices provide an environment favorable for fibroblast-infiltration (Itano et al., 2002; Zoltan-Jones et al., 2003).

HA-induced signaling pathways govern the migratory phenotypes of stromal cells via interaction with HA receptors (Turley et al., 2002). For instance, CD44 and RHAMM, both typical HA receptors, have been implicated in the HA-dependent cell migration and invasion of stromal fibroblasts. Additionally, the interaction of HA and CD44 can activate receptor tyrosine kinases, which in turn induce the activation of downstream Ras/MAPK and PI3K/Akt signaling pathways (Turley et al., 2002). HA binding to CD44 also activates Rac1 signals, which regulate actin assembly associated with membrane ruffling and cell motility (Bourguignon et al., 2000). Thus, HA appears to promote cell motility by acting on intracellular signaling pathways and controlling the assembly of the actin cytoskeleton. Alternatively, HA–CD44 interactions may recruit mesenchymal stem cells (MSCs) (Zhu et al., 2006); in a mouse model of acute renal failure, MSCs injected into mice migrated to the injured kidney, where HA is abundant (Herrera et al., 2007). Renal localization of MSCs is blocked by preincubation with CD44 blocking antibodies or soluble HA. Likewise, MSCs derived from CD44 knockout mice do not localize to the injured kidney, but are rescued by transfection with cDNA encoding CD44. This same mechanism may participate in the recruitment of MSCs to HA-rich tumors.

Lastly for consideration of HA-mediated stromal cell recruitment is the action of HA-binding molecules. Versican (also called PG-M), an HA-binding proteoglycan, is highly expressed in tissue compartments undergoing active cell proliferation and migration (Wight, 2002) and participates in the formation of an HA-rich ECM (Fig. 10.3). In the peripheral invasive areas of infiltrating ductal carcinomas, the most intense staining by a versican-specific antibody is visualized in the mesenchymal

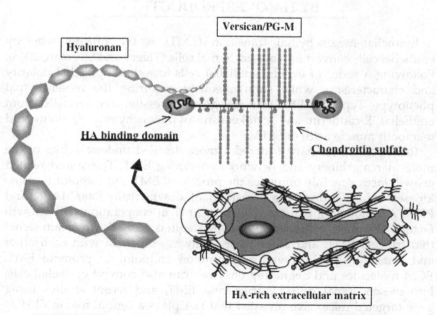

FIGURE 10.3 Scheme of HA-rich extracellular matrix and its constituents.

tissues between carcinoma cell clumps and surrounding tissues, where HA can be demonstrated histochemically (Nara et al., 1997). This cooperative action of HA and versican in mobilizing stromal cells has been demonstrated by the Matrigel plug assay (Koyama et al., 2007). When Matrigel plugs are subcutaneously implanted into nude mice, only trace numbers of stromal cells migrate into the Matrigel plugs containing high molecular weight (HMW) HA alone. However, the infiltration of stromal cells is markedly increased in the presence of HA/versican aggregates. In tumor xenograft models, exogenously added HA/versican aggregates also significantly promote the infiltration of stromal cells within tumors (Koyama et al., personal communication). In concert with versican, HA may therefore allow cells to prepare for migration by enhancing cell detachment from the ECM (Fig. 10.3). The anti-adhesive and motility-promoting effects of versican is evident by combining an earlier observation that versican can inhibit cancer cell attachment to fibronectin, together with the recent finding that formation of an HA/versican pericellular matrix promotes prostate cancer cell motility (Yamagata and Kimata, 1994; Ricciardelli et al., 2007). Current studies have also implied a role of versican in enhancement of cell motility in the assembly of cytoskeletal machinery and transmitting signals.

EPITHELIAL–MESENCHYMAL TRANSITION CAUSED BY HA OVERPRODUCTION

Epithelial–mesenchymal transition (EMT) is the process whereby epithelial cells convert into mesenchymal cells (Thiery and Sleeman, 2006). Following a series of events, epithelial cells lose their epithelial polarity and characteristics while simultaneously acquiring the mesenchymal phenotype. Typically, EMT switches gene expression characteristics from epithelial E-cadherin and cytokeratin to mesenchymal vimentin and α-smooth muscle actin (α-SMA).

Recent advances have fostered a more detailed understanding of the molecular machinery and networks governing EMT. TGF-β and related growth factors mainly influence the process of EMT and receptor tyrosine kinases, which induce the activation of downstream Ras/MAPK and PI3K/Akt signaling pathways, govern EMT in cooperation with growth factors. The nuclear translocation of β-catenin is a key downstream signal that triggers EMT, and all of these pathways crosstalk with each other and transmit signals towards a common endpoint to promote EMT. ECM molecules and degrading enzymes can also convert epithelial cells into mesenchymal cells by triggering EMT, and recent studies using gene-targeted mice have revealed that HA plays a central role in EMT as well (Camenisch et al., 2000); in one report, HAS-2 null mice showed severe cardiac and vascular abnormalities, and died during midgestation due to a lack of transformation of cardiac endothelial cells to mesenchyme.

EMT was originally defined as a morphological conversion during normal development, but has recently gained attention as a central mechanism for carcinoma progression and metastasis (Huber et al., 2005). The progression of carcinoma cells to metastatic cells frequently involves an EMT-like epithelial change towards a migratory fibroblastic phenotype, particularly at the invasive front of tumors. During tumor progression, downregulation of E-cadherin, a hallmark of EMT, aids tumor cells in spreading. In one study, infection of recombinant HAS-2 adenoviral vectors converted normal Madin-Darby canine kidney and MCF-10A human mammary epithelial cells to mesenchyme as assessed by upregulation of vimentin, dispersion of cytokeratin, and loss of E-cadherin at intercellular boundaries (Zoltan-Jones et al., 2003). All of this suggests that increased HA production appears to be sufficient to induce EMT. Our recent observation using HAS-2 conditional Tg mice has also revealed that over-production of HA in mammary carcinoma cells results in the suppression of E-cadherin expression and nuclear translocation of β-catenin, further providing evidence for HA-mediated promotion of EMT (Koyama et al., 2007). Carcinoma cells having undergone EMT may then participate in the formation of intratumoral stroma observed in HA overproducing tumors, such as those seen in human breast cancers (Petersen et al., 2003).

ROLES OF HA-RICH ECM IN TUMOR ANGIOGENESIS

Angiogenesis, the formation of new capillaries from preexisting vessels, is an absolute requirement for tumor growth and metastasis (Carmeliet, 2003) and is controlled by the aberrant production of angiogenic factors expressed by malignant tumor cells, host cells, or both. Among such factors, vascular endothelial growth factor (VEGF) has emerged as a central regulator. In addition, the local composition of the ECM surrounding the vasculature can affect angiogenesis either positively or negatively (Sottile, 2004), and HA oligosaccharides have been implicated in the promotion of angiogenesis (West et al., 1985). Studies in chick chorioallantoic membranes and rat skin have demonstrated that HA degradation products of specific size (3–10 disaccharide units) have the potential to induce neo-vascularization (Sattar et al., 1994; Slevin et al., 1998). Furthermore, HA oligosaccharides, together with angiogenic factors such as VEGF and basic fibroblast growth factor (bFGF), synergistically stimulate endothelial cell proliferation, migration, and capillary formation in vitro. The angiogenic activity of HA depends on its molecular mass; HMW native HA is anti-angiogenic by inhibiting endothelial cell proliferation and migration and capillary formation in a three-dimensional matrix (Feinberg and Beebe,

1983). Because angiogenesis is the result of complex interactions between positive and negative regulators of angiogenesis (Slevin et al., 1998), the balance of regulatory HMW HA and effecter HA oligosaccharides may be important for controlling the angiogenic response.

HAS gene manipulation provides an opportunity to assess the role of HA during angiogenesis *in vivo*. The significance of HA has been highlighted in HAS-2 deficient mice having vascular defects, implicating a critical function of HA in embryonic vasculogenesis (Camenisch et al., 2000). HAS-2 Tg mouse models of breast cancer have also shown that overproduction of HA in tumor cells accelerates formation of intratumoral neovasculature (Koyama et al., 2007). This altered formation in genetically manipulated mice may be explained by the well-known fact that HA degradation products induce an angiogenic response. Indeed, HAS-2-overexpressing tumors contained significant amounts of small HA oligosaccharides, as assessed by gel filtration chromatography of tumor homogenates, whereas control tumors contained mostly HMW HA (Koyama et al., 2007). This supports the conventional notion that HA oligosaccharides influence tumor-induced angiogenesis. Although the physiological significance of HA oligosaccharides in the promotion of angiogenesis is well-established, it is still open to debate whether the ECM consisting of HMW HA and HA-binding molecules has any role in angiogenesis. Interestingly, in HAS-2-overexpressing mammary tumors, most neovasculature is predominantly found penetrating into the intratumoral stroma where HA is abundant as a constituent of ECM (Koyama et al., 2007). The constituents of stromal ECM therefore likely provide a supporting framework for easy penetration of endothelial cells and subsequent neovascularization.

Versican is abundant in both the perivascular elastic tissues of blood vessels and stromal ECM (Nara et al., 1997). In our recent study, administration of HA-versican aggregates, but not native HMW HA alone, promote the infiltration of endothelial cells within Matrigel plugs containing angiogenic bFGF (Koyama et al., 2007), suggesting the potency of HMW HA to accelerate angiogenesis in the presence of versican. Currently, one can only speculate as to the function of HA/versican aggregates in the context of angiogenesis (Fig. 10.4). Since HA/versican complexes can stimulate cell migration, their possible role would be one that enhances migration and invasion of endothelial cells. As an alternative explanation, the HA-rich matrix can be proposed to serve as a reservoir for various growth factors involved in vessel development; degradation of the matrix results in the release of various growth factors sequestered within, which in turn promotes an angiogenic response (Fig. 10.4). Further investigation is being conducted to clarify both the functional aspect of HA/versican aggregates in angiogenesis, as well as the relationship between such aggregates and HA oligosaccharides.

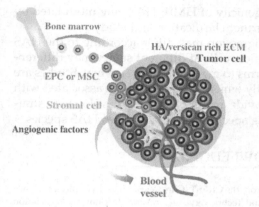

FIGURE 10.4 A model of stroma-induced angiogenesis.

Until recently, new blood vessels in adults were thought to grow exclusively through the sprouting of preexisting vasculature (Hillen and Griffionen, 2007). However, emerging evidence suggests that bone marrow-derived endothelial progenitor cells (EPCs) contribute to tissue vascularization during both embryonic and pathogenetic conditions; circulating bone marrow-derived EPCs are mobilized from the bone marrow and recruited to the foci of neovascularization where they form new *in situ* blood vessels through vasculogenesis. However, the homing process of EPCs remains unclear. Similarly to the recruitment of MSCs, HA-rich matrices may provide a stem cell niche for recruitment and retention of circulating EPCs (Fig. 10.4). In the future, a greater understanding of the mechanisms regulating selective cell movement and recruitment will lead to the development of novel anticancer therapeutic agents targeting reparative progenitor cells.

CONCLUSION AND PERSPECTIVES

This review focused on the role of HA in cancer progression with respect to its biosynthesis and function. A wealth of data has been accumulated on HA function in the promotion of malignancies, showing that enhanced cancer invasion and dissemination may be partly dependent on the mesenchymal conversion of cancer cells by HA overexpression. Furthermore, recent studies have enabled postulation of reliable mechanisms by which HA influences tumor growth and invasion by modulating the tumor microenvironment to recruit stromal cells and vasculature. Although the angiogenic function of HA oligosaccharides has been well-established, the anti-angiogenicity of HMW HA being modulated by HA-binding molecules needs further clarification and study.

The roles of HA in cancer progression may differ according to the HAS isoforms expressed, meaning cancer cells at different stages may differentially utilize the three HAS isoforms to maximize their survival. Studies are now in progress to identify exactly which HAS proteins are associated with cancer progression. This will provide an opportunity to develop new strategies for cancer therapy targeting specific cancer-associated HAS species.

ACKNOWLEDGMENTS

This work was supported by grants from the CREST (Core Research for Evolutional Science and Technology) of JST (Japan Science and Technology Agency) and the Shinshu Association for the Advancement of Medical Sciences (N.I). The help of Dr. Shun'ichiro Taniguchi (Shinshu University Graduate School of Medicine) is greatly appreciated.

References

Adamia, S., Reiman, T., Crainie, M., et al. (2005). Intronic splicing of hyaluronan synthase 1 (HAS-1): a biologically relevant indicator of poor outcome in multiple myeloma. *Blood* **105**, 4836–4844.

Auvinen, P., Tammi, R., Parkkinen, J., et al. (2000). Hyaluronan in peritumoral stroma and malignant cells associates with breast cancer spreading and predicts survival. *Am J Pathol* **156**, 529–536.

Bharadwaj, A. G., Rector, K., and Simpson, M. A. (2007). Inducible hyaluronan production reveals differential effects on prostate tumor cell growth and tumor angiogenesis. *J Biol Chem* **282**, 20561–20572.

Bhowmick, N. A. and Moses, H. L. (2005). Tumor–stroma interactions. *Curr Opin Genet Dev* **15**, 97–101.

Bhowmick, N. A., Chytil, A., Plieth, D., et al. (2004). TGF-beta signaling in fibroblasts modulates the oncogenic potential of adjacent epithelia. *Science* **303**, 848–851.

Bourguignon, L. Y., Zhu, H., Shao, L., and Chen, Y. W. (2000). CD44 interaction with tiam1 promotes Rac1 signaling and hyaluronic acid-mediated breast tumor cell migration. *J Biol Chem* **275**, 1829–1838.

Camenisch, T. D., Spicer, A. P., Brehm-Gibson, T., et al. (2000). Disruption of hyaluronan synthase-2 abrogates normal cardiac morphogenesis and hyaluronan-mediated transformation of epithelium to mesenchyme. *J Clin Invest* **106**, 349–360.

Carmeliet, P. (2003). Angiogenesis in health and disease. *Nat Med* **9**, 653–660.

Desmouliere, A., Guyot, C., and Gabbiani, G. (2004). The stroma reaction myofibroblast: a key player in the control of tumor cell behavior. *Int J Dev Biol* **48**, 509–517.

Dong, J., Grunstein, J., Tejada, M., et al. (2004). VEGF-null cells require PDGFR alpha signaling-mediated stromal fibroblast recruitment for tumorigenesis. *Embo J* **23**, 2800–2810.

Enegd, B., King, J. A., Stylli, S., et al. (2002). Overexpression of hyaluronan synthase-2 reduces the tumorigenic potential of glioma cells lacking hyaluronidase activity. *Neurosurgery* **50**, 1311–1318.

Feinberg, R. N. and Beebe, D. C. (1983). Hyaluronate in vasculogenesis. *Science* **220**, 1177–1179.

Fraser, J. R. and Laurent, T. C. (1989). Turnover and metabolism of hyaluronan. Ciba Found Symp 143, 41–53. *discussion* **53–59**, 281–285.

Hamerman, D., Todaro, G. J., and Green, H. (1965). The production of hyaluronate by spontaneously established cell lines and viral transformed lines of fibroblastic origin. *Biochim Biophys Acta* **101**, 343–351.

Herrera, M. B., Bussolati, B., Bruno, S., et al. (2007). Exogenous mesenchymal stem cells localize to the kidney by means of CD44 following acute tubular injury. *Kidney Int* **72**, 430–441.

Hillen, F. and Griffioen, A. W. (2007). Tumour vascularization: sprouting angiogenesis and beyond. *Cancer Metastasis Rev* **26**, 489–502.

Hopwood, J. J. and Dorfman, A. G. (1977). Lycosaminoglycan synthesis by cultured human skin fibroblasts after transformation with simian virus 40. *J Biol Chem* **252**, 4777–4785.

Huber, M. A., Kraut, N., and Beug, H. (2005). Molecular requirements for epithelial–mesenchymal transition during tumor progression. *Curr Opin Cell Biol* **17**, 548–558.

Itano, N. and Kimata, K. (2002). Mammalian hyaluronan synthases. *IUBMB Life* **54**, 195–199.

Itano, N., Atsumi, F., Sawai, T., et al. (2002). Abnormal accumulation of hyaluronan matrix diminishes contact inhibition of cell growth and promotes cell migration. *Proc Natl Acad Sci USA* **99**, 3609–3614.

Itano, N., Sawai, T., Atsumi, F., et al. (2004). Selective expression and functional characteristics of three mammalian hyaluronan synthases in oncogenic malignant transformation. *J Biol Chem* **279**, 18679–18687.

Itano, N., Sawai, T., Yoshida, M., et al. (1999a). Three isoforms of mammalian hyaluronan synthases have distinct enzymatic properties. *J Biol Chem* **274**, 25085–25092.

Itano, N., Sawai, T., Miyaishi, O., and Kimata, K. (1999b). Relationship between hyaluronan production and metastatic potential of mouse mammary carcinoma cells. *Cancer Res* **59**, 2499–2504.

Kalluri, R. and Zeisberg, M. (2006). Fibroblasts in cancer. *Nat Rev Cancer* **6**, 392–401.

Kennedy, C. I., Diegelmann, R. F., Haynes, J. H., and Yager, D. R. (2000). Proinflammatory cytokines differentially regulate hyaluronan synthase isoforms in fetal and adult fibroblasts. *J Pediatr Surg* **35**, 874–879.

Knudson, W. (1996). Tumor-associated hyaluronan. Providing an extracellular matrix that facilitates invasion. *Am J Pathol* **148**, 1721–1726.

Knudson, W., Biswas, C., and Toole, B. P. (1984). Interactions between human tumor cells and fibroblasts stimulate hyaluronate synthesis. *Proc Natl Acad Sci USA* **81**, 6767–6771.

Knudson, W., Biswas, C., Li, X. Q., Nemec, R. E., and Toole, B. P. (1989). The role and regulation of tumour-associated hyaluronan. Ciba Found Symp 143, 150–159. *discussion* **159–169**, 281–285.

Kosaki, R., Watanabe, K., and Yamaguchi, Y. (1999). Overproduction of hyaluronan by expression of the hyaluronan synthase HAS-2 enhances anchorage-independent growth and tumorigenicity. *Cancer Res* **59**, 1141–1145.

Koyama, H., Hibi, T., Isogai, Z., et al. (2007). Hyperproduction of hyaluronan in neu-induced mammary tumor accelerates angiogenesis through stromal cell recruitment: possible involvement of versican/PG-M. *Am J Pathol* **170**, 1086–1099.

Laurent, T. C. and Fraser, J. R. (1992). Hyaluronan. *Faseb J* **6**, 2397–2404.

Leonard, J. G., Hale, A. H., Roll, D. E., Conrad, H. E., and Weber, M. J. (1978). Turnover of cellular carbohydrates in normal and Rous sarcoma virus-transformed cells. *Cancer Res* **38**, 185–188.

Li, Y. and Heldin, P. (2001). Hyaluronan production increases the malignant properties of mesothelioma cells. *Br J Cancer* **85**, 600–607.

Liu, N., Gao, F., Han, Z., et al. (2001). Hyaluronan synthase 3 overexpression promotes the growth of TSU prostate cancer cells. *Cancer Res* **61**, 5207–5214.

Mantovani, A., Schioppa, T., Porta, C., Allavena, P., and Sica, A. (2006). Role of tumor-associated macrophages in tumor progression and invasion. *Cancer Metastasis Rev* **25**, 315–322.

Merrilees, M. J. and Finlay, G. J. (1985). Human tumor cells in culture stimulate glycosaminoglycan synthesis by human skin fibroblasts. *Lab Invest* **53**, 30–36.

Nara, Y., Kato, Y., Torii, Y., et al. (1997). Immunohistochemical localization of extracellular matrix components in human breast tumours with special reference to PG-M/versican. Histochem J **29**, 21–30.

Nishida, Y., Knudson, W., Knudson, C. B., and Ishiguro, N. (2005). Antisense inhibition of hyaluronan synthase-2 in human osteosarcoma cells inhibits hyaluronan retention and tumorigenicity. Exp Cell Res **307**, 194–203.

Orimo, A., Gupta, P. B., Sgroi, D. C., et al. (2005). Stromal fibroblasts present in invasive human breast carcinomas promote tumor growth and angiogenesis through elevated SDF-1/CXCL12 secretion. Cell **121**, 335–348.

Petersen, O. W., Nielsen, H. L., Gudjonsson, T., et al. (2003). Epithelial to mesenchymal transition in human breast cancer can provide a nonmalignant stroma. Am J Pathol **162**, 391–402.

Pienimaki, J. P., Rilla, K., Fulop, C., et al. (2001). Epidermal growth factor activates hyaluronan synthase 2 in epidermal keratinocytes and increases pericellular and intracellular hyaluronan. J Biol Chem **276**, 20428–20435.

Ponting, J., Howell, A., Pye, D., and Kumar, S. (1992). Prognostic relevance of serum hyaluronan levels in patients with breast cancer. Int J Cancer **52**, 873–876.

Recklies, A. D., White, C., Melching, L., and Roughley, P. J. (2001). Differential regulation and expression of hyaluronan synthases in human articular chondrocytes, synovial cells and osteosarcoma cells. Biochem J **354**, 17–24.

Ricciardelli, C., Russell, D. L., Ween, M. P., et al. (2007). Formation of hyaluronan- and versican-rich pericellular matrix by prostate cancer cells promotes cell motility. J Biol Chem **282**, 10814–10825.

Ropponen, K., Tammi, M., Parkkinen, J., et al. (1998). Tumor cell-associated hyaluronan as an unfavorable prognostic factor in colorectal cancer. Cancer Res **58**, 342–347.

Sattar, A., Rooney, P., Kumar, S., et al. (1994). Application of angiogenic oligosaccharides of hyaluronan increases blood vessel numbers in rat skin. J Invest Dermatol **103**, 576–579.

Setala, L. P., Tammi, M. I., Tammi, R. H., et al. (1999). Hyaluronan expression in gastric cancer cells is associated with local and nodal spread and reduced survival rate. Br J Cancer **79**, 1133–1138.

Shekhar, M. P., Pauley, R., and Heppner, G. (2003). Host microenvironment in breast cancer development: extracellular matrix–stromal cell contribution to neoplastic phenotype of epithelial cells in the breast. Breast Cancer Res **5**, 130–135.

Simpson, M. A., Wilson, C. M., Furcht, L. T., et al. (2002a). Manipulation of hyaluronan synthase expression in prostate adenocarcinoma cells alters pericellular matrix retention and adhesion to bone marrow endothelial cells. J Biol Chem **277**, 10050–10057.

Simpson, M. A., Wilson, C. M., and McCarthy, J. B. (2002b). Inhibition of prostate tumor cell hyaluronan synthesis impairs subcutaneous growth and vascularization in immunocompromised mice. Am J Pathol **161**, 849–857.

Slevin, M., Krupinski, J., Kumar, S., and Gaffney, J. (1998). Angiogenic oligosaccharides of hyaluronan induce protein tyrosine kinase activity in endothelial cells and activate a cytoplasmic signal transduction pathway resulting in proliferation. Lab Invest **78**, 987–1003.

Sottile, J. (2004). Regulation of angiogenesis by extracellular matrix. Biochim Biophys Acta **1654**, 13–22.

Sugiyama, Y., Shimada, A., Sayo, T., Sakai, S., and Inoue, S. (1998). Putative hyaluronan synthase mRNA are expressed in mouse skin and TGF-beta upregulates their expression in cultured human skin cells. J Invest Dermatol **110**, 116–121.

Thiery, J. P. and Sleeman, J. P. (2006). Complex networks orchestrate epithelial–mesenchymal transitions. Nat Rev Mol Cell Biol **7**, 131–142.

Toole, B. P., Munaim, S. I., Welles, S., and Knudson, C. B. (1989). Hyaluronate–cell interactions and growth factor regulation of hyaluronate synthesis during limb development. Ciba Found Symp 143, 138–145. discussion **145–149**, 281–285.

Toole, B. P. (2004). Hyaluronan: from extracellular glue to pericellular cue. *Nat Rev Cancer* **4**, 528–539.

Turley, E. A., Noble, P. W., and Bourguignon, L. Y. (2002). Signaling properties of hyaluronan receptors. *J Biol Chem* **277**, 4589–4592.

Udabage, L., Brownlee, G. R., Waltham, M., et al. (2005). Antisense-mediated suppression of hyaluronan synthase 2 inhibits the tumorigenesis and progression of breast cancer. *Cancer Res* **65**, 6139–6150.

Weigel, P. H., Hascall, V. C., and Tammi, M. (1997). Hyaluronan synthases. *J Biol Chem* **272**, 13997–14000.

West, D. C., Hampson, I. N., Arnold, F., and Kumar, S. (1985). Angiogenesis induced by degradation products of hyaluronic acid. *Science* **228**, 1324–1326.

Wight, T. N. (2002). Versican: a versatile extracellular matrix proteoglycan in cell biology. *Curr Opin Cell Biol* **14**, 617–623.

Yabushita, H., Noguchi, M., Kishida, T., et al. (2004). Hyaluronan synthase expression in ovarian cancer. *Oncol Rep* **12**, 739–743.

Yamada, Y., Itano, N., Narimatsu, H., et al. (2004). Elevated transcript level of hyaluronan synthase1 gene correlates with poor prognosis of human colon cancer. *Clin Exp Metastasis* **21**, 57–63.

Yamagata, M. and Kimata, K. (1994). Repression of a malignant cell-substratum adhesion phenotype by inhibiting the production of the anti-adhesive proteoglycan, PG-M/versican. *J Cell Sci* **107** (Pt 9), 2581–2590.

Zhu, H., Mitsuhashi, N., Klein, A., et al. (2006). The role of the hyaluronan receptor CD44 in mesenchymal stem cell migration in the extracellular matrix. *Stem Cells* **24**, 928–935.

Zoltan-Jones, A., Huang, L., Ghatak, S., and Toole, B. P. (2003). Elevated hyaluronan production induces mesenchymal and transformed properties in epithelial cells. *J Biol Chem* **278**, 45801–45810.

HYALURONAN DEGRADATION, THE HYALURONIDASES, AND THE PRODUCTS OF DEGRADATION

CHAPTER

11

Hyaluronidase: Both a Tumor Promoter and Suppressor

Vinata B. Lokeshwar and Marie G. Selzer

INTRODUCTION

Originally termed as the "spreading factor," hyaluronidases (HAases) are present in a variety of toxins and venoms. For example, HAase is the virulent factor of β-hemolytic *Streptococci* and it is also present in the venoms of snake, bee, wasp, scorpion, etc, where it aids in the spread of these venoms in the body (Markovic-Housley et al., 2000; Girish and Kemparaju, 2006; Morey et al., 2006; Nagaraju et al., 2006; Skove et al., 2006). In mammals, testicular HAase present in the sperm acrosome is necessary for the fertilization of the ovum (Gould and Bernstein, 1975). Despite a lot of work on bacterial, invertebrate and testicular HAases, a connection between HAase and cancer was unequivocally established just over a decade ago and the functional significance of HAases in cancer was demonstrated just about a year ago (Lokeshwar et al., 1999; 2005a, b; Jacobsen et al., 2002; Simpson, 2006). In this part of the review, we will focus on the recent advances in our understanding of the role of HAases in cancer.

HYALURONIDASES

HAases are a class of enzymes that predominantly degrade hyaluronic acid (HA). However, HAases can also degrade chondroitin sulfate and chondroitin, albeit at a slower rate (Stern and Jedrzejas, 2006). HAases are endoglycosidases, as they degrade the β-N-acetyl-D-glucosaminidic linkages in the HA polymer. Six HAase genes are present in the human genome and these occur in two linked triplates. HYAL-1, -2 and -3 genes are clustered in the chromosome 3p21.3 locus, whereas HYAL-4, HYAL-P1 and PH-20 (encodes testicular HAase) reside in the chromosome 7q31.3 locus (Csoka et al., 2001). It is likely that the six mammalian HAase genes must have arisen through gene duplication events, since they share a significant amino acid identity. For example, HYAL-1, -2, -3, -4 and PH-20 share ~40% amino acid identity (Stern and Jedrzejas, 2006). Based on their pH activity profiles, HAases are divided into two categories. HYAL-1, -2, and -3 are considered as acidic HAases because they are active at acidic pH. For example, HYAL-1 has a pH optimum around 4.0–4.2 and the enzyme is inactive above pH 5.0 (Lokeshwar et al., 2001). On the contrary, PH-20 is a neutral active HAase as it is active at pH 7.0 (pH activity profile 3.0–9.0) (Franzmann et al., 2003).

Among the six mammalian HAases, HYAL-1, -2 and PH-20 are well-characterized. As described above, PH-20 is necessary for ovum fertilization and several natural and synthetic HAase inhibitors have been tested for their use as contraceptives (Hardy et al., 2004; Suri, 2004; Garg et al., 2005a, b). PH-20, as well as HYAL-2 are glycosyl phosphatidyl-inositol (GPI)-linked proteins. HYAL-2 degrades HA into ~20 kDa oligosaccharides (~25 disaccharide units). HYAL-1 is the serum HAase and is expressed in several somatic tissues (Lepperdinger et al., 2001; Miller, 2003; Chow and Knudson, 2005; Stern and Jedrzejas, 2006). HYAL-1 has also been purified from human urine, where it is expressed as two molecular forms (Csoka et al., 1997). Although HYAL-1 has high specific activity for degrading HA, its concentration in human serum is low (60 ng/ml) (Stern and Jedrzejas, 2006).

Site directed mutagenesis of PH-20, identification of naturally occurring mutations in HYAL-1 and alternatively spliced variants of HYAL-1 and HYAL-3, crystal structure of bee venom HAase and 3-D X-ray structure of bovine PH-20 have revealed the catalytic site of HAases involved in HA degradation (Markovic-Housley et al., 2000; Triggs-Raine et al., 1999; Arming et al., 1997; Lokeshwar et al., 2002; Botzski et al., 2004). The crystal structure of the bee HAase and X-ray structure of bovine PH-20 show that HAases have a classical $(\beta/\alpha)_8$ TIM barrel structure. The dominant feature of the HAase structure is a large grove that extends perpendicular to the barrel axis. In bee HAase, the loops following the β strands 2, 3, and 4 form one wall of the groove, and those of 1, 5, 7 and 7 form the other wall. This groove is large enough to accommodate a hexasaccharide. In bee HAase, the catalytic site that cleaves the glycosaminidic bond between N-acetyl-D-glucosamine and D-glucuronic acid lies in amino acid residues Asp[111] and Glu[113] (Stern and Jedrzejas, 2006). In a substrate-assisted acid-base catalytic mechanism Glu[113] acts as the proton donor, and the N-acetyl group of the substrate acts as the nucleophile. In all six mammalian HAases, this Glu residue is conserved along with the Asp and is believed to be responsible for the substrate cleavage. For example, site directed mutagenesis has identified Glu[148] and Asp[146] in human PH-20 as the important residues involved in the actual catalysis of the glucosaminidic linkage. In HYAL-1 the equivalent residues are Glu[131] and Asp[129]. In addition to the active site, a 30 amino acid sequence that is conserved in all six mammalian HAases and also in the bee HAase, appears to be necessary for HAase activity (Lokeshwar et al., 2002). In HYAL-1, this sequence appears in amino acids 301 to 330. Based on the bee HAase crystal structure, the 30 amino acid sequence (amino acid 313 to 342 in the bee HAase sequence), forms β sheets 6 and 7, α-helix 8 and the loops in between (Markovik-Housley et al., 2000). Thus, this 30 amino acid sequence is an integral part of one of the walls of the substrate binding groove. In addition, in this 30 amino acid sequence, a Trp residue (Trp[333], bee HAase, Trp321 HYAL-1) is conserved in all mammalian and bee HAases and in chitinolytic enzymes and is involved in hydrophobic interaction with the N-acetyl side chain (Markovik-Housley et al., 2000). It is noteworthy that in HYAL-1 and HYAL-3 transcripts, this 30 amino acid sequence is encoded by a separate exon that is alternatively spliced (Lokeshwar et al., 2002).

Among the six mammalian HAases, HYAL-1 is the major tumor-derived HAase and is expressed by a variety of tumor cells. HYAL-1 was initially purified from the urine of patients with high-grade bladder cancer and was shown to be expressed in epithelial cells of bladder and prostate tumors and in head and neck squamous cell carcinoma cells (Lokeshwar et al., 1999; Franzmann et al., 2003).

HAase EXPRESSION IN TUMOR CELLS

Detection and measurement of HAase activity in tissues, body fluids, and cell conditioned media became possible because of an HAase ELISA-like assay developed by Stern and Stern (1992). A modified version of this assay was used by Lokeshwar et al. to measure HAase levels in prostate and bladder carcinoma tissues, cells, and in the urine of bladder cancer patients (Lokeshwar et al., 1996; 1999; 2000; 2001; 2002; Pham et al., 1997; Franzmann et al., 2003; Hautmann et al., 2004; Schroeder et al., 2004). The modified HAase ELISA-like assay is called the HAase test, which involves incubation of tissue extracts, urine, or cell conditioned media on HA-coated microtiter well plates in an HAase assay buffer. Following incubation at 37°C for ~16 hours, the degraded HA is washed off and the HA remaining on the HA-coated plate is detected using a biotinylated bovine nasal cartilage HA binding protein. The HAase present in biological specimens is determined from a standard graph, plotted as HAase (mU/ml) versus O.D.$_{405nm}$. The HAase activity is then normalized to total protein concentration (mg/ml) or to cell number (if assaying cell conditioned media). Using the HAase test and also a substrate (HA)-gel assay, Lokeshwar et al. found that HAase levels are elevated in prostate cancer tissues, when compared to normal prostate and benign prostatic hyperplasia tissues (Lokeshwar et al., 1996). This study also linked for the first time, HAase levels to tumor progression. In that study, HAase levels were found to be elevated 3–7-fold in high-grade (Gleason ≥7) prostate cancer tissues when compared to low-grade (Gleason 5–7) prostate cancer tissues. Metastatic prostate cancer lesions were found to have even higher HAase levels than the high-grade primary tumor (Lokeshwar et al., 1996; 2001). HAase levels are also elevated in high-grade bladder tumor tissues and in the urine of patients with high-grade bladder cancer. HAase levels in low-grade bladder tumor tissues and urine are comparable to those found in normal bladder tissues and urine (Pham et al., 1997; Lokeshwar et al., 2000; 2002; Hautmann et al., 2004; Schroeder et al., 2004). These studies have established a link between HAase and the tumor invasive/metastatic phenotype. In addition to bladder and prostate carcinomas, HAase levels have also been shown to be elevated in the urine of children with Wilms' tumor (Stern et al., 1991). In addition to genito-urinary tumors, HAase levels are elevated in head and neck squamous cell carcinoma, breast tumors, metastatic tumors and glioma cells (Bertrand et al., 1997; Madan et al., 1999; Victor et al., 1999; Godin et al., 2000; Beech et al., 2002; Delpech et al., 2002; Enegd et al., 2002; Franzmann et al., 2003; Junker et al., 2003; Bertrand et al., 2005; Udabage et al., 2005; Paiva et al., 2005; Christopolous et al., 2006).

RT-PCR and cDNA cloning, protein purification, immunoblotting, pH activity profile, and immunohistochemistry have revealed that HYAL-1 is the major tumor-derived HAase expressed in prostate and bladder carcinoma cells. HYAL-1 is an ~55–60 kDa protein consisting of 435 amino acids. In fact HYAL-1 was the first HAase to be recognized as being expressed by tumor cells and its expression correlates with their invasive/metastatic potential (Lokeshwar et al., 1999; 2001). No HYAL-1 expression is observed in the tumor-associated stroma, although HYAL-1 expression appears to correlate and perhaps induce HA production in the tumor-associated stroma (Lokeshwar et al., 2005a, b).

Patients with head and neck squamous cell carcinomas have been shown to have elevated HAase levels in their saliva and HYAL-1 is the major HAase that is expressed in these tumor tissues (Franzmann et al., 2003). However, in addition to HYAL-1, RT-PCR analysis has revealed PH-20 expression in head and neck carcinoma, especially laryngeal carcinoma (Victor et al., 1999; Christopolous et al., 2006; Godin et al., 2000). Interestingly, the pH activity profile of the HAase activity expressed in tumor tissues is similar to that of HYAL-1, and not PH-20 (Franzmann et al., 2003; Christopoulos et al., 2006). HAase levels are also shown to be elevated in breast tumors and RT-PCR analysis has detected the expression of PH-20, HYAL-2 and HYAL-3 in breast cancer tissues (Junker et al., 2003; Udabage et al., 2005). As in the case of prostate and bladder carcinomas, HAase levels in metastatic breast tumors are found to be 4-fold higher than those expressed in primary tumors (Bertrand et al., 1997). Similarly, HAase levels were higher in brain metastatic lesions of carcinomas other than primary glioblastomas (Delpech et al., 2002). Furthermore, there is some evidence that while less invasive breast cancer cells express HAS-3 and HYAL-3, highly invasive cells express HAS-2 and HYAL-2 (Udabage et al., 2005). However, how and why the HA production by HAS-2 and HA degradation by HYAL-2 promote tumor cell invasion, but HA production by HAS-3 and HA degradation by HYAL-3 associates with low-invasive phenotype is unclear. It is noteworthy that in these studies, the expression of HAS and HYAL isoforms was studied only at the transcript level by real time RT-PCR. Given that functionally inactive splice variants of HYAL-1 and HYAL-3 are previously reported (as discussed below), the expression of HYAL genes at the transcript level does not necessarily translate into HAase activity produced by breast cancer or any other cell type. Similar observations regarding HYAL-2 and HYAL-3 expression were reported for endometrial carcinoma. In a relatively small number of endometrial carcinoma specimens ($n = 13$), HYAL-2 and HYAL-3 mRNA expression, determined by real time RT-PCR was found to be >1000- and >30-fold more than HYAL-1, respectively (Paiva et al., 2005).

Contrary to the findings regarding elevated expression of one or more HAases in tumors, it has been shown that the chromosome locus 3p21.3, where HYAL-1, -2 and -3 genes are clustered, is deleted in lung and some breast carcinomas at a higher frequency; however, the tumor suppressor gene in this region is RASSF1 and not a HAase gene (Junker et al., 2003; Csoka et al., 1998; Ji et al., 2002). Nonetheless, it was previously believed that HYAL-1 is a tumor suppressor gene (Csoka et al., 1998; Frost et al., 2000; Stern, 2005). Interestingly, again based on the real time RT-PCR studies, Bertrand et al. reported that HYAL-2 expression correlates with lymphoma diagnosis, but the expression actually decreases in high-grade lymphomas, when compared to low-grade lymphomas (Bertrand et al., 2005).

Taken together, HAase expression appears to be elevated in many carcinomas and the expression correlates with tumor invasiveness. However, in some carcinomas HAase expression depends on the status of the chromosome 3p21.3 locus and may inversely correlate with tumor grade.

HAase FUNCTIONS IN GENITO-URINARY TUMORS

HAase as a Tumor Promoter

Extensive digestion of HA by HAase generates tetrasaccharides, whereas limited digestion generates HA fragments, some of which are angiogenic (3–25 disaccharide units). HA fragments of 10–15 disaccharide units have been shown to stimulate endothelial cell proliferation, adhesion and capillary formation (Lokeshwar and Selzer, 2000; Takahashi et al., 2005). Such angiogenic HA fragments are found in the urine of patients with high-grade bladder cancer, in the tissue extracts of high-grade prostate tumors, and in the saliva of patients with head and neck squamous cell carcinoma, suggesting that the HA–HAase system is active in high-grade invasive tumors (Lokeshwar et al., 1997; 2001; Franzmann et al., 2003).

Recent evidence based on cDNA transfection studies shows that HYAL-1 is involved in tumor growth, muscle infiltration by tumor, and tumor angiogenesis (Lokeshwar et al., 2005a, b; Simpson et al., 2006). Lokeshwar et al. have shown that blocking HYAL-1 expression in bladder and prostate cancer cells decreases tumor cell proliferation by ~4-fold, due to cell cycle arrest in the G2-M phase and decreases their invasive activity. In xenografts, inhibition of HYAL-1 expression resulted in a decrease in tumor growth by 9–17-fold. While HYAL-1 expressing tumors infiltrated muscle and blood vessels, tumors lacking HYAL-1 expression resembled benign neoplasm and had 4–9-fold less microvessel density and smaller capillaries (Lokeshwar et al., 2005a, b). The contribution of HYAL-1 expression to muscle invasion by a bladder tumor has been observed in bladder cancer patients. Aboughalia has shown that HYAL-1 expression in tumor cells exfoliated in urine correlates with tumor invasion into the bladder muscle and beyond (Aboughalia, 2006). It is noteworthy that patients with muscle invasive bladder cancer have poor prognosis, as 60%

of the patients with muscle invasive bladder cancer will have metastasis within 2 years and two-thirds will die within 5 years. Interestingly, HA production by the tumor stroma correlates with HYAL-1 levels in tumor cells, suggesting crosstalk between the tumor and the tumor-associated stroma (Lokeshwar et al., 2001; 2005a, b). Such crosstalk between HA and HYAL-1, with respect to tumor growth and angiogenesis, was recently confirmed by Simpson who tested tumor growth and angiogenesis following the expression of HAS-2 and HYAL-1, either individually or together, in a non-invasive prostate cancer cell line. While HAS-2 or HYAL-1 when expressed individually in a prostate cancer cell line, increased tumor growth and angiogenesis, their co-expression had a synergistic effect on this increase (Simpson, 2006). Expression of HYAL-1 in a human prostate cancer cell line also causes a slight increase in its ability to form lung metastasis in xenograft (Patel et al., 2002).

HAase as a Tumor Suppressor

Contrary to the tumor promoting effects of HYAL-1, a prevalent concept has been that, in general, HAases are tumor suppressors (Csoka et al., 1998; Frost et al., 2000; Stern, 2005). The origin of this concept lies in the observation that in some epithelial carcinomas, the 3p21.3 locus is deleted and although the tumor suppressor gene in this locus was shown not to be an HYAL gene (i.e., HYAL-1, -2, or -3), the concept continued (Csoka et al., 1998; Junker et al., 2003; Stern, 2005; Stern and Jedrzejas, 2006). Perhaps this concept became popular because HA is known to promote tumor metastasis, and therefore, conceptually it was easier to explain that an enzyme that degrades HA was a tumor suppressor. In support of this concept, Jacobson et al. reported that while HAS-2 expression in a rat colon carcinoma line promoted tumor growth, the over-expression of HYAL-1, at levels (220–360 mu/10^6 cells) that are not found in tumor tissues and tumor cells, inhibited tumor growth and generated necrotic tumors (Jacobson et al., 2002). Furthermore, Shuster et al. showed that administration of super high concentrations of bovine testicular HAase (300 units) caused an ~50% regression in breast tumor xenografts (Shuster et al., 2002). The controversy whether HAase is a tumor promoter or a suppressor was recently resolved, when Lokeshwar et al. showed that while HYAL-1 levels that are expressed in tumor tissues and cells promote tumor growth, invasion, and angiogenesis, HAase levels exceeding 100 milliunits/10^6 cells), i.e., at levels that are not naturally expressed by tumor cells, significantly reduce tumor incidence and growth due to induction of apoptosis (Lokeshwar et al., 2005a). Therefore, the function of HAase as a tumor promoter or a suppressor is a concentration-dependent phenomenon, but in tumor tissues, the tumor cell-derived HAase acts mainly as a tumor promoter.

Regulation of HAase Activity

One of the mechanisms to control cellular HAase expression is the loss of the chromosome 3p21.3 locus, which occurs at a higher frequency in some epithelial tumors (Marsit et al., 2004; Hilbe et al., 2006; Pizzi et al., 2005). Alternative mRNA splicing is another mechanism by which HAase activity is regulated. A common internal splicing event occurs in the 5′ untranslated region present in exon 1 (Junker et al., 2003; Frost et al., 2000). This splicing event joins nucleotides 109 and 597. Frost et al. and Junker et al. reported that HYAL-1 protein levels and HAase activity in tumor cells correlate with a HYAL-1 transcript in which this 5′ untranslated region is spliced. Furthermore, HYAL-1 protein is not detected in tumor cells which express a HYAL-1 transcript that retains the 5′ untranslated region. Based on these findings, Frost et al. and Junker et al. concluded that the HYAL-1 transcript containing the 5′ untranslated region is not translated (Frost et al., 2000; Junker et al., 2003). However, it is unclear how and why the 5′-untranslated region in the HYAL-1 mRNA prevents translation. Using normal and bladder tumor tissues and bladder and prostate cancer cells, Lokeshwar et al. have reported several alternatively spliced variants of HYAL-1 and HYAL-3 transcripts. These variants are generated by alternative splicing occurring in the coding regions of HYAL-1 and HYAL-3 transcripts which encode truncated proteins that lack HAase activity (Lokeshwar et al., 2002). For example, five alternatively spliced variants of the HYAL-1 transcript that affect the coding region have been reported. HYAL-1-v1 protein lacks a 30 amino acid stretch between amino acids 300 and 3001 and is generated by alternative splicing of exon 2. The HYAL-1-v2 protein sequence from amino acids 183 to 435 is identical to HYAL-1 and the HYAL-1-v3 protein contains the first 207 amino acids of the HYAL-1 wild type protein. HYAL-1-v4 and HYAL-1-v5 proteins consist of amino acids 260–435 and 340–435, respectively, that are present in the wild type protein. Among the HYAL-3 splice variants, HYAL-3-v1 lacks a 30 amino acid sequence present in the wild type protein and this truncation joins amino acid 298 to 329. HYAL-3-v1 is generated by alternative splicing of exon 3. HYAL-3-v2 encodes a 168 amino acid protein, and this is identical to amino acids 249–417 in the HYAL-3 wild type protein. HYAL-3-v3 protein encodes a 138 amino acid protein that is 100% identical to amino acids 249–417 except that it also lacks the 30 amino acid sequence from 299 to 328. As discussed above, although various splicing events maintain the open reading frame of the HYAL-1 and HYAL-3 proteins, none of these variants are functionally active (Lokeshwar et al., 2002).

Recent data on one of the HYAL-1 variants, HYAL-1-v1, show that the expression of HYAL-1-v1 is higher in normal bladder tissues than in bladder tumor tissues. Furthermore, HYAL-1-v1 expression reduces HAase activity secreted by bladder cancer cells because of a complex formation between HYAL-1 and HYAL-1-v1. HYAL-1-v1 expression induces apoptosis in bladder cancer cells and reduces tumor growth, infiltration and angiogenesis (Lokeshwar et al., 2006). This suggests that a critical balance between the levels of HYAL-1 and HYAL-1 variants may regulate HYAL-1 function in cancer.

REGULATION OF HAase GENE EXPRESSION

Although HAases are important, the regulation of HAase gene expression in normal and cancer cells was unknown until recently. The minimal promoter region for HYAL-2 has been identified; however, the regulation of HYAL-2 expression is unknown (Chow and Knudson, 2005). By deletion analyses Chow and Knudson demonstrated that the region between nucleotides +959 and +1158 (within intron-1) contains the basal promoter elements and the region between nucleotides +224 and +958 contain negative elements that may control the basal expression level of HYAL-2. HYAL-2 promoter lacks a TATA-binding site but contains a GATA-binding region. Recently, Lokeshwar et al. mapped the minimal promoter region of HYAL-1 and showed that HYAL-1 expression in normal and tumor cells is regulated by promoter methylation (Lokeshwar et al., 2008). They identified the minimal promoter region between nucleotides −93 and −38 upstream of the transcription start site. HYAL-1 promoter contains a TACAAA sequence and nucleotides −73 to −50, which contain overlapping binding consensus sites for SP1, Egr-1, and AP-2, are important for promoter activity. In addition $C^{-71}pG$ and $C^{-59}pG$ dinucleotides and a NFκB binding site (at position −15), also appear to be necessary for promoter activity. Although, HYAL-1 promoter lacks a CpG island, methylation at C^{-71} and C^{-59} and differential binding of SP1 to the methylated promoter and Egr-1/Ap-2 binding to the unmethylated promoter appear to regulate HYAL-1 promoter activity. Specifically in non-HYAL-1 expressing cells both C^{-71} and C^{-59} are methylated and SP1 binds to the promoter. But in HYAL-1 expressing cells, C^{-71} and C^{-59} are unmethylated and Egr-1/AP-2 binds to the Regulation of HYAL-1 promoter activity by methylation raises an interesting question about cancer therapeutics involving DNA demethylating agents. Hypermethylation of tumor suppressor genes has been extensively investigated for developing cancer markers and therapeutics. However, if DNA hypomethylation turns on the genes such as HYAL-1 (and also heparanase, uPA, MMP-2, [Ehrlich, 2002]) that function in tumor growth and metastasis, then DNA hypo-methylation-inducing therapies may only have short-term efficacy, as they could speed up the progression of surviving cancer cells (Ehrlich, 2002).

HAase AND SIGNALING

HAase and Cell Cycle Progression

As discussed above, blocking HYAL-1 expression in bladder and prostate cancer cells induces cell cycle arrest in the G2-M phase. G2-M arrest results from the down regulation of the positive regulators of G2-M transition. For example, stable HYAL-1 anti-sense transfectants show down regulation of cdc25c, cyclin B1, and cdk1 levels, as well as, cdk1 kinase activity (Lokeshwar et al., 2005a, b). In HSC3 oral carcinoma cells, HYAL-1 expression caused a 145% increase in the S-phase fraction, with a concomitant decrease in the G0-G1 phase (Lin and Stern, 2001).

The mechanism by which HYAL-1-induces cell cycle transition and up-regulates the levels of positive regulators of G2-M transition is unknown. However, testicular HAase has been shown to induce phosphorylation of c-jun N-terminal kinases (JNK)-1 and -2 and p44/42 ERK in murine fibroblasts cells L929 (Chang, 2001). ERK is required for G2-M and G1-S transitions (Liu et al., 2004). Lokeshwar et al. have previously shown that cell surface interaction between HA oligosaccharides and RHAMM stimulates phosphorylation of p42/p44 ERK (activated p42/44ERK) and focal adhesion kinase in human endothelial cells (Lokeshwar and Selzer, 2000). RHAMM co-immunoprecipitates with src and ERK and contains recognition sequences for these kinases, suggesting a direct interaction (Zhang et al., 1988; Hall et al., 1996). Activated FAK also activates ERK through Grb2 and Shc and PI3 kinase through a direct interaction (McLean et al., 2005; Mitra et al., 2005). It is noteworthy that angiogenic HA fragments are detected in high-grade tumor tissues and in body fluids (e.g., urine and saliva) of cancer patients (Lokeshwar et al., 1994; 2001; Franzmann et al., 2003). In addition to ERK activity, transient activation of JNKs is required for G2-M transition. For example, activated JNK may phosphorylate cdc25c and modulate its activity (Mingo-Sion et al., 2004). Furthermore, activated JNKs phosphorylate c-jun, which then increases cdc2 expression (Goss et al., 2003). However, at the present time it is unknown whether hyaluronidase-mediated regulation of the cell cycle involves JNK and/or ERK pathways.

HAase and Apoptosis

As discussed above, the super high expression of HYAL-1 induces apoptosis in prostate cancer cells. The apoptosis induction by HYAL-1 involves mitochondrial depolarization and induction of a pro-apoptotic protein, WOX1. WOX1 is a ww-domain containing oxidoreductase that contains a nuclear localization signal, a mitochondrial localization signal, and an alcohol dehydrogenase domain (Chang, 2002). Chang has shown that transient transfection of the murine fibroblast line L929, by HYAL-1 or HYAL-2 cDNA or ectopic addition of bovine testicular HAase (100 U/ml) enhances TNF-induced cytotoxicity, which is mediated by increased WOX1 expression and prolonged NK_{KB} activation (Chang, 2001; Chang et al., 2001; 2003). WOX1 is known to induce apoptosis in a p53 independent manner, which involves WOX1 activation (i.e. WOX1-^{P}Tyr33), its translocation to mitochondria and down regulation of anti-apoptotic proteins bcl2 and bclx$_L$ (Chang et al., 2003). Although the kinase, which phosphorylates WOX1, is unknown, JNK1 directly interacts with WOX1 (Chang et al., 2003). JNK is also associated with the mitochondria-mediated apoptotic pathway, as it phosphorylates bcl-2 and bclx$_{XL}$, and suppresses their anti-apoptotic activity (Basu and Haldar, 2003; Deng et al., 2001).

Recently, Lokeshwar et al. have shown that the expression of HYAL-1-v1 in bladder cancer cells, that express wild-type HYAL-1, induces G2-M arrest and apoptosis. HYAL-1 and HYAL-1-v1 form a non-covalent complex, which is enzymatically inactive. The HYAL-1-v1 induced apoptosis involves the extrinsic pathway, since HYAL-1-v1 expression induces activation of caspases -8, -9 and -3, Fas and FADD (Fas associated death domain) upregulation, and BID activation. Moreover, inhibition of Fas expression by Fas siRNA inhibits HYAL-1-v1 induced apoptosis (Ehrlich, 2002). These reports suggest that HYAL-1 and its variants are capable of inducing apoptotic pathways, the understanding of which has only recently begun.

HAase as a Diagnostic and Prognostic Indicator

The diagnostic potential of HAase, either alone or together with HA has been extensively explored in bladder cancer. For example, urinary HAase levels, measured using the HAase test, have been shown to be 3–7-fold elevated among patients with intermediate (G2) and high (G3)-grade bladder cancer when compared to normal individuals, patients with one of the many benign urologic conditions, patients with a history of bladder cancer, and patients with low-grade bladder cancer (Pham et al., 1997). In a study of 513 urine specimens, the HAase test had 81.5% sensitivity, 83.8% specificity, and 82.9% accuracy to detect G2/G3 patients. When the HAase test was combined with the HA test, which measures urinary HA levels, the combined HA–HAase test had higher sensitivity (91.2%) and accuracy (88.3%), and comparable specificity (84.4%) to detect bladder cancer, regardless of the tumor grade and stage (Lokeshwar et al., 2000). In another study, where 70 bladder cancer patients were prospectively followed for a period of 4 years to monitor bladder cancer recurrence, the HA–HAase test had 91% sensitivity and 70% specificity to detect bladder cancer recurrence (Lokeshwar et al., 2002b). More importantly, a patient with a false-positive HA–HAase test had a 10-fold increased risk for developing bladder cancer within 5 months. In a side-by-side comparison, the HA–HAase test was also superior to a variety of FDA-approved bladder tumor markers (Hautmann et al., 2004; Schroeder et al., 2004). Hautmann et al. have shown a correlation between increased tumor-associated HYAL-1 and HA in tumor tissues and a positive HA–HAase test (Chang et al., 2001). This suggests that tumor-associated HYAL-1 and HA are released into the urine when it comes in contact with a tumor in the bladder. In addition to urinary HAase levels, measurement of HYAL-1 mRNA levels in exfoliated cells found in urine also appears to be a marker for bladder cancer. For example, Eissa et al. found that HYAL-1 mRNA expression determined by RT-PCR has >90% accuracy in detecting bladder cancer (Eissa et al., 2005). Furthermore, HYAL-1 mRNA levels measured in exfoliated cells are elevated in patients with invasive and poorly differentiated carcinoma (Aboughalia, 2006). These studies show that HAase is a highly accurate marker for detecting high-grade bladder cancer, and when it is combined with HA, it detects both low-grade and high-grade bladder cancer with ~90% accuracy.

The prognostic potential of HYAL-1 has been explored in prostate cancer. Standard clinical and pathological parameters provide very limited information to clinicians regarding which prostate cancers will progress, and/or have a poor prognosis, and as a result, it is difficult to predict which patients need aggressive treatment, from those for whom watchful waiting would be sufficient. By performing immunohistochemistry on radical prostatectomy specimens, on whom there was a minimum 5-year follow-up, Posey et al. and Ekici et al. found that HYAL-1 is highly expressed in specimens from patients who later had a biochemical recurrence (Posey et al., 2003; Ekici et al., 2004). Biochemical recurrence is defined as increasing serum prostate specific antigen (PSA) levels following radical prostatectomy and is an indicator of disease progression (i.e. either local recurrence or metastasis to distant sites). HYAL-1 staining in radical prostatectomy specimens appears to be an independent predictor of biochemical recurrence. Furthermore, HYAL-1 staining when combined with HA staining has 87% accuracy in predicting disease progression (Ekici et al., 2004). It is noteworthy that in prostate cancer specimens while HYAL-1 is exclusively expressed by tumor cells, HA is mostly expressed by the tumor-associated stroma (Posey et al., 2003). These results show that consistent with the function of HYAL-1 in tumor growth, infiltration, and angiogenesis, it is most likely a prognostic indicator for disease progression.

In a limited number of studies, hyaluronidase expression has also been studied in other carcinomas. For example, there is some evidence that HYAL-1 may be an accurate marker for head and neck squamous cell carcinomas and that salivary HAase levels are elevated in head and neck cancer patients (Franzmann et al., 2003). In addition to HYAL-1, PH-20 mRNA levels have been shown to be elevated in primary and lymph node metastatic lesions of laryngeal carcinoma when compared to normal laryngeal tissues (Christopoulos et al., 2006; Godin et al., 2000; Victor et al., 1999). In contrast to the observations in many other carcinomas, increased HYAL-2 expression inversely correlates with invasion in B-cell lymphomas and may serve as a prognostic indicator (Bertrand et al., 2005).

HAase and Cancer Therapeutics

Testicular HAase has been added in cancer chemotherapy regimens to improve drug penetration. Tumor cells growing in 3-dimensional multi-cellular masses, such as spheroids *in vitro* and solid tumors acquire resistance to chemotherapeutic drugs (i.e. multicellular resistance) (Green et al., 2004). The resistance of multicellular spheroids of EMT-6 to 4-hydro-peroxycyclophosphamide (4-HC) can be abolished by treatment of these spheroids by HAase (St Croix et al., 1988; Kerbel et al., 1996; Croix et al., 1996). Consistent with the findings that HAase is necessary for cell cycle progression (Lokeshwar et al., 2005a, b; Lin and Stern, 2001), HAase treatment increases recruitment of disaggregated cells into the cycling pool, and thus renders them more sensitive to a cell cycle dependent drug (St Croix et al., 1998; Croix et al., 1996; Kerbel et al., 1996). In limited clinical studies, HAase has been used to enhance the efficacy of vinblastin in the treatment of malignant melanoma and Kaposi's sarcoma (Spruss et al., 1995; Smith et al., 1997), boron neutron therapy of glioma (Haselberger et al., 1998; Hobart et al., 1992), intravesical mitomycin treatment for bladder cancer (Hobart et al., 1992; Maier and Baumgartner, 1989), and chemotherapy involving cisplatin and vindesine in the treatment of head and neck squamous cell carcinoma (Klocker et al., 1994; 1998). It is note-worthy that the HAase concentrations ($1 \times 10^5 - 2 \times 10^5$ IU) used in these clinical studies far exceed the amount of HAase present in tumor tissues, and therefore, it is unlikely that at these concentrations the infused HAase will act as a tumor promoter.

In summary, HAase is an endoglycosidase that functions in tumor growth, infiltration, and angiogenesis. At concentrations that are present in tumor tissues, HAase acts as a tumor promoter. However, artificially increasing these concentrations, results in HAase functioning as a tumor suppressor. HYAL-1 type HAase regulates cell cycle progression and apoptosis, and therefore, may regulate tumor growth and angiogenesis. The regulation of HAase in cancer appears to be controlled at the tran-scription level. HAases either alone, or together with HA are potentially accurate diagnostic and prognostic indicators for cancer detection and tumor metastasis. We are only beginning to understand the complex role that this enzyme plays in cancer. In the future because of its role in tumor growth and progression, this enzyme may be targeted for developing novel cancer therapeutics and diagnostics.

ACKNOWLEDGMENT

This research had the following grant support: NIH/NCI RO1 072821-06 (VBL), NIH R01 DK66100 (VBL).

References

Aboughalia, A. H. (2006). Elevation of hyaluronidase-1 and soluble intercellular adhesion molecule-1 helps select bladder cancer patients at risk of invasion. *Arch Med Res* **37**, 109–116.

Arming, S., Strobl, B., Wechselberger, C., and Kreil, G. (1997). In vitro mutagenesis of PH-20 hyaluronidase from human sperm. *Eur J Biochem* **247**, 810–814.

Basu, A. and Haldar, S. (2003). Identification of a novel Bcl-xL phosphorylation site regulating the sensitivity of taxol- or 2-methoxyestradiol-induced apoptosis. *FEBS Lett* **538**, 41–47.

Beech, D. J., Madan, A. K., and Deng, N. (2002). Expression of PH-20 in normal and neoplastic breast tissue. *J Surg Res* **103**, 203–207.

Bertrand, P., Courel, M. N., Maingonnat, C., et al. (2005). Expression of HYAL-2 mRNA, hyaluronan and hyaluronidase in B-cell non-Hodgkin lymphoma: relationship with tumor aggressiveness. *Int J Cancer* **113**, 207–212.

Bertrand, P., Girard, N., Duval, C., et al. (1997). Increased hyaluronidase levels in breast tumor metastases. *Int J Cancer* **73**, 327–346.

Botzki, A., Rigden, D. J., Braun, S., et al. (2004). L-Ascorbic acid 6-hexadecanoate, a potent hyaluronidase inhibitor. X-ray structure and molecular modeling of enzyme-inhibitor complexes. *J Biol Chem* **279**, 45990–45997.

Chang, N. S. (2001). Hyaluronidase activation of c-Jun N-terminal kinase is necessary for protection of L929 fibrosarcoma cells from staurosporine-mediated cell death. *Biochem Biophys Res Commun* **283**, 278–286.

Chang, N. S. (2002). Transforming growth factor-beta1 blocks the enhancement of tumor necrosis factor cytotoxicity by hyaluronidase HYAL-2 in L929 fibroblasts. *BMC Cell Biol* **3**, 8.

Chang, N. S., Doherty, J., and Ensign, A. (2003). JNK1 physically interacts with WW domain-containing oxidoreductase (WOX1) and inhibits WOX1-mediated apoptosis. *J Biol Chem* **278**, 9195–9202.

Chang, N. S., Pratt, N., Heath, J., et al. (2001). Hyaluronidase induction of a WW domain-containing oxidoreductase that enhances tumor necrosis factor cytotoxicity. *J Biol Chem* **276**, 3361–3370.

Chow, G. and Knudson, W. (2005). Characterization of promoter elements of the human HYAL-2 gene. *J Biol Chem* **280**, 26904–26912.

Christopoulos, T. A., Papageorgakopoulou, N., Theocharis, D. A., et al. (2006). Hyaluronidase and CD44 hyaluronan receptor expression in squamous cell laryngeal carcinoma. *Biochim Biophys Acta* **1760**, 1039–1045.

Croix, B. S., Rak, J. W., Kapitain, S., et al. (1996). Reversal by hyaluronidase of adhesion-dependent multicellular drug resistance in mammary carcinoma cells. *J Natl Cancer Inst* **88**, 1285–1296.

Csoka, A. B., Frost, G. I., and Stern, R. (2001). The six hyaluronidase-like genes in the human and mouse genomes. *Matrix Biol* **20**, 499–508.

Csoka, A. B., Frost, G. I., Heng, H. H., et al. (1998). The hyaluronidase gene HYAL-1 maps to chromosome 3p21.2-p21.3 in human and 9F1–F2 in mouse, a conserved candidate tumor suppressor locus. *Genomics* **48**, 63–70.

Csoka, A. B., Frost, G. I., Wong, T., and Stern, R. (1997). Purification and microsequencing of hyaluronidase isozymes from human urine. *FEBS Lett* **417**, 307–310.

Delpech, B., Laquerriere, A., Maingonnat, C., Bertrand, P., and Freger, P. (2002). Hyaluronidase is more elevated in human brain metastases than in primary brain tumours. *Anticancer Res* **22**, 2423–2427.

Deng, X., Xiao, L., Lang, W., et al. (2001). Novel role for JNK as a stress-activated Bcl2 kinase. *J Biol Chem* **276**, 23681–23688.

Ehrlich, M. (2002). DNA methylation in cancer, too much, but also too little. *Oncogene* **21**, 5400–5413.

Eissa, S., Kassim, S. K., Labib, R. A., et al. (2005). Detection of bladder carcinoma by combined testing of urine for hyaluronidase and cytokeratin 20 RNAs. *Cancer* **103**, 1356–1362.

Ekici, S., Cerwinka, W. H., Duncan, R., et al. (2004). Comparison of the prognostic potential of hyaluronic acid, hyaluronidase (HYAL-1), CD44v6 and microvessel density for prostate cancer. *Int J Cancer* **112**, 121–129.

Enegd, B., King, J. A., Stylli, S., et al. (2002). Overexpression of hyaluronan synthase-2 reduces the tumorigenic potential of glioma cells lacking hyaluronidase activity. *Neurosurgery* **50**, 1311–1318.

Franzmann, E. J., Schroeder, G. L., Goodwin, W. J., et al. (2003). Expression of tumor markers hyaluronic acid and hyaluronidase (HYAL-1) in head and neck tumors. *Int J Cancer* **106**, 438–445.

Frost, G. I., Mohapatra, G., Wong, T. M., et al. (2000). HYAL-1LUCA-1, a candidate tumor suppressor gene on chromosome 3p21.3, is inactivated in head and neck squamous cell carcinomas by aberrant splicing of pre-mRNA. *Oncogene* **19**, 870–877.

Garg, A., Anderson, R. A., Zaneveld, L. J., and Garg, S. (2005a). Biological activity assessment of a novel contraceptive antimicrobial agent. *J Androl* **26**, 414–421.

Garg, S., Vermani, K., Garg, A., et al. (2005b). Development and characterization of bioadhesive vaginal films of sodium polystyrene sulfonate (PSS), a novel contraceptive antimicrobial agent. *Pharm Res* **22**, 584–595.

Girish, K. S. and Kemparaju, K. (2006). Inhibition of *Naja naja* venom hyaluronidase: role in the management of poisonous bite. *Life Sci* **78**, 1433–1440.

Godin, D. A., Fitzpatrick, P. C., Scandurro, A. B., et al. (2000). PH-20: a novel tumor marker for laryngeal cancer. *Arch Otolaryngol Head Neck Surg* **126**, 402–404.

Goss, V. L., Cross, J. V., Ma, K., et al. (2003). SAPK/JNK regulates cdc2/cyclin B kinase through phosphorylation and inhibition of cdc25c. *Cell Signal* **15**, 709–718.

Gould, S. F. and Bernstein, M. H. (1975). The localisation of bovine sperm hyaluronidase. *Differentiation* **3**, 123–132.

Green, S. K., Francia, G., Isidoro, C., and Kerbel, R. S. (2004). Antiadhesive antibodies targeting E-cadherin sensitize multicellular tumor spheroids to chemotherapy in vitro. *Mol Cancer Ther* **3**, 149–159.

Hall, C. L., Lange, L. A., Prober, D. A., Zhang, S., and Turley, E. A. (1996). pp60(c-src) is required for cell locomotion regulated by the hyaluronan receptor RHAMM. *Oncogene* **13**, 2213–2224.

Hardy, C. M., Clydesdale, G., Mobbs, K. J., et al. (2004). Assessment of contraceptive vaccines based on recombinant mouse sperm protein PH-20. *Reproduction* **127**, 325–334.

Haselsberger, K., Radner, H., and Pendl, G. (1998). Boron neutron capture therapy for glioblastoma: improvement of boron biodistribution by hyaluronidase. *Cancer Lett* **131**, 109–111.

Hautmann, S. H., Lokeshwar, V. B., Schroeder, G. L., et al. (2001). Elevated tissue expression of hyaluronic acid and hyaluronidase validates the HA–HAase urine test for bladder cancer. *J Urol* **165**, 2068–2074.

Hautmann, S., Toma, M., Lorenzo-Gomez, M. F., et al. (2004). Immunocyte and the HA–HAase urine tests for the detection of bladder cancer, a side-by-side comparison. Eur Urol 46, 466–471.

Hilbe, W., Auberger, J., Dirnhofer, S., et al. (2006). High rate of molecular alteration in histologically tumour-free bronchial epithelium of NSCLC patients detected by multicolour fluorescence in situ hybridisation. Oncol Rep 15, 1233–1240.

Hobart, K., Maier, U., and Marberger, M. (1992). Topical chemoprophylaxis of superficial bladder cancer with mitomycin C and adjuvant hyaluronidase. Eur Urol 21, 206–210.

Jacobson, A., Rahmanian, M., Rubin, K., and Heldin, P. (2002). Expression of hyaluronan synthase 2 or hyaluronidase 1 differentially affect the growth rate of transplantable colon carcinoma cell tumors. Int J Cancer 102, 212–219.

Ji, L., Nishizaki, M., Gao, B., et al. (2002). Expression of several genes in the human chromosome 3p21.3 homozygous deletion region by an adenovirus vector results in tumor suppressor activities in vitro and in vivo. Cancer Res 62, 2715–2720.

Junker, N., Latini, S., Petersen, L. N., and Kristjansen, P. E. (2003). Expression and regulation patterns of hyaluronidases in small cell lung cancer and glioma lines. Oncol Rep 10, 609.

Kerbel, R. S., St Croix, B., Florenes, V. A., and Rak, J. (1996). Induction and reversal of cell adhesion-dependent multicellular drug resistance in solid breast tumors. Hum Cell 9, 257–264.

Klocker, J., Sabitzer, H., Raunik, W., Wieser, S., and Schumer, J. (1995). Combined application of cisplatin, vindesine, hyaluronidase and radiation for treatment of advanced squamous cell carcinoma of the head and neck. Am J Clin Oncol 18, 425–428.

Klocker, J., Sabitzer, H., Raunik, W., Wieser, S., and Schumer, J. (1998). Hyaluronidase as additive to induction chemotherapy in advanced squamous cell carcinoma of the head and neck. Cancer Lett 131, 113–115.

Lepperdinger, G., Mullegger, J., and Kreil, G. (2001). HYAL-2 – less active, but more versatile? Matrix Biol 20, 509–514.

Lin, G. and Stern, R. (2001). Plasma hyaluronidase (HYAL-1) promotes tumor cell cycling. Cancer Lett 163, 95–101.

Liu, X., Yan, S., Zhou, T., Terada, Y., and Erikson, R. L. (2004). The MAP kinase pathway is required for entry into mitosis and cell survival. Oncogene 23, 763–776.

Lokeshwar, V. B. and Selzer, M. G. (2000). Differences in hyaluronic acid-mediated functions and signaling in arterial, microvessel, and vein-derived human endothelial cells. J Biol Chem 275, 27641–27649.

Lokeshwar, V. B., Cerwinka, W. H., and Lokeshwar, B. L. (2005a). HYAL-1 hyaluronidase: a molecular determinant of bladder tumor growth and invasion. Cancer Res 65, 2243–2250.

Lokeshwar, V. B., Cerwinka, W. H., Isoyama, T., and Lokeshwar, B. L. (2005b). HYAL-1 hyaluronidase in prostate cancer: a tumor promoter and suppressor. Cancer Res 65, 7782–7789.

Lokeshwar, V. B., Estrella, V., Lopez, L., et al. (2006). HYAL-1-v1, an alternatively spliced variant of HYAL-1 hyaluronidase: a negative regulator of bladder cancer. Cancer Res 66, 11219–11227.

Lokeshwar, V. B., Gomez, P., Kramer, M., et al. (2008). Epigenetic regulation of HYAL-1 hyaluronidase expression, identification of HYAL-1 promoter. J Biol Chem. Oct 24;283 (43): 29215–29227. Epub 2008 Aug 21.

Lokeshwar, V. B., Lokeshwar, B. L., Pham, H. T., and Block, N. L. (1996). Association of elevated levels of hyaluronidase, a matrix-degrading enzyme, with prostate cancer progression. Cancer Res 56, 651–657.

Lokeshwar, V. B., Obek, C., Pham, H. T., et al. (2000). Urinary hyaluronic acid and hyaluronidase: markers for bladder cancer detection and evaluation of grade. J Urol 163, 348–356.

Lokeshwar, V. B., Obek, C., Soloway, M. S., and Block, N. L. (1997). Tumor-associated hyaluronic acid: a new sensitive and specific urine marker for bladder cancer. Cancer Res 57, 773–777. Erratum in: Cancer Res (1998). 58, 3191.

Lokeshwar, V. B., Rubinowicz, D., Schroeder, G. L., et al. (2001). Stromal and epithelial expression of tumor markers hyaluronic acid and HYAL-1 hyaluronidase in prostate cancer. *J Biol Chem* **276**, 11922–11932.

Lokeshwar, V. B., Schroeder, G. L., Carey, R. I., Soloway, M. S., and Iida, N. (2002a). Regulation of hyaluronidase activity by alternative mRNA splicing. *J Biol Chem* **277**, 33654–33663.

Lokeshwar, V. B., Schroeder, G. L., Selzer, M. G., et al. (2002b). Bladder tumor markers for monitoring recurrence and screening comparison of hyaluronic acid–hyaluronidase and BTA-Stat tests. *Cancer* **95**, 61–72.

Lokeshwar, V. B., Young, M. J., Goudarzi, G., et al. (1999). Identification of bladder tumor-derived hyaluronidase: its similarity to HYAL-1. *Cancer Res* **59**, 4464–4470.

Madan, A. K., Yu, K., Dhurandhar, N., Cullinane, C., Pang, Y., and Beech, D. J. (1999). Association of hyaluronidase and breast adenocarcinoma invasiveness. *Oncol Rep* **6**, 607–609.

Maier, U. and Baumgartner, G. (1989). Metaphylactic effect of mitomycin C with and without hyaluronidase after transurethral resection of bladder cancer: randomized trial. *J Urol* **141**, 529–530.

Markovic-Housley, Z., Miglierini, G., et al. (2000). Crystal structure of hyaluronidase, a major allergen of bee venom. *Structure* **8**, 1025–1035.

Marsit, C. J., Hasegawa, M., Hirao, T., et al. (2004). Loss of heterozygosity of chromosome 3p21 is associated with mutant TP53 and better patient survival in non-small-cell lung cancer. *Cancer Res* **64**, 8702–8707.

McLean, G. W., Carragher, N. O., Avizienyte, E., et al. (2005). The role of focal-adhesion kinase in cancer – a new therapeutic opportunity. *Nat Rev Cancer* **5**, 505–515.

Miller, A. D. (2003). Identification of HYAL-2 as the cell-surface receptor for jaagsiekte sheep retrovirus and ovine nasal adenocarcinoma virus. *Curr Top Microbiol Immunol* **275**, 179–199.

Mingo-Sion, A. M., Marietta, P. M., Koller, E., Wolf, D. M., and Van Den Berg, C. L. (2004). Inhibition of JNK reduces G2/M transit independent of p53, leading to endo-reduplication, decreased proliferation, and apoptosis in breast cancer cells. *Oncogene* **23**, 596–604.

Mitra, S. K., Hanson, D. A., and Schlaepfer, D. D. (2005). Focal adhesion kinase, in command and control of cell motility. *Nat Rev Mol Cell Biol* **6**, 56–68.

Morey, S. S., Kiran, K. M., and Gadag, J. R. (2006). Purification and properties of hyaluronidase from *Palamneus gravimanus* (Indian black scorpion) venom. *Toxicon* **47**, 188–195.

Nagaraju, S., Mahadeswaraswamy, Y. H., Girish, K. S. and Kemparaju, K. (2006). Venom from spiders of the genus Hippasa: biochemical and pharmacological studies. *Comp Biochem Physiol C Toxicol Pharmacol*. Sep; 144(1): 1–9. Epub 2006 May 22.

Paiva, P., Van Damme, M. P., Tellbach, M., et al. (2005). Expression patterns of hyaluronan, hyaluronan synthases and hyaluronidases indicate a role for hyaluronan in the progression of endometrial cancer. *Gynecol Oncol* **98**, 193–202.

Patel, S., Turner, P. R., Stubberfield, C., et al. (2002). Hyaluronidase gene profiling and role of HYAL-1 overexpression in an orthotopic model of prostate cancer. *Int J Cancer* **97**, 416–424. Erratum in: *Int J Cancer* **98**, 957.

Pham, H. T., Block, N. L., and Lokeshwar, V. B. (1997). Tumor-derived hyaluronidase: a diagnostic urine marker for high-grade bladder cancer. *Cancer Res* 57, 778–783. Erratum in: *Cancer Res* **57**, 1622.

Pillwein, K., Fuiko, R., Slavc, I., et al. (1998). Hyaluronidase additional to standard chemotherapy improves outcome for children with malignant brain tumors. *Cancer Lett* **131**, 101–108.

Pizzi, S., Azzoni, C., Bottarelli, L., et al. (2005). RASSF1A promoter methylation and 3p21.3 loss of heterozygosity are features of foregut, but not midgut and hindgut, malignant endocrine tumours. *J Pathol* **206**, 409–416.

Posey, J. T., Soloway, M. S., Ekici, S., et al. (2003). Evaluation of the prognostic potential of hyaluronic acid and hyaluronidase (HYAL-1) for prostate cancer. *Cancer Res* **63**, 2638–2644.

Schroeder, G. L., Lorenzo-Gomez, M. F., Hautmann, S. H., et al. (2004). A side by side comparison of cytology and biomarkers for bladder cancer detection. *J Urol* **172**, 1123–1126.

Shuster, S., Frost, G. I., Csoka, A. B., Formby, B., and Stern, R. (2002). Hyaluronidase reduces human breast cancer xenografts in SCID mice. *Int J Cancer* **102**, 192–197.

Simpson, M. A. (2006). Concurrent expression of hyaluronan biosynthetic and processing enzymes promotes growth and vascularization of prostate tumors in mice. *Am J Pathol* **169**, 247–257.

Skov, L. K., Seppala, U., Coen, J. J., et al. (2006). At 2.0 Angstrom resolution: structural analysis of an allergenic hyaluronidase from wasp venom. *Acta Crystallogr D Biol Crystallogr* **62**. *Structure of recombinant Ves* **v. 2**, 595–604.

Smith, K. J., Skelton, H. G., Turiansky, G., and Wagner, K. F. (1997). Hyaluronidase enhances the therapeutic effect of vinblastine in intralesional treatment of Kaposi's sarcoma. Military Medical Consortium for the Advancement of Retroviral Research (MMCARR). *J Am Acad Dermatol* **6** (2 Pt 1), 239–242.

Spruss, T., Bernhardt, G., Schonenberger, H., and Schiess, W. (1995). Hyaluronidase significantly enhances the efficacy of regional vinblastine chemotherapy of malignant melanoma. *J Cancer Res Clin Oncol* **121**, 193–202.

St Croix, B., Man, S., and Kerbel, R. S. (1998). Reversal of intrinsic and acquired forms of drug resistance by hyaluronidase treatment of solid tumors. *Cancer Lett* **131**, 35–44.

Stern, M. and Stern, R. (1992). An ELISA-like assay for hyaluronidase and hyaluronidase inhibitors. *Matrix* **12**, 397–403.

Stern, M., Longaker, M. T., Adzick, N. S., Harrison, M. R., and Stern, R. (1991). Hyaluronidase levels in urine from Wilms' tumor patients. *J Natl Cancer Inst* **83**, 1569–1574.

Stern, R. (2005). Hyaluronan metabolism: a major paradox in cancer biology. *Pathol Biol (Paris)* **53**, 372–382.

Stern, R. and Jedrzejas, M. J. (2006). Hyaluronidases: their genomics, structures, and mechanisms of action. *Chem Rev* **106**, 818–839.

Suri, A. (2004). Sperm specific proteins – potential candidate molecules for fertility control. *Reprod Biol Endocrinol* **10** (2), 10.

Takahashi, Y., Li, L., Kamiryo, M., et al. (2005). Hyaluronan fragments induce endothelial cell differentiation in a CD44- and CXCL1/GRO1-dependent manner. *J Biol Chem* **280**, 24195–24204.

Triggs-Raine, B., Salo, T. J., Zhang, H., Wicklow, B. A., and Natowicz, M. R. (1999). Mutations in HYAL-1, a member of a tandemly distributed multigene family encoding disparate hyaluronidase activities, cause a newly described lysosomal disorder, mucopolysaccharidosis IX. *Proc Natl Acad Sci USA* **96**, 6296–6300.

Udabage, L., Brownlee, G. R., Nilsson, S. K., and Brown, T. J. (2005). The over-expression of HAS-2, HYAL-2 and CD44 is implicated in the invasiveness of breast cancer. *Exp Cell Res* **310**, 205–217.

Victor, R., Chauzy, C., and Girard, N. (1999). Human breast-cancer metastasis formation in a nude-mouse model, studies of hyaluronidase, hyaluronan and hyaluronan-binding sites in metastatic cells. *Int J Cancer* **82**, 77–83.

Zhang, S., Chang, M. C., Zylka, D., et al. (1998). The hyaluronan receptor RHAMM regulates extracellular-regulated kinase. *J Biol Chem* **273**, 11342–11348.

CHAPTER

12

Hyaluronidases in Cancer Biology

Robert Stern

INTRODUCTION

Given the constant synthesis and degradation of hyaluronan (HA) in tissues, it is remarkable that the body maintains precise levels of HA as tightly as it does. Surprisingly, elevated amounts of HA correlate with several types of malignancies. Levels of HA within cancer cells, on the surface of tumor cells, and in their surrounding stroma can increase with malignant progression. On the other hand, levels of hyaluronidases, the enzymes that degrade HA, are variable in cancer; in some cases elevated and in others suppressed, relative to normal tissues.

A relatively simple explanation is that the mere presence of a lysosomal hyaluronidase enzyme is not sufficient for it to degrade its substrate. Older studies on lysosomal hyaluronidases in fetal development conclude that the substrate must be presented to the enzyme in an acidic environment (<pH 4.5) in order to support enzymatic depolymerization. While the presence of a lysosomal hyaluronidase such as HYAL-1 is found in circulation at higher levels than most tissues (4–6 U/ml) it is clear that this is insufficient to tip the balance for local HA catabolism. Thus, a fundamental question remains towards just what level of acid active enzyme production is necessary to modulate local HA catabolism, and what additional tools exist to modulate HA levels, such as HA synthase (HAS) levels, receptor activation, the presence of enzyme inhibitors, or the immobilization of extracellular HA with aggregating proteoglycans.

While studies in mice have clearly demonstrated that an HAS-2 is the *sine qua non* of the fetal HA system based upon embryonic lethality, no such smoking gun has been identified on the catabolic side with regard to the confirmed human hyaluronidases, HYAL-1, HYAL-2 and PH-20. Rather, these enzyme knockouts are all viable with phenotypes that do not explain the high rate of turnover of HA. The mechanisms involved in HA catabolism in development homeostasis and malignancy are just beginning to be uncovered.

The hyaluronidase enzymes can correlate with cancer progression, but can also function as tumor suppressors. This anomaly can be explained in several ways. The HA exists in a high molecular weight form as well as in a myriad of lower molecular size intra- and extracellular sizes. While high molecular weight HA is a reflection of intact, healthy tissues, the fragmented forms, indicators of distress signals, occur in abundance in malignancies. They promote angiogenesis, stimulate production of inflammatory cytokines, and activate signaling pathways that are critical for cancer progression. These fragments may be truncated products of the synthetic reaction, but may also be the result of hyaluronidase activities. How the levels of these enzymes are modulated is not known. Free, unfettered hyaluronidase activity would create great havoc in tissues and cells. It is obvious that such activities must be finely controlled. But the means by which this is accomplished is not known. Potent hyaluronidase inhibitory activities have been detected in tissue extracts, as well as inhibitors that are unique to cancer. Regulation of enzyme transcription, translation, including the production of various splices variants, and the competition between these various forms are additional sites of potential control. A review of the various aspects of this important class of enzymes, including what is known of their modulation, is presented here.

HYALURONIDASES AND THEIR STRUCTURE

The HYAL enzymes degrade predominantly HA. The term is a misnomer, since they have the limited ability to also degrade chondroitin and chondroitin sulfates. In vertebrate tissues, they are present in exceedingly low concentrations, occurring, e.g. at 60 ng/ml in human serum, but possess extraordinarily high specific activities. The first somatic hyaluronidase to be isolated was purified from out-dated human plasma, and was named HYAL-1 (Frost et al., 1997). From the EST (expressed sequence tag) database, it was established that there are six such sequences in the human genome, three at chromosome 3p21.3, and another three at 7q31.3 (Csoka et al., 1999; 2001). Intriguingly, both of these loci occur at sites of putative tumor suppressor genes (TSGs) (Csoka et al., 1998; Edelson et al., 1997; Mateo et al., 1999). That at position 3p21.3 was located initially by positional cloning. The one termed HYAL-1 had identity to a sequence previously termed LuCa1 (Lung Cancer 1), so named because of loss of heterozygosity or homozygous deletion that occur in most lung cancers. But subsequent work documents that the HYAL's are not the critical or the only tumor suppressor gene products at that locus (Zabarovsky et al., 2002). Multiple 3p21.3 TSGs occur there, and tumor acquired promoter DNA methylation is an epigenetic mechanism for inactivating the expression of many of these genes in human malignancy. Thus, both genetic and epigenetic abnormalities of several genes residing in chromosome region 3p21.3 are important for the development of cancer, but it is still obscure how many of them exist and which of the numerous candidate TSGs are the key players in cancer pathogenesis.

Loss of hyaluronidase does provide the cancer cell, however, with an HA-rich environment that can stimulate growth and motility, and facilitate metastatic spread. The HYAL-1 protein as well as the mRNA exists in multiple forms, and these various isomers may compete with each other in the net expression of activity. The protein can exist in two major isoforms, one being the product of two endoproteolytic cleavage reactions that eliminate 99 amino acids from the molecule (Csoka et al., 1997).

Two species of HYAL-1 transcripts can be identified by RT-PCR when primers are used that include the 5' untranslated region. The predominant mRNA species does not correlate with protein translation and contains a retained intron. The second spliced form lacking this intron occurs that produces HYAL-1 protein. Inactivation of HYAL-1 in those cases is a result of incomplete splicing of its pre-mRNA that appears to be epigenetic in nature. Multiple stop codons are contained in the retained intron that prevent translation. Both isoforms occurs in most tissues, but the relative ratios between them vary widely. This inactivation of HYAL-1 occurs at the RNA level, indicating that not all TSGs are of DNA origin (Frost et al., 2000; Junker et al., 2003).

Lokeshwar and colleagues (2002) report several additional alternatively spliced variants of both HYAL-1 and HYAL-3 occurring in the coding region, including proteins that lack enzymatic activity. All of these enzyme protein variants may be competing with each other for effective expression of activity.

Several mammalian species are reported to lack HYAL-1 activity (Fiszer-Szarfarz et al., 1990). This would suggest that HYAL-1 is not an important enzyme, and that other hyaluronidases may be able to substitute for the enzyme. However, an inactive higher molecular weight form can be observed in the circulation of such animals by immuno-staining on SPAGE using HYAL-1 specific polyclonal antibodies (R. Stern, unpublished observations). This suggests that incomplete processing of circulating enzyme occurs in some mammalian species. Intracellular expression in endothelium rather than liver secretion has also been reported in these species (bovine, rabbit, etc.)

A CATABOLIC SCHEME FOR HYALURONAN

Hyaluronan has an extraordinarily high rate of turnover in vertebrates. In the human, 5 g of the 15 g of total body HA turnover daily, mostly the result of the HYAL enzymes, through three catabolic pathways. At the cellular level, the two predominant HYALs are HYAL-1 and HYAL-2, degrading the HA to progressively smaller fragments. Hyaluronan deposition and turnover is even more abundant and more rapid in malignant tissues. The proportion of low molecular weight (LMW) fragments of HA is greater in tumors and tumor patients than in the normal (Kumar et al., 1989; Lokeshwar et al., 1997).

A putative cellular scheme for HA catabolism was recently formulated (Stern, 2003; 2004a). A very simplified version of this scheme is presented in Fig. 12.1, provided originally in the Glycoforum website, Science of Hyaluronan www.glycoforum.gr.jp (Stern, 2004b).

The polymer is degraded in a series of steps generating ever-smaller fragments. The variously sized fragments have a wide and occasionally opposing spectrum of biological activities. Cell surface HYAL-2 makes an initial cut of high molecular weight (HMW) matrix HA generating fragments of 50 to 100 saccharides. These intermediate-sized fragments enter early endosomes, become lysosomal, and degrade further to predominantly tetrasaccharides by lysosomal, acid-active, HYAL-1. The controls that permit accumulation of particular sized fragments are entirely unknown.

THE WARBURG EFFECT

The ability of cancer cells to use anaerobic metabolism and to generate lactate, even in the presence of adequate oxygen, is known as the Warburg

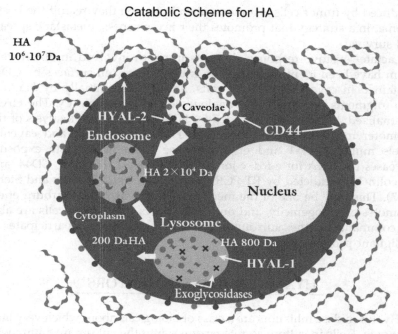

FIGURE 12.1 High molecular weight HA from the extracellular matrix (ECM) is bound to cell surfaces by the combined effects of HYAL-2 and CD44. The clustered complex is guided into caveolae-rich lipid rafts. There, in association with Na^+-H^+ Exchange Protein-1 (not shown), the HA is internalized, and the HA is cleaved to LMW fragments. Such fragments are delivered into early endosomes, and finally to lysosomes. The combined effects of the acid-active β-endoglycosidase HYAL-1, and the two acid-active β-exoglycosidases, β-glucuronidase and β-N-acetylglucosaminidase degrade the HA to tetrasaccharides, and ultimately to individual sugars. Such sugars can then exit lysosomes to enter the cytoplasm where they participate in other cellular reactions. Considering that 5 g of HA are turned over daily in the average 70 Kg individual, the amounts of sugar so processed are not trivial. It is not clear at what point the HA from the ECM becomes intracellular. Additionally, a certain amount of synthesized HA remains intracellular without ever becoming a component of the ECM. How this second HA compartments catabolized is unknown. (Courtesy of Glycoforum website. Science of Hyaluronan. www.glycoforum.gr.jp)

effect (Warburg et al., 1924; Warburg, 1954). Lactate added to cultured fibroblasts increases their HA production (Stern et al., 2002). The phenomenon is dose-dependent at physiological levels, between zero and 10 mM lactate. Lactate also increases expression of CD44, a transmembrane glycoprotein, and the predominant HA receptor on cell surfaces.

The stroma that surrounds carcinomas has increased HA, providing an environment that promotes the growth and motility of cancer cells. Levels of HA often correlate with degree of malignancy. The lactate

V. HYALURONAN DEGRADATION, THE HYALURONIDASES

produced by tumor cells is a mechanism by which they recruit the host's stroma, in a strategy that promotes their invasiveness, metastatic spread, and survival.

Lactate-sensitive response elements in genes involved in HA metabolism have been identified. The HA is tethered to cell surfaces by CD44 contained in caveolin-rich lipid rafts, and cleaved initially by HYAL-2 into fragments that are highly angiogenic and inflammatory. The HA is internalized, and further degraded by HYAL-1. Sequence analysis of the promoter regions of genes for CD44, HYAL-2, HYAL-1, and caveolin revels multiple AP-1 and ets-1 response elements. Lactate exposure increases the RNA for c-fos, c-jun, c-ets-1, HYAL-1, HYAL-2, CD44, and caveolin-1, as indicted by RT-PCR (Stern et al., 2002; Formby and Stern, 2003). This may be one of the mechanisms by which the Warburg effect promotes carcinogenicity, and one of the ways that cancer cells are able to commandeer the surrounding normal stroma to participate in malignant progression.

HYALURONIDASE INHIBITORS

Hyaluronidase inhibitors are a class of biologicals about which very little is known. Cells in culture secrete enzymes into the culture medium away from the cells, but this does not occur in tissues. The production of unopposed hyaluronidase activity would be very destructive in tissues. Potent hyaluronidase inhibitors can be detected in tissue extracts. These are obvious during the process of enzyme purification, the apparent number of total units of activity rising significantly after the initial steps of isolation. The hyaluronidase inhibitors may parallel the careful control of the matrix metalloproteinases (MMPs), another class of matrix-degrading enzymes, by their tissue inhibitors (TIMPs) (Nagase and Visse, 2006). The hyaluronidase inhibitors were first detected over 60 years ago in blood (Haas, 1946; Dorfman et al., 1948). These activities are magnesium dependent, are acute phase substances synthesized by the liver, some of which are members of the Kunitz type of inter-alpha-inhibitor family (Mio and Stern, 2000; Mio et al., 2000).

An ever-present inhibitor of HYAL-2 activity on cell surfaces would have to be invoked in normal, healthy tissues, to preserve ECM integrity. A combination of CD44, HYAL-2, and the $Na^+ H^-$ exchanger 1 (NHE1) are complexed within lipid rafts on cell surfaces for binding HMW HA (Bourguignon et al., 2004). Alternatively, a block of any one of these components could function as a potential inhibitor, for preserving HA integrity.

An entirely different class of inhibitors of hyaluronidases is found in the circulation of cancer patients (Kiriluk et al., 1950; Kolárová, 1975), activities that are independent of magnesium (Fiszer-Szafarz, 1968). They have not been further characterized, but are clearly important for cancer progression. Using the theoretical scheme for HA metabolism, invoking an inhibitor of HYAL-1 would permit accumulation of intermediate-sized HA fragments, the products of HYAL-2 cleavage. These stimulate the angiogenesis necessary for cancer viability after a primary tumor or its metastases have grown past a critical stage. Such fragments also induce the inflammatory reaction that facilitates tumor growth (Coussens and Werb, 2002).

HYALURONIDASE MODULATES CD44 EXPRESSION

CD44 has multiple isoforms generated by alternative exon splicing of a single gene. CD44 and many of its variants are expressed on cancer cells, but the mechanisms by which splice variant exons are selected are unknown. The presence of HA in the environment of the cell appears to influence that selection process. The expression of particular splice variants of CD44 as well as the simultaneous presence of HA is important for motility, invasion, and the metastatic spread of some tumors. The influence of hyaluronidase digestion on the expression of CD44 was examined in a number of human cancer cell lines. CD44 isoforms containing alternatively spliced exons are sensitive to hyaluronidase digestion in all the cell lines examined. However expression of CD44s, the standard form, is resistant to digestion in most cancer cell lines. A tentative model proposes that CD44 isoform variants are unstable, requiring the continuous presence of the HA ligand for expression. On the other hand CD44s are relatively more stable, not requiring the continuous presence of HA or its expression (Stern et al., 2001). A diagrammatic presentation is given in the accompanying figure. Evidence indicates that CD44 variant expression and stabilization requires phosphorylation events catalyzed by casein kinase II (Stern et al., 2001). Stabilization of CD44 occurs through phosphorylation of ser 323 and 325 on the intracellular portion of the transmembrane molecule (Formby and Stern, 1998).

HYALURONAN FRAGMENTATION

Hyaluronan exists not only in an HMW form but also in a number of discreet LMW sizes that have a wide variety of biological activities (Stern, 2006). Tumor cells secrete abundant levels of HYAL-1. The enzyme generates HA fragments that in turn, induce proteases that cleave the tumor cells own CD44 while also stimulating motility. This is an autocrine, paracrine process that increases the tumor cells own malignant potential (Sugahara et al., 2003; 2006). This re-enforces the concept that HA

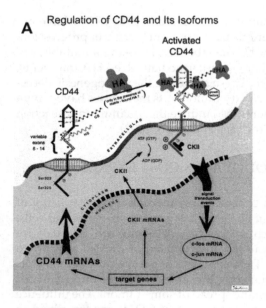

A Regulation of CD44 and Its Isoforms

B Steps in the Downregulation of CD44 in the Presence of Hyaluronidase

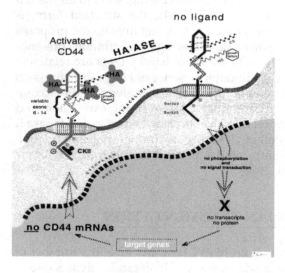

FIGURE 12.2 CD44 exists in a great number of isofoms, in part, the result of expression of ten variable exons within the single *CD44* gene. CD44s, the standard form which contains none of the variable exons, is a stable isoform that does not require the continuous presence of extracellular HA for expression. Other CD44 isoforms, and in particular some that are cancer-associated, are highly unstable transcripts requiring the continuous presence of HA for their expression. Ser-323 and Ser-325 of the intracellular component of transmembrane CD44 are sites of phosphorylation (Neame and Isacke, 1992). Such phosphorylation is catalyzed in part by casein kinase II. Additional kinases have also been detected (B. Formby and R. Stern, unpublished observations). Signal transduction events lead to increased levels of transcripts for c-fos, c-jun, and c-ets mRNAs (A).

In the presence of hyaluronidase activity, HA of the ECM is degraded. The unstable CD44 isoforms requiring the continuous presence of HA for their expression become downregulated when HA is absent or degraded. There is no phosphorylation of the Ser-323 or Ser-325 sites, no signal transduction or transcription of target genes, including the genes for certain CD44 transcripts containing critical exon variants (B).

fragments facilitate cancer progression. Additionally, the autocrine mechanism described above indicates that this malignant progression is self-stimulating without the necessity of external signals.

The size range of HA oligosaccharides that induce cleavage in this cultured pancreatic tumor cell system was carefully investigated. Added exogenously, 6–14 saccharides induced maximal cleavage in a dose-dependent manner. The CD44 theoretically released into the circulation might be part of a strategy for the cancer to become independent of CD44-related control mechanisms, providing a circulating CD44 that would compete with cell-bound CD44. Enhanced CD44 cleavage has been demonstrated in gliomas, breast, colon, and ovarian cancers, and in non-small cell carcinomas of the lung (Okamoto et al., 2002).

Human tumors inoculated into immunodeficient (SCID) mice regress when treated with PH-20. The chromosomal loci at 3p21.3 and 7q31.3 are both TSG sites. It is tempting to assume that HYALs are TSG products, and must be eliminated for tumor progression. Together with the observation that 100% of human small cell (oat cell) carcinomas, and 80% of bronchogenic carcinomas carry deletions at the 3p21.3 site makes the argument all the more attractive. However, 3p21.3 is a gene-rich locus, and more recent data indicate that the HYALs are not the identity of the TSG products as was assumed initially (Wong et al., 2006; Oh et al., 2007).

Highly invasive bladder cancers generate HA fragments in the 30–50 saccharide range, sizes that are highly mitogenic for endothelial cells, and therefore very angiogenic (Lokeshwar, 1997). This may be the mechanism that supports their invasiveness. Similar size fragments also activate tumor integrins that enhance cell binding (Fujisaki et al., 1999).

By contrast, the very small HA oligosaccharides have the ability to inhibit a variety of tumors. In the 6–7 saccharide range, they inhibit anchorage-independent growth of tumor cells (Ghatak et al., 2002). Even the smallest tetrasaccharides, the limit product of hyaluronidase digestion (Stern and Jedrzejas, 2006), have biological functions. They induce heat shock proteins, and prevent apoptosis (Xu et al., 2002). The very small oligomers appear to ameliorate the effect of intermediate-size HA fragments, and may even be a mechanism of host defense against the tumor. Oligosaccharides of HA thus appear to be able to either promote or inhibit tumor progression, depending on size.

HYALURONIDASES AS ANTI-CANCER CHEMOTHERAPEUTIC AGENTS

Hyaluronidases have long been added to anti-cancer regimens, particularly in Europe. Tumors previously resistant to chemotherapy become sensitive when hyaluronidase is added (Klocker et al., 1998; Baumgartner et al., 1998; Baumgartner and Hamilton, this volume). The enzyme may decrease intratumoral pressure, permitting drugs to penetrate the malignancy. However, studies are available suggesting that hyaluronidase has intrinsic anti-tumor activity.

Anti-Cancer Properties of Hyaluronidase

Evidence for the anti-cancer effects of hyaluronidase come from experimental model systems. The enzyme enhances the anti-cancer effects of adriamycin *in vitro* (Beckenlehner et al., 1992). Human cancers grown in SCID mice regress dramatically following administration of purified testicular hyaluronidase (PH-20) (Shuster et al., 2002). Over expression of HYAL-1 suppresses tumorigenicity in a model for colon carcinoma (Jacobson et al, 2002). Hyaluronidase administration delays the appearance of carcinogen-induced tumors (Pawlowski et al., 1979). Hyaluronidase treatment also prevents lymph node invasion in a murine model for T-cell lymphoma (Zahalka et al., 1995; Naor et al., this volume). The mouse has several alleles for HYAL-1, while there is only one in the human. The murine alleles have different levels of circulating hyaluronidase activity. The growth rates of murine malignancies correlate inversely with enzyme levels (DeMaeyer and DeMaeyer-Guignard, 1992). Hyaluronidase also blocks TNF-mediated cancer cell death, reverses multidrug resistance (Chang, 1998), and alters cell cycle kinetics of chemo-resistant carcinomas (St. Croix et al., 1996).

HYALURONIDASES IN CANCER PROGRESSION

However, data have also accumulated that hyaluronidase correlates with tumor progression. This has been particularly documented in tumors of the male genito-urinary tract, in prostate (Lokeshwar et al., 1996; Madan et al., 1999a; Kovar et al., 2006) and urinary bladder cancers (Lokeshwar et al., 2000; Lokeshwar et al., 2005). Aggressiveness of other human cancers also correlates with hyaluronidase, including breast (Madan et al., 1999b; Beech et al., 2002) and laryngeal cancer (Godin et al., 2000). At the same time levels of hyaluronidase sufficient to remove HA are reportedly anti-tumorigenic in prostate tumors. Whether this dose-dependent effect is a result of reaching a threshold of enzyme necessary to remove HA is still not clear. On the other, overexpression of HYAL-2 in murine astrocytoma cells accelerates tumor formation (Novak et al., 1999). The inconsistencies that abound suggest that additional work must be performed, before comprehensive understanding emerges.

THE CONUNDRUMS INVOLVING HYALURONIDASES IN MALIGNANCY

The confusion and inconsistencies of HA and the HYALs in tumor biology can be attributed to the concept that different tumors do different things, and the same tumor can do different things at different times. The phenomenon of the angiogenic switch (Folkman, 2002) may be applicable here. Early in the course of a malignancy, HMW HA may be required to open up tissue spaces, for the flow of nutrients at the primary site. But when simple diffusion no longer suffices, the action of hyaluronidase, particularly HYAL-2, can provide the HA fragments that induce angiogenesis and the neo-vascularization to support the malignancy (West et al., 1985). There is also a dose effect, as outlined by Lokeshwar and Selzer in this volume, in which opposite effects of hyaluronidase enzymes are achieved at different doses. However, HA has a very rapid rate of turnover in vertebrate tissues, and increased amounts of HA in human cancers may merely reflect simultaneous enhanced synthetic and degradative enzyme levels (Jacobson et al., 2002; Simpson, 2006).

Instability of the tumor genome, and the constant pressure of the Darwinian selection process of metastatic tumor cells underscore the resilience and ingenuity of malignant cells in their ability to survive and thrive. This appears to be applicable also to the cancer cells ability to generate different size of HA fragments at various points in the life of the cancer, with profoundly different effects achieved by large, intermediate, and small HA fragments.

References

Baumgartner, G., Gomar-Höss, C., Sakr, L., Ulsperger, E., and Wogritsch, C. (1998). The impact of extracellular matrix on the chemoresistance of solid tumors – experimental and clinical results of hyaluronidase as additive to cytostatic chemotherapy. *Cancer Lett* **131**, 85–99.

Beckenlehner, K., Bannke, S., Spruss, T., et al. (1992). Hyaluronidase enhances the activity of adriamycin in breast cancer models in vitro and in vivo. *J Cancer Res Clin Oncol* **118**, 591–596.

Beech, D. J., Madan, A. K., and Deng, N. (2002). Expression of PH-20 in normal and neoplastic breast tissue. *J Surg Res* **103**, 203–207.

Bourguignon, L. Y., Singleton, P. A., Diedrich, F., Stern, R., and Gilad, E. (2004). CD44 interaction with Na+-H+ exchanger (NHE1) creates acidic microenvironments leading to hyaluronidase-2 and cathepsin B activation and breast tumor cell invasion. *J Biol Chem* **279**, 26991–27007.

Chang, N. S. (1998). Transforming growth factor-beta protection of cancer cells against tumor necrosis factor cytotoxicity is counteracted by hyaluronidase. *Int J Mol Med* **2**, 653–659.

Coussens, L. M. and Werb, Z. (2002). Inflammation and cancer. *Nature* **420**, 860–867.

Csóka, A. B., Frost, G. I., Heng, H. H., et al. (1998). The hyaluronidase gene HYAL-1 maps to chromosome 3p21.2-p21.3 in human and 9F1-F2 in mouse, a conserved candidate tumor suppressor locus. *Genomics* **48**, 63–70.

Csoka, A. B., Frost, G. I., and Stern, R. (2001). The six hyaluronidase-like genes in the human and mouse genomes. *Matrix Biol* **20**, 499–508.

Csóka, A. B., Frost, G. I., Wong, T., and Stern, R. (1997). Purification and microsequencing of hyaluronidase isozymes from human urine. *FEBS Lett* **417**, 307–310.

Csóka, A. B., Scherer, S. W., and Stern, R. (1999). Expression analysis of six paralogous human hyaluronidase genes clustered on chromosomes 3p21 and 7q31. *Genomics* **60**, 356–361.

De Maeyer, E. and De Maeyer-Guignard, J. (1992). The growth rate of two transplantable murine tumors, 3LL lung carcinoma and B16F10 melanoma, is influenced by HYAL-1, a locus determining hyaluronidase levels and polymorphism. *Int J Cancer* **51**, 657–660.

Dorfman, A., Ott, M. L., and Whitney, R. (1948). The hyaluronidase inhibitor of human blood. *J Biol Chem* **174**, 621–629.

Edelson, M. I., Scherer, S. W., Tsui, L. C., et al. (1997). Identification of a 1300 kilobase deletion unit on chromosome 7q31.3 in invasive epithelial ovarian carcinomas. *Oncogene* **14**, 2784–2979.

Fiszer-Szafarz, B. (1968). Demonstration of a new hyaluronidase inhibitor in serum of cancer patients. *Proc Soc Exp Biol Med* **129**, 300–302.

Fiszer-Szafarz, B., Szafarz, D., and Vannier, P. (1990). Polymorphism of hyaluronidase in serum from man, various mouse strains and other vertebrate species revealed by electrophoresis. *Biol Cell* **68**, 95–100.

Folkman, J. (2002). Role of angiogenesis in tumor growth and metastasis. *Semin Oncol* **29**, 15–18.

Formby, B. and Stern, R. (1998). Phosphorylation stabilizes alternatively spliced CD44 mRNA transcripts in breast cancer cells. Inhibition by antisense complementary to casein kinase II mRNA. *Mol Cell Biochem* **187**, 23–31.

Formby, B. and Stern, R. (2003). Lactate-sensitive response elements in genes involved in hyaluronan catabolism. *Biochem Biophys Res Commun* **305**, 203–208.

Frost, G. I., Csóka, A. B., Wong, T., and Stern, R. (1997). Purification, cloning, and expression of human plasma hyaluronidase. *Biochem Biophys Res Commun* **236**, 10–15.

Frost, G. I., Mohapatra, G., Wong, T. M., et al. (2000). HYAL-1,[LUCA-1] a candidate tumor suppressor gene on chromosome 3p21.3, is inactivated in head and neck squamous cell carcinomas by aberrant splicing of pre-mRNA. *Oncogene* **19**, 870–877.

Fujisaki, T., Tanaka, Y., Fujii, K., et al. (1999). CD44 stimulation induces integrin-mediated adhesion of colon cancer cell lines to endothelial cells by up-regulation of integrins and c-Met and activation of integrins. *Cancer Res* **59**, 4427–4434.

Ghatak, S., Misra, S., and Toole, B. P. (2002). Hyaluronan oligosaccharides inhibit anchorage-independent growth of tumor cells by suppressing the phosphoinositide 3-kinase/Akt cell survival pathway. *J Biol Chem* **277**, 38013–38020.

Godin, D. A., Fitzpatrick, P. C., Scanduro, A. B., et al. (2000). PH-20: a novel tumor marker for laryngeal cancer. *Arch Otolaryngol Head Neck Surg* **126**, 402–404.

Haas, E. (1946). On the mechanism of invasion. I. Anti-invasin, an enzyme in plasma. *J Biol Chem* **163**, 63–88.

Jacobson, A., Rahmanian, M., Rubin, K., and Heldin, P. (2002). Expression of hyaluronan synthase 2 or hyaluronidase 1 differentially affect the growth rate of transplantable colon carcinoma cell tumors. *Int J Cancer* **102**, 212–219.

Junker, N., Latini, S., Petersen, L. N., and Kristjansen, P. E. (2003). Expression and regulation patterns of hyaluronidases in small cell lung cancer and glioma lines. *Oncol Rep* **10**, 609–616.

Kiriluk, L. B., Kremen, A. J., and Glick, D. (1950). Mucolytic enzyme systems XII. Hyaluronidase in human and animal tumors, and further studies on the serum hyaluronidase inhibitor in human cancer. *J Natl Cancer Inst* **10**, 993–1000.

Klocker, J., Sabitzer, H., Raunik, W., Wieser, S., and Schumer, J. (1998). Hyaluronidase as additive to induction chemotherapy in advanced squamous cell carcinoma of the head and neck. *Cancer Lett* **131**, 113–115.

Kolárová, M. (1975). Host-tumor relationship XXXIII. Inhibitor of hyaluronidase in blood serum of cancer patients. *Neoplasma* **22**, 435–439.

Kovar, J. L., Johnson, M. A., Volcheck, W. M., Chen, J., and Simpson, M. A. (2006). Hyaluronidase expression induces prostate tumor metastasis in an orthotopic mouse model. *Am J Pathol* **169**, 1415–1426.

Kumar, S., West, D. C., Ponting, J. M., and Gattamaneni, H. R. (1989). Sera of children with renal tumours contain low-molecular-mass hyaluronic acid. *Int J Cancer* **44**, 445–448.

Lokeshwar, V. B., Cerwinka, W. H., and Lokeshwar, B. L. (2005). HYAL-1 hyaluronidase: a molecular determinant of bladder tumor growth and invasion. *Cancer Res* **65**, 2243–2250.

Lokeshwar, V. B., Lokeshwar, B. L., Pham, H. T., and Block, N. L. (1996). Association of elevated levels of hyaluronidase, a matrix-degrading enzyme, with prostate cancer progression. *Cancer Res* **56**, 651–657.

Lokeshwar, V. B., Obek, C., Pham, H. T., et al. (2000). Urinary hyaluronic acid and hyaluronidase, markers for bladder cancer detection and evaluation of grade. *J Urol* **163**, 348–356.

Lokeshwar, V. B., Schroeder, G. L., Carey, R. I., Soloway, M. S., and Iida, N. (2002). Regulation of hyaluronidase activity by alternative mRNA splicing. *J Biol Chem* **277**, 33654–33663.

Madan, A. K., Pang, Y., Wilkiemeyer, M. B., Yu, D., and Beech, D. J. (1999a). Increased hyaluronidase expression in more aggressive prostate adenocarcinoma. *Oncol Rep* **6**, 1431–1433.

Madan, A. K., Yu, K., Dhurandhar, N., et al. (1999b). Association of hyaluronidase and breast adenocarcinoma invasiveness. *Oncol Rep* **6**, 607–609.

Mateo, M., Mollejo, M., Villuendas, R., et al. (1999). 7q31-32 allelic loss is a frequent finding in splenic marginal zone lymphoma. *Am J Pathol* **154**, 1583–1589.

Mio, K. and Stern, R. (2002). Inhibitors of the hyaluronidases. *Matrix Biol* **21**, 31–37.

Mio, K., Carrette, O., Maibach, H. I., and Stern, R. (2000). Evidence that the serum inhibitor of hyaluronidase may be a member of the inter-alpha-inhibitor family. *J Biol Chem* **275**, 32413–32421.

Nagase, H., Visse, R., and Murphy, G. (2006). Structure and function of matrix metalloproteinases and TIMPs. *Cardiovasc Res* 15; **69**, 562–573.

Neame, S. J. and Isacke, C. M. (1992). Phosphorylation of CD44 in vivo requires both Ser-323 and Ser-325 but does not regulate membrane localization or cytoskeletal interaction in epithelial cells. *EMBO J* **11**, 4733–4738.

Novak, U., Stylli, S. S., Kaye, A. H., and Lepperdinger, G. (1999). Hyaluronidase-2 overexpression accelerates intracerebral but not subcutaneous tumor formation of murine astrocytoma cells. *Cancer Res* **59**, 6246–6250.

Oh, J.J., Boctor, B.N., Jimenez, C.A., et al. (2007). Promoter methylation study of the H37/ RBM5 tumor suppressor gene from the 3p21.3 human lung cancer tumor suppressor locus. *Hum Genet* **123**, 55–64.

Okamoto, I., Tsuiki, H., Kenyon, L. C., et al. (2002). Proteolytic cleavage of the CD44 adhesion molecule in multiple human tumors. *Am J Pathol* **160**, 441–447.

Pawlowski, A., Haberman, H. F., and Menon, I. A. (1979). The effects of hyalurodinase upon tumor formation in BALB/c mice painted with 7,12-dimethylbenz-(a)anthracene. *Int J Cancer* **23**, 105–109.

Shuster, S., Frost, G. I., Csoka, A. B., Formby, B., and Stern, R. (2002). Hyaluronidase reduces human breast cancer xenografts in SCID mice. *Int J Cancer* **102**, 192–197.

Simpson, M. A. (2006). Concurrent expression of hyaluronan biosynthetic and processing enzymes promotes growth and vascularization of prostate tumors in mice. *Am J Pathol* **169**, 247–257.

St Croix, B., Man, S., and Kerbel, R. S. (1998). Reversal of intrinsic and acquired forms of drug resistance by hyaluronidase treatment of solid tumors. *Cancer Lett* **131**, 35–44.

Stern, R. (2003). Devising a pathway for hyaluronan catabolism: are we there yet? *Glycobiology* **13** (12), 105R–115R.

Stern, R. (2004a). Hyaluronan catabolism: a new metabolic pathway. *Eur J Cell Biol* **83**, 317–325.

Stern, R. (2004b). Mammalian hyaluronidases, an update. Glycoforum website. www.glycoforum.gr.jp.

Stern, R., Asari, A. A., and Sugahara, K. N. (2006). Hyaluronan fragments: an information-rich system. *Eur J Cell Biol* **85**, 699–715.

Stern, R. and Jedrzejas, M. (2006). Hyaluronidases: their genomics, structures, and mechanisms of action. *Chem Rev* **106**, 818–839.

Stern, R., Shuster, S., Neudecker, B. A., and Formby, B. (2002). Lactate stimulates fibroblast expression of hyaluronan and CD44: the Warburg effect revisited. *Exp Cell Res* **276**, 24–31.

Stern, R., Shuster, S., Wiley, T. S., and Formby, B. (2001). Hyaluronidase can modulate expression of CD44. *Exp Cell Res* **266**, 167–176.

Sugahara, K. N., Hirata, T., Hayasaka, H., et al. (2006). Tumor cells enhance their own CD44 cleavage and motility by generating hyaluronan fragments. *J Biol Chem* **281**, 5861–5868.

Sugahara, K. N., Murai, T., Nishinakamura, H., et al. (2003). Hyaluronan oligosaccharides induce CD44 cleavage and promote cell migration in CD44-expressing tumor cells. *J Biol Chem* **278**, 32259–32265.

Warburg, O. (1956). On the origin of cancer cells. *Science* **129**, 309–314.

Warburg, O., Posener, K., and Negelein, E. (1924). Ueber den Stoffwechsel der Carcinomzelle. *Biochem Z* **152**, 309–344.

West, D. C., Hampson, I. N., Arnold, F., and Kumar, S. (1985). Angiogenesis induced by degradation products of hyaluronic acid. *Science* **228**, 1324–1326.

Wong, M. L., Tao, Q., Fu, L., Wong, K. Y., et al. (2006). Aberrant promoter hypermethylation and silencing of the critical 3p21 tumour suppressor gene, RASSF1A, in Chinese oesophageal squamous cell carcinoma. *Int J Oncol* **28**, 767–773.

Xu, H., Ito, T., Tawada, A., et al. (2002). Effect of hyaluronan oligosaccharides on the expression of heat shock protein 72. *J Biol Chem* **277**, 17308–17314.

Zabarovsky, E. R., Lerman, M. I., and Minna, J. D. (2002). Tumor suppressor genes on chromosome 3p involved in the pathogenesis of lung and other cancers. *Oncogene* **21**, 6915–6935.

CHAPTER

13

Hyaluronan Fragments: Informational Polymers Commandeered by Cancers

Kazuki N. Sugahara

INTRODUCTION

Hyaluronan (HA), a high molecular weight polymer ($\sim 10^7$ Da) consisting of repeating disaccharide units of D-glucuronic acid and N-acetyl-D-glucosamine (Laurent and Fraser, 1992), was first isolated from the vitreous body of the eye by Karl Meyer in 1934 (Meyer and Palmer, 1934), and was shown to act as a viscous space-filler of the extracellular cavity due to its high capacity to hold water (Meyer et al., 1939). The involvement of HA in inflammatory diseases related to synovial joints, such as rheumatoid arthritis (RA), was widely investigated in the first few decades after its discovery (Balazs et al., 1967; Ragan and Meyer, 1949). In the 1950s and 1960s, it was reported that a high amount of HA is present in the body fluids of patients bearing tumors such as mesothelioma (Blix, 1951; Truedsson, 1951) and sarcomas (Marcante, 1965), and that HA is produced by Rous sarcoma *in vitro* (Grossfeld, 1962). In the late 1970s and 1980s, it became clear that HA plays a critical role in the development and the remodeling of tissues (Toole and Gross, 1971). It was found that not only the accumulation of HA in such tissues was critical for the events, but also the degradation and removal of the HA that accumulated in the tissues, were crucial for the following maturation steps (Feinberg and Beebe, 1983; Toole and Gross, 1971). The findings by West et al. in 1985 added a twist to these understandings (West et al., 1985). They demonstrated that the degraded forms of HA mainly consisting of HA 8–50-mers, but not the native high molecular weight HA (HMW-HA), induced angiogenesis in a chick chorioallantoic membrane (CAM) and in a porcine heart. They called the HA fragments "angiogenic oligosaccharides." This finding supported the profound significance of the degradation of HA during tissue remodeling, and also showed that even the degradation products had an active role in such events.

Accumulating evidence suggests that such degraded forms of HA have distinct and diverse functions from HMW-HA, and are involved in the pathology of cancer and inflammation (Stern et al., 2006; Sugahara et al., 2004; Toole, 2004). It is suggested that these HA fragments are generated by direct synthesis or enzymatic degradation of HMW-HA under such conditions. The HA fragments interact with various HA binding proteins such as CD44 (Aruffo et al., 1990), receptor for HA-mediated motility (RHAMM) (Hardwick et al., 1992), and toll-like receptors (TLRs) (Akira et al., 2006) that mediate their signals. The functions of HA fragments with respect to their receptors will be discussed here in detail.

PRODUCTION OF HA FRAGMENTS

A variety of reports demonstrates the accumulation of low molecular weight HA fragments *in vivo* when cancer or inflammation is present, suggesting that such HA fragments contribute to the pathology of the diseases. For instance, HA fragments similar to the "angiogenic oligosaccharides" in size have been found in tumor tissues and body fluids of patients with cancer (Kumar et al., 1989; Lokeshwar et al., 1997; Lokeshwar et al., 2001). A high amount of low polymerized HA was found in the synovial fluids of patients with RA (Balazs et al., 1967; Ragan and Meyer, 1949). HA fragments ranging from 5.5–146 kDa (approximately 28–760-mers) were found in transgenic mice that produce a high amount of HA in the tumor tissues while developing spontaneous breast tumors (Koyama et al., 2007). Intermediate-sized HA fragments ranging from 20–160 kDa (approximately 100–830-mers) were present in mice with non-infectious pulmonary inflammation (Teder et al., 2002). Various tumor cells were shown to produce HA fragments similar in size as the "angiogenic oligosaccharides" (Lokeshwar et al., 2001; Sugahara et al., 2006).

Two mechanisms can be considered for the production of HA fragments. One is the direct synthesis of HA fragments by HA synthases (HAS), and the other is the degradation of HMW-HA by enzymes and other means. The three synthases known in mammals, HAS-1, HAS-2, and HAS-3 (Itano et al., 1999b), have slightly different features from each other. HAS-2 and HAS-3 have higher abilities to synthesize HA than HAS-1 (Itano et al., 1999b). HAS-1 and HAS-3 synthesize and secrete HA of broad size distributions ranging from 200–2000 kDa (approximately 1000–10,000-mers), while HAS-2 synthesizes extremely large HA mainly over 2000 kDa (approximately 10,000-mers) (Itano et al., 1999b). Therefore, theoretically, intermediate-sized HA fragments around 200 kDa (approximately 1000-mers) can be synthesized by HAS-1 and HAS-3. A number of reports demonstrate the correlation between high HAS-1 expression and progression of tumors such as endometrial cancer (Yabushita et al., 2005), mesothelioma (Kanomata et al., 2005), ovarian cancer (Yabushita et al., 2004), and colon cancer (Yamada et al., 2004). It has also been shown experimentally that HAS-1 has an important role in tumor progression and metastases. For example, stable transfection of HAS-1 in a mouse mammary carcinoma cell line defective in HA synthesis rescued the ability of the cancer cells to metastasize to the lungs (Itano et al., 1999a), and a HAS-1 stable transfectant of transformed 3Y1 rat fibroblast showed growth promotion *in vitro* and *in vivo* (Itano et al., 2004).

Besides the contribution of HA synthases in the generation of HA fragments, it is very likely that the degradation of HA takes place in this process as well, especially when HA fragments of small sizes are produced. Such degradation can be achieved by various means including hyaluronidases and reactive oxygen species (ROS) (Stern, 2004; Stern et al., 2006). While ROS-directed degradation of HA is considered to be the major pathway to generate HA fragments in inflammatory and autoimmune diseases such as RA and osteoarthritis, hyaluronidase-mediated HA degradation is often involved in tumor progression (Stern, 2004; Stern et al., 2006). Among the five known human hyaluronidases, HYAL-1 and HYAL-2 play a central role in HA catabolism (Stern, 2004). HYAL-2 first degrades HMW-HA tethered on cell surfaces by HA receptors such as

V. HYALURONAN DEGRADATION, THE HYALURONIDASES

CD44 into intermediate-sized HA fragments around 20 kDa (approximately 100-mers). Then, HYAL-1 further degrades the 100-mer fragments into 4-mers in lysosomes. These hyaluronidases are often found to be active in tumors. For instance, HYAL-1 was present in high-grade human prostate cancer tissues that also contained biologically active HA fragments ranging from 20–30-mers (Lokeshwar et al., 2001). Because the expression levels of HYAL-1 correlate with the stages and metastases of human bladder cancer, it has been suggested that HYAL-1 can be used as a tumor marker for bladder cancer (Lokeshwar et al., 2005). We have found that both HYAL-1 and HYAL-2 were expressed in pancreatic tumor cells and were secreted into the culture supernatant (Sugahara et al., 2006). The supernatant was active in degrading HA, and contained HA fragments ranging from approximately 10–40-mers.

Although it is becoming clear that HA synthases and hyaluronidases contribute to tumor growth and metastases, it is often unclear how their products come into the stories except for the cases in which the sizes and biological functions of the actual HA found in the systems were analyzed (Koyama et al., 2007; Lokeshwar et al., 1997; Lokeshwar et al., 2001; Sugahara et al., 2006). Since there must be a tight connection and a balance between HA synthesis and degradation as part of the HA turnover process, it is very likely that there are large quantities of HA fragments of various sizes in tumors that over express HA synthases, hyaluronidases, or both. Therefore, it is of high importance to understand the HA profiles in such tumors and to dissect out the principal actor(s) in each event. In addition, the receptors that mediate the signals of such HA fragments have to be investigated as well.

RECEPTORS OF HA FRAGMENTS

Various molecules have been reported to function as the receptor for HA fragments (Stern et al., 2006). Such receptors, and the sizes of HA that they are able to recognize, are summarized in Table 13.1. In the text, three major molecules will be discussed.

CD44

CD44 is a widely distributed transmembrane cell adhesion molecule that is probably the most well-characterized receptor for HA (Aruffo et al., 1990). CD44–HA interaction is involved in various biological events, such as lymphocyte rolling, tumor cell migration, and invasion (Toole, 2004). Recently, CD44 has regained attention because it is now recognized as one of the most important markers for cancer stem cells in various tumors such as prostate, colon, breast, and pancreatic cancers (Ailles and Weissman, 2007).

TABLE 13.1 HA Binding Proteins and the Sizes of HA Fragments That Interact with Them

HA binding proteins	Smallest size of HA that efficiently binds to the HA binding proteins	Additional information	References
CD44	6–10-mers 10-mers (keratinocytes)	HA 4-mers interacted with CD44, but with a much lower affinity than 6-mers.	Lesley et al., 2000 (Tammi et al., 1998)
RHAMM	14-mers	HA 2–4-mers and 4–6-mers did not interact with RHAMM.	Manzaranes et al., 2007 Lokeshwar et al., 2000
TLR2	135–200 kDa	HA 4–16-mers did not interact with TLR2.	Jiang et al., 2005 Tesar et al., 2006 Termeer et al., 2002
TLR4	4–6-mers	HA 4–6-mers up to HMW-HA interacted with TLR4.	Voelcker et al., 2008 See text for further references
TSG-6	6–8-mers	HA 6–16-mers bound with a similar affinity. HA 4-mers did not bind to TSG-6.	Kahmann et al., 2000
SHAP	10-mers	HA 8-mers competed less with HA. HA 6-mers failed to do so.	Yoneda et al., 1990
Aggrecan	10-mers	HA 4–8-mers weakly interacted with aggrecan.	Hascall and Heinegård, 1974
Versican	10-mers	HA 6–8-mers competed less with HA. HA 24-mers induced a ternary complex.	Seyfried et al., 2005
Link protein	10-mers	HA 6–8-mers competed less with HA. HA 24-mers induced a ternary complex.	Seyfried et al., 2005

Refer to text for abbreviations.

The interaction between CD44 and HA is of very low affinity as described by Kd ranging from 5–150 µM (Lesley et al., 2000). When a CD44-expressing cell attaches to an HA substrate, multiple receptor-

V. HYALURONAN DEGRADATION, THE HYALURONIDASES

ligand bonds are formed, resulting in an increase in the avidity of the interaction. This is enabled by the repeating disaccharide structure and the huge size (up to 10^7 Da) of HA. Therefore, when a CD44-expressing cell binds an HA of a much smaller size, it is not surprising that the avidity is much weaker than the interaction with native HMW-HA.

Lesley et al. showed that HA 6-mer was the minimum oligomer size required for efficiently occupying the HA binding site of CD44 (Lesley et al., 2000). In addition, they showed that beginning at HA 20-mers, there was a dramatic and progressive increase in the avidity with increasing oligomer size up to 38-mers, suggesting that HA 20-mers or larger can occupy more than one HA binding site of CD44 (Lesley et al., 2000). This differential binding pattern of CD44 and HA might partially explain why HA fragments ranging from 6–36-mers exhibit distinct effects compared to intermediate-sized HA or native HMW-HA (refer to "HA fragments of 6–36-mers" later in this chapter). Additional speculations can also be made to explain the distinct functions of HA fragments of this size range. For example, it is possible that these HA fragments, especially those larger than 20-mers, have the capability to induce cross-linking of CD44 molecules, an event that initiates CD44-mediated intracellular signaling (Fujii et al., 2001). It is also possible that due to the small size, these HA fragments have a high accessibility to the HA binding sites of CD44, or are able to tickle a specific epitope within the HA binding site. These HA fragments might be internalized much more easily than HMW-HA (Thankamony and Knudson, 2006) and alter cellular functions.

Takeda et al. observed an interesting phenomenon when CD44 interacts with HA fragments (Takeda et al., 2006). They found that the HA binding domain of CD44 showed a rearrangement of the β-strands in the extended lobe when the HA binding domain interacted with HA 6-mers. In addition, they observed that the C-terminal region of the HA binding domain became disordered and was released from the structural domain upon interaction with HA fragments, and that this structural change resulted in a higher susceptibility to proteolytic cleavage of the CD44 molecule by trypsin (Figure 13.1). These findings suggest that HA fragments may directly affect

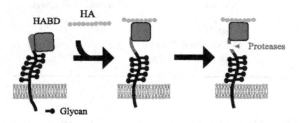

FIGURE 13.1 Proposed schematic model for the structural change-induced CD44 proteolytic cleavage upon HA binding. Binding of HA 6-mers to CD44 induces a conformational change in CD44 that makes it more susceptible to protease cleavage. HABD stands for hyaluronan binding domain. Figure adopted from Takeda et al., 2006.

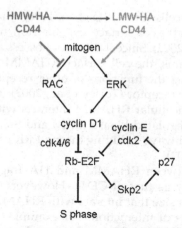

FIGURE 13.2 Differential regulation of cell proliferation signals by HMW-HA and HA fragments (LMW-HA; low molecular weight hyaluronan). HMW-HA suppresses cell proliferation by inhibiting Rac dependent signaling to cyclin D1, whereas HA fragments activate ERK and stimulate cell proliferation. Figure adopted from Kothapalli et al., 2008.

the structure of CD44 that might lead to proteolytic cleavage of the molecule. It is well-accepted that CD44 cleavage is deeply related to the migration of tumor cells (Okamoto et al., 1999; Sugahara et al., 2003; Sugahara et al., 2006), tumor progression (Okamoto et al., 2002), and the turnover of CD44 molecules (Okamoto et al., 2001). Therefore, it is an attractive hypothesis that HA fragments affect tumor cell functions by modifying the conformation of CD44 and altering cellular functions such as the up-regulation of CD44 cleavage on cell surfaces and cell motility. Opposing effects of HMW-HA and HA fragments on cell proliferation through CD44 appear to be strictly regulated by signaling pathways (Kothapalli et al., 2008). After binding to CD44 on vascular smooth muscle cells, HMW-HA suppresses cell cycle progression, while HA fragments stimulate it. This difference results from a differential regulation of signaling pathways to cyclin D1, the common signaling target for both HMW-HA and HA fragments. HMW-HA suppresses Rac dependent signaling to cyclin D1 by inhibiting GTP loading of Rac, whereas HA fragments activate ERK and stimulate ERK-dependent cyclin D1 gene expression (Figure 13.2). Similar differential regulation of intracellular signaling may be expected for other effects induced by HA fragments but not by HMA-HA.

Receptor for HA-Mediated Motility (RHAMM)

RHAMM was cloned as an HA binding protein required for cell locomotion regulated by a *ras*-oncogene (Hardwick et al., 1992). It exists in almost all cell types such as fibroblasts, macrophages, endothelial cells, and breast carcinoma cells (Entwistle et al., 1996), and in multiple cellular components including the cell surface, cytoskeleton, mitochondria, and the nucleus (Turley et al., 2002). Since it does not possess any signal proteins or transmembrane domains, the cell surface RHAMM initiates intracellular signaling by modifying the functions of other receptors such as platelet-derived growth factor receptor (Turley et al., 2002) upon interaction with HA. In addition, intracellular RHAMM interacts with molecules such as extra cellular signal-regulated kinase (ERK) and Src, leading to the regulation of signaling pathways including the ras-ERK pathways (Slevin et al., 2007; Turley et al., 2002).

The interaction between RHAMM and HA fragments has not been studied as extensively as that of CD44. However, reports cover a wide range of HA fragment size that interact with RHAMM, thus allowing one to speculate the mode of interaction. For example, it was demonstrated that native HA and HA 20–30-mers induced mitogenic responses in human pulmonary artery endothelial cells (ECs) and human lung microvessel ECs, while HA 4–6-mers or HA 2–4-mers did not (Lokeshwar and Selzer, 2000). Both native HA and HA 20–30-mers were able to induce tyrosine phosphorylation of molecules in the ECs, although for an unknown reason, HA 20–30-mers had a stronger effect than the native HA (Lokeshwar and Selzer, 2000). In addition, it was shown that an HA preparation of an average molecular mass of 200 kDa (approximately 1000-mers) stimulated ciliary activity in airway epithelial cells through an RHAMM-dependent manner (Forteza et al., 2001), and that HA 14-mers and 20–30 kDa HA (approximately 100–150-mers) were able to do so, whereas HA larger than 500 kDa (approximately 2600-mers) were not (Manzanares et al., 2007). Therefore, this information may suggest that the minimum HA size required to bind to RHAMM is larger than HA 6-mers, and that the size of the HA fragments that can efficiently induce an effect through RHAMM might range somewhere from 8–14-mers up to 200 kDa (approximately 1000-mers) and less than 500 kDa (approximately 2600-mers). Although this is not a fair assumption because the systems and cell types used in the experiments were different, some reports support this notion. For example, it has been reported that the up-regulation of RHAMM-mediated cell motility is induced by HA fragments ranging from 250–300 kDa (approximately 1300–1500-mers) (Hamilton et al., 2007) and a mixture of 10 kDa HA (approximately 50-mers) and 276.7 kDa HA (approximately 1440-mers) (Tolg et al., 2006).

Collectively, RHAMM seems to favor a certain size range of HA fragments when mediating effects, as does CD44. Although the mechanism of such regulation is ill-defined as is with CD44–HA fragment interaction, several reports provide us clues to understand the mechanism. It was demonstrated that RHAMM, CD44, and ERK1,2 formed a complex in invasive breast cancer cell lines and such complex formation sustained the high motility of the cancer cells (Hamilton et al., 2007). It has also been shown that RHAMM was required for cell surface display of CD44 and the formation of CD44-ERK1,2 complex (Tolg et al., 2006). These reports clearly show that RHAMM and CD44 co-operate to mediate the signals of HA fragments in tumor cells. Investigating the role of HA size in inducing such complex formation might guide us to the mechanism of the differential function of HA fragments mediated by CD44 and RHAMM.

Toll-Like Receptor (TLR)

TLRs are pattern-recognition receptors that regulate innate immunity (Akira et al., 2006). They are type I integral membrane glycoproteins with extracellular domains containing leucine-rich repeats and cytoplasmic domains homologous to that of the interleukin 1 (IL-1) receptor (Akira et al., 2006). So far, 12 members of TLRs have been identified in mammals. TLR1, TLR2, and TLR6 mainly recognize lipids, while TLR7, TLR8, and TLR9 recognize nucleic acids (Akira et al., 2006). TLR4 is unusual in a way that it recognizes structurally unrelated ligands including lipopolysaccharides (LPS), virus glycoproteins, and fibronectin (Akira et al., 2006).

Among the members of the TLR family, TLR2 and TLR4 have been shown to serve as the receptors for HA (Chang et al., 2007; Jiang et al., 2005; Scheibner et al., 2006; Taylor et al., 2004; Termeer et al., 2002; Tesar et al., 2006). Although they both are able to recognize HA fragments, they appear to have different tastes in the size of the fragments, judging from published reports. For example, TLR4 has been shown to mediate signals of HA 4–6-mers (Voelcker et al., 2008), HA 4–8-mers (Taylor et al., 2004), HA 4–16-mers (Termeer et al., 2002), HA 30-mers, 23 kDa (approximately 120-mers), and 230 kDa (approximately 1200-mers) (Chang et al., 2007), 135–200 kDa HA (approximately 700–1000-mers) (Jiang et al., 2005; Tesar et al., 2006), 500–800 kDa HA (approximately 2600–4200-mers) (Wang et al., 2006), and 1300 kDa HA (approximately 6800-mers) and larger (Chang et al., 2007). In contrast, TLR2 has been reported to be involved in events initiated by 135–200 kDa HA (approximately 700–1000-mers) (Jiang et al., 2005; Tesar et al., 2006) and 200 kDa HA (approximately 1000-mers) (Scheibner et al., 2006), and its involvement in mediating signals of HA 4–16-mers was denied (Termeer et al., 2002). Although the optimal HA sizes that TLR4 and TLR2 recognize might possibly differ between cell types, it can be roughly estimated that TLR4 might be able to recognize a broad range of HA sizes beginning from HA 4–6-mers up to HMW-HA, whereas TLR2 prefers intermediate-sized HA fragments. This is in line with the fact that TLR4 recognizes a variety of structurally unrelated pathogen-associated molecular patterns, while TLR2 recognizes limited ligands such as lipids and some glycoproteins (Akira et al., 2006). Structural analyses of TLR4 and

TLR2 with or without association with various sizes of HA fragments will provide us further information to draw a conclusion.

How TLRs recognize such HA fragments is unclear. Recently, Taylor et al. (2007) reported an interesting finding that is similar to what is known about the collaboration between CD44 and RHAMM (Hamilton et al., 2007; Tolg et al., 2006). When they treated macrophages with HA, they found that TLR4 formed a complex with CD44 and MD-2, an adapter molecule required for TLR4 signaling, and that this complex was responsible for HA recognition. In addition, this complex formation was not observed under LPS treatment, suggesting that the recognition pattern of HA and LPS by TLR4 is different. Considering the reports regarding the complex formation between CD44 and RHAMM (Hamilton et al., 2007; Tolg et al., 2006), it is possible that different sizes of HA fragments are able to tie up different receptors resulting in unique receptor complexes depending on the cell types.

It is now becoming clearer that TLRs serve as sensors on macrophages and dendritic cells (DCs) that mediate danger signals in response to HA fragments under non-infectious inflammations (Akira et al., 2006; Jiang et al., 2007). A similar scenario can be envisaged under tumor-bearing conditions because of the accumulation of HA fragments (Lokeshwar et al., 1997; Sugahara et al., 2006) and the presence of tumor-associated macrophages and DCs in such conditions (Chang et al., 2007). However, an emerging concept demonstrates that various immune cells within the tumor microenvironment such as macrophages (Giraudo et al., 2004) and neutrophils (Pahler et al., 2008) co-operate with the tumor cells to help them grow. Therefore, the interaction of HA fragments with the immune cells through TLRs may be a "double-sided blade" that can evoke a danger signal to the host, but helps the tumor cells to make their comfortable beds at the same time. In addition, TLRs appear to be directly involved in tumor cell functions (Voelcker et al., 2008). It has been reported that a series of TLRs are expressed on various tumor cells such as colon cancer, breast cancer, prostate cancer, lung cancer, and melanoma, and that TLR4 expressed on colon cancer cells mediates the suppression of T cell proliferation and natural killer cell activity enabling the cancer cells to evade immune surveillance (Huang et al., 2005). Therefore, TLRs might not only be regulating innate immunity, but also are involved in cellular functions of non-immune cells including tumor cells when they interact with certain ligands such as HA fragments.

DIFFERENT SIZES OF HA HAVE DIFFERENT BIOLOGICAL FUNCTIONS

The various functions of HA fragments are discussed below in three groups according to the sizes of the HA fragments (Tables 13.2–13.4).

TABLE 13.2 Biological Functions of HA 4–6-mers

MW (kDa)	Number of saccharides	Biological functions	Responsible receptors	References
1–4	N.D.	Blood flow was increased in cryoinjured grafts. Number of blood vessels was increased in the grafts.	N.D	Lees et al., 1995[**]
N.D.	4-mers	Heat shock protein 72 expression was upregulated.	CD44[*]	Xu et al., 2002
N.D.	4–6 mers	The internalization and intracellular catabolism of HA by cartilage explant cultures were inhibited.	N.D.	Ng et al., 1995[**]
		Proliferation of melanoma cells enhanced by HMW-HA was inhibited.	CD44	Ahrens et al., 2001[**]
		Mmp-9 and Mmp-13 expression was enhanced in tumor cells.	CD44[*] RHAMM[*] TLR[*]	Fieber et al., 2004
		NF-κB, MMP-2, and IL-8 expression and cell migration was enhanced in melanoma cells via TLR4.	TLR4	Voelcker et al., 2008
	6-mers	Transcription factors were activated in articular chondrocytes. Expression of MMP-3 and collagen II was upregulated.	N.D.	Ohno et al., 2005
		MMP-13 expression was induced in articular chondrocytes through activation of NF-κB and p38 MAPK.	CD44 and U.R.	Ohno et al., 2006
		Concormational change in the HA binding domain of CD44 was induced.	CD44	Takeda et al., 2006

The abbreviations used are: N.D., not described; U.R., unknown receptor. Refer to text for other abbreviations. Receptors with asterisks indicate those that were found to be NOT responsible for the events. References that do not appear in the text (double asterisks) are as follows.

Ahrens, T., Assmann, V., Fieber, C., et al. (2001). CD44 is the principal mediator of hyaluronic-acid-induced melanoma cell proliferation. *J Invest Dermatol* **116**, 93–101.

Lees, V.C., Fan, T.P., and West, D.C. (1995) Angiogenesis in a delayed revascularization model is accelerated by angiogenic oligosaccharides of hyaluronan. *Lab Invest* **73**, 259–266.

Ng, C.K., Handley, C.J., Preston, B.N. et al. (1995). Effect of exogenous hyaluronan and hyaluronan oligosaccharides on hyaluronan and aggrecan synthesis and catabolism in adult articular cartilage explants. *Arch Biochem Biophys* **316**, 596–606.

V. HYALURONAN DEGRADATION, THE HYALURONIDASES

TABLE 13.3   Biological Functions of HA 6–36-mers (Approximately 1.35–6.9 kDa)

MW (kDa)	Saccharide units	Biological functions	Responsible receptors	References
N.D.	4–8-mers	IL-8 release from human ECs was enhanced. MIP-2 and KC levels were increased in mouse serum.	TLR4	Taylor et al., 2004
		NO production was increased through the activation of NO synthase in articular chondrocytes.	CD44	Iacob and Kundson, 2006**
N.D. PIP	4–14-mers	PIP$_3$ production and Akt phosphorylation levels were reduced. Apotosis of T lymphoma cells lines were induced.	N.D.	Alaniz et al., 2006**
		Drug resistant T cell lymphoma cells were sensitized to vincristine.	CD44	Cordo Russo et al., 2008
N.D.	4–16-mers	Maturation of human blood derived DCs was induced. IL-1 β, TNF- α and IL-12 production from DCs was increased.	CD44* RHAMM*	Termeer et al., 2000
		Human and murine DCs were activated *via* TLR4 but not TLR2.	TLR4 TLR2*	Termeer et al., 2002
1.35–4.5	6–20-mer	HA fragments were found in the sera of patients with Wilms' tumors and bone metastaszing renal tumors of childhood.	N.D.	Kumar et al., 1989
		Angiogenesis was induced (chick CAM assay). Sythesis of collagen type I and type VIII was promoted.	N.D.	Rooney et al., 1993

TABLE 13.3 (*continued*)

MW (kDa)	Saccharide units	Biological functions	Responsible receptors	References
		Blood vessel numbers were increased in rat skin. EC migration was promoted.	N.D.	Sattar et al., 1994
		Angiogenesis was induced synergistically with VEGF. Components of the PA-plasmin system were activated.	N.D.	Montesano et al., 1996
		Gene expression was upregulated in ECs. (*c-fos, c-jun, jun-B, Krox-20, Krox-24*).	CD44	Deed et al., 1997**
		CD44 phosphorylation was induced leading to increased tyrosine phosphorylation.	CD44	Slevin et al., 1998
		Phosphorylation and translocation of PLC wound recovery, ERK1/2 activation were induced.	N.D.	Selvin et al., 2002**
		PI3K activity and Akt phosphorylation were inhibited. PTEN expression was stimulated in tumor cells.	CD44	Ghatak et al., 2002
		HA oligosaccharides were found in the post-mortem tissues and serum of patients with acute-stage stroke injury.	N.D.	A'Qteishat et al., 2006**
N.D.	6–24-mers	Growth of B16F10 melanoma cells was inhibited.	CD44(?)	Zeng et al., 1998
N.D.	6–32-mers	Proliferation of bovine aortic ECs was induced.	N.D.	West and Kumar, 1989

(*continued*)

V. HYALURONAN DEGRADATION, THE HYALURONIDASES

TABLE 13.3 (*continued*)

MW (kDa)	Saccharide units	Biological functions	Responsible receptors	References
1.35–6.9	6–36-mers	CD44 cleavage was induced in tumor cells. CD44-dependent tumor cell migration was promoted.	CD44	Sugahara et al., 2003
~500	6-mers ~	Chemokine gene expression was induced in murine alveolar ECs. (*crg-2, MCP-1, RANTES, MIP-1 a, MIP-1 β*).	CD44	Mckee et al., 1996
N.D.	8-mers	Tumor cell motility and invasion were inhibited *in vitro*. Number of lung metastases was decreased *in vivo*.	CD44	Hosono et al., 2007
N.D.	8–50-mers	Angiogenesis was induced (chick CAM assay).	N.D.	West et al., 1985
N.D.	~10–40-mers	HA oligosaccharides generated by tumor cells enhanced. CD44 cleavage and tumor cell motility.	CD44	Sugahara et al., 2006
N.D.	12-mers	Tube formation of brain ECs was induced in collagen gels.	CD44 (?)	Rahmanian et al., 1997
		Ec differentiation and CXCL1 upregulation were induced.	CD44	Takahashi et al., 2005
2.675 and 20	14-mers and N.D.	CBF was stimulated in tracheal epithelial cells.	RHAMM	Manzanares et al., 2007
N.D.	20–30-mers	Found in high-grade TCC patient urine specimens. Mitogenic respose in HMVEC-L was induced.	N.D.	Lokeshwar et al., 1997

V. HYALURONAN DEGRADATION, THE HYALURONIDASES

TABLE 13.3 (*continued*)

MW (kDa)	Saccharide units	Biological functions	Responsible receptors	References
		Tyrosine phosphorylation of p 125 [FAK], paxillin, and p42/44 ERK was induced in HPAEC and HMVEC-L.	RHAMM	Lokeshwar and Selzer, 2000
		Found in high-grade human prostate cancer specimens.	N.D.	Lokeshwar et al., 2001
5.5–146	N.D.	HA fragments were detected in HAS-2 overexpressing breast tumor transgenic mice.	N.D.	Koyama et al., 2007
N.D.–230	30-mers–N.D.	Osteoclastogenesis was induced in bone marrow Mø.	TLR4 (?)	Chang et al., 2007
6.8	32-mers	bFGF-induced *in vivo* angiogenesis was promoted.	N.D.	Koyama et al., 2007
		Erk 1/2 phosphorylation and migration of aortic smooth muscle cellls were inhibited.	CD44	Vigetti et al., 2007**
6.9	36-mers	LFA-1 was activated in tumor cells. Tumor cell binding to ICAM-1 was enhanced.	CD44 (?)	Fujisaki et al., 1999**
		Fas expression was enhanced. Fas-mediated early apoptotic change of RA synovial cells was increased.	CD44	Fujii et al., 2001

The abbreviations used are: N.D., not described; KC, keratinocute chemoattractan; NO, nitric oxide; PIP3, phosphatidylinositol-3,4,5-triphosphate; PA, plasminogen activator; PLC 1, phospholipase C 1; PTEN, phosphatase and tensin homolog deleted on chromosome 10; CBF, ciliary beat frequency; HMVEC-L, human lung microvessel endothelial cell; HPAEC, human pulmonary artey EC; Mø, macrophage; LFA-1, lymphocyte function-associated antigen-1. Refer to text for other abbreviations. Receptors with asterisks indicate those that were found to be resposible for the events.(?) indicates the receptors that were suggested to be involved in the events but not experimentally proved.

V. HYALURONAN DEGRADATION, THE HYALURONIDASES

References that do not appear in the text (double asterisks) are as follows.
Alaniz, L., Garcia, M. G., Gallo-Rodriguez, C., et al. (2006). Hyaluronan oligosaccharides induce cell death through PI3-K/Akt pathway independently of NF-kappaB transcription factor. *Glycobiology* **16**, 359–367.
A'Qteishat, A., Gaffney, J., Krupinski, J., et al. (2006). Changes in hyaluronan production and metablism following ischaemic stroke in man. *Brain* **1229**, 2158–2176.
Deed, R., Rooney, P., Kumar, P., et al. (1997). Early-response gene signalling is induced by angiogenic oligosaccharides of hyaluronan in endothelial cells. Inhibition by non-angiogenic, high-molecular-weight hyaluronan. *Int J Cancer*. **71**, 251–256.
Fujisaki, T., Tanaka, Y., Fujii, K., et al. (1999). CD44 stimulation induces integrin-mediated adhesion of colon cancer cell lines to endothlial cells by up-regulation of integrins and c-Met and activation of integrins. *Cancer Res* **59**, 4427–4434.
Iacob, S. and Knudson, C.B. (2006). Hyaluronan fragments activate nitric oxide synthase and the production of nitric oxide by articular chondrocytes. *Int J Biochem Cell Biol* **38**, 123–133.
Slevin, M., Kumar, S. and Gaffney, J. (2002). Angiogenic oligosaccharides of hyaluronan induce multiple signaling pathways affecting vascular endothelial cell mitogenic and wond healing responses. *J Biol Chem* **277**, 41046–41059.
Vigetti, D., Viola, M., Karousou, E., et al. (2008). Hyaluronan-CD44-ERK1/2 regulate human aortic smooth muscle cell motility during aging. *J Biol Chem* **283**, 4448–4458.

HA Fragments of 4–6-mers ($\approx$1.35 kDa) (Table 13.2)

HA fragments as small as 4–6-mers are special in a sense that they do not efficiently bind to known HA binding proteins such as RHAMM (Lokeshwar and Selzer, 2000), TSG-6 (Kahmann et al., 2000), serum-derived HA-associated protein (SHAP) (Yoneda et al., 1990), aggrecan (Hascall and Heinegård, 1974), versican (Seyfried et al., 2005), and link protein (Seyfried et al., 2005) (refer to "Receptors of HA fragments" and Table 13.1). The functions of HA 6-mers can be partially explained through its interaction with CD44 because it is able to bind to CD44, although with a relatively low affinity (Lesley et al., 2000). For instance, HA 6-mers up-regulated transcription factors and several mRNAs including *matrix metalloprotease 3* (*Mmp3*) (Ohno et al., 2005), and also MMP13 protein through a CD44-dependent pathway as well as through a CD44-independent pathway (Ohno et al., 2006). However, regarding HA 4-mers, it is difficult to explain its mode of function through conventional HA receptors. It was reported that HA 4-mers up-regulated the expression of heat shock protein 72 in a chronic myelogeneous leukemia cell line, K562, while HA 2-mers and 6–12-mers did not, and that CD44 was not the responsible receptor for these events (Xu et al., 2002). Then, what can be the receptor for these very small HA fragments? One receptor that should be considered as a candidate is TLR4. It has been demonstrated that HA 4–6-mers activate NF-κB and enhance the expression of IL-8 and MMP2 in melanoma cells through interaction with TLR4 (Voelcker et al., 2008). However, some reports exclude the involvement of TLR4 in mediating the signals of HA 4–6-mers, suggesting the presence of a yet-unknown HA fragment receptor. HA 4–6-mers but not 60–200 kDa HA (approximately 300–1000-mers) or HA larger

TABLE 13.4 Biological Functions of HA Fragments of Approximately 100-mers (20 kDa) or Larger

MW (kDa)	Saccharide units	Biological functions	Responsible receptors	References
<10	N.D.	Receptor-mediated uptake of DNA/PEI was facilitated.	CD44	Hornof et al., 2008
10 + 276.7	N.D.	Random motility of fibroblasts was stimulated.	RHAMM	Tolg et al., 2006
~20–160	N.D.	Accumulation of HA fragments was observed in bleomycin-induced lung injury in CD44-/- mice.	CD44	Teder et al., 2002
40–80	N.D.	IL-1 β, TNF α, IGF-1 expression in murine bone marrow-derived Mø was induced.	CD44	Nobel et al., 1993
80–600	N.D.	ICAM-1 and VCAM-1 expression through activation of NF- B and AP-1 was up-regulated.	N.D.	Oertli et al., 1998
		COX-2 expression and TXA production was increased in renal tubular epithelial cells and ^{2}Mø.	N.D.	Sun et al., 2001
80–800	N.D.	MCP-1 expression was up-regulated in cortical tubular cells. Chemokine production was stimulated in Mø.	CD44	Beck-Schimmer et al., 1998
135–200	N.D.	An inflammatory response was initiated to promote recovery in acute lung injury.	TLR2 TLR4	Jiang et al., 2005
		DC maturation was induced and alloimmunity was initiated.	TLR4 TLR2	Tesar et al., 2006

(continued)

V. HYALURONAN DEGRADATION, THE HYALURONIDASES

TABLE 13.4 (*continued*)

MW (kDa)	Saccharide units	Biological functions	Responsible receptors	References
200	N.D.	iNOS mRNA and protein expression was induced, alone or synergistically with IFN- in murine Mø.	CD44	Mckee et al., 1997
		IP-10 mRNA and protein expression was induced synergistically with IFN- in murine Mø.	N.D.	Horton et al., 1998
		IL-10 and IFN-independently inhibited HA fragment-induced chemokine gene expression in murine Mø.	N.D.	Horton et al., 1998**
		iNOS expression was induced *via* the NF-κB pathway in rat sinusoidal ECs and Kupffer cells.	N.D.	Rockey et al., 1998
		MME expression and enzyme activity were upregulated in alveolar Mø.	CD44*	Horton et al., 1999
		HA fragment-induced VCAM-1 expression was inhibited by a PKC inhibitor.	N.D.	Schwalder et al., 1999
		PAI activity was augmented and uPA was inhibited in alveolar Mø.	N.D.	Horton et al., 2000
		CBF in tracheal epithelial cells was stimulated. RHAMM.		Forteza et al., 2001
		Innate immune response was activated *via* TLR-2.	TLR2 CD44* TLR4*	Scheibner et al., 2006
250–300	N.D.	Motility of MDA-MB231 but not MCF-7 was stimulated. RHAMM.		Hamilton et al., 2007

TABLE 13.4 (*continued*)

MW (kDa)	Saccharide units	Biological functions	Responsible receptors	References
267–513	N.D.	*NF- B/I- B^a* system was activated in murine Mø.	N.D.	Noble et al., 1996
280	N.D.	RANTES, MIP-1 a, and MIP-1 ß secretion and active IL-12 production were induced in elicited Mø.	CD44	Hodge-Dufour et al., 1997
400	N.D.	Activities of collagenase, gelatinase, and IL-8 production in human myometrial tissue cultures were stimulated.	N.D.	EI Maradny et al., 1997**
500–800	N.D.	TNF- a production was elevated in Mø.	CD44 RHAMM TLR4	Wang et al., 2006

The abbreviations used are: N.D., not described; Mø, macrophage. Refer to text for other abbreviations. Receptors with asterisks indicate those that were found to be NOT responsible for the events.
References that do not appear in the text (double asterisks) are as follows.
Horton, M.R., Burdick, M.D., Strieter, R.M., Bao, C and Noble, P.W. (1998). Regulation of hyaluronan-induced chemokine gene expression by IL-10 and IFN-gamma in mouse macrophages. *J Immunol* **160**, 3023–3030.
El Maradny, E., Kanayama, N., Kobayashi, H., et al. (1997). The role of hyaluronic acid as a mediator and regulator of cervical ripening. *Hum Reprod* **12**, 1080–1088.

than 600 kDa (approximately 3000-mers) induced the expression of *Mmp9* and *Mmp13* mRNA through the activation of nuclear factor (NF)-κB in 3LL Lewis lung carcinoma cells and normal murine embryonic fibroblasts (Fieber et al., 2004). In these events, CD44, RHAMM, and also TLR4 were all experimentally excluded as candidates for the responsible receptor.

Collectively, although the pathways are not fully uncovered, it is clear that even these very small HA fragments are capable of initiating signals such as the activation of NF-κB that leads to increased expression of various MMPs. These effects can promote tumor development, as it is well-established that MMPs are crucially involved in migration of tumor cells, tissue remodeling, and cell–cell communication (McCawley and Matrisian, 2001). However, it is also possible that such HA fragments act as a tumor suppressor because MMPs may also participate in tumor suppression (López-Otín and Matrisian, 2007) and the HA fragments might affect the immune system as well as

V. HYALURONAN DEGRADATION, THE HYALURONIDASES

tumor cell functions through interaction with TLR4. In any case, identification of the receptor for HA 4–6-mers will allow us to understand further the characteristics of these very small HA fragments.

HA Fragments of 6–36-mers (≈ 1.35–6.9 kDa) (Table 13.3)

Most of the reports that deal with the biological functions of HA fragments fall into this category. Biological functions of the HA fragments of this size include their roles in angiogenesis, tumor cell biology, and the immune system.

Angiogenesis

As discussed earlier, the discovery of "angiogenic oligosaccharides" by West et al. (1985) opened the door to the research field of HA fragments. Later, they supported their own findings by demonstrating that HA 6–32-mers stimulated the proliferation of bovine aortic ECs but not fibroblasts or smooth muscle cells (West and Kumar, 1989). They also found that HMW-HA but not HA 6–32-mers disrupted a confluent monolayer of the ECs (West and Kumar, 1989) suggesting the importance of HA fragments not only in angiogenesis but also in maintaining vascular integrity. A number of reports followed this discovery further supporting their notion. For example, HA 6–20-mers induced angiogenesis and also the deposition of collagen type I and VIII at the sites of angiogenesis (Rooney et al., 1993), and HA 12-mers induced tube formation of brain capillary ECs (Rahmanian et al., 1997; Takahashi et al., 2005). The angiogenic effects of HA 6–20-mers were also observed *in vivo* by injecting the HA fragments into the skin of rats (Sattar et al., 1994).

HA fragments appear to evoke such responses by regulating several key molecules involved in angiogenesis. It was shown that HA fragments induce the expression of vascular endothelial growth factor (VEGF), one of the key molecules that promote angiogenesis (Adams and Alitalo, 2007) in endocardial cushion cultures (Rodgers et al., 2006), and synergistically co-operate with VEGF in bovine microvascular ECs to induce angiogenesis (Montesano et al., 1996). HA 6–20-mers up-regulated the expression levels of active urokinase type plasminogen activator (uPA) and plasminogen activator inhibitor type 1 (PAI-1) (Montesano et al., 1996) that might contribute to the invasiveness of the ECs. In addition, HA fragments were able to initiate signaling cascades that include molecules such as ERK and protein kinase C (PKC) (Slevin et al., 1998) that might lead to the proliferation of the ECs. These signals induced by HA fragments are often mediated by CD44 (Slevin et al., 1998), but some reports show that RHAMM is also involved in these events. It was demonstrated that HA 20–30-mers bound to human pulmonary artery ECs, human lung microvessel ECs, and human umbilical vein ECs through RHAMM, and such interaction lead to tyrosine phosphorylation of p125^{FAK}, paxillin, and p42/44 ERK (Lokeshwar and Selzer, 2000).

These reports are in line with the notion that HA fragments ranging from approximately 6–20-mers induce angiogenesis while HMW-HA does not. However, Koyama et al. (2007) appear to challenge this concept by demonstrating that not only HA oligosaccharides but also certain forms of HMW-HA are able to induce angiogenesis. They crossed a line of conditional transgenic mice expressing murine HAS-2 with the mouse mammary tumor virus-Neu transgenic model of spontaneous breast cancer to make a spontaneous breast tumor model that produces high amount of HA (HAS-2$^{\Delta Neo}$). They found that the breast tumors of HAS-2$^{\Delta Neo}$ showed an increased number of angiogenic blood vessels, and that degraded forms of HA ranging from 5.5–146 kDa (approximately 28–760-mers) were present in the tumors suggesting that these HA fragments contributed to the angiogenic response. Indeed, they were able to confirm the well-accepted notion that HA fragments but not HMW-HA induce angiogenesis in a matrigel plug assay using HA 32-mers and HMW-HA derived from *Streptococcus*, but surprisingly, when they tested the effect of HMW-HA from human umbilical cord, they noticed that the HMW-HA was able to induce angiogenic responses as well as HA 32-mers. In attempt to explain the discrepancy, they found a high amount of versican enriched in the human umbilical cord HA preparation forming an aggregate with HMW-HA. This aggregate induced angiogenesis in the matrigel plug assay while HMW-HA alone had no effect. The receptor responsible for this effect has not been revealed yet, but CD44 seems to be excluded because the angiogenic response still persisted in CD44-deficient mice. These results suggest that HA behaves distinctively by changing its size and also by interacting with molecules that co-exist with it in the microenvironment, adding a new concept to understand the functions of HA *in vivo*.

Recently, Koyama et al. found another interesting phenomenon induced by HA using the HAS-2$^{\Delta Neo}$ mice (Koyama et al., 2008). In the tumors of the transgenic mice, they found an increased number of lymphangiogenic vessels that were accompanied by an increased expression of VEGF-C and VEGF-D, suggesting the role of HA not only in angiogenesis but also in lymphangiogenesis. They confirmed in a xenograft breast tumor model that again HA-versican aggregates induced this response. In the report, the size-dependent effects of HA in lymphangiogenesis were not examined, but since HAS-2$^{\Delta Neo}$ produces degraded forms of HA (Koyama et al., 2007), it is possible that HA fragments as well as HA-versican aggregates contributed to the induction of lymphangiogenesis.

Effects on Tumor Cells

Similar to what Koyama et al. observed in the HAS-2$^{\Delta Neo}$ mice (Koyama et al., 2007), a variety of reports demonstrate the presence of HA

V. HYALURONAN DEGRADATION, THE HYALURONIDASES

fragments in tumor tissues or in body fluids of cancer patients. Kumar et al. found "angiogenic oligosaccharides" in the sera of patients with Wilms' tumors and bone metastasizing renal tumors of the childhood (Kumar et al., 1989). Lokeshwar et al. found that urine from healthy people contained a small fraction of intermediate-sized HA, while a large amount of HA that contained both HMW-HA and HA 20–30-mers was detected in the urine specimens of patients with transient cell carcinoma (TCC) of the bladder (Lokeshwar et al., 1997). The HA 20–30-mers isolated from the urine of high-grade TCC patients were shown to have mitogenic responses in human lung microvessel ECs, suggesting the active role of HA fragments in transient cell carcinoma progression. They also found a similar HA profile in human prostate cancer tissues (Lokeshwar et al., 2001). These findings point to the fact that HA fragments are generated under tumor-bearing conditions, and somehow contribute to tumor progression.

Our findings substantiate this notion and add one way to explain the mechanism of HA fragment-induced tumor progression. We found that tumor cells produce HA fragments at cellular levels, and that such HA fragments activate the tumor cells through a CD44-dependent mechanism (Sugahara et al., 2003; 2006). The cell surface CD44 is proteolytically cleaved at the extracellular domain by proteases such as membrane type 1-MMP (Kajita et al., 2001), a disintegrin and metalloproteinase (ADAM)-10 and ADAM-17 (Nagano et al., 2004), followed by further cleavage at the intracellular domain that leads to increased expression of CD44 (Okamoto et al., 2001). Elevated CD44 cleavage is observed in various human tumors such as glioma, breast carcinoma, non-small cell lung carcinoma, colon carcinoma, and ovarian carcinoma (Okamoto et al., 2002) suggesting its critical role in tumor progression, and has been experimentally shown to be involved in tumor cell migration along extracellular matrices (Kajita et al., 2001; Okamoto et al., 1999). We found that HA 6–12-mers and a 6.9 kDa HA preparation that mainly contained HA 36-mers enhanced CD44 cleavage in tumor cells, while larger HA preparations such as 36 kDa, 200 kDa, and 1000 kDa HA did not (Sugahara et al., 2003). This phenomenon was induced by the interaction of the HA fragments and the cell surface CD44 molecules as demonstrated by the inhibition of the HA-fragment-induced CD44 cleavage by an anti-CD44 blocking antibody. The 6.9 kDa HA enhanced not only CD44 cleavage but also the migration of tumor cells in a CD44-dependent manner. We further found that MIA PaCa-2 human pancreatic cancer cell line expressed and secreted at least two hyaluronidases, HYAL-1 and HYAL-2, and that the culture supernatant of MIA PaCa-2 was able to degrade HA (Sugahara et al., 2006). Indeed, fragmented HA ranging from approximately 10–40-mers was found in the culture supernatant of tumor cells such as MIA PaCa-2. These purified tumor-associated HA fragments successively enhanced CD44 cleavage and cell migration in MIA PaCa-2 cells through a CD44-dependent manner. Collectively, these results suggest the presence of an autocrine/ paracrine activation mechanism in tumor cells which is mediated by HA fragments produced by the tumor cells themselves.

In contrast to the reports that demonstrate the contribution of HA fragments in tumor progression, some claim that the fragments counteract tumor development. Ghatak et al. reported that the *in vivo* tumor growth of LX-1 human lung carcinoma and TA3/St murine mammary carcinoma was inhibited and that their anchorage-dependent growth was inhibited as well by HA 6–20-mers (Ghatak et al., 2002). They further demonstrated that the survival pathway necessary for anchorage-independent growth charac- terized by the activity of phosphoinositide 3-kinase (PI3K) and the phos- phorylation of Akt, was inhibited by the fragments. Such growth inhibitory effects of HA fragments were also observed in B16F10 murine melanoma subcutaneous tumors (Zeng et al., 1998). In addition, it has been reported that HA fragments exert pro-apoptotic effects on tumor cells (Cordo Russo et al., 2008; Hosono et al., 2007). Two possibilities can be considered to explain the suppressive effects of the HA fragments in tumor development. First, it is possible that the HA fragments actively initiate signals in the tumor cells that negatively regulate the cellular functions. Another possi- bility is that the effects observed (Cordo Russo et al., 2008; Ghatak et al., 2002; Hosono et al., 2007; Zeng et al., 1998) may have been due to the perturbation of the interaction between HMW-HA and CD44 expressed on the tumor cells, as discussed by Ghatak et al. (Ghatak et al., 2002). In fact, the latter possibility is valid because the HA fragments used in these reports include HA 20-mers (Cordo Russo et al., 2008; Ghatak et al., 2002; Hosono et al., 2007; Zeng et al., 1998) that is able to occupy more than one HA binding site of CD44 and efficiently inhibit the interaction between HA and CD44 (Lesley et al., 2000) that is essential to maintain the signaling cascade for tumor cell survival (Kosaki et al., 1999). In either explanation, it is not surprising even if HA fragments of a same size showed both tumor- promoting and tumor-regressing effects because of the differential binding patterns of such HA fragments (refer to "CD44" under "Receptors for HA fragments"). The simple repeating disaccharide structure of HA enables such complicated behavior which is an important and unique aspect of HA.

Immune Response and Inflammation

Termeer et al. reported an important observation in which they showed for the first time that a member of the TLR family serves as one of the receptors for HA fragments. They demonstrated that HA fragments ranging from 6–16-mers induced the maturation of both human (Termeer et al., 2000) and murine (Termeer et al., 2002) DCs. This was accompanied by the production of various cytokines from the DCs, and was mediated by the activation of the mitogen-activated protein (MAP) kinase pathway and NF-κB (Termeer et al., 2002). Upon treatment of the DCs with an anti-TLR4 blocking antibody prior to the HA fragment treatment, the DC maturation did not occur, suggesting that the HA fragments induced the maturation by interacting with TLR4. This was further confirmed by using bone marrow-derived DCs from TLR4-deficient mice. In contrast, treatment of TLR2-deficient DCs with the HA fragments still resulted in the maturation of the DCs, indicating that TLR4 but not TLR2 serves as the receptor for the HA fragments in DCs. Similarly, it was shown in human microvessel dermal ECs that HA fragments up-regulate IL-8 through TLR4 (Taylor et al., 2004). These reports suggest that certain sizes of HA fragments initiate a systemic self-defensive response by eliciting innate immunity under certain conditions and also a local inflammation perhaps at the sites of injuries by interacting with ECs to promote wound healing.

Opposed to such protective functions, HA fragments appear to cause an undesired consequence in some cases by inducing immunological reactions. For instance, it was demonstrated that HA 36-mers up-regulate Fas expression on synovial cells and augment Fas-mediated early apoptosis of the cells which might be one of the mechanisms of the pathogenesis of RA. Indeed, it was already noticed about 50 years ago that in spite of the elevated concentration of HA, the viscosity of the synovial fluid of RA patients was lower than that of normal synovial fluids (Ragan and Meyer, 1949), and that it was due to the low degree of polymerization of HA (Balazs et al., 1967; Ragan and Meyer, 1949). Perhaps, HA fragments actively participate in the pathogenesis of immunological disorders and inflammatory diseases as well.

HA Fragments of 100-mers or Larger ($\approx$ 20 kDa) (Table 13.4)

HA fragments of this size range may be described as intermediate-sized HA fragments. They have been shown to be capable of up-regulating the expression of adhesion molecules such as intracellular adhesion molecule (ICAM)-1 and vascular cell adhesion molecule (VCAM)-1 (Oertli et al., 1998; Schawalder et al., 1999), enhancing ciliary functions in tracheal epithelial cells (Forteza et al., 2001), and elevating random motility of tumor cells (Hamilton et al., 2007) and fibroblasts (Tolg et al., 2006). Among the various functions of intermediate-sized HA fragments, their roles in regulating macrophages and DCs are most well-characterized. For example, sonicated HA preparations mainly consisting of HA fragments of 438 kDa (approximately HA 2300-mers), 474 kDa (approximately HA 2500-mers), and 513 kDa (approximately HA 2700-mers) in size induced the expression of I-κBα mRNA in mouse alveolar macrophages while native HMW-HA of 6000 kDa (approximately HA 31,000-mers) did not (Noble et al., 1996). HA fragments within a similar size range as the sonicated HA such as 40–80 kDa (approximately HA 200–400-mers) (Noble et al., 1993), those smaller than 500 kDa (approximately HA 2600-mers) (McKee et al., 1996), and 200 kDa (approximately HA 1000-mers)

(Hodge-Dufour et al., 1997; Horton et al., 1998) were also biologically active and able to induce the expression of chemokines. Such effects are often mediated by NF-κB (Noble et al., 1993; Noble et al., 1996; Rockey et al., 1998). In addition, a 280 kDa HA preparation (approximately HA 1500-mers) was demonstrated to be much more potent than HA 6-mers in inducing chemokines such as macrophage inflammatory protein (MIP)-1α, MIP-1β, cytokine-responsive gene (crg)-2, regulated on activation normal T cell expressed and secreted (RANTES), and monocyte chemoattractant protein (MCP)-1 (McKee et al., 1996). These findings suggest that HA fragments within the size range of approximately 40–500 kDa (approximately HA 200–2600-mers) are particularly potent in activating macrophages.

HA fragments of such size range also induce key molecules in inflammatory responses other than chemokines. HA fragments of 200 kDa (approximately HA 1000-mers) but not HMW-HA enhanced the expression of cytokines such as IL-12 that induce a Th1-type immune response in macrophages (Hodge-Dufour et al., 1997). Such HA fragments also increased the expression of inducible nitric oxide synthase (iNOS) in macrophages (McKee et al., 1997), Kupffer cells (Rockey et al., 1998), and ECs in the liver (Rockey et al., 1998), and cyclooxygenase type 2 (COX-2) and thromboxane A2 (TXA2) in macrophages and renal tubular cells (Sun et al., 2001). In addition, they induced the expression of macrophage metalloelastase (MME), an MMP that degrades various extracellular matrix (ECM) components such as elastin, collagen, and fibronectin, and plays an integral role in the development of emphysema (Horton et al., 1999). They were even shown to modulate the balance of fibrinolysis. The 200 kDa HA preparation (approximately HA 1000-mers) but not HA 2-mers or HA 12–16-mers augmented the expression and inhibitory activity of PAI-1 and diminished the expression and activity of uPA in macrophages, tilting the balance of fibrinolysis to a lesser extent (Horton et al., 2000).

Many of the effects of HA fragments discussed above were found to be mediated by CD44 (Beck-Schimmer et al., 1998; Hodge-Dufour et al., 1997; McKee et al., 1997; Noble et al., 1993; Wang et al., 2006). The involvement of CD44 in HA fragment-induced inflammatory responses was verified *in vivo* by Teder et al. (2002). They found that CD44-deficient mice succumb to unremitting pulmonary inflammation after bleomycin treatment. It was characterized by the accumulation of HA fragments ranging from 20–160 kDa in size (approximately HA 100–830-mers) and a decreased activity of transforming growth factor-β (TGF-β). The survival defect and the histologic changes were significantly reversed by reconstitution with CD44-positive bone marrow, indicating that CD44 plays a crucial role in resolving inflammatory responses. Although it is clear that CD44 is involved in the clearance of the HA fragments under such inflammatory conditions, one question rises from this report. How did the HA fragments

V. HYALURONAN DEGRADATION, THE HYALURONIDASES

that accumulated in the injured lungs impact the inflammation under a CD44-negative circumstance, and what was the responsible receptor? The findings reported by Jiang et al. give a possible answer to this question (Jiang et al., 2005). These authors demonstrated that a 135 kDa HA preparation (approximately HA 700-mers) and HA fragments isolated from the serum of patients with acute lung injury (approximately HA 1000-mers) induced MIP-2 expression in macrophages through both TLR2 and TLR4 by using macrophages that lack TLR2, TLR4, or both. The involvement of CD44, TLR1, TLR3, TLR5, and TLR9 in these events was excluded. In contrast to roles of intermediate-sized HA fragments in eliciting inflammatory responses, they found that HMW-HA was protective against acute lung injury. Therefore, they concluded that HA degradation products generated after tissue injury initiate inflammatory responses by stimulating macrophages while HMW-HA maintain epithelial integrity, and that these effects are mediated by TLR-2 and TLR-4.

Although it is clear that TLR2 and TLR4 mediate the signals induced by intermediate-sized HA that initiate inflammatory responses, further discussion is required to understand their division of labor in these events. Similar to the findings by Jiang et al. (2005), it was shown that 135 kDa HA (approximately HA 700-mers) activated bone marrow-derived DCs mainly through TLR4 and partial synergy with TLR2 (Tesar et al., 2006). On the other hand, it was reported that 200 kDa HA (approximately HA 1000-mers) but not HMW-HA was able to induce a series of inflammatory chemokines, cytokines, and enzymes in thioglycollate-elicited peritoneal macrophages and MIP-1α in bone marrow-derived DCs through TLR2 but not TLR4, CD44, IL-1R, or IL-10R (Scheibner et al., 2006). It is currently difficult to explain how TLR2 and TLR4 co-operate. However, taking these reports into account, it appears that both TLR2 and TLR4 mediate the inflammatory responses induced by intermediate-sized HA fragments, and under some conditions, TLR2 and TLR4 compensate each other. It is of particular interest if different TLRs recognize different sizes of HA, or different cell types respond to HA fragments using different TLRs depending on their location and type of inflammation even if they were categorized as the same cell type such as macrophages.

CONCLUSIONS AND FUTURE PROSPECTS

HA fragments of various sizes have a variety of functions on different cell types, such as endothelial cells, tumor cells, and immune cells. The effects are mediated by different receptors including CD44, RHAMM, and TLRs, and also by unknown molecules. The effects look diverse, however have tight links to each other. From tumor biology point of view, perhaps such effects can be summarized as shown in Fig. 13.3. HA fragments, most

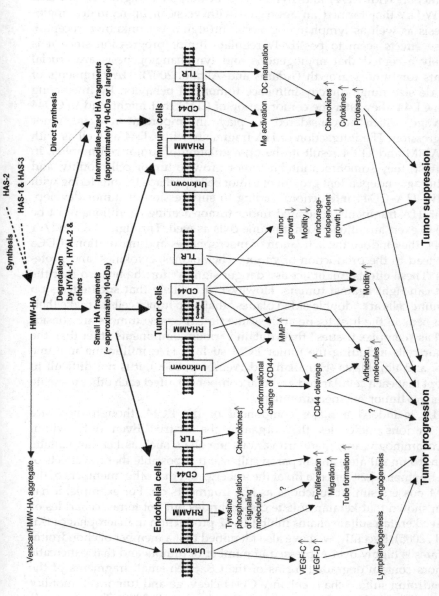

FIGURE 13.3 Effects of HA fragments on tumor development. HA fragments are generated through direct synthesis by HA synthases and/or degradation of HMW-HA by hyaluronidases and other means. The HA fragments cause various changes in the cellular functions of endothelial cells, tumor cells, and immune cells through receptors such as CD44, RHAMM, TLR, and a yet unknown receptor(s) that lead to tumor progression or suppression. "?" indicates connections that have been suggested but not experimentally proven. Refer to text for abbreviations.

of them within the range of 6–36-mers, promote angiogenesis through interaction with CD44 and RHAMM. Not only HA fragments, but also HMW-HA that formed an aggregate with versican up-regulates angiogenesis as well as lymphangiogenesis through a yet unknown receptor. These effects seem to positively regulate tumor progression since it is widely accepted that angiogenesis and lymphangiogenesis are crucial events for tumors growth (Adams and Alitalo, 2007). HA fragments of a wide size range directly influence tumor cell behavior. By interacting with CD44, they alter the conformation of CD44 that might lead to CD44 cleavage which is considered to play an important role in tumor progression. The interaction of HA fragments with CD44 as well as with RHAMM and TLR4 result in the up-regulation of tumor cell motility. In contrast, they sometimes inhibit tumor growth, tumor cell motility, and anchorage-independent growth of tumor cells, perhaps by interfering with HMW–HA–CD44 interaction, leading to suppression of tumor development. HA fragments generated under tumor-bearing conditions must be able to exert an influence on immune cells as well. Through TLRs and/or CD44, they induce the activation of macrophages and maturation of DCs that lead to the production of various chemokines, cytokines, and proteases. These effects might act as "danger signals" for the host so that the host can fight against tumors. However, it is likely that such effects on immune cells are "double-sided blades" that help tumor cells to make their own beds in the host tissue at the same time. Because tumors are considered as "complex tissues" that contain various components other than the tumor cells including the tumor ECM such as HA, infiltrating immune cells, and fibroblasts (Hanahan and Weinberg, 2000), it is not difficult to accept that each cellular and acellular component affect each other to create a perfect tumor microenvironment.

HA is indeed a major component of the ECM, though there are still various molecules that organize the matrix. Even only within glycosaminoglycans, there are various members such as keratan sulfate, chondroitin sulfates, and heparan sulfates. It is possible that the effects of HA fragments just reflect a tip of the iceberg, and that other members of the ECM possess similar functions as HA fragments do. For example, it has been shown that keratan sulfate disaccharides but not tetrasaccharides or native keratan sulfate chains inhibit IL-12 production in macrophages (Xu et al., 2005). Recently, we have also identified that a member of chondroitin sulfates is highly over-expressed in number of tumors and that pancreatic tumors contain degraded forms of the CS. Such small fragments of the chondroitin sulfate chain enhance CD44 cleavage and tumor cell motility in a fashion similar to HA fragments (Sugahara et al., 2008). Therefore, it is tempting to think that, as tumors develop, tumor-associated ECM becomes degraded and some of the degradation products within the tumor microenvironment start to send signals to the surrounding cells.

V. HYALURONAN DEGRADATION, THE HYALURONIDASES

By understanding such mechanisms, it is possible to make use of them. Some attempts are already made with HA fragments. For instance, different sizes of HA fragments were compared for their ability to enhance elastin matrix synthesis and generate a scaffolding material for the regeneration of vascular elastin matrices (Ibrahim et al., 2007; Joddar and Ramamurthi, 2006). In addition, HA fragments were tested for use to facilitate DNA uptake by cells (Hornof et al., 2008). It was demonstrated that when DNA/polyethylenimine (PEI) was coated with HA fragments smaller than 10 kDa (approximately HA 50-mers), the specific uptake of DNA by CD44 was enhanced in human corneal epithelial cells, whereas coating with larger HA polymers did not allow such enhancement. In addition to the use of HA fragments themselves, hyaluronidases that generate HA fragments are being tested as tumor markers in bladder cancers (Lokeshwar and Selzer, 2006). Such utilization of our knowledge regarding HA fragments in diagnostic and therapeutic technologies must be the next step that we should take to fight against tumors and inflammatory diseases.

ACKNOWLEDGMENTS

I thank Drs. Robert Stern (Department of Pathology, School of Medicine, University of California, San Francisco, CA), Vincent C. Hascall (Department of Biomedical Engineering and Orthopaedic Research Center/ND20, The Cleveland Clinic Foundation, Cleveland, OH), Masayuki Miyasaka (Laboratory of Immunodynamics, Department of Microbiology and Immunology, Osaka University Graduate School of Medicine, Osaka, Japan), Toshiyuki Tanaka (Laboratory of Immunobiology, School of Pharmacy, Hyogo University of Health Sciences, Kobe, Japan), Eva Engvall (Burnham Institute for Medical Research, La Jolla, CA), Tambet Teesalu (Vascular Mapping Center, Burnham Institute for Medical Research at UCSB, Santa Barbara, CA), Yukihiko Ebisuno (Department of Molecular Genetics, Institute of Biomedical Science, Kansai Medical University, Osaka, Japan), and Kazuyuki Sugahara (Laboratory of Proteoglycan Signaling and Therapeutics, Frontier Research Center for Post-Genomic Science and Technology, Hokkaido University Graduate School of Life Science, Hokkaido, Japan) for critically reading this manuscript. I also thank Dr. Erkki Ruoslahti (Vascular Mapping Center, Burnham Institute for Medical Research at UCSB) for his generous support.

References

Adams, R. H. and Alitalo, K. (2007). Molecular regulation of angiogenesis and lymphangiogenesis. *Nat Rev Mol Cell Biol* **8**, 464–478.

Ailles, L. E. and Weissman, I. L. (2007). Cancer stem cells in solid tumors. *Curr Opin Biotechnol* **18**, 460–466.

Akira, S., Uematsu, S., and Takeuchi, O. (2006). Pathogen recognition and innate immunity. *Cell* **124**, 783–801.

Aruffo, A., Stamenkovic, I., Melnick, M., Underhill, C. B., and Seed, B. (1990). CD44 is the principal cell surface receptor for hyaluronate. *Cell* **61**, 1303–1313.

Balazs, E. A., Watson, D., Duff, I. F., and Roseman, S. (1967). Hyaluronic acid in synovial fluid. I. Molecular parameters of hyaluronic acid in normal and arthritis human fluids. *Arthritis Rheum* **10**, 357–376.

Beck-Schimmer, B., Oertli, B., Pasch, T., and Wüthrich, R. P. (1998). Hyaluronan induces monocyte chemoattractant protein-1 expression in renal tubular epithelial cells. *J Am Soc Nephrol* 9, 2283–2290.

Blix, G. (1951). Hyaluronic acid in the pleural and peritoneal fluids from a case of mesothelioma. *Acta Soc Med Ups* 56, 47–50.

Chang, E. J., Kim, H. J., Ha, J., et al. (2007). Hyaluronan inhibits osteoclast differentiation via Toll-like receptor 4. *J Cell Sci* 120, 166–176.

Cordo Russo, R. I., Garcia, M. G., Alaniz, L., et al. (2008). Hyaluronan oligosaccharides sensitize lymphoma resistant cell lines to vincristine by modulating P-glycoprotein activity and PI3K/Akt pathway. *Int J Cancer* 122, 1012–1018.

Entwistle, J., Hall, C. L., and Turley, E. A. (1996). HA receptors: regulators of signalling to the cytoskeleton. *J Cell Biochem* 61, 569–577.

Feinberg, R. N. and Beebe, D. C. (1983). Hyaluronate in vasculogenesis. *Science* 220, 1177–1179.

Fieber, C., Baumann, P., Vallon, R., et al. (2004). Hyaluronan-oligosaccharide-induced transcription of metalloproteases. *J Cell Sci* 117, 359–367.

Forteza, R., Lieb, T., Aoki, T., et al. (2001). Hyaluronan serves a novel role in airway mucosal host defense. *Faseb J* 15, 2179–2186.

Fujii, K., Fujii, Y., Hubscher, S., and Tanaka, Y. (2001). CD44 is the physiological trigger of Fas up-regulation on rheumatoid synovial cells. *J Immunol* 167, 1198–1203.

Ghatak, S., Misra, S., and Toole, B. P. (2002). Hyaluronan oligosaccharides inhibit anchorage-independent growth of tumor cells by suppressing the phosphoinositide 3-kinase/Akt cell survival pathway. *J Biol Chem* 277, 38013–38020.

Giraudo, E., Inoue, M., and Hanahan, D. (2004). An amino-bisphosphonate targets MMP-9-expressing macrophages and angiogenesis to impair cervical carcinogenesis. *J Clin Invest* 114, 623–633.

Grossfeld, H. (1962). Production of hyaluronic acid in tissue culture of Rous sarcoma. *Nature* 196, 782–783.

Hamilton, S. R., Fard, S. F., Paiwand, F. F., et al. (2007). The hyaluronan receptors CD44 and Rhamm (CD168) form complexes with ERK1,2 that sustain high basal motility in breast cancer cells. *J Biol Chem* 282, 16667–16680.

Hanahan, D. and Weinberg, R. A. (2000). The hallmarks of cancer. *Cell* 100, 57–70.

Hardwick, C., Hoare, K., Owens, R., et al. (1992). Molecular cloning of a novel hyaluronan receptor that mediates tumor cell motility. *J Cell Biol* 117, 1343–1350.

Hascall, V. C. and Heinegård, D. (1974). Aggregation of cartilage proteoglycans. II. Oligosaccharide competitors of the proteoglycan-hyaluronic acid interaction. *J Biol Chem* 249, 4242–4249.

Hodge-Dufour, J., Noble, P. W., Horton, M. R., et al. (1997). Induction of IL-12 and chemokines by hyaluronan requires adhesion-dependent priming of resident but not elicited macrophages. *J Immunol* 159, 2492–2500.

Hornof, M., de la Fuente, M., Hallikainen, M., Tammi, R. H., and Urtti, A. (2008). Low molecular weight hyaluronan shielding of DNA/PEI polyplexes facilitates CD44 receptor mediated uptake in human corneal epithelial cells. *J Gene Med* 10, 70–80.

Horton, M. R., McKee, C. M., Bao, C., et al. (1998). Hyaluronan fragments synergize with interferon-gamma to induce the C-X-C chemokines Mig and interferon-inducible protein-10 in mouse macrophages. *J Biol Chem* 273, 35088–35094.

Horton, M. R., Shapiro, S., Bao, C., Lowenstein, C. J., and Noble, P. W. (1999). Induction and regulation of macrophage metalloelastase by hyaluronan fragments in mouse macrophages. *J Immunol* 162, 4171–4176.

Horton, M. R., Olman, M. A., Bao, C., et al. (2000). Regulation of plasminogen activator inhibitor-1 and urokinase by hyaluronan fragments in mouse macrophages. *Am J Physiol Lung Cell Mol Physiol* 279, L707–715.

Hosono, K., Nishida, Y., Knudson, W., et al. (2007). Hyaluronan oligosaccharides inhibit tumorigenicity of osteosarcoma cell lines MG-63 and LM-8 *in vitro* and *in vivo* via perturbation of hyaluronan-rich pericellular matrix of the cells. *Am J Pathol* **171**, 274–286.

Huang, B., Zhao, J., Li, H., et al. (2005). Toll-like receptors on tumor cells facilitate evasion of immune surveillance. *Cancer Res* **65**, 5009–5014.

Ibrahim, S., Joddar, B., Craps, M., and Ramamurthi, A. (2007). A surface-tethered model to assess size-specific effects of hyaluronan (HA) on endothelial cells. *Biomaterials* **28**, 825–835.

Itano, N., Sawai, T., Miyaishi, O., and Kimata, K. (1999a). Relationship between hyaluronan production and metastatic potential of mouse mammary carcinoma cells. *Cancer Res* **59**, 2499–2504.

Itano, N., Sawai, T., Yoshida, M., et al. (1999b). Three isoforms of mammalian hyaluronan synthases have distinct enzymatic properties. *J Biol Chem* **274**, 25085–25092.

Itano, N., Sawai, T., Atsumi, F., et al. (2004). Selective expression and functional character-istics of three mammalian hyaluronan synthases in oncogenic malignant transformation. *J Biol Chem* **279**, 18679–18687.

Jiang, D., Liang, J., Fan, J., et al. (2005). Regulation of lung injury and repair by Toll-like receptors and hyaluronan. *Nat Med* **11**, 1173–1179.

Jiang, D., Liang, J., and Noble, P. W. (2007). Hyaluronan in tissue injury and repair. *Annu Rev Cell Dev Biol* **23**, 435–461.

Joddar, B. and Ramamurthi, A. (2006). Fragment size- and dose-specific effects of hyaluronan on matrix synthesis by vascular smooth muscle cells. *Biomaterials* **27**, 2994–3004.

Kahmann, J. D., O'Brien, R., Werner, J. M., et al. (2000). Localization and characterization of the hyaluronan-binding site on the link module from human TSG-6. *Structure* **8**, 763–774.

Kajita, M., Itoh, Y., Chiba, T., et al. (2001). Membrane-type 1 matrix metalloproteinase cleaves CD44 and promotes cell migration. *J Cell Biol* **153**, 893–904.

Kanomata, N., Yokose, T., Kamijo, T., et al. (2005). Hyaluronan synthase expression in pleural malignant mesotheliomas. *Virchows Arch* **446**, 246–250.

Kosaki, R., Watanabe, K., and Yamaguchi, Y. (1999). Overproduction of hyaluronan by expression of the hyaluronan synthase HAS-2 enhances anchorage-independent growth and tumorigenicity. *Cancer Res* **59**, 1141–1145.

Kothapalli, D., Flowers, J., Xu, T., et al. (2008). Differential activation of ERK and Rac mediates the proliferative and anti-proliferative effects of hyaluronan and CD44. *J Biol Chem* **283**, 31823–31829.

Koyama, H., Hibi, T., Isogai, Z., et al. (2007). Hyperproduction of hyaluronan in neu-induced mammary tumor accelerates angiogenesis through stromal cell recruitment: possible involvement of versican/PG-M. *Am J Pathol* **170**, 1086–1099.

Koyama, H., Kobayashi, N., Harada, M., et al. (2008). Significance of tumor-associated stroma in promotion of intratumoral lymphangiogenesis: pivotal role of a hyaluronan-rich tumor microenvironment. *Am J Pathol* **172**, 179–193.

Kumar, S., West, D. C., Ponting, J. M., and Gattamaneni, H. R. (1989). Sera of children with renal tumours contain low-molecular-mass hyaluronic acid. *Int J Cancer* **44**, 445–448.

Laurent, T. C. and Fraser, J. R. (1992). Hyaluronan. *Faseb J* **6**, 2397–2404.

Lesley, J., Hascall, V. C., Tammi, M., and Hyman, R. (2000). Hyaluronan binding by cell surface CD44. *J Biol Chem* **275**, 26967–26975.

Lokeshwar, V. B., Öbek, C., Soloway, M. S., and Block, N. L. (1997). Tumor-associated hy-aluronic acid: a new sensitive and specific urine marker for bladder cancer. *Cancer Res* **57**, 773–777.

Lokeshwar, V. B. and Selzer, M. G. (2000). Differences in hyaluronic acid-mediated functions and signaling in arterial, microvessel, and vein-derived human endothelial cells. *J Biol Chem* **275**, 27641–27649.

V. HYALURONAN DEGRADATION, THE HYALURONIDASES

Lokeshwar, V. B., Rubinowicz, D., Schroeder, G. L., et al. (2001). Stromal and epithelial expression of tumor markers hyaluronic acid and HYAL-1 hyaluronidase in prostate cancer. *J Biol Chem* **276**, 11922–11932.

Lokeshwar, V. B., Cerwinka, W. H., and Lokeshwar, B. L. (2005). HYAL-1 hyaluronidase: a molecular determinant of bladder tumor growth and invasion. *Cancer Res* **65**, 2243–2250.

Lokeshwar, V. B. and Selzer, M. G. (2006). Urinary bladder tumor markers. *Urol Oncol* **24**, 528–537.

López-Otín, C. and Matrisian, L. M. (2007). Emerging roles of proteases in tumour suppression. *Nat. Rev Cancer* **7**, 800–808.

Manzanares, D., Monzon, M. E., Savani, R. C., and Salathe, M. (2007). Apical oxidative hyaluronan degradation stimulates airway ciliary beating via RHAMM and RON. *Am J Respir Cell Mol Biol* **37**, 160–168.

Marcante, M. L. (1965). Properties of hyaluronic acid from a human myxosarcoma. *Nature* **208**, 896.

McCawley, L. J. and Matrisian, L. M. (2001). Matrix metalloproteinases: they're not just for matrix anymore!. *Curr Opin Cell Biol* **13**, 534–540.

McKee, C.M., Penno, M.B., Cowman, M., et al. Hyaluronan (HA) fragments induce chemokine gene expression in alveolar macrophages. The role of HA size and CD44. *J Clin Invest* **98**, 2403–2413.

McKee, C. M., Lowenstein, C. J., Horton, M. R., et al. (1997). Hyaluronan fragments induce nitric-oxide synthase in murine macrophages through a nuclear factor kappaB-dependent mechanism. *J Biol Chem* **272**, 8013–8018.

Meyer, K. and Palmer, J. W. (1934). The polysaccharide of the vitreous humor. *J Biol Chem* **107**, 629–634.

Meyer, K., Smyth, E. M., and Dawson, M. H. (1939). The isolation of a mucopolysaccharide from synovial fluid. *J Biol Chem* **128**, 319–327.

Montesano, R., Kumar, S., Orci, L., and Pepper, M. S. (1996). Synergistic effect of hyaluronan oligosaccharides and vascular endothelial growth factor on angiogenesis in vitro. *Lab Invest* **75**, 249–262.

Nagano, O., Murakami, D., Hartmann, D., et al. (2004). Cell-matrix interaction via CD44 is independently regulated by different metalloproteinases activated in response to extracellular Ca(2+) influx and PKC activation. *J Cell Biol* **165**, 893–902.

Noble, P. W., Lake, F. R., Henson, P. M., and Riches, D. W. (1993). Hyaluronate activation of CD44 induces insulin-like growth factor-1 expression by a tumor necrosis factor-alpha-dependent mechanism in murine macrophages. *J Clin Invest* **91**, 2368–2377.

Noble, P. W., McKee, C. M., Cowman, M., and Shin, H. S. (1996). Hyaluronan fragments activate an NF-kappa B/I-kappa B alpha autoregulatory loop in murine macrophages. *J Exp Med* **183**, 2373–2378.

Oertli, B., Beck-Schimmer, B., Fan, X., and Wüthrich, R. P. (1998). Mechanisms of hyaluronan-induced up-regulation of ICAM-1 and VCAM-1 expression by murine kidney tubular epithelial cells: hyaluronan triggers cell adhesion molecule expression through a mechanism involving activation of nuclear factor-kappa B and activating protein-1. *J Immunol* **161**, 3431–3437.

Ohno, S., Im, H. J., Knudson, C. B., and Knudson, W. (2005). Hyaluronan oligosaccharide-induced activation of transcription factors in bovine articular chondrocytes. *Arthritis Rheum* **52**, 800–809.

Ohno, S., Im, H. J., Knudson, C. B., and Knudson, W. (2006). Hyaluronan oligosaccharides induce matrix metalloproteinase 13 via transcriptional activation of NFkappaB and p38 MAP kinase in articular chondrocytes. *J Biol Chem* **281**, 17952–17960.

Okamoto, I., Kawano, Y., Tsuiki, H., et al. (1999). CD44 cleavage induced by a membrane-associated metalloprotease plays a critical role in tumor cell migration. *Oncogene* **18**, 1435–1446.

Okamoto, I., Kawano, Y., Murakami, D., et al. (2001). Proteolytic release of CD44 intracellular domain and its role in the CD44 signaling pathway. *J Cell Biol* **155**, 755–762.

Okamoto, I., Tsuiki, H., Kenyon, L. C., et al. (2002). Proteolytic cleavage of the CD44 adhesion molecule in multiple human tumors. *Am J Pathol* **160**, 441–447.

Pahler, J. C., Tazzyman, S., Erez, N., et al. (2008). Plasticity in tumor-promoting inflammation: impairment of macrophage recruitment evokes a compensatory neutrophil response. *Neoplasia* **10**, 329–340.

Ragan, C. and Meyer, K. (1949). The hyaluronic acid of synovial fluid in rheumatoid arthritis. *J Clin. Invest* **28**, 56–59.

Rahmanian, M., Pertoft, H., Kanda, S., et al. (1997). Hyaluronan oligosaccharides induce tube formation of a brain endothelial cell line in vitro. *Exp Cell Res* **237**, 223–230.

Rockey, D. C., Chung, J. J., McKee, C. M., and Noble, P. W. (1998). Stimulation of inducible nitric oxide synthase in rat liver by hyaluronan fragments. *Hepatology* **27**, 86–92.

Rodgers, L. S., Lalani, S., Hardy, K. M., et al. (2006). Depolymerized hyaluronan induces vascular endothelial growth factor, a negative regulator of developmental epithelial-to-mesenchymal transformation. *Circ Res* **99**, 583–589.

Rooney, P., Wang, M., Kumar, P., and Kumar, S. (1993). Angiogenic oligosaccharides of hyaluronan enhance the production of collagens by endothelial cells. *J Cell Sci* **105**, 213–218.

Sattar, A., Rooney, P., Kumar, S., et al. (1994). Application of angiogenic oligosaccharides of hyaluronan increases blood vessel numbers in rat skin. *J Invest Dermatol* **103**, 576–579.

Schawalder, A., Oertli, B., Beck-Schimmer, B., and Wüthrich, R. P. (1999). Regulation of hyaluronan-stimulated VCAM-1 expression in murine renal tubular epithelial cells. *Nephrol Dial Transplant* **14**, 2130–2136.

Scheibner, K. A., Lutz, M. A., Boodoo, S., et al. (2006). Hyaluronan fragments act as an endogenous danger signal by engaging TLR2. *J Immunol* **177**, 1272–1281.

Seyfried, N. T., McVey, G. F., Almond, A., et al. (2005). Expression and purification of functionally active hyaluronan-binding domains from human cartilage link protein, aggrecan and versican: formation of ternary complexes with defined hyaluronan oligosaccharides. *J Biol Chem* **280**, 5435–5448.

Slevin, M., Krupinski, J., Kumar, S., and Gaffney, J. (1998). Angiogenic oligosaccharides of hyaluronan induce protein tyrosine kinase activity in endothelial cells and activate a cytoplasmic signal transduction pathway resulting in proliferation. *Lab Invest* **78**, 987–1003.

Slevin, M., Krupinski, J., Gaffney, J., et al. (2007). Hyaluronan-mediated angiogenesis in vascular disease: uncovering RHAMM and CD44 receptor signaling pathways. *Matrix Biol* **26**, 58–68.

Stern, R. (2004). Hyaluronan catabolism: a new metabolic pathway. *Eur J Cell Biol* **83**, 317–325.

Stern, R., Asari, A. A., and Sugahara, K. N. (2006). Hyaluronan fragments: an information-rich system. *Eur J Cell Biol* **85**, 699–715.

Sugahara, K. N., Hirata, T., Murai, T., and Miyasaka, M. (2004). Hyaluronan oligosaccharides and tumor progression. *Trends Glycosci Glycotech* **16**, 187–197.

Sugahara, K. N., Murai, T., Nishinakamura, H., et al. (2003). Hyaluronan oligosaccharides induce CD44 cleavage and promote cell migration in CD44-expressing tumor cells. *J Biol Chem* **278**, 32259–32265.

Sugahara, K. N., Hirata, T., Hayasaka, H., et al. (2006). Tumor cells enhance their own CD44 cleavage and motility by generating hyaluronan fragments. *J Biol Chem* **281**, 5861–5868.

Sugahara, K. N., Hirata, T., Tanaka, T., et. al. (2008). Chondroitin sulfate E fragments enhance CD44 cleavage and CD44-dependent motility in tumor cells. *Cancer Res* **6**, 7191–7199.

Sun, L. K., Beck-Schimmer, B., Oertli, B., and Wüthrich, R. P. (2001). Hyaluronan-induced cyclooxygenase-2 expression promotes thromboxane A2 production by renal cells. *Kidney Int* **59**, 190–196.

Takahashi, Y., Li, L., Kamiryo, M., et al. (2005). Hyaluronan fragments induce endothelial cell differentiation in a CD44- and CXCL1/GRO1-dependent manner. *J Biol Chem* **280**, 24195–24204.

Takeda, M., Ogino, S., Umemoto, R., et al. (2006). Ligand-induced structural changes of the CD44 hyaluronan-binding domain revealed by NMR. *J Biol Chem* **281**, 40089–40095.

V. HYALURONAN DEGRADATION, THE HYALURONIDASES

Taylor, K. R., Trowbridge, J. M., Rudisill, J. A., et al. (2004). Hyaluronan fragments stimulate endothelial recognition of injury through TLR4. *J Biol Chem* **279**, 17079–17084.

Taylor, K. R., Yamasaki, K., Radek, K. A., et al. (2007). Recognition of hyaluronan released in sterile injury involves a unique receptor complex dependent on Toll-like receptor 4, CD44, and MD-2. *J Biol Chem* **282**, 18265–18275.

Teder, P., Vandivier, R. W., Jiang, D., et al. (2002). Resolution of lung inflammation by CD44. *Science* **296**, 155–158.

Termeer, C., Benedix, F., Sleeman, J., et al. (2002). Oligosaccharides of Hyaluronan activate dendritic cells via toll-like receptor 4. *J Exp Med* **195**, 99–111.

Termeer, C. C., Hennies, J., Voith, U., et al. (2000). Oligosaccharides of hyaluronan are potent activators of dendritic cells. *J Immunol* **165**, 1863–1870.

Tesar, B. M., Jiang, D., Liang, J., et al. (2006). The role of hyaluronan degradation products as innate alloimmune agonists. *Am J Transplant* **6**, 2622–2635.

Thankamony, S. P. and Knudson, W. (2006). Acylation of CD44 and its association with lipid rafts are required for receptor and hyaluronan endocytosis. *J Biol Chem* **281**, 34601–34609.

Tolg, C., Hamilton, S. R., Nakrieko, K. A., et al. (2006). Rhamm−/− fibroblasts are defective in CD44-mediated ERK1,2 motogenic signaling, leading to defective skin wound repair. *J Cell Biol* **175**, 1017–1028.

Toole, B. P. and Gross, J. (1971). The extracellular matrix of the regenerating newt limb: synthesis and removal of hyaluronate prior to differentiation. *Dev Biol* **25**, 57–77.

Toole, B. P. (2004). Hyaluronan: from extracellular glue to pericellular cue. *Nat Rev Cancer* **4**, 528–539.

Truedsson, E. (1951). A case of mesothelioma of the pleura and peritoneum producing hyaluronic acid. *Acta Soc Med Ups* **56**, 39–45.

Turley, E. A., Noble, P. W., and Bourguignon, L. Y. (2002). Signaling properties of hyaluronan receptors. *J Biol Chem* **277**, 4589–4592.

Voelcker, V., Gebhardt, C., Averbeck, M., et al. (2008). Hyaluronan fragments induce cytokine and metalloprotease upregulation in human melanoma cells in part by signalling via TLR4. *Exp Dermatol* **17**, 100–107.

Wang, M. J., Kuo, J. S., Lee, W. W., et al. (2006). Translational event mediates differential production of tumor necrosis factor-alpha in hyaluronan-stimulated microglia and macrophages. *J Neurochem* **97**, 857–871.

West, D. C., Hampson, I. N., Arnold, F., and Kumar, S. (1985). Angiogenesis induced by degradation products of hyaluronic acid. *Science* **228**, 1324–1326.

West, D. C. and Kumar, S. (1989). The effect of hyaluronate and its oligosaccharides on endothelial cell proliferation and monolayer integrity. *Exp Cell Res* **183**, 179–196.

Xu, H., Ito, T., Tawada, A., et al. (2002). Effect of hyaluronan oligosaccharides on the expression of heat shock protein 72. *J Biol Chem* **277**, 17308–17314.

Xu, H., Kurihara, H., Ito, T., et al. (2005). The keratan sulfate disaccharide Gal(6S03) beta1,4-GlcNAc(6S03) modulates interleukin 12 production by macrophages in murine Thy-1 type autoimmune disease. *J Biol Chem* **280**, 20879–20886.

Yabushita, H., Noguchi, M., Kishida, T., et al. (2004). Hyaluronan synthase expression in ovarian cancer. *Oncol Rep* **12**, 739–743.

Yabushita, H., Kishida, T., Fusano, K., et al. (2005). Role of hyaluronan and hyaluronan synthase in endometrial cancer. *Oncol Rep* **13**, 1101–1105.

Yamada, Y., Itano, N., Narimatsu, H., et al. (2004). Elevated transcript level of hyaluronan synthase1 gene correlates with poor prognosis of human colon cancer. *Clin Exp Metastasis* **21**, 57–63.

Yoneda, M., Suzuki, S., and Kimata, K. (1990). Hyaluronic acid associated with the surfaces of cultured fibroblasts is linked to a serum-derived 85 kDa protein. *J Biol Chem* **265**, 5247–5257.

Zeng, C., Toole, B. P., Kinney, S. D., Kuo, J. W., and Stamenkovic, I. (1998). Inhibition of tumor growth in vivo by hyaluronan oligomers. *Int J Cancer* **77**, 396–401.

HYALURONAN IN CANCER EPITHELIAL-STROMAL INTERACTIONS

CHAPTER

14

Hyaluronan in Human Tumors: Importance of Stromal and Cancer Cell-Associated Hyaluronan

*Raija H. Tammi, Anne H. Kultti,
Veli-Matti Kosma, Risto Pirinen,
Päivi Auvinen, and Markku I. Tammi*

© 2009 Elsevier Inc.

HYALURONAN ACCUMULATION IN MALIGNANT TUMORS

Hyaluronan, a huge linear sugar polymer, composed of alternating N-acetyl-glucosamine (GlcNAc) and glucuronic acid (GlcUA) units, is common in many vertebrate extracellular matrices (Fraser et al., 1997). The number of the repeating disaccharide units varies, but can reach at least 25,000, corresponding to a molecular mass of 10 million Daltons, and an extended length of 22.5 μm. Its unique physicochemical properties, like the capacity to bind large amounts of water and form viscous gels at relatively low concentrations, has been suggested to provide a pliable matrix for tissue remodeling during embryonic development, and wound healing (Toole, 1997). It also acts like a filter, facilitating the diffusion of small, but excluding large molecules (Fraser et al., 1997). In contrast to other glycosaminoglycans, hyaluronan does not contain sulfate groups, is not covalently attached to a core protein, but can bind to proteoglycans and other proteins to organize pericellular and extracellular matrix. More recently, its capacity to activate various intracellular signaling routes through specific plasma membrane receptors has gained wide interest. Hyaluronan has turned out to be an active regulator of cell behavior, rather than solely an inert extracellular matrix component (see sections II and III in this volume).

The high concentration of hyaluronan in many embryonic tissues correlates with their rates of cell migration and proliferation. In addition, hyaluronan accumulation signals for epithelial to mesenchymal transformation, for example at the future valve sites of the embryonic heart tube (Camenisch et al., 2000). Because of the similarities between developing and tumor tissues, it was suggested decades ago that hyaluronan might be important for cancer growth. In line with this idea, biochemical analyses of glycosaminoglycans have shown accumulation of hyaluronan in tumors originating from connective tissues but also in those of epithelial origin, like carcinomas of breast, head and neck, ventricle, colon, and pancreas (reviewed by Knudson et al., 1989 and references in this paper, and Theocharis et al., 2003; Garcia et al., 2000; Llaneza et al., 2000).

The question whether hyaluronan resided in the tumor parenchyma, or the peritumoral stroma, remained obscure, as it could not be addressed with biochemical assays, and no suitable histological methods were available. During the last ten years, taking advantage of the high affinity binding between hyaluronan and the cartilage proteoglycan aggrecan (Hardingham and Muir, 1972), using either whole aggrecan (Knudson and Toole, 1985) or its isolated hyaluronan binding region (Ripellino et al., 1985) has enabled the acquisition of detailed information on the localization and relative content of hyaluronan in the different compartments of tumor tissues (Wang et al., 1996). Today, such analyses performed on a number of patient materials and tumor types allows us to recognize distinct patterns and trends. Overall, it is now well-established that hyaluronan levels both in the cancer epithelium and peritumoral stroma relate to tumor progression, and often indicate an unfavorable outcome of the disease. Moreover, recent reports suggest that hyaluronan is not only a prognostic indicator, but an active participant in the disease, and a novel target of therapy (Nakazawa et al., 2006). Therefore, in the future we may see hyaluronan inhibitors as one asset in the battle against cancer.

HYALURONAN ASSOCIATED WITH CANCER CELLS

There is a large body of evidence from experimental animals and *in vitro* models suggesting that the production of hyaluronan by tumor cells is important for their malignant behavior. Carcinoma cells isolated from different human or animal tumors produce variable amounts of hyaluronan under *in vitro* conditions, and although not universally true, high hyaluronan synthesis rate correlates with an aggressive, metastatic behavior in many cancer cell types (Kimata et al., 1983; van Muijen et al., 1995). Experimental overexpression or underexpression of hyaluronan synthases (HASes) supports the hypothesis that hyaluronan production by malignant cells enhances tumor spreading. Thus, HAS overexpression enhances local tumor growth (Kosaki et al., 1999; Liu et al., 2002; Jacobson et al., 2002; Itano et al., 2004), and distant metastases (Itano et al., 1999; Simpson et al., 2002), while antisense inhibition of HAS expression reportedly inhibits the growth of the primary tumor (Udabage et al., 2005) and distant metastasis (Simpson et al., 2002). Analyses of clinical materials has supported the relevance of the experimental studies cited above, by showing that hyaluronan on cancer cells may indeed contribute to high spreading potential of human tumors *in vivo*. However, opposite trends are seen in tumors arising from simple or stratified epithelia as described below.

Hyaluronan in Tumor Cells Arising from Simple Epithelia

The normal simple epithelia, and the transitional epithelium of the bladder show either a low signal for hyaluronan staining, or none at all (Fig. 14.1) (de la Torre et al., 1993; Wang et al., 1996; Auvinen et al., 1997; Pirinen et al., 1998; Lokeshwar et al., 2001; Hautmann et al., 2001), while cancer cells arising from these epithelia are often decorated by hyaluronan. Neoplastic growth starting from simple epithelia is thus associated with acquisition of hyaluronan expression (Fig. 14.1, Table 14.1). In most cases hyaluronan is localized on the plasma membranes, but sometimes also in the cytoplasm, and occasionally even in the nucleus (Ropponen et al., 1998;

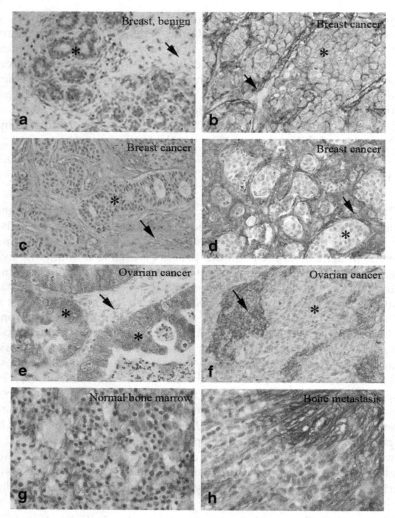

FIGURE 14.1 Stromal hyaluronan versus cancer cell-associated hyaluronan. Tumors archived in paraffin were sectioned and stained with a probe specific for hyaluronan (brown), nuclei were visualized with hematoxylin (blue). In benign breast tissue the normal tubular cells are hyaluronan negative and the stromal component is faintly hyaluronan-positive (a). In invasive breast carcinomas (b–d) cancer-cell-associated hyaluronan is some-times present close to plasma membrane (b) and the stromal staining varies from faint (c), to strong (d). An ovarian carcinoma with tumor cell-associated hyaluronan and very low stromal staining is shown in (e), and another case with strong stromal hyaluronan signal without epithelial staining in (f). Bone metastasis of a breast carcinoma shows a strong stromal hyaluronan staining (h) while the adjacent normal bone marrow is less intensely stained (g). Stars point the tumor epithelial cells and the arrows the stromal compartment. (See Page 6 in Color Section at the back of the book).

Auvinen et al., 2000). Some hyaluronan-positive cells are found in about half of breast and prostate tumors, while in ovarian, colorectal, and gastric cancers more than 80 and 90% of the tumors can be positive (Ropponen et al., 1998; Setälä et al., 1999; Auvinen et al., 2000; Anttila et al., 2000; Lipponen et al., 2001). However, all tumors are internally heterogeneous in the expression of cancer cell-associated hyaluronan, with areas of high and low frequency of positive cells in a single tumor. In general, the number of positive cells is typically higher in colon and gastric cancers than for example in ovarian cancers (Ropponen et al., 1998; Setälä et al., 1999; Anttila et al., 2000). Different histological subtypes of cancers arising from a certain epithelial tissue can display distinct hyaluronan expression profiles, as shown by examples of lung (Pirinen et al., 2001) (Fig. 14.2), gastric (Setälä et al., 1999), salivary gland (Xing et al., 1998), and thyroid cancers (Böhm et al., 2002). Hyaluronan content also correlates with other biological parameters of cancer cells, such as missing estrogen and progesterone receptors in breast cancer cells (Auvinen et al., 2000). This must be considered when comparing tumor sets with different histological subtype distributions.

Several findings support the idea that an ectopic expression of hyaluronan in the malignant cells arising from simple epithelia contributes to an aggressive behavior of these tumors (Table 14.1). Tumor grades, or degrees of cellular dedifferentiation, show direct correlations with the proportion of hyaluronan-positive cancer cells, or the intensity of hyaluronan staining on those cells (Table 14.1). Cell-associated hyaluronan in cancers arising from simple epithelia also correlates with local invasive growth, spreading to local lymph nodes, and even with distant metastasis, as in salivary gland and colorectal carcinomas (Table 14.1). Interestingly, metastases often contain cancer cells with a high level of hyaluronan expression and correlate with those of the original tumors (Anttila et al., 2000; Böhm et al., 2002).

The correlations described above, suggesting that tumors with hyaluronan-positive cells are more aggressive, have been studied further by analyzing patient prognosis. A high level of hyaluronan on tumor cells is a strong indicator of unfavorable outcome, i.e. shortened overall survival (OS) and/or the time of disease free survival (DFS) in breast, gastric, and colorectal cancers (Table 14.1). In colon cancer, only 20% of patients survived if a large fraction of their tumor cells were hyaluronan-positive, while 80% survived when there were just few hyaluronan expressing cells (Ropponen et al., 1998) (Fig. 14.3). In gastric cancer the corresponding survival levels were 25 and 60% in cases with high and low hyaluronan expression, respectively (Setälä et al., 1999), and in breast cancer 55 and 80% (Auvinen et al., 2000). In a Cox multivariate analysis of colon cancer, including all conventional prognostic factors, tumor cell hyaluronan is an independent prognostic factor for both OS and DFS (Ropponen et al., 1998; Köbel et al., 2004).

TABLE 14.1 Cancer Cell-Associated Hyaluronan, Tumor Characterics, and Patient Prognosis

| Tumor | HA content | HA correlation with | | | | HA as a prognostic factor | | | | Reference |
| | | Grade | Local invasion | Nodal invasion | Distant metastases | Univariate*** | | Multivariate*** | | |
						OS	RFS/DFS	OS	RFS/DFS	
Simple epithelia										
Bladder	Up	+								Hautmann et al. (2001)
Breast	Up	+		+		+	–		–	Auvinen et al. (2000)
Breast*	Up	+					+		–	Suwiwat et al. (2004)
Colon	Up	+	+			+	+	+	+	Ropponen et al. (1998)
Colon	Up	–		+	+	+		–/+**		Köbel et al. (2004)
Gastric	Up	+	+	–	–	+		–		Setälä et al. (1999)
Lung, adenocarcinoma	Up	+				–				Pirinen et al. (2001)
Ovarian	Up	+				–				Anttila et al. (2000)
Prostate	Up		Perineural invasion			–	–			Lipponen et al. (2001)
Salivary	Up				+					Wein et al. (2005)
Thyroid	Up		+							Böhm et al. (2002)
Stratified epithelia										
Esophagus	Down	+			+		+			Wang et al. (1996)
Larynx	Down	+					+		–	Hirvikoski et al. (1999)
Lung, SCC + adenocarcinoma	Down	–						–		Pirinen et al. (2001)

Cancer		Breslow thickness						Reference
Melanoma	Down			+	+	−	−	Karjalainen et al. (2000)
Oral	Down	+	+?	+	+	+	+	Kosunen et al. (2004)
Skin	Down	+						Karvinen et al. (2003a)

OS, overall survival; DFS/RFS, disease free/recurrence free survival.
SCC, squamous cell carcinoma.
*Node negative patients only.
**Combined with CD44v6 expression.
***+ correlates with short OS/RFS, – no correlation.

VI. HYALURONAN IN CANCER EPITHELIAL–STROMAL INTERACTIONS

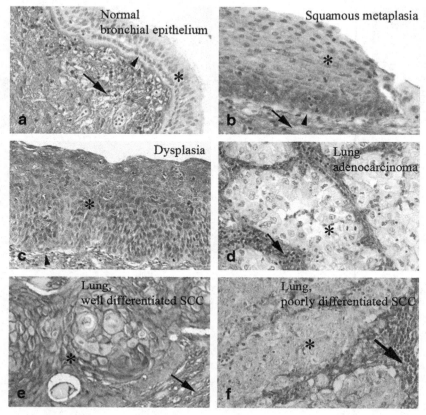

FIGURE 14.2 Association of epithelial hyaluronan expression with squamous cell differentiation. Tumors archived in paraffin were sectioned and stained with a probe specific for hyaluronan (brown), nuclei were visualized with hematoxylin (blue). Basal cells (arrowhead) may show a faint hyaluronan signal, while most of the normal bronchial epithelium does not stain for hyaluronan (a). Metaplastic bronchial epithelium is associated with a strong hyaluronan expression, especially in the superficial cell layers (b). Epithelial hyaluronan expression is also seen in dysplastic bronchial epithelium (c) and in well differentiated SCCs (e). In the latter case the staining pattern is very similar to that in cancers originating from squamous epithelia, like skin SCC or oral SCC. In contrast, a poorly differentiated lung SCC shows just a faint hyaluronan staining on the cancer cells (f). Well differentiated lung adenocarcinoma shows very little hyaluronan among the cancer cells, but the stroma is intensely stained (d). The stars point the (tumor) epithelial cells and the arrows the stromal compartment. (See Page 7 in Color Section at the back of the book).

Furthermore, hyaluronan is the only significant prognostic factor in the Dukes B subgroup of patients (Ropponen et al., 1998). The histological findings are in agreement with radioimmunometric assays of hyaluronan concentration in a large series of colon and gastric carcinomas (Llaneza et al., 2000; Vizoso et al., 2004).

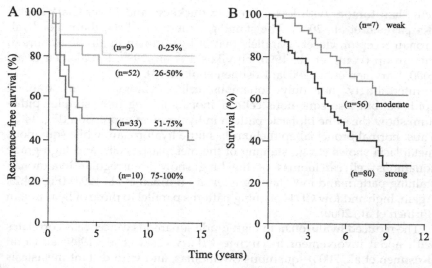

FIGURE 14.3 Prognosis of cancer patients – correlations with cell-associated and stromal hyaluronan. Sections of colorectal carcinoma were stained for hyaluronan with a specific probe, scored for the proportion of tumor cells positive for hyaluronan, and grouped into the indicated ranges of percentages for Kaplan–Maier curves of cumulative recurrence-free survival (**A**). Sections of breast cancer were scored for the intensity of hyaluronan staining in the peritumoral stroma (weak, moderate, strong) to present curves for survival as above (**B**). The numbers of patients in each group are shown in parenthesis. The data have been compiled from Ropponen et al. (1998) and Auvinen et al. (2000), respectively.

Hyaluronan in Tumor Cells Arising from Stratified Epithelia

Stratified epithelial cells display a fair amount of surface hyaluronan already in their native state (Tammi et al., 2004). Therefore, it is difficult to tell whether hyaluronan increases in the early grade neoplasms arising from stratified epithelia, while it is quite obvious that cell-associated hyaluronan is decreased in high-grade as compared to low-grade squamous cancers (Table 14.1). Anyway, several premalignant or early malignant lesions, like dysplasias and the *in situ* carcinomas of oral, laryngeal, esophageal and skin epithelium, all show a tendency to increased hyaluronan expression compared to normal epithelium (Wang et al., 1996; Hirvikoski et al., 1999; Karvinen et al., 2003a; Kosunen et al., 2004). Hyaluronan expression remains high in well-differentiated carcinomas of these tissues, while in high-grade (poorly differentiated) squamous cell carcinomas hyaluronan staining becomes irregular, and its overall intensity is reduced (Table 14.1). Skin melanomas show a hyaluronan expression pattern similar to that of squamous cell carcinomas (Karjalainen et al., 2000). Thus, aggressive melanomas contain less hyaluronan-positive cells than tumors with lower Breslow thickness and lower Clark's level (Karjalainen et al., 2000). The staining pattern of CD44, the major hyaluronan receptor, closely parallels that of hyaluronan in all the tumors in this group (Wang et al., 1996; Hirvikoski et al., 1999; Karjalainen et al., 2000; Karvinen et al., 2003a; Kosunen et al., 2004).

Interestingly, not only squamous cell carcinomas from stratified epithelia, but also squamous cell carcinomas arising from simple epithelium show the same biphasic pattern in hyaluronan expression (Fig. 14.2). Thus, normal bronchial epithelium has little hyaluronan, while squamous metaplasia shows strong staining of the metaplastic cells, and high-grade squamous cell carcinomas of the lung show an irregular hyaluronan staining pattern and low staining intensity (Pirinen et al., 1998) (Fig. 14.2). Again, high and low CD44 staining patterns parallel to those of hyaluronan (Pirinen et al., 2000).

The reduced hyaluronan in high-grade squamous cancer cells correlates with nodal involvement in laryngeal (Hirvikoski et al., 1999) and oral (Kosunen et al., 2004) squamous carcinomas, and with distant metastasis in laryngeal cancers (Hirvikoski et al., 1999) (Table 14.1). The importance of the loss of hyaluronan in squamous cell carcinoma progression was indicated by the findings that it was associated with poor patient outcome both in laryngeal and oral carcinomas (Hirvikoski et al., 1999; Kosunen et al., 2004), as well as in skin melanomas (Karjalainen et al., 2000). Only 50% of oral carcinoma patients showing irregular and low hyaluronan staining at the time of diagnosis were alive after 5 years, as compared to 80% survival among those with normal hyaluronan staining pattern (Kosunen et al., 2004). The loss of hyaluronan from the oral cancer cells turned out to be an independent prognostic factor in Cox multivariant analysis of (Kosunen et al., 2004) (Table 14.1).

The analyses of tumor biopsies do not tell whether the decline of hyaluronan in the high-grade squamous carcinomas results from reduced synthesis, enhanced rate of degradation, or loss from cell surface due to CD44 depletion. Increased and ectopic hyaluronidase expression (Franzmann et al., 2003; Christopoulos et al., 2006), and increased expression of MT-MMP1 and ADAM proteases, capable of CD44 stripping from cell surface, were reported in head and neck squamous cell carcinomas (Nakamura et al., 2004; Zhai et al., 2005), suggesting that hyaluronan turnover was increased. In addition to hyaluronan and CD44, other adhesion molecules like cadherins and catenins are often downregulated or their function is impaired (Thomas and Speight, 2005). Thus, the decrease of hyaluronan and CD44 may reflect the more general loss of adhesion molecules in poorly differentiated SCCs. Although the cell–cell contacts based on cadherin probably dominate in normal conditions, the loss of the low affinity contacts mediated by CD44 and hyaluronan may become significant in the absence of cadherin.

In summary, cell-associated hyaluronan has a completely different role in the progress of adenocarcinomas and squamous cell carcinomas, originating from monolayered and stratified epithelia, respectively.

STROMAL ACCUMULATION OF HYALURONAN

Even the first histochemical stainings with the hyaluronan probe derived from cartilage aggrecan suggested that stromal hyaluronan is increased in lung and breast tumors (Knudson et al., 1989; Bertrand et al., 1992; Ponting et al., 1993), and that the increase is more pronounced in the invasion front as compared with the central tumor areas (Kudson et al., 1989; Bertrand et al., 1992; Ponting et al., 1993). The increased stromal hyaluronan staining in breast cancer has been later confirmed in several studies, and is also present in prostate, ovarian, bladder, endometrial, and thyroid carcinomas, and lung adenocarcinomas (Table 14.2) (Fig. 14.1). The accumulation of stromal hyaluronan can be drastic in tissues with naturally low hyaluronan concentration. For example, in grade three ovarian cancer the median increase of hyaluronan concentration was 49-fold (Hiltunen et al., 2002) (Fig. 14.1).

Poorly differentiated or high-grade tumors have typically more stromal hyaluronan than well-differentiated tumors, examples including breast, prostate, and ovarian cancers (Table 14.2). Stromal hyaluronan accumulation also associates with cancer cell penetration of capsules, lymph vessels, and nerves in prostate carcinomas, local lymph node infiltration in breast cancers, and with distant metastasis in thyroid, ovarian, and prostate cancers (Table 14.2).

In line with its association with invasion, strong stromal hyaluronan staining indicates poor patient outcome in breast, ovarian, prostate, and thyroid cancers (Table 14.2). In breast cancer, half of the patients showing strong stromal hyaluronan signal were alive after 5 years follow-up, while all patients with low stromal hyaluronan signal survived (Auvinen et al., 2000). In ovarian cancers the corresponding values were 25 and 45% (Anttila et al., 2000). Furthermore, strong hyaluronan signal is an independent prognostic factor in breast and ovarian tumors in Cox multivariate analysis (Table 14.2). In ovarian cancers, which present a rather bleak outlook even in early disease, strong stromal hyaluronan signal is an independent prognostic factor for short DFS and OS also when FIGO I and II stages are examined separately (Anttila et al., 2000). The lack of a similar statistically significant correlation in a node negative subgroup of breast cancers (Suwiwat et al., 2004) with a generally favorable prognosis, may just reflect the low number of cases (disease recurrences or deaths) in these patient groups during a limited follow-up period. Figure 14.1 shows hyaluronan staining of a metastatic lesion of a breast cancer in bone

TABLE 14.2   Stromal Hyaluronan Expression, Tumor Characteristics, and the Patient Prognosis

Tumor	HA content	HA correlation with				HA as a prognostic factor				Reference
						Univariate		Multivariate		
		Grade	Local invasion	Nodal invasion	Distant metastases	OS	RFS/DFS	OS	RFS/DFS	
Bladder	Up	+								Hautmann et al. (2001)
Breast	Up	+		+		+		+		Auvinen et al. (2000)
Breast	Up	+	+				+			Bertrand et al. (1992)
Breast	Up	−	+					−		Pontig et al. (1993)
Breast*	Up	+						−	−	Suwiwat et al. (2004)
Breast*	Up	+		+						Wernicke et al. (2003)
Endometrium	Up	−					+			Afify et al. (2005)
Lung, adenocarcinoma	Up	−					+		+	Pirinen et al. (2001)
Ovarian	Up	+			+	+		+	+	Anttila et al. (2000)
Ovarian	Up	+								Hiltunen et al. (2002)
Prostate	Up	Gleason	+		+	+		−	−	Lipponen et al. (2001)
Prostate	Up		+						−	Aaltomaa et al. (2002)
Prostate	Up		+							Lokeshwar et al. (2001)

				Posey et al. (2003)
		–/+**	–	Ekici et al. (2004)
		+	+	Böhm et al. (2002)

Prostate	Up	+	+***	
Prostate	Up	+		
Thyroid	Up	–		

OS, overall survival; DFS/RFS, disease free/recurrence free survival.
*Node negative patients.
**If HYAL-1 is not included in the model.
***Cancer related mortality.
+ Correlates with short OS/RFS/DFS; – no correlation.

marrow. The strong hyaluronan staining highlights the cancer in the host tissue. Metastases of thyroid and ovarian carcinomas show a similar hyaluronan-rich peritumoral stroma which correlates with the expression level in the original tumor (Anttila et al., 2000; Böhm et al., 2002). Invasive lesions of prostate cancer into adjacent tissues like seminal vesicle often contain a high level of hyaluronan (Lokeshwar et al., 2001). The data suggest that tumors able to induce stromal hyaluronan accumulation have a growth advantage in the new environment as compared to those without this capacity.

The staining intensity of stromal hyaluronan does not associate with the progression of tumors from the gastrointestinal tract (Ropponen et al., 1998; Setälä et al., 1999; Köbel et al., 2004), or those arising from stratified epithelia (Hirvikoski et al., 1999; Karjalainen et al., 2000; Karvinen et al., 2003a), perhaps because hyaluronan is naturally abundant in the stroma of these tissues. The strong prognostic power of stromal hyaluronan in breast and ovarian cancers may be due to the ease of detection of the increased signal in these hyaluronan poor tissues, but probably also because of the more profound change in tissue structure induced by hyaluronan accumulation (de la Torre et al., 1993; Auvinen et al., 1997; Anttila et al., 2000) (Fig. 14.1).

The hyaluronan-rich peritumoral stroma is also enriched in versican, a hyaluronan binding proteoglycan (Voutilainen et al., 2003), the expression level of which can further define the prognostic value of hyaluronan (Suwiwat et al., 2004).

MECHANISMS OF HYALURONAN ACCUMULATION IN TUMORS

The assays on tumor biopsies do not give information about the dynamic processes that control the synthesis and catabolism of hyaluronan. Currently, we have not much data on the relative importance of factors that determine the content of hyaluronan in malignant tissues, but some of the alternatives are discussed below.

Enhanced Synthesis of Hyaluronan in Tumors

In a given cell, hyaluronan is produced by one or more of the three homologous hyaluronan synthases (HAS1, HAS2, and HAS3). Each of these enzymes can take care of all the tasks required for hyaluronan secretion, including initiation of the chain, the alternating transfer of glucuronic acid (GlcUA) and N-Acetyl-glucosamine (GlcNAc) from their UDP-activated forms, holding the growing polysaccharide, and creating a plasma membrane pore for its continuous extrusion (Weigel et al., 1997).

Therefore, the rate of hyaluronan synthesis is mostly determined by the activity of HAS.

While experimental UDP-GlcUA depletion may limit the rate of hyaluronan synthesis in some cases (Kakizaki et al., 2004; Vigetti et al., 2006), there is limited evidence that increasing the cellular UDP-GlcNAc and UDP-GlcUA pool size could markedly stimulate hyaluronan synthesis. Instead, most of the studies currently available indicate that hyaluronan synthesis correlates with the level of HAS mRNA (Pienimäki et al., 2001; Karvinen et al., 2003b; Pasonen-Seppänen et al., 2003), suggesting that transcriptional regulation is a major determinant of the net HAS activity. The importance of HAS transcription as a major determinant of hyaluronan synthesis is in line with the short turnover time of the HAS protein (Rilla et al., 2005), and the rapid and large fluctuations observed in HAS expression (Fülop et al., 1997; Pienimäki et al., 2001). However, most of these studies have been conducted *in vitro*, leaving some uncertainty, for instance, in the HAS protein turnover times *in vivo* (Anggiansah et al., 2003).

The genomic region containing the HAS2 gene is frequently amplified in prostate cancer (Tsuchiya et al., 2002), but it is not known whether such amplifications correlate with Has mRNA levels or hyaluronan synthesis. In fact, it seems more likely that up-regulation of HAS expression is mediated by the frequently mutated signaling proteins such as those of the ErbB family (Normanno et al., 2006) and Ras (Shaw and Cantley, 2006) with downstream targets in the HAS promoter (Saavalainen et al., 2005).

Various cellular signaling pathways are likely to contribute to transcriptional up-regulation of HASes, since growth factor exchange between epithelial and stromal cells is common (Smola et al., 1993). It was suggested long ago that the cancer (epithelial) cells have adopted an ability to stimulate the adjacent stromal cells to constantly change the tissue structure to enable tumor growth, hyaluronan being an important component in this molecular remodeling package (Knudson et al., 1988). The idea of growth factor-induced hyaluronan accumulation in cancer is consistent with the findings that overexpression of EGF and EGFR (Normanno et al., 2006), TGF-beta (Glick, 2004), PDGF (Pietras et al., 2003), IGF-1 (Giovannucci, 2001), and FGF (Kwabi-Addo et al., 2004) is common in cancers, and that HAS2 expression is often increased by the same growth factors (Jacobson et al., 2000; Pienimäki et al., 2001; Karvinen et al., 2003b). PDGF and TGFβ are good candidates for the stimulation of hyaluronan synthesis in stromal cells (Mueller et al., 2004).

To resolve the issue of HAS involvement in hyaluronan accumulations, analyses of both HAS mRNA and protein levels in tissues would be highly desirable, using real-time RT-PCR and Western blotting. Unfortunately, good HAS antibodies have not been generally available and the level of HAS proteins is low, which has delayed progress in the field. Using immunohistochemistry, expression of all three HAS proteins was recently reported in tissue sections of mesotheliomas (Kanomata et al., 2005), ovarian cancers (Yabushita et al., 2004), and endometrial cancers (Yabushita et al., 2005). However, due to general problems in the immunohistochemical reactions, these findings need confirmation by more specific assays like western blotting, and support from quantitative mRNA analyses of the corresponding HASes.

As discussed above, a reasonable hypothesis is that increased hyaluronan content in the malignant tumors is associated with transcriptional up-regulation of one or more of the HASes. However, at the moment relatively limited data are available on HAS mRNA levels in malignant tumors. HAS1 has been associated with unfavorable prognosis in colon cancer (Yamada et al., 2004) and myeloma (Calabro et al., 2002; Adamia et al., 2005), while our unpublished real-time PCR data indicate that there is hardly any HAS1 mRNA in serous ovarian cancers (Nykopp et al., submitted for publication), and that HAS2 and HAS3 messages are not statistically significantly elevated as compared with normal ovaries and benign ovarian tumors.

Catabolism as a Determinant of Tumor Hyaluronan Content?

The uncertainty concerning biosynthesis as the only or main determinant of hyaluronan accumulation in cancer tissue makes examination of hyaluronan catabolism warranted. Unlike synthesis, which is mostly controlled by HAS activity, the rate of hyaluronan catabolism depends on multiple reaction chains that occur sequentially and in parallel with each other.

Hyaluronan, once released from HAS, may remain bound to cell surface receptors like CD44, or get immobilized in the extracellular matrix due to its large size, enhanced by complex formation with hyaluronan binding proteoglycans like versican and aggrecan (Iozzo, 1998), or specific cross-linking systems like TSG6/IαI (Milner et al., 2006).

It is reasonable to assume that the catabolism of native, high molecular mass hyaluronan can be triggered by a few initial cleavages by hyaluronidases or oxygen free radicals (Ågren et al., 1997; Tammi et al., 2001; Stern, 2005). Hyaluronan thus released may either become endocytosed and degraded by local cells (Tammi et al., 2001; Knudson et al., 2002) or diffuse into lymph vessels which carry it for specific, high capacity uptake receptors in the lymph nodes and liver endothelial cells (Harris et al., 2004).

Except PH-20, the enzyme on the surface of spermatozoa, most hyaluronidases have their pH optima below 4, suggesting that they are only active in lysosomes (Stern and Jedrzejas, 2006). Therefore, it is currently unknown, whether the more commonly expressed HYAL-1-3 hyaluronidases can degrade hyaluronan in the extracellular pH, usually above 6. If hyaluronidases mainly work within endosomes or lysosomes, they are unlikely to control tissue hyaluronan degradation, which becomes more dependent on the rate of endocytosis and the initial cleavages that facilitate its diffusion into lymphatics.

Increased HYAL-1 expression and hyaluronidase activity occur in prostate and bladder cancers, and form a strong, unfavorable prognostic indicator (Lokeshwar et al., 2001; 2002). On the other hand, hyaluronidase activity is low in advanced ovarian cancer, as compared with normal ovary (Hiltunen et al., 2002), suggesting that in different tumors, hyaluronidase may have opposite effects. Moreover, a recent study suggests that both elevated and lowered levels of HYAL-1 enhance malignant growth in a single cancer cell type (Lokeshwar et al., 2005). In general, the role of hyaluronidases in cancer is contradictory, since by degrading hyaluronan they would counteract the formation of environment favorable for tumor progression, while the hyaluronan oligosaccharides created by its activity stimulate angiogenesis, important for cancer growth (Rooney et al., 1995; Takahashi et al., 2005).

Changes, defects, or allelic imbalance in chromosome 3p21.3 containing HYAL-1, HYAL-2 and HYAL-3 hyaluronidases, are frequent in neoplasms, suggesting tumor suppressor function (Stern, 2005). However, allelic imbalance in this genomic region does not correlate with patient survival or hyaluronan content in ovarian cancer (Tuhkanen et al., 2005), which may reflect the inaccuracy of the LOH technique, and involvement of other genes in the region (Ito et al., 2005). Nevertheless, the expression of HYAL-1 is significantly reduced, and correlates with reduced hyaluronidase activity and increased hyaluronan concentration in ovarian cancer (Nykopp et al., submitted for publication). At this time, it is not known whether chromosomal defects or transcriptional down-regulation account for the reduced HYAL-1 gene expression. Nevertheless, the lower HYAL-1 expression, concomitantly with unchanged expression of HASes, suggests that hyaluronidase activity somehow contributes to hyaluronan accumulation in ovarian cancer (Nykopp et al. submitted for publication).

Genomic Changes in the Tumor Stroma

As described above, tumor stroma is an unusual type of connective tissue with hyaluronan accumulation of as a specific feature. While the factors responsible for the accumulation of stromal hyaluronan remain to be defined, recent analyses of the stromal cells have revealed a novel alternative to explain its distinct nature. Several chromosomal probes revealed frequent loss of heterozygosity (LOH) in the stromal cells adjacent to the malignant tumor epithelium. Some of the defective chromosomal sites were identical, some different from those in the actual cancer cells (Tuhkanen et al., 2004). Therefore, the accumulation of hyaluronan may not be just due to stimulation of the normal stroma of the organ, but reflect the existence of a novel, transformed connective tissue which serves the needs of the tumor. Peritumoral stromal cells in breast cancer express epithelial keratins, suggesting that they are probably tumor cells that have undergone epithelial-mesenchymal transition (Petersen et al., 2003). While these myofibroblast type cells cannot form tumors on their own, they are crucial in supporting their malignant epithelial companions. The exact origin of these genetically altered stromal cells is puzzling. Have they differentiated from a common ancestral cancer cell? Do they reflect a continuous epithelial-mesenchymal transition in the malignant cells? Are both cell types derived from resident tumor stem cells? Moreover, the role of stromal cells in tumorigenesis was stressed by the recent finding that a genomic defect in the TGF receptor of stromal cells can initiate tumor in the prostate epithelial cells (Bhowmick et al., 2004). At this time, it looks highly likely that many of the most common carcinomas have a transformed stromal cell type that determines the progression of the tumor.

HOW DOES HYALURONAN CONTRIBUTE TO CANCER PROGRESSION?

Hyaluronan Support for Migration and Survival of Malignant Cells

The invasive front of a malignant tumor often shows a very strong hyaluronan signal (Knudson et al., 1989; Bertrand et al., 1992; Ponting et al., 1993), suggesting that hyaluronan is involved in the migration of the cancer cells into the adjacent tissue. Indeed, there is a large body of evidence for the stimulatory effect of hyaluronan on the locomotory activity of cells. Hyaluronan as a gel-forming, extremely hydrophilic polymer may, owing to its swelling pressure, open up extracellular spaces for cells to migrate into.

Mesenchymal type migratory properties are induced in carcinoma cells that gain active hyaluronan synthesis (Zoltan-Jones et al., 2003), perhaps through stimulation of lamellipodial extensions and Rac activity (Bakkers et al., 2004) or microvillous extensions (Kultti et al., 2006) (Fig. 14.4). Cell migration was inhibited by blocking HAS2 expression (Bakkers et al., 2004) or hyaluronan synthesis through UDP-GlcUA depletion (Kakizaki et al., 2004; Rilla et al., 2004). The stimulatory effect of hyaluronan on cell migration is at least partly mediated through its receptors on cell surface, and the activation of signaling pathways within the cells (Thorne et al., 2004). The role of hyaluronan as a pro-invasion agent is supported by the finding that hyaluronan content predicts local recurrence of breast cancer better than any of the common risk factors (Dr. Elda Tagliabue, Milan, Italy, personal communication).

In addition to the signals related to migration, hyaluronan binding to its receptors also activates protein phosphorylation cascades that prevent apoptosis, a threat to cells that have escaped their regular adhesive restraints and endogenous growth restrictions (Misra et al., 2006). This is exemplified by the ability of hyaluronan to stimulate PI-3-kinase/Akt signaling that keeps cancer cell lines alive in the presence of cytotoxic drugs (Misra et al., 2005) and in soft agar (Ghatak et al., 2002), a medium that tests the ability to survive without regular matrix support. These functions of hyaluronan may also operate *in vivo*, as suggested by the ability of hyaluronan oligomers to inhibit the growth of cancer cells inoculated subcutaneously, perhaps by a mechanism involving displacement of the endogenous high molecular mass hyaluronan from its CD44 receptor (Zeng et al., 1998).

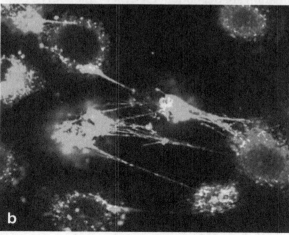

FIGURE 14.4 Hyaluronan dependent microvilli and hyaluronan "cables" on GFP-HAS3 overexpressing breast cancer cells. Cultured breast cancer cells (MCF7) overexpressing GFP-HAS3 show long microvillous extensions (green) on its dorsal cell surface (a). These microvilli are covered with extensive hyaluronan coats (red), which exclude the red blood cells from the vicinity of the cells (a). A part of the cell-associated hyaluronan extends relatively far out, and is visible as hyaluronan "cables" after methanol fixation (b). (See Page 8 in Color Section at the back of the book).

Hyaluronan and Inflammation

The inflammatory reaction has a dual role in malignant tumors. Cells of the immune system can attack and destroy the transformed cells, but they can also enhance tumor growth when inflammation becomes chronic (reviewed in Ben-Baruch, 2006) (Fig. 14.5). The development of cancer at sites of chronic inflammation is well-known in several organs, including skin (reviewed in Ben-Baruch, 2006; Mueller, 2006). On the other hand, a growing malignant tumor often maintains the tissue in an inflammatory state, as demonstrated by the residence of large numbers of leukocytes (Mueller and Fusenig, 2004). The inflammatory cells in tumors are often immunologically suppressed, i.e. unable to destroy the cancer cells (Ben-Baruch, 2006), but still secrete growth factors that contribute to tumor growth (Pollard, 2004).

Hyaluronan deposition in epithelium is a general feature of inflammatory reactions, for instance that associated with skin wounds (Tammi et al., 2005) and the immunological injury of celiac disease (Kemppainen et al., 2005). Recent studies on colon (de la Motte et al., 2003), kidney (Wang and Hascall, 2004), and arterial wall (Wilkinson et al., 2006) indicate that an excess of free hyaluronan, in the form of hyaluronan "cables," is synthesized by the local cells during the inflammatory reaction (de la Motte et al., 2003). This hyaluronan can immobilize monocyte (macrophage) type of cells patrolling in the tissue (Fig. 14.5). Although monocytes may acutely secrete pro-inflammatory signals that may injure cancer cells, it seems that upon binding to high molecular mass hyaluronan, they are deactivated. The deactivation means that they reduce the synthesis of cytokines, start hyaluronan clearance, and increase the synthesis of growth factors and matrix molecules (Day and de la Motte, 2005). Accordingly, the down-regulation of the acute inflammatory reaction and the start of the healing process in experimental lung inflammation are dependent on CD44+ monocytic cells (Teder et al., 2002). It has also been suggested that the deactivation of monocytes and the resulting immunosuppression is mediated via binding of high molecular mass hyaluronan to CD44 on tumor cells (Mytar et al., 2003; del Fresno et al., 2005). On the other hand, long-term stimulation of Toll-like receptors by degradation products of matrix molecules like hyaluronan may cause deactivation of the immune system (reviewed in Tsan, 2006). The binding of leukocytes to hyaluronan "cables" may prevent their association with the cancer cell surface, necessary for a proper immune reaction (Day and de la Motte, 2005). This kind of immunological protection had been shown already decades ago in cancer cells surrounded by large hyaluronan coats (Dick et al., 1983). Accumulation of hyaluronan is thus a feature common to inflamed and malignant tissues, and by gathering monocytes hyaluronan may help to catch the monocytes in the first place, but also to suppress their inflammatory activity and facilitate the secretion

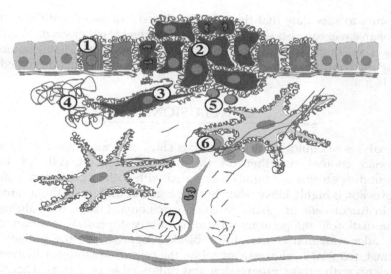

FIGURE 14.5   Localizations and forms of tumor hyaluronan (red), and suggested roles in the growth and spreading of cancers derived from simple epithelia.

1. Hyaluronan is not present in normal simple epithelia, but emerges in epithelial inflammation or injury, presumably to enhance cell proliferation and migration to cover the defect.
2. Hyaluronan is up-regulated when the epithelial cells undergo malignant transformation. It supports cell proliferation, prevents apoptosis, maintains intercellular space to facilitate nutrient diffusion, and enhances cell locomotion which stimulates invasion.
3. Hyaluronan synthesis enhances epithelial to mesenchymal transition in the cancer cell phenotype which releases the cells from their epithelial compartment for invasion.
4. Accumulation of hyaluronan in the stroma opens the fibrillar matrix for cell migration.
5. A coat of hyaluronan on cancer cells shields them from the cytotoxic effects of T-lymphocytes.
6. Free hyaluronan, arranged as "cables" in fixed cell preparations, bind tissue macrophages and modulate their activity to favor tumor growth.
7. Fragments of hyaluronan (oligosaccharides) stimulate endothelial cell proliferation and budding of new capillaries that allow tumor expansion. (See Page 9 in Color Section at the back of the book)

of growth factors and matrix molecules that support the formation of a tumor-friendly host tissue.

Recently, we found that HAS2 and HAS3 overexpressing cells form unique plasma membrane extensions, which form the skeleton of an extensive hyaluronan coat (Fig. 14.4; Kultti et al., 2006). Similar structures are also seen in untransfected cells which secrete large amounts of hyaluronan (Rilla et al., 2008) and bind monocytes (Jokela et al., 2008). It is tempting to speculate that these HAS-induced extensions with their hyaluronan cover can also act as immunological shields either via mechanical basis or via immunomodulatory actions. A hypothetical scheme summarizing the contribution of hyaluronan to cancer spreading is presented in Fig. 14.5.

CONCLUSIONS

Analyses of clinical patient materials show that alterations in the hyaluronan content, whether on the actual malignant cells or their surrounding stroma are tightly associated with patient prognosis. Tumor progression is highly likely when hyaluronan is abundant on the surface or within tumor cells of gastric or colon carcinoma. Likewise, hyaluronan accumulation in the peritumoral stroma of breast, prostate, ovarian, and lung adenocarcinomas indicate a bleak prognosis for the patient. In contrast, reduced hyaluronan signal on the cells of squamous carcinomas correlates with cancer progression and unfavorable prognosis. There are a number of ways hyaluronan can be involved in the regulation of cancer growth and spreading, as suggested by experiments on animals, and studies *in vitro*. However, understanding the relative importance of the various aspects of hyaluronan functions and metabolism in human cancers *in vivo* is still lacking, and warrants more research on clinical materials. Nevertheless, there is reason to believe that hyaluronan will be an important therapeutic target in several human cancers.

ACKNOWLEDGMENTS

The authors gratefully acknowledge the contributions of the many graduate students and collaborators to the original papers published by the team during the past ten years. The authors have received financial support from The Academy of Finland # 107173 and 127169 (MT), Finnish Cancer Foundation (RT & V-MK), Kuopio University Hospital EVO funds (V-MK & MT), North Savo Cancer Foundation (AK), Emil Aaltonen Foundation (AK), Paavo Koistinen Foundation (AK), Finnish Cultural Foundation, North Savo and Southern Ostrobotnia Regional Funds (AK), Kuopio University Foundation (AK), The Mizutani Foundation (MT), and Sigrid Juselius Foundation (MT).

References

Aaltomaa, S., Lipponen, P., Tammi, R., et al. (2002). Strong stromal hyaluronan expression is associated with PSA recurrence in local prostate cancer. *Urol Int* **69**, 266–272.

Adamia, S., Reiman, T., Crainie, M., et al. (2005). Intronic splicing of hyaluronan synthase 1 (HAS-1): a biologically relevant indicator of poor outcome in multiple myeloma. *Blood* **105**, 4836–4844.

Afify, A. M., Craig, S., Paulino, A. F., and Stern, R. (2005). Expression of hyaluronic acid and its receptors, CD44s and CD44v6, in normal, hyperplastic, and neoplastic endometrium. *Ann Diagn Pathol* **9**, 312–318.

Ågren, U. M., Tammi, R. H., and Tammi, M. I. (1997). Reactive oxygen species contribute to epidermal hyaluronan catabolism in human skin organ culture. *Free Radic Biol Med* **23**, 996–1001.

Anggiansah, C. L., Scott, D., Poli, A., et al. (2003). Regulation of hyaluronan secretion into rabbit synovial joints in vivo by protein kinase C. *J Physiol* **550**, 631–640.

Anttila, M. A., Tammi, R. H., Tammi, M. I., et al. (2000). High levels of stromal hyaluronan predict poor disease outcome in epithelial ovarian cancer. *Cancer Res* **60**, 150–155.

Auvinen, P., Tammi, R., Parkkinen, J., et al. (2000). Hyaluronan in peritumoral stroma and malignant cells associates with breast cancer spreading and predicts survival. *Am J Pathol* **156**, 529–536.

Auvinen, P. K., Parkkinen, J. J., Johansson, R. T., et al. (1997). Expression of hyaluronan in benign and malignant breast lesions. *Int J Cancer* **74**, 477–481.

Bakkers, J., Kramer, C., Pothof, J., et al. (2004). HAS-2 is required upstream of Rac1 to govern dorsal migration of lateral cells during zebrafish gastrulation. *Development* **131**, 525–537.

Ben-Baruch, A. (2006). Inflammation-associated immune suppression in cancer: the roles played by cytokines, chemokines and additional mediators. *Semin Cancer Biol* **16**, 38–52.

Bertrand, P., Girard, N., Delpech, B., et al. (1992). Hyaluronan (hyaluronic acid) and hyaluronectin in the extracellular matrix of human breast carcinomas: comparison between invasive and non-invasive areas. *Int J Cancer* **52**, 1–6.

Bhowmick, N. A., Neilson, E. G., and Moses, H. L. (2004). Stromal fibroblasts in cancer initiation and progression. *Nature* **432**, 332–337.

Böhm, J., Niskanen, L., Tammi, R., et al. (2002). Hyaluronan expression in differentiated thyroid carcinoma. *J Pathol* **196**, 180–185.

Calabro, A., Oken, M. M., Hascall, V. C., and Masellis, A. M. (2002). Characterization of hyaluronan synthase expression and hyaluronan synthesis in bone marrow mesenchymal progenitor cells: predominant expression of HAS-1 mRNA and up-regulated hyaluronan synthesis in bone marrow cells derived from multiple myeloma patients. *Blood* **100**, 2578–2585.

Camenisch, T. D., Spicer, A. P., Brehm-Gibson, T., et al. (2000). Disruption of hyaluronan synthase-2 abrogates normal cardiac morphogenesis and hyaluronan-mediated transformation of epithelium to mesenchyme. *J Clin Invest* **106**, 349–360.

Christopoulos, T. A., Papageorgakopoulou, N., Theocharis, D. A., et al. (2006). Hyaluronidase and CD44 hyaluronan receptor expression in squamous cell laryngeal carcinoma. *Biochim Biophys Acta* **1760**, 1039–1045.

Day, A. J. and de la Motte, C. A. (2005). Hyaluronan cross-linking: a protective mechanism in inflammation? *Trends Immunol* **26**, 637–643.

de la Motte, C. A., Hascall, V. C., Drazba, J., et al. (2003). Mononuclear leukocytes bind to specific hyaluronan structures on colon mucosal smooth muscle cells treated with polyinosinic acid, polycytidylic acid: inter-alpha-trypsin inhibitor is crucial to structure and function. *Am J Pathol* **163**, 121–133.

de la Torre, M., Wells, A. F., Bergh, J., and Lindgren, A. (1993). Localization of hyaluronan in normal breast tissue, radial scar, and tubular breast carcinoma. *Hum Pathol* **24**, 1294–1297.

del Fresno, C., Otero, K., Gomez-Garcia, L., et al. (2005). Tumor cells deactivate human monocytes by up-regulating IL-1 receptor associated kinase-M expression via CD44 and TLR4. *J Immunol* **174**, 3032–3040.

Dick, S. J., Macchi, B., Papazoglou, S., et al. (1983). Lymphoid cell–glioma cell interaction enhances cell coat production by human gliomas: novel suppressor mechanism. *Science* **220**, 739–742.

Ekici, S., Cerwinka, W. H., Duncan, R., et al. (2004). Comparison of the prognostic potential of hyaluronic acid, hyaluronidase (HYAL-1), CD44v6 and microvessel density for prostate cancer. *Int J Cancer* **112**, 121–129.

Franzmann, E. J., Schroeder, G. L., Goodwin, W. J., et al. (2003). Expression of tumor markers hyaluronic acid and hyaluronidase (HYAL-1) in head and neck tumors. *Int J Cancer* **106**, 438–445.

Fraser, J. R., Laurent, T. C., and Laurent, U. B. (1997). Hyaluronan: its nature, distribution, functions and turnover. *J Intern Med* **242**, 27–33.

Fülop, C., Salustri, A., and Hascall, V. C. (1997). Coding sequence of a hyaluronan synthase homologue expressed during expansion of the mouse cumulus–oocyte complex. *Arch Biochem Biophys* **337**, 261–266.

Garcia, I., Vizoso, F., Suarez, C., et al. (2000). Relationship of tumoral hyaluronic acid and cathepsin D contents with histological type of gastric carcinoma. *Int J Biol Markers* **15**, 215–218.

Ghatak, S., Misra, S., and Toole, B. P. (2002). Hyaluronan oligosaccharides inhibit anchorage-independent growth of tumor cells by suppressing the phosphoinositide 3-kinase/Akt cell survival pathway. *J Biol Chem* **277**, 38013–38020.

Giovannucci, E. (2001). Insulin, insulin-like growth factors and colon cancer: a review of the evidence. *J Nutr* **131** (11 Suppl), 3109S–3120S.

Glick, A. B. (2004). TGFbeta1, back to the future: revisiting its role as a transforming growth factor. *Cancer Biol Ther* **3**, 276–283.

Hardingham, T. E. and Muir, H. (1972). The specific interaction of hyaluronic acid with cartilage proteoglycans. *Biochim Biophys Acta* **279**, 401–405.

Harris, E. N., Weigel, J. A., and Weigel, P. H. (2004). Endocytic function, glycosaminoglycan specificity, and antibody sensitivity of the recombinant human 190-kDa hyaluronan receptor for endocytosis (HARE). *J Biol Chem* **279**, 36201–36209.

Hautmann, S. H., Lokeshwar, V. B., Schroeder, G. L., et al. (2001). Elevated tissue expression of hyaluronic acid and hyaluronidase validates the HA–HAase urine test for bladder cancer. *J Urol* **165**, 2068–2074.

Hiltunen, E. L., Anttila, M., Kultti, A., et al. (2002). Elevated hyaluronan concentration without hyaluronidase activation in malignant epithelial ovarian tumors. *Cancer Res* **62**, 6410–6413.

Hirvikoski, P., Tammi, R., Kumpulainen, E., et al. (1999). Irregular expression of hyaluronan and its CD44 receptor is associated with metastatic phenotype in laryngeal squamous cell carcinoma. *Virchows Arch* **434**, 37–44.

Iozzo, R. V. (1998). Matrix proteoglycans: from molecular design to cellular function. *Annu Rev Biochem* **67**, 609–652.

Itano, N., Sawai, T., Atsumi, F., et al. (2004). Selective expression and functional characteristics of three mammalian hyaluronan synthases in oncogenic malignant transformation. *J Biol Chem* **279**, 18679–18687.

Itano, N., Sawai, T., Miyaishi, O., and Kimata, K. (1999). Relationship between hyaluronan production and metastatic potential of mouse mammary carcinoma cells. *Cancer Res* **59**, 2499–2504.

Ito, M., Ito, G., Kondo, M., et al. (2005). Frequent inactivation of RASSF1A, BLU, and SEMA3B on 3p21.3 by promoter hypermethylation and allele loss in non-small cell lung cancer. *Cancer Lett* **225**, 131–139.

Jacobson, A., Brinck, J., Briskin, M. J., Spicer, A. P., and Heldin, P. (2000). Expression of human hyaluronan synthases in response to external stimuli. *Biochem J* **348**, 29–35.

Jacobson, A., Rahmanian, M., Rubin, K., and Heldin, P. (2002). Expression of hyaluronan synthase 2 or hyaluronidase 1 differentially affect the growth rate of transplantable colon carcinoma cell tumors. *Int J Cancer* **102**, 212–219.

Jokela, T. A., Lindgren, A., Rilla, K., et al. (2008). Induction of hyaluronan cables and monocyte adherence in epidermal keratinocytes. *Connect Tissue Res* **49**, 115–119, 2008.

Kakizaki, I., Kojima, K., Takagaki, K., et al. (2004). A novel mechanism for the inhibition of hyaluronan biosynthesis by 4-methylumbelliferone. *J Biol Chem* **279**, 33281–33289.

Kanomata, N., Yokose, T., Kamijo, T., et al. (2005). Hyaluronan synthase expression in pleural malignant mesotheliomas. *Virchows Arch* **446**, 246–250.

Karjalainen, J. M., Tammi, R. H., Tammi, M. I., et al. (2000). Reduced level of CD44 and hyaluronan associated with unfavorable prognosis in clinical stage I cutaneous melanoma. *Am J Pathol* **157**, 957–965.

Karvinen, S., Kosma, V. M., Tammi, M. I., and Tammi, R. (2003a). Hyaluronan, CD44 and versican in epidermal keratinocyte tumours. *Br J Dermatol* **148**, 86–94.

Karvinen, S., Pasonen-Seppänen, S., Hyttinen, J. M. T., et al. (2003b). Keratinocyte growth factor stimulates migration and hyaluronan synthesis in the epidermis by activation of keratinocyte hyaluronan synthases 2 and 3. *J Biol Chem* **278**, 49495–49504.

Kemppainen, T., Tammi, R., Tammi, M., et al. (2005). Elevated expression of hyaluronan and its CD44 receptor in the duodenal mucosa of coeliac patients. *Histopathology* **46**, 64–72.

Kimata, K., Honma, Y., Okayama, M., et al. (1983). Increased synthesis of hyaluronic acid by mouse mammary carcinoma cell variants with high metastatic potential. *Cancer Res* **43**, 1347–1354.

Knudson, C. B. and Toole, B. P. (1985). Fluorescent morphological probe for hyaluronate. *J Cell Biol* **100**, 1753–1758.

Knudson, W. and Toole, B. P. (1988). Membrane association of the hyaluronate stimulatory factor from LX-1 human lung carcinoma cells. *J Cell Biochem* **38**, 165–177.

Knudson, W., Biswas, C., Li, X. Q., Nemec, R. E., and Toole, B. P. (1989). The role and regulation of tumour-associated hyaluronan. *Ciba Found Symp* **143**, 150–159.

Knudson, W., Chow, G., and Knudson, C. B. (2002). CD44-mediated uptake and degradation of hyaluronan. *Matrix Biol* **21**, 15–23.

Köbel, M., Weichert, W., Cruwell, K., et al. (2004). Epithelial hyaluronic acid and CD44v6 are mutually involved in invasion of colorectal adenocarcinomas and linked to patient prognosis. *Virchows Arch* **445**, 456–464.

Kosaki, R., Watanabe, K., and Yamaguchi, Y. (1999). Overproduction of hyaluronan by expression of the hyaluronan synthase HAS-2 enhances anchorage-independent growth and tumorigenicity. *Cancer Res* **59**, 1141–1145.

Kosunen, A., Ropponen, K., Kellokoski, J., et al. (2004). Reduced expression of hyaluronan is a strong indicator of poor survival in oral squamous cell carcinoma. *Oral Oncol* **40**, 257–263.

Kultti, A., Rilla, K., Tiihonen, R., et al. (2006). Hyaluronan synthesis induces microvillus-like cell surface protrusions. *J Biol Chem* **281**, 15821–15828.

Kwabi-Addo, B., Ozen, M., and Ittmann, M. (2004). The role of fibroblast growth factors and their receptors in prostate cancer. *Endocr Relat Cancer* **11**, 709–724.

Lipponen, P., Aaltomaa, S., Tammi, R., et al. (2001). High stromal hyaluronan level is associated with poor differentiation and metastasis in prostate cancer. *Eur J Cancer* **37**, 849–856.

Liu, N., Gao, F., Han, Z., et al. (2001). Hyaluronan synthase 3 overexpression promotes the growth of TSU prostate cancer cells. *Cancer Res* **61**, 5207–5214.

Llaneza, A., Vizoso, F., Rodriguez, J. C., et al. (2000). Hyaluronic acid as prognostic marker in resectable colorectal cancer, 871690–871696.

Lokeshwar, V. B., Schroeder, G. L., Selzer, M. G., et al. (2002). Bladder tumor markers for monitoring recurrence and screening comparison of hyaluronic acid–hyaluronidase and BTA–Stat tests. *Cancer* **95**, 61–72.

Lokeshwar, V. B., Cerwinka, W. H., Isoyama, T., and Lokeshwar, B. L. (2005). HYAL-1 hyaluronidase in prostate cancer: a tumor promoter and suppressor. *Cancer Res* **65**, 7782–7789.

Lokeshwar, V. B., Rubinowicz, D., Schroeder, G. L., et al. (2001). Stromal and epithelial expression of tumor markers hyaluronic acid and HYAL-1 hyaluronidase in prostate cancer. *J Biol Chem* **276**, 11922–11932.

Milner, M., Higman, V. A., and Day, A. J. (2006). a pluripotent inflammatory mediator? *Biochem Soc Trans* **34**, 446–450.

Misra, S., Ghatak, S., and Toole, B. P. (2005). Regulation of MDR1 expression and drug resistance by a positive feedback loop involving hyaluronan, phosphoinositide 3-kinase, and ErbB2. *J Biol Chem* **280**, 20310–20315.

Misra, S., Toole, B. P., and Ghatak, S. (2006). Hyaluronan constitutively regulates activation of multiple receptor tyrosine kinases in epithelial and carcinoma cells. *J Biol Chem* **281**, 34936–34941.

Mueller, M. M. (2006). Inflammation in epithelial skin tumours: old stories and new ideas. *Eur J Cancer* **42**, 735–744.

Mueller, M. M. and Fusenig, N. E. (2004). Friends or foes – bipolar effects of the tumour stroma in cancer. *Nat Rev Cancer* **4**, 839–849.

Mytar, B., Woloszyn, M., Szatanek, R., et al. (2003). Tumor cell-induced deactivation of human monocytes. *J Leukoc Biol* **74** (6), 1094–1101.

Nakamura, H., Suenaga, N., Taniwaki, K., et al. (2004). Constitutive and induced CD44 shedding by ADAM-like proteases and membrane-type 1 matrix metalloproteinase. *Cancer Res* **64**, 876–882.

Nakazawa, H., Yoshihara, S., Kudo, D., et al. (2006). 4-methylumbelliferone, a hyaluronan synthase suppressor, enhances the anticancer activity of gemcitabine in human pancreatic cancer cells. *Cancer Chemother Pharmacol* **57**, 165–170.

Normanno, N., De Luca, A., Bianco, C., et al. (2006). Epidermal growth factor receptor (EGFR) signaling in cancer. *Gene* **366** (1), 2–16.

Nykopp, T. K., Sironen, R., Tammi, M. I., Tammi, R. H., Hämäläinen, K., Heikkinen, A.-M., Komulainen, M., Kosma, V.-M., and Anttila, M.: Increased accumulation of hyaluronan in serous ovarian carcinoma correlates with decreased expression of HYAL 1 hyaluronidase. Submitted for publication.

Pasonen-Seppänen, S., Karvinen, S., Törrönen, K., et al. (2003). EGF up-regulates, whereas TGF-beta downregulates, the hyaluronan synthases HAS-2 and HAS-3 in organotypic keratinocyte cultures: correlations with epidermal proliferation and differentiation. *J Invest Dermatol* **120**, 1038–1044.

Petersen, O. W., Nielsen, H. L., Gudjonsson, T., et al. (2003). Epithelial to mesenchymal transition in human breast cancer can provide a nonmalignant stroma. *Am J Pathol* **162**, 391–402.

Pienimäki, J. P., Rilla, K., Fülop, C., et al. (2001). Epidermal growth factor activates hyaluronan synthase 2 in epidermal keratinocytes and increases pericellular and intracellular hyaluronan. *J Biol Chem* **276**, 20428–20435.

Pietras, K., Sjöblom, T., Rubin, K., Heldin, C. H., and Ostman, A. (2003). PDGF receptors as cancer drug targets. *Cancer Cell* **3**, 439–443.

Pirinen, R., Hirvikoski, P., Böhm, J., et al. (2000). Reduced expression of CD44v3 variant isoform is associated with unfavorable outcome in non-small cell lung carcinoma. *Hum Pathol* **31**, 1088–1095.

Pirinen, R., Tammi, R., Tammi, M., et al. (2001). Prognostic value of hyaluronan expression in non-small-cell lung cancer: increased stromal expression indicates unfavourable outcome in patients with adenocarcinoma. *Int J Cancer* **95**, 12–17.

Pirinen, R. T., Tammi, R. H., Tammi, M. I., et al. (1998). Expression of hyaluronan in normal and dysplastic bronchial epithelium and in squamous cell carcinoma of the lung. *Int J Cancer* **79**, 251–255.

Pollard, J. W. (2004). Tumour-educated macrophages promote tumour progression and metastasis. *Nat Rev Cancer* **4**, 71–78.

Ponting, J., Kumar, S., and Pye, D. (1993). Co-localization of hyaluronan and hyaluronectin in normal and neoplastic breast tissues. *Int J Oncol* **2**, 889–893.

Posey, J. T., Soloway, M. S., Ekici, S., et al. (2003). Evaluation of the prognostic potential of hyaluronic acid and hyaluronidase (HYAL-1) for prostate cancer. *Cancer Res* **63**, 2638–2644.

Rilla, K., Pasonen-Seppänen, S., Rieppo, J., Tammi, M., and Tammi, R. (2004). The hyaluronan synthesis inhibitor 4-methylumbelliferone prevents keratinocyte activation and epidermal hyperproliferation induced by epidermal growth factor. *J Invest Dermatol* **123**, 708–714.

Rilla, K., Siiskonen, H., Spicer, A. P., et al. (2005). Plasma membrane residence of hyaluronan synthase is coupled to its enzymatic activity. *J Biol Chem* **280**, 31890–31897.

Rilla, K., Tiihonen, R., Kultti, A., et al. (2008). Pericellular hyaluronan coat visualized in live cells with a fluorescent probe is scaffolded by plasma membrane protrusions. *J Histochem Cytochem* **56**, 901–910.

Ripellino, J. A., Klinger, M. M., Margolis, R. U., and Margolis, R. K. (1985). The hyaluronic acid binding region as a specific probe for the localization of hyaluronic acid in tissue sections. Application to chick embryo and rat brain. *J Histochem Cytochem* **33**, 1060–1066.

Rooney, P., Kumar, S., Ponting, J., and Wang, M. (1995). The role of hyaluronan in tumour neovascularization (review). *Int J Cancer* **60**, 632–636.

Ropponen, K., Tammi, M., Parkkinen, J., et al. (1998). Tumor cell-associated hyaluronan as an unfavorable prognostic factor in colorectal cancer. *Cancer Res* **58**, 342–347.

Saavalainen, K., Pasonen-Seppänen, S., Dunlop, T. W., et al. (2005). The human hyaluronan synthase 2 gene is a primary retinoic acid and epidermal growth factor responding gene. *J Biol Chem* **280**, 14636–14644.

Setälä, L. P., Tammi, M. I., Tammi, R. H., et al. (1999). Hyaluronan expression in gastric cancer cells is associated with local and nodal spread and reduced survival rate. *Br J Cancer* **79**, 1133–1138.

Shaw, R. J. and Cantley, L. C. (2006). Ras, PI(3)K and mTOR signalling controls tumour cell growth. *Nature* **441**, 424–430.

Simpson, M. A., Wilson, C. M., Furcht, L. T., et al. (2002). Manipulation of hyaluronan synthase expression in prostate adenocarcinoma cells alters pericellular matrix retention and adhesion to bone marrow endothelial cells. *J Biol Chem* **277**, 10050–10057.

Smola, H., Thiekotter, G., and Fusenig, N. E. (1993). Mutual induction of growth factor gene expression by epidermal–dermal cell interaction. *J Cell Biol* **122**, 417–429.

Stern, R. (2005). Hyaluronan metabolism: a major paradox in cancer biology. *Pathol Biol (Paris)* **53**, 372–382.

Stern, R. and Jedrzejas, M. J. (2006). Hyaluronidases: their genomics, structures, and mechanisms of action. *Chem Rev* **106**, 818–839.

Suwiwat, S., Ricciardelli, C., Tammi, R., et al. (2004). Expression of extracellular matrix components versican, chondroitin sulfate, tenascin, and hyaluronan, and their association with disease outcome in node-negative breast cancer. *Clin Cancer Res* **10**, 2491–2498.

Takahashi, Y., Li, L., Kamiryo, M., et al. (2005). Hyaluronan fragments induce endothelial cell differentiation in a CD44- and CXCL1/GRO1-dependent manner. *J Biol Chem* **280**, 24195–24204.

Tammi, R., Pasonen-Seppänen, S., Kolehmainen, E., and Tammi, M. (2005). Hyaluronan synthase induction and hyaluronan accumulation in mouse epidermis following skin injury. *J Invest Dermatol* **124**, 898–905.

Tammi, R., Rilla, K., Pienimäki, J. P., et al. (2001). Hyaluronan enters keratinocytes by a novel endocytic route for catabolism. *J Biol Chem* **276**, 35111–35122.

Tammi, R. H. and Tammi, M. I. (2004). Hyaluronan in epidermis and other epithelial tissues. Chemistry and Biology of Hyaluronan. In: Garg, H. G. & Hales, C. A. (Eds.). Kidlington: Elsevier, pp. 395–407.

Teder, P., Vandivier, R. W., Jiang, D., et al. (2002). Resolution of lung inflammation by CD44. *Science* **296**, 155–158.

Theocharis, A. D., Vynios, D. H., Papageorgakopoulou, N., Skandalis, S. S., and Theocharis, D. A. (2003). Altered content composition and structure of glycosaminoglycans and proteoglycans in gastric carcinoma. *Int J Biochem Cell Biol* **35**, 376–390.

Thomas, G. J. and Speight, P. M. (2001). Cell adhesion molecules and oral cancer. *Crit Rev Oral Biol Med* **12**, 479–498.

Thorne, R. F., Legg, J. W., and Isacke, C. M. (2004). The role of the CD44 transmembrane and cytoplasmic domains in co-ordinating adhesive and signalling events. *J Cell Sci* **117**, 373–380.

Toole, B. P. (1997). Hyaluronan in morphogenesis. *J Intern Med* **242**, 35–40.

VI. HYALURONAN IN CANCER EPITHELIAL–STROMAL INTERACTIONS

Tsan, M. F. (2006). Toll-like receptors, inflammation and cancer. *Semin Cancer Biol* **16**, 32–37.

Tsuchiya, N., Kondo, Y., Takahashi, A., et al. (2002). Mapping and gene expression profile of the minimally overrepresented 8q24 region in prostate cancer. *Am J Pathol* **160**, 1799–1806.

Tuhkanen, H., Anttila, M., Kosma, V. M., et al. (2004). Genetic alterations in the peritumoral stromal cells of malignant and borderline epithelial ovarian tumors as indicated by allelic imbalance on chromosome 3p. *Int J Cancer* **109**, 247–252.

Udabage, L., Brownlee, G. R., Waltham, M., et al. (2005). Antisense-mediated suppression of hyaluronan synthase 2 inhibits the tumorigenesis and progression of breast cancer. *Cancer Res* **65**, 6139–6150.

van Muijen, G. N., Danen, E. H., Veerkamp, J. H., et al. (1995). Glycoconjugate profile and CD44 expression in human melanoma cell lines with different metastatic capacity. *Int J Cancer* **61**, 241–248.

Vigetti, D., Ori, M., Viola, M., et al. (2006). Molecular cloning and characterization of UDP-glucose dehydrogenase from the amphibian Xenopus laevis and its involvement in hyaluronam synthesis. *J Biol Chem* **281**, 8254–8263.

Vizoso, F. J., del Casar, J. M., Corte, M. D., et al. (2004). Significance of cytosolic hyaluronan levels in gastric cancer. *Eur J Surg Oncol* **30**, 318–324.

Voutilainen, K., Anttila, M., Sillanpää, S., et al. (2003). Versican in epithelial ovarian cancer: relation to hyaluronan, clinicopathologic factors and prognosis. *Int J Cancer* **107**, 359–364.

Wang, A. and Hascall, V. C. (2004). Hyaluronan structures synthesized by rat mesangial cells in response to hyperglycemia induce monocyte adhesion. *J Biol Chem* **279**, 10279–10285.

Wang, C., Tammi, M., Guo, H., and Tammi, R. (1996). Hyaluronan distribution in the normal epithelium of esophagus, stomach, and colon and their cancers. *Am J Pathol* **148**, 1861–1869.

Weigel, P. H., Hascall, V. C., and Tammi, M. (1997). Hyaluronan synthases. *J Biol Chem* **272**, 13997–14000.

Wein, R. O., McGary, C. T., Doerr, T. D., et al. (2006). Hyaluronan and its receptors in mucoepidermoid carcinoma. *Head Neck* **28**, 176–181.

Wernicke, M., Pineiro, L. C., Caramutti, D., et al. (2003). Breast cancer stromal myxoid changes are associated with tumor invasion and metastasis: a central role for hyaluronan. *Mod Pathol* **16**, 99–107.

Wilkinson, T. S., Bressler, S. L., Evanko, S. P., Braun, K. R., and Wight, T. N. (2006). Over-expression of hyaluronan synthases alters vascular smooth muscle cell phenotype and promotes monocyte adhesion. *J Cell Physiol* **206**, 378–385.

Xing, R., Regezi, J. A., Stern, M., Shuster, S., and Stern, R. (1998). Hyaluronan and CD44 expression in minor salivary gland tumors. *Oral Dis* **4**, 241–247.

Yabushita, H., Kishida, T., Fusano, K., et al. (2005). Role of hyaluronan and hyaluronan synthase in endometrial cancer. *Oncol Rep* **13**, 1101–1105.

Yabushita, H., Noguchi, M., Kishida, T., et al. (2004). Hyaluronan synthase expression in ovarian cancer. *Oncol Rep* **12**, 739–743.

Yamada, Y., Itano, N., Narimatsu, H., et al. (2004). Elevated transcript level of hyaluronan synthase1 gene correlates with poor prognosis of human colon cancer. *Clin Exp Metastasis* **21**, 57–63.

Zeng, C., Toole, B. P., Kinney, S. D., Kuo, J. W., and Stamenkovic, I. (1998). Inhibition of tumor growth in vivo by hyaluronan oligomers. *Int J Cancer* **77**, 396–401.

Zhai, Y., Hotary, K. B., Nan, B., et al. (2005). Expression of membrane type 1 matrix metalloproteinase is associated with cervical carcinoma progression and invasion. *Cancer Res* **65**, 6543–6550.

Zoltan-Jones, A., Huang, L., Ghatak, S., and Toole, B. P. (2003). Elevated hyaluronan production induces mesenchymal and transformed properties in epithelial cells. *J Biol Chem* **278**, 45801–45810.

CHAPTER

15

The Oncofetal Paradigm Revisited: MSF and HA as Contextual Drivers of Cancer Progression

Seth L. Schor, Ana M. Schor, Ian R. Ellis, Sarah J. Jones, Margaret Florence, Jacqueline Cox, and Anne-Marie Woolston

OUTLINE

INTRODUCTION

Cell differentiation during fetal development is driven by the sequential transcription of certain genes and the reciprocal transcriptional silencing of others. Such changes in gene expression are stably transmitted to daughter cells and ultimately result in the emergence of functionally restricted (specialized) cell lineages. Recent technological advances have fueled an explosion in our detailed understanding of the epigenetic mechanisms controlling this process. Changes in gene expression also characterize cancer pathogenesis and are similarly responsible for the emergence of distinct lineages of neoplastic cells. Advances in molecular biology during the latter half of the 20th century made it possible to document the multiple mutational and chromosomal alterations associated with cancer. The impressive accomplishments of this venture have resulted in the now widely accepted view that genetically regulated changes in the expression of oncogenes and tumor suppressor genes provide the necessary and sufficient impetus to drive cancer progression.

Cancer progression is also invariably accompanied by the re-expression of fetal isoforms of various structural and regulatory proteins not normally present in healthy adult tissues. Different models involving both genetic and epigenetic mechanisms have been proposed to account for the observed "oncofetal" pattern of gene expression. *Migration stimulating factor* (MSF) is an oncofetal regulatory molecule constitutively expressed by epithelial and stromal cells during fetal development. Although not expressed in the majority of healthy adult tissues, it is transiently re-expressed during wound healing and persistently re-expressed by both tumor and tumor-associated stromal cells in common human cancers. MSF is also systemically re-expressed at distant, uninvolved sites in patients with cancer. MSF exhibits a number of potent bioactivities of direct relevance to both fetal development and disease progression, including the stimulation of hyaluronan (HA) synthesis, cell motility, and angiogenesis. In this chapter we shall review MSF in terms of (i) its molecular characterization and spectrum of bioactivities, including the role of HA in mediating the motogenic response of certain target cells; (ii) the oncofetal pattern of its local and systemic expression; (iii) the tissue-level control of target cell response to MSF; and (iv) the postulated epigenetic control of MSF expression within the context of an "extended" oncofetal model of cancer inception and progression. We conclude with a brief discussion of the potential clinical implications of these concepts for improving the management of patients with cancer.

MSF: ITS MOLECULAR CHARACTERIZATION AND SPECTRUM OF BIOACTIVITIES

MSF was first identified in studies comparing the behavior of different cell types in a then recently developed collagen gel migration assay (Schor, 1980; Schor et al., 1985a; 1988a). As part of this work, we reported that (i) fetal fibroblasts migrated into a 3-dimensional collagen matrix to a significantly greater extent than did their normal adult counterparts; (ii) this behavioral difference resulted from the production of a soluble "migration stimulating factor" (MSF) by the fetal, but not adult cells; and (iii) tumor-derived and skin fibroblasts obtained from patients with various types of cancer resembled fetal cells with respect to their elevated migration and production of MSF (Durning et al., 1984; Schor et al., 1985b; 1988b; Haggie et al., 1987). Normal adult fibroblasts did not produce MSF or an inhibitor of MSF; the migration of these cells was, however, stimulated by conditioned media from fetal and cancer patient fibroblasts, thereby providing a sensitive bioassay for the subsequent purification of MSF. MSF was eventually cloned and, most unexpectedly, shown to be a genetically truncated isoform of fibronectin (Schor and Schor, 2001; Schor et al., 2003).

Fibronectin is a ubiquitously distributed macromolecule, present as an insoluble constituent of the extracellular matrix (ECM) and as a soluble component of serum (Hynes, 1990). Both forms consist of two similar, but not necessarily identical, protein chains (with individual molecular masses in the region of 260 kDa) covalently linked by disulfide bonds at their respective C-termini. Each protein chain consists of a tandem array of "functional domains" defined on the basis of their binding proclivities for other matrix molecules and integrin receptors on the cell surface (Fig. 15.1). Starting at the N-terminus, these include: the *Fib1/Hep1-binding domain* (exhibiting affinity for fibrin and heparin), the *Gel-binding domain* (affinity for collagen/gelatin), the *Cell-binding domain* (affinity for certain cell surface integrins, such as $\alpha5\beta1$), the *Hep2-binding domain* (affinity for heparin), and the C-terminal *Fib2-binding domain* (affinity for fibrin). Each of these functional domains consists of different combinations of three "structural homology modules" designated types I, II, and III. This modular structure is reflected by a corresponding modularity in the fibronectin gene, in which types I and II modules are each coded by a single exon, while the majority of type III modules are coded by two exons (designated "a" and "b"). There are approximately 20 previously described "full-length" fibronectin isoforms, all having molecular masses in the region of 260 kDa. These are generated by alternative splicing of the primary fibronectin gene transcript involving the retention or deletion of two particular type III exons (EDA and EDB), as well as a more complex splicing repertoire within the downstream IIICS region (Hynes, 1990).

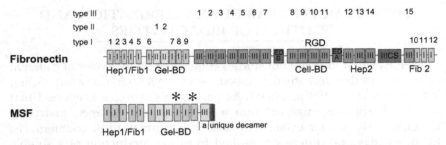

FIGURE 15.1   MSF is a truncated isoform of fibronectin. Fibronectin is a modular glycoprotein consisting of the following functional domains: Hep-1/Fib-1 (N-terminal low affinity binding to heparin and fibrin), Gel-BD (binding to gelatin/collagen), Cell-BD (RGD-mediated binding to cell surface integrins), Hep-2 (high affinity heparin binding), and Fib-2 (C-terminal fibrin binding site). Each functional domain is composed of three possible structural modules: types I, II, and III. MSF is a 70 kDa truncated isoform of fibronectin, identical to its N-terminus, up to and including the amino acid sequence coded by exon III-1a. MSF terminates in a unique intron-derived 10 amino acid sequence. Asterisks mark the location of the two IGD motifs that mediate the bioactivity of MSF on fibroblasts.

MSF message is transcribed from the fibronectin gene by an unusual "failure" of normal alternative splicing involving read-through of the intron separating exons III-1a and III-1b, and subsequent premature transcript cleavage (Schor et al., 2003). The retained intronic sequence contains a 30 bp in-frame coding sequence (immediately contiguous with exon III-1a), followed by several in-frame stop codons and a cleavage/poly-adenylation signal. The resultant MSF protein is consequently a truncated isoform of fibronectin, identical to its 70 kDa N-terminus, up to and including the amino acid sequence coded by exon III-1a, and ends in a unique, intron-derived, 10 amino acid sequence not present in any previously identified "full-length" fibronectin isoform (Fig. 15.1). As is the case with many cytokines and stress response molecules, MSF message exhibits an extremely short half-life as a consequence of an AU rich instability element in its 3'-UTR (Schor et al., 2003; Bakheet et al., 2001; Chen et al., 1994).

Zhao et al. (2001) have identified a similarly foreshortened fibronectin message in Zebrafish embryos and tissues of the adult fish which is identical to the 5'-end of Zebrafish fibronectin, up to and including exon III-3, and, like MSF, terminates in an intron-derived sequence coding for a unique peptide not present in any full-length Zebrafish fibronectin. Similar truncated isoforms of fibronectin have been detected in cDNA libraries from (i) fetal and adult human liver, prostate, ovary, and spleen; (ii) mouse liver, spleen, intestine, brain, heart, thymus, kidney, testis, stomach, lung and muscle; as well as (iii) goldfish and rainbow trout (Liu et al., 2003). Taken together, these various observations make it apparent that genetically truncated isoforms of fibronectin are widely expressed across the phylogenetic tree. The functionality of these truncated fibro-nectins remains to be determined. Initial results indicate that, unlike MSF (terminating at exon III-1), the larger truncated isoforms (terminating at exon III-3) promote adhesion and appear to inhibit basal and MSF-stim-ulated migration (Zhao 2001; our unpublished observations).

Recombinant MSF exhibits a broad spectrum of bioactivities, including the (i) stimulation of fibroblast, epithelial, and endothelial cell migration (Fig. 15.2); (ii) up-regulation of HA synthesis by target fibroblasts (Fig. 15.2); (iii) induction of endothelial cell activation ("sprouting") *in vitro* (Fig. 15.3); (iv) induction of angiogenesis *in vivo* (Fig. 15.4) and (v) prote-olysis (Schor and Schor; 2001; Schor et al., 1989; 2003; Houard, 2005). These activities are unusually potent, commonly being manifest *in vitro* and *in vivo* at femtomolar concentrations. *In vitro* mutagenesis studies (Schor et al., 2003; Millard et al., 2007) indicate that MSF stimulation of fibroblast migration is mediated by two of its constituent IGD tripeptide motifs located in structural modules I-7 and I-9 (Fig. 15.1). The IGD motif is a highly conserved feature of fibronectin type I modules (Hynes, 1990), although no biological functionality had previously been ascribed to it. Significantly, synthetic tri- and tetra-peptides containing the IGD motif mimic all MSF bioactivities (Schor et al., 1999; Schor et al., 2003; unpub-lished data). Initial data indicate that cellular response to MSF/IGD requires maintenance of integrin $\alpha v \beta 3$ functionality and is mediated, at least in part, by the PI-3 kinase signal transduction pathway (Schor et al., 1999, unpublished observations).

Full-length fibronectin isoforms do not express any of MSF's potent bioactivities, presumably as a consequence of steric hindrance of their constitutive bioactive motifs. The functional "unmasking" of these in MSF is accordingly postulated to result from appropriate alterations in its tertiary structure (higher order folding) resulting from truncation. Analogously,

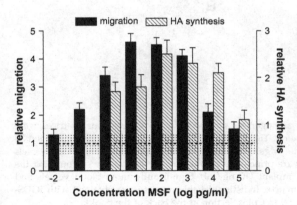

FIGURE 15.2 MSF stimu-lation of cell migration and HA synthesis by adult skin fibro-blasts. The effect of MSF on migration and HA synthesis by confluent adult skin fibro-blasts growing on a 3D native type I collagen matrix were determined as previously described (Ellis et al., 1992). Data are expressed relative to control cultures incubated in the absence of MSF.

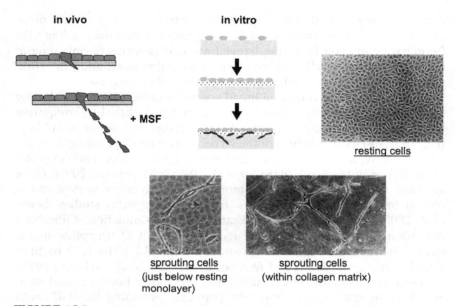

FIGURE 15.3 Induction of endothelial cell sprouting *in vitro* by MSF. The induction of new blood vessel formation by an angiogenic factor *in vivo* commences with the migration of "activated" endothelial cells from the vessel lumen into the surrounding tissue stroma where they adopt an elongated "sprouting cell" morphology. Endothelial cells *in vitro* may be induced to mimic this process. In this example, confluent endothelial cells growing on the surface of 3D collagen gel form a monolayer of resting "cobblestone" cells reminiscent of cells lining the vessel lumen. This resting monolayer remains stable in culture in the absence of exogenous angiogenic factor. Sprouting cells are induced by the addition of MSF (as well as any other angiogenic factor). In analogy with angiogenesis *in vivo*, these sprouting cells migrate down into the 3D collagen matrix where they form an interconnected network of cells, both just below the cobblestone monolayer and deeper within the collagen matrix.

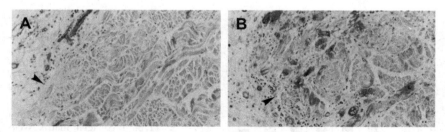

FIGURE 15.4 Induction of angiogenesis *in vivo* by MSF. Collagenous matrices were implanted subcutaneously in the rat. Tissues were excised, fixed, and paraffin-embedded 28 days later. Arrowheads mark the edge of the implants. **A**, Control implant. Blood vessels (stained brown) and fibroblasts are principally confined to the host tissue surrounding the implant. **B**, MSF impregnated implant (10 ng/ml). Significantly more blood vessels and fibroblasts have infiltrated the matrix. Indistinguishable results were obtained with IGDS-impregnated implants. (See Page 9 in Color Section at the back of the book).

proteolytically generated fragments of fibronectin display a host of "neo-activities" not expressed by the parental molecule (Fukai et al., 1995), including the stimulation of monocyte migration (Clark et al., 1998), the inhibition of cell proliferation (Muir and Manthorpe, 1992), the induction of protease gene expression by adherent synovial fibroblasts (Werb et al., 1989), adipocyte differentiation (Fukai et al., 1993) and an RGDS-independent mediation of cell migration (Fukai et al., 1991). The two active IGD motifs in MSF (in I-7 and I-9) reside in the gelatin-binding domain (Gel-BD), and as predicted, Gel-BD generated by proteolytic cleavage of fibro-nectin mimics all MSF bioactivities (Schor et al., 1996; 2003). Interestingly, the proteolytically generated cell-binding fragment of fibronectin (con-taining the RGD motif) inhibits the motogenic activity of MSF, Gel-BD and IGD synthetic peptide (Schor et al., 1999; unpublished observations).

Another bioactive motif has been identified in MSF. The putative zinc-binding motif (HEEGH), located in module I-8, is required for fibronectin–proteinase activity (Houard 2005). These workers further demonstrated that this motif is required to promote the migration of a breast tumor cell line (MCF-7). Mutagenesis of the two histidine residues to phenylalanine (FEEGF) abolished both the proteinase and motogenic activities of MSF. We have recently observed that the HEEGH motif functions in addition to the bioactive IGD motifs in stimulating the migration of target endothelial cells: i.e. both the HEEGH and IGD motifs in MSF must be mutated in order to abolish the motogenic response of these cells, whereas only the IGD motifs appear to be required to stimulate fibroblast migration.

Available evidence suggests that the motogenic response of target fibroblasts to MSF is mediated by its stimulation of HA synthesis: i.e. (i) exogenously provided HA stimulates the migration of adult-derived fibroblasts; and (ii) the motogenic response of adult fibroblasts to MSF is abrogated by co-incubation with *Streptococcal* hyaluronidase (Schor et al., 1989). MSF is not unique in this regard, as other, but not all, motogenic cytokines also up-regulate HA synthesis which similarly appears to mediate cytokine-stimulated migration (Schor, 1994; Ellis et al., 1992; 1997; 2007). As predicted, fetal fibroblasts produce significantly more HA and migrate to a greater extent compared to adult cells; hyaluronidase and MSF function-neutralizing antibody reduces the elevated migration of fetal cells down to that of their adult counterparts. In contrast, endothelial cells *in vitro* do not synthesize HA, either in the presence or absence of MSF (Winterbourne et al., 1983; Amanuma and Mitsui, 1991; unpublished observations); the stimulation of endothelial cell migration and adoption of a sprouting cell phenotype by MSF *in vitro* must accordingly be mediated by an HA-independent mechanism. The situation *in vivo* is clearly more complex. In view of the well-documented effect of HA and HA fragments on angiogenesis, it is likely that the pro-angiogenic activity of MSF in animal model systems may also be indirectly mediated by its effect on HA synthesis by other cell populations (e.g. fibroblasts, carcinoma cells) within a multi-component tissue environment.

Section Summary and Significance

- MSF is a 70 kDa truncated isoform of fibronectin.
- MSF exhibits a number of potent bioactivities, including the stimulation of cell migration, angiogenesis, and the synthesis of HA. These bioactivities are, at least in part, mediated by two IGD tri-peptide motifs respectively located within the I-7 and I-9 type I structural modules.
- Another bioactive motif, HEEGH, appears to complement IGD in stimulating the migration of endothelial cells.
- MSF bioactivities are cryptic within all full-length fibronectin isoforms, probably as a consequence of steric hindrance. Fragments of fibronectin generated by proteolytic degradation also exhibit bioactivities which are held cryptic within the full-length parental molecule.
- The transcriptional control of fibronectin truncation provides a tightly controlled mechanism for regulating the spatial and temporal expression of potent bioactivities without the concomitant generation of potentially competing/confounding activities expressed by the full range of fibronectin proteolytic fragments.

THE ONCOFETAL PATTERN OF MSF EXPRESSION: CONTEXTUAL CONTROL BY ECM AND SOLUBLE FACTORS

MSF exhibits an "oncofetal" pattern of expression as defined by its (i) constitutive production by keratinocytes, fibroblasts, and endothelial cells in fetal skin; (ii) reduced or undetectable expression in skin and other tissues in healthy adults; and (iii) re-expression by both carcinoma and stromal cells in the majority of common human tumors, including those of the breast, lung, colon, prostate, and oral mucosa (Schor et al., 2003; unpublished observations). As is the case with other oncofetal regulatory molecules, including full-length fibronectin isoforms containing the EDA and/or EDB type III modules (ffrench-Constant et al., 1989), MSF is *transiently* re-expressed during acute wound healing (Fig. 15.5). This is first apparent approximately three days after wounding, when MSF is re-expressed by keratinocytes in the migrating epithelial tongue. In this experimental protocol, the epithelium was resealed by day 7, at which point MSF was exuberantly expressed by both the new epithelium and underlying granulation tissue (fibroblasts and endothelial cells). MSF expression declined to low or undetectable levels of expression by 21–28 days. MSF may play

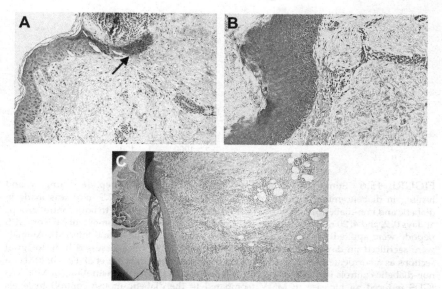

FIGURE 15.5 The transient re-expression of MSF during acute wound healing. **A**, Day 3: MSF expression is first apparent in the epithelial tongue migrating beneath the wound scab (arrow). **B**, Day 7: MSF is abundantly expressed by keratinocytes in the newly re-epithelized cell sheet covering the wound and by fibroblasts and endothelial cells in the subjacent granulation tissue. **C**, Day 21: Undetectable or low levels of MSF expression. (See Page 10 in Color Section at the back of the book).

a hitherto unrecognized role in orchestrating the spatial and temporal control of tissue repair (Schor et al., 2005), as topically applied MSF and IGD synthetic peptides to full-thickness wounds in diabetic mice significantly increase both the kinetics of wound closure and neovascularization (Fig. 15.6). Pilot studies have additionally demonstrated that MSF is re-expressed in association with other common pathologies, including scleroderma and rheumatoid arthritis (unpublished observations).

Expression of MSF protein by adult skin fibroblasts is regulated by an unusual post-transcriptional mechanism involving the initial generation of a 5.9 kb MSF pre-message by read-through of the majority of the intron separating exons III-1a and III-1b of the fibronectin gene (Kay et al., 2005). This precursor remains sequestered within the nucleus where it exhibits a short half-life. In response to an interdependent array of regulatory signals (as discussed below), the intron-derived 3'-UTR of the precursor mRNA is cleaved to produce the shorter 2.1 kb mature MSF message, which is then exported to the cytoplasm for translation (Kay et al., 2005). This post-transcriptional control mechanism, coupled with the aforementioned message instability (see section on p. 287), allows the cell to respond rapidly to microenvironmental cues, thereby controlling the temporal and spatial

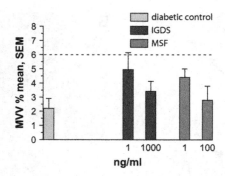

FIGURE 15.6 Stimulation of angiogenesis by MSF and IGD-peptide during wound healing in diabetic mice. A standard full-thickness wound (7.5 × 7.5 mm) was made in diabetic and non-diabetic mice and 20 microliters of PBS were applied to both control groups at days 0, 2, and 4; 20 microliters of the indicated concentrations of recombinant MSF or IGDS peptide were applied to the wounds in diabetic mice at the same time intervals. Animals were sacrificed on day 10 and microvascular volume (MVV) was assessed in histological sections as a surrogate measure of angiogenesis as described in Schor et al. (1998). MMV in non-diabetic controls is indicated by the dashed line. The lower concentrations of MSF and IGDS induced an increase in MMV (compared to the diabetic mouse control) to levels approaching that of the non-diabetic control. The study utilized eight mice per experimental group (unpublished data).

expression of MSF bioactivities. In contrast to this post-transcriptional mechanism, epithelial and endothelial cells appear to constitutively express MSF protein and regulate the manifestation of its potent bioactivities by the variable co-expression of selective inhibitors (as discussed below).

In addition to its *local* expression by tumor cells and tumor-associated stromal cells, patients with cancer may also exhibit an aberrant *systemic* expression of MSF. For example, MSF is inappropriately expressed by skin fibroblasts obtained from distant uninvolved sites in patients with cancer; these aberrant "fetal-like" cells further differ from their normal adult counterparts in terms of their migratory phenotype and elevated expression of HA (Fig. 15.7). We have also reported that the tumor-free, histologically normal, tissue margin adjacent to resected breast carcinomas commonly contain MSF-expressing intra-lobular fibroblasts (Schor et al., 1994; unpublished observations). The presence of such a functionally perturbed peritumor field may carry as yet unrecognized prognostic significance, as well as provide useful insight into the factors (systemic and local) which contribute to cancer pathogenesis (to be discussed below).

MSF bioactivity has also been detected systemically in the serum of approximately 90% of patients with breast cancer, as compared to only 10% of age- and sex-matched healthy controls (Fig. 15.8) (Picardo et al., 1991; additional unpublished data). This systemic expression of MSF does not appear to be a measure of tumor burden (as is the case with other oncofetal tumor markers), as it may persist for decades after resection of

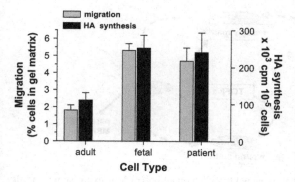

FIGURE 15.7 Fetal-like pattern of migration and HA synthesis by skin fibroblasts from patients with breast cancer. Cell migration and HA synthesis by confluent skin fibroblasts growing on 3D type I collagen gels were ascertained as described in Ellis et al. (1997) and unpublished data. Comparative data (mean ± SD) are presented for four lines each of fetal-derived, healthy adult and breast cancer patients.

the primary tumor in patients with no evidence of recurrent disease. The presence of bioactive MSF in the serum of cancer patients stands in marked contrast to results obtained in a corresponding study in patients recovering from post surgical trauma; in these individuals, MSF bioactivity was not present in serum, in spite of being clearly detected locally in wound fluid (Picardo et al., 1992).

A number of lines of evidence indicate that the post transcriptional control of MSF expression by adult skin fibroblasts may be manipulated by the concerted action of cytokine and matrix signals. For example, a brief (6–8 hour) exposure of adult skin fibroblasts growing on matrices characteristic of "wounded" tissue to certain cytokines, such as TGFβ1 (Kay et al., 2005), results in the switch-on of MSF expression. Significantly, these "activated" cells *continue to express MSF for the entire duration of their* in vitro *lifespan* when subsequently cultured under standard conditions *in the absence of inducing matrix molecule or cytokine* (Fig. 15.9). Remarkably, a second transient exposure of these activated fibroblasts growing on a matrix characteristic of healthy tissue *to the same cytokine* results in the equally persistent "switch-off" of MSF expression (Fig. 15.9). This inducible and persistent switching on and off of MSF expression may be repeated numerous times.

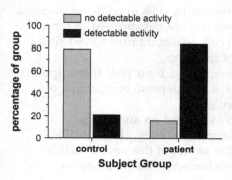

FIGURE 15.8 MSF bioactivity present in the serum of a majority of patients with breast cancer. The presence of detectable levels of MSF bioactivity in the serum of breast cancer patients ($n = 30$) and age- and sex-matched healthy controls ($n = 30$) was ascertained as described in Picardo et al. (1991). Data are presented as the percentage of positive (detectable) and negative (not detectable) samples in the two subject groups.

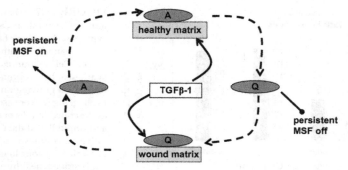

FIGURE 15.9   The persistent and reversible switch on and off of MSF expression by adult skin fibroblasts. Quiescent (Q) adult skin fibroblasts do not produce MSF. They may be induced to do so ("switched-on") by a single transient (8 h) exposure to TGF-β1 when cultured on a "wound" matrix, such as fibrin. These activated (A) cells continue to express MSF for the entire duration of their *in vitro* lifespan when cultured under standard tissue culture conditions. MSF expression may, however, be persistently switched-off by a subsequent transient exposure of such activated cells, this time when growing on a "healthy tissue" matrix (such as native type I collagen), to TGF-β1. This switch on and off remains completely reversible and may be repeated several times.

Our data further indicate that exposure to either matrix or cytokine alone is not sufficient to switch MSF expression either on or off, the concerted signaling of both is required. The inducible, heritable, and reversible nature of this on-off switch is most consistent with the underlying involvement of epigenetic control mechanisms (see section on p. 300).

Section Summary and Significance

- MSF is constitutively expressed during fetal development and is re-expressed in association with cancer and other common pathologies in the adult.
- Significantly, MSF may be expressed by both epithelial and stromal cells (such as fibroblasts and endothelium) within a given tissue.
- In addition to such localized expression by cells in fetal and diseased adult tissues, MSF may also be systemically re-expressed in cancer patients by epithelial and stromal cells in distant uninvolved sites, as well as a circulating component in serum.
- MSF expression by fibroblasts is regulated by a post transcriptional regulatory mechanism involving the interdependent signaling of cytokines and matrix macromolecules.
- MSF expression may be repeatedly switched on and off by the concerted action of microenvironmental cues. The inducible, persistent (heritable), and reversible nature of this on/off switch suggests that epigenetic control mechanisms may be involved.

THE CONTEXTUAL CONTROL OF TARGET CELL RESPONSE TO MSF BY ECM AND SOLUBLE FACTORS

The migratory response of target cells to putative motogenic factors is most commonly studied *in vitro* with the transmembrane (Boyden chamber) assay. This experimental protocol may be used to distinguish between *chemokinesis* (i.e. stimulation of non-directionally biased cell motility in response to an isotropic field of soluble factor) and *chemotaxis* (the stimulation of directional motility in response to a concentration gradient of soluble factor) (Zigmond and Hirsch, 1973). The polycarbonate membrane used in the transmembrane assay does not support cell attachment and must consequently be coated with an adhesive compound, gelatin (*denatured* type I collagen) being a popular choice. Such adhesion-promoting coatings have generally been assumed, either explicitly or tacitly, to provide a "neutral" substratum which does not influence target cell response to the evaluated motogenic factor.

In order to study cell behavior in a more "tissue-like" environment *in vitro*, we developed means for obtaining quantitative data regarding cell migration and proliferation on and within 3-dimensional gels of *native* type I collagen fibers (Schor, 1980). The production of MSF by fetal and cancer patient fibroblasts was first demonstrated using this assay. The critical role played by the substratum in determining cell migratory response to MSF became evident in later studies utilizing the transmembrane assay. In this assay, adult fibroblasts adherent to membranes coated with *native* type I collagen (as used in the collagen gel assay) responded to MSF, whereas the same target cells attached to *denatured* collagen (as typically used in the transmembrane assay) were completely unresponsive (Fig. 15.10). Matrices may accordingly be classified as either "permissive" (as is the case with native type I collagen) or "non-permissive" (as with denatured collagen) with respect to the manifestation of MSF motogenicity. This

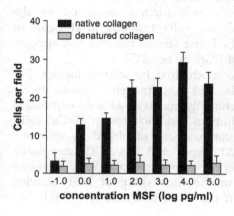

FIGURE 15.10 Matrix-dependence of MSF stimulation of fibroblast migration. Fibroblast migration was assessed in the transmembrane or Boyden chamber assay using membranes coated with either native type I collagen or denatured type I collagen (i.e. gelatin) as described in Schor et al. (1999). Fibroblasts adherent to a native collagen substratum exhibited a motogenic response to MSF (as well as IGD peptide), whereas those adherent to denatured collagen were unresponsive.

matrix-dependency was also apparent in studies of the induction of an early signal transduction response, as generically indicated by the enhancement of plasma membrane proton pump activity (Schor et al., 1999). IGD synthetic peptides exhibit an identical matrix-dependence with respect to both early signaling events and stimulation of cell migration (Schor et al., 1999). Insight into the molecular mechanisms responsible for such matrix-modulation has been provided by observations that (i) the ligation of RGD peptide by integrin $\alpha5\beta1$ inhibits the motogenic response of adult fibroblasts to MSF/IGD; and (ii) the α-chains of type I collagen contain cryptic RGD motifs which are rendered accessible by denaturation (Schor et al., 1999).

The permissive or non-permissive nature of the substratum is not invariant, but may vary with respect to the particular target cell population. For example, although denatured type I collagen is non-permissive for fibroblast response to MSF, it is permissive for endothelial cells (unpublished observations). Similarly, small amounts of added HA or thrombospondin-1 render a native type I collagen matrix non-permissive for endothelial cells, but do not alter the permissive nature of this substratum for target fibroblasts.

Cellular response to other motogenic factors is also modulated by the precise nature of the ECM (Schor, 1994). In this regard, we have recently reported that although the migration of target fibroblasts into a 3-dimensional native collagen matrix is equally stimulated by both EGF and TGF-α in serum-containing medium, this substratum is only permissive for the expression of EGF motogenic activity under serum-free conditions (Ellis et al., 2007). This finding is particularly unexpected, as the cellular response to these two structurally homologous cytokines is contingent upon their respective binding to the same cell surface EGF receptor (EGFR1). Interestingly, the addition of vitronectin (a principal ligand of integrin $\alpha v\beta3$) to serum-free medium is sufficient to restore cell motogenic response to TGF-α. The differential requirement for $\alpha v\beta3$ in the manifestation of TGF-α motogenicity was further suggested by the observed ability of function-neutralizing $\alpha v\beta3$ antibody to abrogate of cellular response to TGF-α, but not EGF. In a similar manner, HA appears to mediate the motogenic activity of EGF, by not TGF-α.

Cellular response to MSF is also modulated by soluble factors. For example, resting endothelial cells and keratinocytes constitutively express MSF which is rendered functionally inactive by the co-expression of an MSF inhibitor. In the case of human keratinocytes (or HaCaT cells, a keratinocyte cell line), this inhibitor has been identified as *neutrophil gelatinase-associated lipocalin*, or NGAL (Jones et al., 2007). The neoplastic transformation of HaCaT cells, either spontaneously or as a consequence of oncogene transfection, results in the unmasking of MSF activity, apparently as a result of a shift in the relative activities of MSF and

NGAL. MSF expressed by resting ("cobblestone") endothelial cells is similarly rendered functionally inactive by co-expression of an as yet unidentified inhibitor. Activation of these cells by angiogenic factors (i.e. as manifest *in vitro* by the adoption of a "sprouting" cell phenotype) is similarly accompanied by the unmasking of MSF functionality.

The complex contextual control of MSF expression and target cell response to it by the concerted action of matrix and soluble factors is summarized in Fig. 15.11.

Section Summary and Significance

- Cellular response to MSF is matrix-dependent: certain matrices are permissive for manifestation of a motogenic response, whereas other matrices are not.
- Cellular response to MSF is also modulated by the presence of soluble factors.
- As a consequence of these various modulatory influences we conclude that the particular bioactivities manifest by MSF *in vivo* are not invariant and are likely to change during the temporal course of disease progression, especially in pathologies characterized by extensive matrix remodeling. A corollary of this dynamic control of MSF functionality is that the presence of MSF in tissues

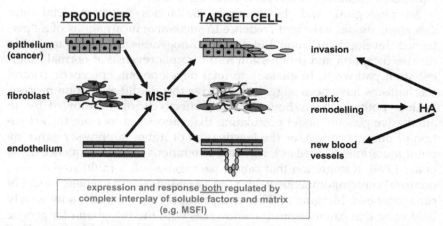

FIGURE 15.11   Contextual modulation of MSF expression and bioactivity. MSF expression by epithelial cells, fibroblasts, and endothelial cells is modulated by a complex and inter-dependent network of regulatory cues, including soluble factors and matrix constituents. The response of target cells to MSF is similarly controlled by soluble factors and matrix macro-molecules. MSF-induced up-regulation of HA synthesis by target fibroblasts may influence the behavior of target epithelial and endothelial cells. This hierarchical regulatory circuitry may contribute to the temporal and spatial control of developmental and pathological processes, such as wound healing and cancer progression.

(as demonstrated by immunohistochemical localization) does not in itself indicate which, if any, of its multiple bioactivities are being expressed nor which of its potential target cells are responding to it.

AN EXTENDED ONCOFETAL PARADIGM OF CANCER PATHOGENESIS: THE INTER DEPENDENT CONTRIBUTION OF GENETIC AND EPIGENETIC MECHANISMS

The term "oncofetal" is used to denote those protein isoforms which are constitutively expressed during fetal development, are generally not expressed in healthy adult tissues, but are re-expressed in association with cancer and other pathological processes, such as wound healing and chronic inflammation. Available data suggest that oncofetal isoforms contribute to the control of relevant tissue-level events, including cell migration and angiogenesis, which are shared features of both fetal development and disease progression. Interestingly, disease-associated stromal cells also display a number of characteristic morphological features variously referred to by pathologists as "fetal-like," "activated," or "plump" when explicitly distinguishing them from their "resting" or "quiescent" healthy counterparts (McNeal, 1984).

Morphological and biochemical similarities between fetal and cancerous tissues have lent credence to successive incarnations of a "perturbed development" model of cancer pathogenesis, according to which, disease inception and progression result from a reversal of normal differentiation pathways. In analogy to fetal development, epigenetic control mechanisms have been suggested to drive this de-differentiation process. Such hypotheses have, however, been effectively overshadowed by an alternative genetic model postulating that cancer results from the activation of oncogenes and/or the inactivation of tumor suppressor genes by point mutations, as well as larger scale alterations to DNA sequence (Ilyas et al., 1999). Recognition that cancer pathogenesis is a multi step process occurred contemporaneously with the ascendance of the genetic model of carcinogenesis. Merging of these two precepts resulted in the now widely held view that carcinogenesis commences with the inception of a genetic lesion affecting the expression of a cancer-critical gene. This "initiating" event is further posited to confer a relative growth advantage to the progeny of the affected cell. Subsequent disease "progression" results from the random accumulation of complementary genetic lesions within the initiated clonal population and the selection of derivative sub-clones displaying ever increasing growth advantage. Successive iterations of this clonal selection process ultimately results in the emergence of progressively dysfunctional populations of pre-neoplastic and overtly neoplastic cells, culminating in the appearance of a clinically identifiable malignancy. A corollary of this genetic model is that cancer inception and progression are "deterministic" processes inexorably driven by essentially irreversible genomic lesions.

Cancer-relevant genetic lesions have also been detected in epithelial cells within histologically normal peritumor tissue. According to the "field cancerization" hypothesis, such lesions are postulated to have arisen from prior exposure to environmental genotoxins, such as tobacco smoke in the case of oral cancer (Nees et al., 1993). The formation of subsequent independent tumors from these initiated cells is speculated to account for the elevated propensity of affected individuals to develop multiple, nonsynchronous primary cancers following the successful ("free margins") resection of the presenting lesion.

Related studies have definitively demonstrated the local and systemic presence of host stromal cells displaying both biochemical and behavioral features (such as anchorage-independent growth and formation of multilayered regions) routinely used to define neoplastic transformation *in vitro* (reviewed in Schor and Schor, 1997; Schor et al., 1987; see also Azzarone et al., 1984; Kopelovich, 1988). These aberrant cells were non-tumorigenic *in vivo* and therefore have commonly been referred to as "partially-transformed." Several genetic mechanisms have been invoked to account for the systemic disposition of these cells, including whole body exposure to environmental genotoxins (in analogy to local field cancerization) and the inheritance of germ-line mutations (as in patients with familial cancers). Seen within the confines of the then prevailing genetic model of cancer pathogenesis, these aberrant host cells were generally not considered to make a direct contribution to cancer development. Instead, their presence was interpreted as telltale indicators of an inherited or incurred genetic lesion which *only* leads to tumor formation when also present within a suitable (usually epithelial) progenitor cell population.

It is important to note, however, that the same phenotypic characteristics used to define "partially-transformed" host cells are also well recognized attributes of fetal cells. We have consequently elected to employ the descriptor "fetal-like" when referring to such aberrant host cells in patients with cancer (Schor and Schor, 1987, 2001; Schor et al., 1987). This is not a mere semantic distinction, as explicit use of the term "fetal" invites speculation that an inappropriate deployment of epigenetic mechanisms may also contribute to the genesis of these cells. This possibility is strongly supported by recent reports documenting the presence of aberrantly methylated CpG islands in the promoter regions of cancer-critical genes in histologically unremarkable peritumor tissue (Eads et al., 2000; Shen et al., 2005). Observations regarding the contextual control of MSF expression are also consistent with the involvement of epigenetic mechanisms. For example, our data indicate that a transient exposure of normal adult fibroblasts to an appropriate combination of cytokine and matrix results in a switch-on of MSF expression. Although this change is heritable (persisting for the entire subsequent *in vitro* lifespan of the treated cells), it may be reversed at any time by a second transient exposure to a different combination of cytokine and matrix. Recent pilot studies further indicate that constitutive MSF expression by both *bona fide* fetal fibroblasts and the fetal-like fibroblasts of cancer patients may similarly be persistently switched-off by a transient exposure to an appropriate combination of cytokine and matrix (unpublished data). Finally, related observations indicate that a transient exposure of normal adult fibroblasts to either fluorodeoxyuridine (FUDR) or 5-azacytidine, two agents which induce changes in gene expression by epigenetic mechanisms, also result in a persistent switch-on of MSF expression which, again, may be switched off by a subsequent exposure to combinations of cytokine and matrix. Taken together, these observations suggest that aberrantly behaving host cells in cancer patients may have arisen by a prior exposure of the individual to agents which, like 5-azacytidine and FUDR, induce persistent (although reversible) epigenetic alterations in gene expression. The key question is whether these "fetal-like" host cells, which may precede the initiation and subsequent progression of an overt malignant lesion, actually contribute to the disease process. In this context, initial observations indicate a statistically significant inverse association between MSF expression by carcinoma cells and the survival of patients with breast cancer (unpublished observations).

Explicit use of the term "fetal-like" also engenders a broader "contextual" framework in which to consider the disease process. In this regard, it has long been recognized that cancer progression can be an extremely indolent process, with many decades elapsing between documented or inferred exposure to an environmental genotoxin and the emergence of a clinically overt malignancy. Indeed, a significant (age-related) proportion of apparently healthy individuals have been shown to harbor pre-malignant and/or early stage malignant lesions prior to their death by non-cancer related causes (Nielsen et al., 1987). We have previously speculated that the inappropriate production of MSF, as well as other oncofetal regulatory molecules, by fetal-like stromal cells may significantly *accelerate* the kinetics of cancer progression by creating a more nurturing (permissive) microenvironment for the clonal expansion of genetically initiated tumor progenitor cells (Schor and Schor, 1997; 2001; Schor et al., 1987). The effects of MSF on cell migration, angiogenesis, and HA synthesis would all be expected to function in this capacity.

In the light of these considerations, we propose an "extended" oncofetal paradigm of cancer pathogenesis according to which (i) both genetic and epigenetic mechanisms contribute to the creation of a perturbed tissue field containing tumor progenitor and/or aberrant host stromal cells; and (ii) the progression of initiated tumor progenitor cells is driven by both genetic (deterministic) and epigenetic (contextual) mechanisms. In agreement with a previously published observation: *"there is nothing new under the sun"* (Ecclesiastes 1:9), this "extended" oncofetal model is essentially a re-statement of the traditional "seed-soil" hypothesis in the contemporary context of clonal selection and epigenetic/genetic mechanisms. Nonetheless, we hope that explicit recognition of the multi factorial (local and systemic, genetic and epigenetic) factors contributing to cancer progression may encourage a more holistic view of patient management strategies. For example, if future studies confirm that MSF and other oncofetal regulatory molecules produced by "fetal-like" host cells do indeed accelerate tumor progression, the potentially reversible nature of their expression may make it possible to develop novel adjuvant therapies designed to switch them off. Such therapies may make it possible to (i) retard the kinetics of disease progression in patients with cancer; (ii) reduce the likelihood of a local recurrence in patients with a resected primary tumor embedded within a histologically normal, but functionally perturbed, field containing fetal-like host cells; and (iii) reduce the risk of cancer development in apparently healthy individuals (approximately 10% of the adult population) who exhibit an aberrant systemic expression of MSF.

Section Summary and Significance

- According to the prevailing clonal selection model, cancer progression is a multi step process driven by the successive emergence of increasingly aberrant clonal populations harboring accumulated genetic lesions. Aberrantly behaving "partially transformed" stromal cells present in patients with cancer have not generally been regarded to make a direct contribution to the disease process.
- We have proposed an "extended oncofetal" model which explicitly acknowledges the additional contribution of epigenetic and contextual control mechanisms to cancer pathogenesis.

ACKNOWLEDGMENTS

This work was supported by funding from Cancer Research UK, Breast Cancer Campaign, National Institutes of Health, USA (grant 1 RO1 DK59144-01), Scottish Enterprise Proof of Concept Programme, Engineering and Physical Sciences Research Council and Biotechnology and Biological Sciences Research Council.

References

Amanuma, K. and Mitsui, Y. (1991). HA synthesis is absent in normal human endothelial cells irrespective of HA synthase inhibitor activity but is significantly high in transformed cells. *Biochem Biophys Acta* **1093**, 336–340.

Azzarone, B., Mareel, M., Billard, C., et al. (1984). Abnormal properties of skin fibroblasts from patients with breast cancer. *Int J Cancer* **33**, 759–764.

Bakheet, T., Frevel, M., Williams, B. R. G., Greer, W., and Khabar, K. S. A. (2001). ARED: human AU-rich element-containing mRNA database reveals an unexpectedly diverse functional repertoire of encoded proteins. *Nucleic Acid Res* **29**, 246–254.

Chen, C.-Y. A., Chen, T.-M., and Shyu, A.-B. (1994). Interplay of two functionally and structurally distinct domains of the c-fos AU-rich element specifies its mRNA-destabilizing function. *Mol Cell Biol* **14**, 416–426.

Chen, W. Y. J., Grant, M. E., Schor, A. M., and Schor, S. L. (1989). Differences between adult and fetal fibroblasts in the regulation of hyaluronate synthesis: correlation with migratory activity. *J Cell Sci* **94**, 577–584.

Clark, R. A. F., Wikner, N. E., Doherty, D. E., and Norris, D. A. (1998). Cryptic chemotactic activity of fibronectin for human monocytes resides in the 120-kDa fibroblastic cell-binding fragment. *J Biol Chem* **263**, 12115–12123.

Durning, P., Schor, S. L., and Sellwood, R. A. (1984). Fibroblasts from patients with breast cancer show abnormal migratory behaviour *in vitro*. *Lancet* **2**, 890–892.

Eads, C. A., Lord, R. V., Kurumboor, S. K., et al. (2000). Fields of aberrant CpG island hypermethylation in Barrett's esophagus and associated adenocarcinoma. *Cancer Res* **60**, 5021–5026.

Ellis, I., Banyard, J., and Schor, S. L. (1997). Differential response of fetal and adult fibroblasts to cytokines, cell migration and hyaluronan synthesis. *Development* **124**, 1593–1600.

Ellis, I., Grey, A. M., Schor, A. M., and Schor, S. L. (1992). Antagonistic effects of TGF-β1 and MSF on fibroblast migration and hyaluronic acid synthesis – possible implications for dermal wound healing. *J Cell Sci* **102**, 447–456.

Ellis, I. R., Schor, A. M., and Schor, S. L. (2007). EGF and TGF-alpha motogenic activities are mediated by the EGF receptor via distinct matrix-dependent mechanisms. *Exp Cell Res* **313**, 732–741.

ffrench-Constant, C., Van de Water, L., Dvorak, H. F., and Hynes, R. O. (1989). Reappearance of an embryonic pattern of fibronectin splicing during wound healing in the adult rat. *J Cell Biol* **109**, 903–914.

Fukai, F., Suzuki, H., Suzuki, K., Tsugita, A., and Katayama, T. (1991). Rat plasma fibronectin contains two distinct chemotactic domains for fibroblastic cells. *J Biol Chem* **246**, 8807–8813.

Fukai, F., Iso, T., Sekiguchi, K., et al. (1993). An amino-terminal fibronectin fragment stimulates the differentiation of ST-13 preadipocytes. *Biochemistry* **32**, 5746–5751.

Fukai, F., Ohtaki, M., Fugii, N., et al. (1995). Release of biological activities from quiescent fibronectin by a conformational change and limited proteolysis by matrix metalloproteinases. *Biochemistry* **34**, 11453–11459.

Haggie, J. A., Sellwood, R. A., Howell, A., Birch, J. M., and Schor, S. L. (1987). Fibroblasts from relatives of patients with hereditary breast cancer show fetal-like behaviour *in vitro*. *Lancet* **1**, 1455–1457.

Houard, X., Germain, S., Gervais, M., et al. (2005). Migration-stimulating factor displays HEXXH-dependent catalytic activity important for promoting tumor cell migration. *Int J Cancer* **116**, 378–384.

Ilyas, M., Straub, J., Tomlinson, I. P., and Bodmer, W. P. (1999). Genetic pathways in colorectal and other cancers. *Eur J Cancer* **35**, 1986–2002.

Hynes, R. (1990). Fibronectins. New York: Springer-Verlag.

Jones, S. J., Florence, M. M., Ellis, I. R., et al. (2007). Co-expression by keratinocytes of migration stimulating factor (MSF) and functional inhibitor of its bioactivity (MSFI). *Exp Cell Res* **313**, 4145–4157.

Kay, R. A., Ellis, I. R., Jones, S. J., et al. (2005). The expression of MSF, a potent oncofetal cytokine, is uniquely controlled by 3'-untranslated region dependent nuclear sequestration of its precursor messenger RNA. *Cancer Res* **65**, 10742–10749.

Kopelovich, L. (1988). The transformed (initiated) human cell phenotype: studies on cultured skin fibroblasts from individuals predisposed to cancer. *Mutat Res* **199**, 369–385.

Liu, X., Zhao, Q., and Collodi, P. (2003). A truncated form of fibronectin is expressed in fish and mammals. *Matrix Biol* **22**, 393–396.

McNeal, J.E. (1984). Anatomy of the prostate and morphogenesis of BPH. In: Kimbal, F.A., Buhl, A.E. and Carter, D.B., (eds.). New Approaches to the Study of Benign Prostatic Hyperplasia. New York: A.R. Liss, pp. 27–53.

Millard, C. J., Ellis, I. R., Pickford, A. R., et al. (2007). The role of fibronectin IGD motif in stimulating fibroblast migration. *J Biol Chem* **282**, 35530–35535.

Muir, D. and Manthorpe, M. (1992). Stromelysin generates a fibronectin fragment that inhibits Schwann cell proliferation. *J Cell Biol* **116**, 177–185.

Nees, M., Hormann, N., Discher, H., et al. (1993). Expression of mutated p53 of tumor-distant epithelia of head and neck cancer patients: a possible molecular basis for the development of multiple tumors. *Cancer Res* **53**, 4189–4196.

Nielsen, M., Thomsen, J.L., Primdahl, S., Dyreborg, U. and Andersen JA. (1987). Breast cancer and atypia among young and middle–aged women. *Br J Cancer* **56**, 814–819.

Picardo, M., Schor, S. L., Grey, A. M., et al. (1991). Migration stimulating activity in serum of breast cancer patients. *Lancet* **337**, 130–133.

Picardo, M., Grey, A. M., McGurk, M., Ellis, I., and Schor, S. L. (1992). Detection of migration stimulating activity in wound fluid. *Exp Mol Pathol* **57**, 8–21.

Schor, A. M., Rushton, G., Ferguson, J. E., et al. (1994). Phenotypic heterogeneity in breast fibroblasts – functional anomaly in fibroblasts from histologically normal tissue adjacent to carcinoma. *Int J Cancer* **59**, 25–32.

Schor, A. M., Pendleton, N., Pazouki, S., et al. (1998). Assessment of vascularity in histological sections: effects of methodology and value as an index of angiogenesis in breast tumours. *Histochem J* **30**, 849–856.

Schor, S. L. (1980). Cell proliferation and migration on collagen substrata *in vitro*. *J Cell Sci* **41**, 159–175.

Schor, S. L. (1994). Cytokine control of cell motility: modulation and mediation by the extracellular matrix. *Prog Growth Factor Res* **5**, 223–248.

Schor, S. L. and Schor, A. M. (1997). Stromal acceleration of tumour progression: role of fetal-like fibroblast subpopulations. *Pathol Update* **4**, 75–95.

Schor, S. L. and Schor, A. M. (2001). Tumour-stroma interactions – phenotypic and genetic alterations in mammary stroma: implications for tumour progression. *Breast Cancer Res* **3**, 373–379.

Schor, S. L., Schor, A. M., Rushton, G., and Smith, L. (1985a). Adult, foetal and transformed fibroblasts display different migratory phenotypes on collagen gels – evidence for an isoformic transition during foetal development. *J Cell Sci* **73**, 221–234.

Schor, S. L., Schor, A. M., Durning, P., and Rushton, G. (1985b). Skin fibroblasts from cancer patients display foetal-like migratory behaviour on collagen gels. *J Cell Sci* **73**, 235–244.

Schor, S. L., Schor, A. M., Howell, A., and Crowther, D. (1987). Hypothesis: persistent expression of fetal phenotypic characteristics by fibroblasts is associated with increased susceptibility to malignant disease. *Eur Exp Cell Biol* **55**, 11–17.

Schor, S. L., Schor, A. M., Grey, A. M., and Rushton, G. (1988a). Fetal and cancer patient fibroblasts produce an autocrine migration-stimulating factor not made by normal adult cells. *J Cell Sci* **90**, 391–399.

VI. HYALURONAN IN CANCER EPITHELIAL–STROMAL INTERACTIONS

Schor, S. L., Schor, A. M., and Rushton, G. (1988b). Fibroblasts from cancer patients display a mixture of both fetal and adult-like phenotypic characteristics. *J Cell Sci* **90**, 401–407.

Schor, S. L., Schor, A. M., Grey, A. M., et al. (1989). Mechanism of action of the migration stimulating factor produced by fetal and cancer-patient fibroblasts: effect on hyaluronic acid synthesis. *In vitro Cell Dev Biol* **25**, 737–746.

Schor, S. L., Ellis, I., Dolman, C., et al. (1996). Substratum-dependent stimulation of fibroblast migration by the gelatin-binding domain of fibronectin. *J Cell Sci* **109**, 2581–2590.

Schor, S. L., Ellis, I., Banyard, J., and Schor, A. M. (1999). Motogenic activity of the IGD amino acid motif. *J Cell Sci* **112**, 3879–3888.

Schor, S. L., Ellis, I. R., Jones, S. J., et al. (2003). Migration-stimulating factor: a genetically truncated onco-fetal fibronectin isoform expressed by carcinoma and tumor-associated stromal cells. *Cancer Res* **63**, 8827–8836.

Schor, S.L., Schor, A.M., Keatch, R.P. and Belch, J.F.F. (2005). Role of matrix macromolecules in the aetiology and treatment of chronic ulcers. In: Lee, B.Y., (ed.). The Wound Management Manual. New York: McGraw Hill, Chapter 10, pp. 109–121.

Shen, L., Kondo, Y., Rosner, G. L., et al. (2005). MGMT promoter methylation and field defect in sporadic colorectal cancer. *J Natl Cancer Inst* **97**, 1330–1338.

Werb, Z., Tremble, P. M., Behrendsten, O., Cowley, E., and Damsky, C. H. (1989). Signal transduction through fibronectin receptor induces collagenase and stromelysin gene expression. *J Cell Biol* **109**, 877–889.

Winterbourne, D., Schor, A. M., and Gallagher, J. T. (1983). Synthesis of GAGs by cloned bovine endothelial cells cultured on collagen gels. *Eur J Biochem* **135**, 271–277.

Zhao, Q., Liu, X., and Collodi, P. (2001). Identification and characterization of a novel fibronectin in Zebrafish. *Exp Cell Res* **268**, 211–219.

Zigmond, S. H. and Hirsch, J. G. (1973). Leukocyte locomotion and chemotaxis. New methods for evaluation and demonstration of a cell derived chemotactic factor. *J Exp Med* **137**, 387–410.

HYALURONAN AND INDIVIDUAL CANCERS

CHAPTER

16

Hyaluronan Synthesis and Turnover in Prostate Cancer

Melanie A. Simpson

OVERVIEW OF PROSTATE CANCER PROGRESSION

Despite the prevalence of prostate cancer, prostate carcinogenesis is not fully understood, either from the standpoint of physical causes or initial molecular progression. Several genetic susceptibility loci have been identified that may account for higher risk through reduced ability to cope with environmental insults. Insults may be in the form of chronic or acute inflammation, bacterial infection, or exposure to carcinogens. Inflammation leads to tissue damage, matrix remodeling, fibrosis, hypoxia, selection for cell survival in extreme conditions, and triggering of normal angiogenic mechanisms. The infiltration of immune cells produces reactive oxygen species (ROS) that damage DNA while cells are being demanded to repopulate the damaged areas of tissue, promoting mutagenesis and ultimately neoplasia.

In the prostate, the luminal epithelium is the site of tumorigenesis and normal proliferation is androgen dependent. Tumors detected early are often managed effectively by surgical resection and/or hormone ablation therapy. However, a significant percentage of tumors will resume growth in the absence of androgens (Pienta and Smith, 2005). The transformation from androgen dependent to androgen independent prostate cancer is incompletely understood, and such tumors are typically highly aggressive. Invasive spread, in which tumor cells are no longer confined to the prostate, may include surrounding smooth muscle tissue, seminal vesicles, or rectum. Metastasis may occur in either androgen dependent or androgen independent prostate cancer and most frequently involves lymph nodes and/or bone. Complications of bone metastatic cancer are the cause of ≈80% of prostate cancer mortality (Jemal et al., 2006).

HA SYNTHASES AND HYALURONIDASES

In humans, HA synthases HAS-1 (Shyjan et al., 1996), HAS-2 (Watanabe and Yamaguchi, 1996) and HAS-3 (Spicer and McDonald, 1998) catalyze the synthesis of HA polymers from ≈100–5000 kDa (Weigel et al., 1997). High levels of HA production are maintained by availability of UDP-esterified glucuronate, which is provided by UDP-glucose dehydrogenase (UGDH). The level of UGDH is normally high in the adult prostate relative to other tissues (with the exception of liver), though HAS expression is extremely low, and its main role may be in solubilization of androgens for excretion (Lapointe and Labrie, 1999; Spicer et al., 1998).

HA turnover depends on the activity of the hyaluronidase enzymes HYAL-1-HYAL-4 and PH-20, each of which catalyzes endolytic cleavage of HA polymers (Csoka et al., 2001; Stern, 2005). Sizes of the processed fragments range from tetrasaccharides to ≈20 kDa polymers, depending upon isozymes and microenvironment. HYAL-1 is found in serum and in extracellular spaces, while HYAL-2 and PH-20 are GPI anchored at the extracellular surface. Both HYAL-1 and HYAL-2 also occur in lysosomes, where they catalyze complete hydrolysis of endocytosed HA. Removal of HA from the cell surface is additionally promoted by surface receptors such as CD44, LYVE-1, and RHAMM. These receptors determine the responses of the cell to extracellular HA in a complex fashion that depends in part on their individual regulatory status and partly on the size of the HA fragment.

HA polymers may be fragmented in the extracellular space by inflammation-induced ROS (Al-Assaf et al., 2006; Jiang et al., 2007), such as hydroxyl radicals, through locally acidified microenvironment activation of extracellular matrix (ECM) hyaluronidase, through release from lysosomes during necrotic cell death, or through mechanical disruption of ECM integrity (reviewed in Stern et al., 2006). Fragments may induce cell proliferation, motility, or apoptosis, while polymers are either inert or suppress these effects. In general, HA is a critical stimulus regulating cellular behavior during wound healing, inflammation, and development (Camenisch et al., 2000; Delpech et al., 1997; Evanko et al., 1999; Fujimoto et al., 1989; Gakunga et al., 1997; Jiang et al., 2007; Laurent et al., 1996; Toole, 1997). The following sections discuss the clinical correlation and functional roles of HA synthases, hyaluronidases, and HA receptors in specific aspects of prostate cancer.

GENETIC SUSCEPTIBILITY TO PROSTATE CANCER

Relatively few genetic determinants of prostate cancer have been identified. Recently, chromosome locus 8q24 was found to be over-represented in human prostate cancer and prostate tumor cell lines, as a result of its duplication and translocation to another chromosome (Tsuchiya et al., 2002). This locus contains, among others, the gene encoding the essential isozyme of the HA biosynthetic enzyme family, HAS-2 (Camenisch et al., 2000). Extra gene copies of the biosynthetic enzyme may account for elevation of HA with the progression of prostate cancer in patients with this genetic lesion. Interestingly, single nucleotide polymorphisms in untranslated DNA at chromosomal locus 8q24 were also newly confirmed by three independent groups simultaneously as a genetic correlate of up to 40% of prostate cancers, in particular accounting for early appearance of the disease in certain European lineages and African-American men (Gudmundsson et al., 2007; Haiman et al., 2007; Yeager et al., 2007).

INFLAMMATION RESPONSE AND HYPOXIA

Several recent reviews address the well documented role of inflammation in prostate carcinogenesis (De Marzo et al., 2007; Goldstraw et al., 2007; Wagenlehner et al., 2007). The prostate may become inflamed as a result of physical insults such as urine reflux, chemical components of the diet such as heterocyclic amines, or bacterial infections (Fig. 16.1). The net result is pro-inflammatory cytokine production, which differentially elevates HAS-2 and/or HAS-3 mediated HA synthesis (Mohamadzadeh et al., 1998). HA recruits macrophages, which infiltrate the damaged site and release copious amounts of ROS to induce epithelial cell death and degrade the HA polymers to oligomers (Jiang et al., 2007). HA oligomers

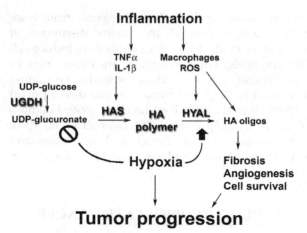

FIGURE 16.1 Flow diagram of inflammation-related prostate carcinogenesis and progression.

promote collagen synthesis, angiogenesis, and epithelial cell proliferation to repair the injured tissue (Mast et al., 1992; Toole, 2004). In a glandular tissue such as the prostate, experiencing chronic inflammation, the new matrix is disorganized and fibrotic (Wernicke et al., 2003). Fibrosis contributes to extreme hypoxia, a hallmark of several cancers including prostate (Movsas et al., 1999), which suppresses proliferation and activates autophagic processes to promote cellular survival (Degenhardt et al., 2006; Jin and White, 2007). These lesions continue to be infiltrated by inflammatory cells and adjacent epithelium may become hyperproliferative as the tissue attempts to regenerate. These lesions are termed Proliferative Inflammatory Atrophy (PIA), and are thought to precede high-grade prostatic intraepithelial neoplasia (HGPIN) and prostate adenocarcinoma.

Areas of PIA are damaged tissues high in mutagenic ROS. The elevated HA polymers in response to the initial damage have a wound healing function, in that HA deposition is observed to preserve cell-free space for organized replacement of basement membrane and polarized epithelium. In the presence of chronically elevated ROS, however, HA degradation to oligomers pathologically causes excess stromal collagen synthesis, reducing oxygen and nutrient permeability, despite concurrently increased cell proliferation. The resulting hypoxia may potentiate rapid aggressive progression of the cancer (Harris, 2002). Hypoxia-induced genes such as HIF-1α and TGF-β1 are then expressed (Liu and Simon, 2004), activating further collagen deposition and initiating angiogenesis cascades to augment blood supply. Hypoxia creates an anti-proliferative microenvironment for the epithelial tumor cells, which then cease efficient aerobic metabolism in favor of lactate production and elevated glycolytic flux (Harris, 2002; Liu and Simon, 2004). Chronically increased glycolysis provides a metabolic advantage by facilitating survival in hypoxic conditions.

Hypoxic regions have been identified with microelectrodes and molecular markers in prostate tumors (Movsas et al., 1999; Wouters et al., 2002). In development and normal wound healing, hypoxia induces angiogenesis, in part because ROS-generated HA fragmentation occurs and because secretion of angiogenic factors such as VEGF is signaled. However, HA polymers are anti-angiogenic and accumulation of HA in hypoxic tumor regions may thereby exacerbate hypoxia and prolong tumor cell survival by potentiating slow proliferation. Tumor cells in a hypoxic microenvironment are highly resistant to chemo- and radio-therapy treatment efforts (Harris, 2002; Wouters et al., 2002) because of their reduced proliferation and autophagic nutrient conservation. Presence of elevated ROS nonetheless leads to mutations in tumor cell DNA of cells residing in these extreme conditions. Ultimately, as the matrix is slowly turned over by activation of proteases and hyaluronidase, the former of which is activated by HIF-1α and the latter by acidification of the local microenvironment, an aggressive prostate tumor may be the outcome.

TUMORIGENESIS

Synthesis and deposition of HA is normally associated with cell division, motility, transformation, and vascular development in embryogenesis (reviewed in Fraser et al., 1997; Toole, 2004). These normal functions continue into adulthood for maintenance of cell and tissue turnover. Additional utility of HA as an adhesion substrate is important for recruitment of circulating cells during wound healing and immune system function. HA-mediated functions are exploited by transformed cells, so the enzymes and receptors which synthesize, process, and signal cellular responses to HA are valuable predictive tools for cancer aggressiveness, and their complex interplay is an active area of research.

Many studies have demonstrated the strong correlation between HA accumulation and malignancy in solid tumors. Prostate histopathology first indicated increased HA content of the stroma in benign prostatic hyperplasia (BPH) relative to normal tissue (De Klerk, 1983), and found the magnitude of the increase directly mirrored level of dedifferentiation within the tumor (De Klerk et al., 1984). Subsequent studies investigated the diagnostic and prognostic potential of HA detection in tumors. In general, HA is detected in tumor stroma if Gleason score is ≥ 5 and does not indicate severity. Tumor cell-associated HA is detected at high Gleason scores and in metastases, each of which correlates with poor prognosis (Aaltomaa et al., 2002; Lipponen et al., 2001; Lokeshwar et al., 2001). In retrospective studies with minimum 5-year follow-up, high HA levels in radical prostatectomy specimens did not independently predict biochemical recurrence (PSA elevation after tumor resection indicating possible local or distant metastasis) (Ekici et al., 2004; Posey et al., 2003). However, combined staining of HA and HYAL-1 is 87% accurate for predicting disease progression (Ekici et al., 2004). It is noteworthy that in early stage prostate cancer specimens, HYAL-1 is exclusively expressed by tumor cells, and HA is primarily in the tumor-associated stroma (Posey et al., 2003), but its cellular origin becomes uncertain as compartmental boundaries erode in progression.

HA production in cultured tumor cells is significantly stimulated by growth factors and proinflammatory cytokines known to be elevated in human prostate cancer specimens. Notably, IGF-1, bFGF (Kuroda et al., 2001), PDGF (Pullen et al., 2001), IL-1β and TNFα (Mohamadzadeh et al., 1998), have all been shown to trigger HA synthesis. PDGF in particular was shown to stimulate HAS-2 expression and HA production in prostate stromal fibroblasts during the transition from normal to BPH. These studies suggest an important link between clinically relevant tumor proliferation signals and inappropriate HA biosynthesis. Targeted disruption of HAS-2 in mice illustrated that HA can signal epithelial to mesenchymal transition during development (Camenisch et al., 2000). This function of HA as a transformation signal has been demonstrated in cancer progression as well, in which case neoplastic transformations may be promoted by HA-mediated activation of receptor tyrosine kinases such as ErbB2 (Ghatak et al., 2005; Misra et al., 2006; Toole et al., 2005).

Despite its documented activation of growth factor receptors, large HA quantities of high molecular mass (i.e., from ≈100 kDa to ≈2 MDa) are antiproliferative (Rooney et al., 1995; West and Kumar, 1989), capable of suppressing growth of both tumor cells (Enegd et al., 2002; Simpson, 2006) and vascular endothelial cells (West and Kumar, 1989). Constitutive engagement of HA receptors by small amounts of newly synthesized HA polymers was shown to stimulate cell cycling through assembly of protein complexes including PI3-kinase, cdc37, and the cytoskeletal adaptor ezrin (Ghatak et al., 2002; Ghatak et al., 2005). HA responses are also context dependent with respect to two-dimensional versus three-dimensional culture conditions (Toole, 2002). Anchorage-independent growth of prostate tumor cells requires HA polymer ligation of cell surface receptors to stabilize cytoskeletal architecture by intracellular association with cytoskeletal proteins and adaptors (Ghatak et al., 2002; Zhu and Bourguignon, 1998). Disruption of these interactions *in vitro* and *in vivo* by administration of HA oligomers induces tumor apoptosis (Ghatak et al., 2002; Zeng et al., 1998).

Among cultured human prostate tumor cell lines, elevated HA production was found specifically in aggressive, metastatic cells, in which HAS-2 and HAS-3 isozymes were upregulated ≈3-fold and ≈30-fold, respectively (Simpson et al., 2001). Suppression of HAS-2 and/or HAS-3 expression by stable antisense RNA reduced the synthesis and cell surface retention of HA (Simpson et al., 2002a), and inhibited primary subcutaneous (Simpson et al., 2002b) or intraprostatic growth (McCarthy et al., 2005). Reduced primary tumor growth was associated with comparable apoptotic and proliferative fractions in culture and in tumors, but virtually no vascularization of tumors. These results implicate HA, and specifically HAS-2 and HAS-3, in tumor angiogenesis, as well as intrinsic growth rate modulation (Simpson et al., 2002b). Interestingly, HA exogenous addition to knock-down cells upon injection restored subcutaneous tumor growth and angiogenesis, implying the existence of a tumor or stromal factor (i.e. a hyaluronidase) that could modulate effects of HA in trans, with the same malignant outcome. HAS-2 overexpression in prostate cells impacted tumor initiation but did not significantly affect size of tumors that formed subcutaneously (Simpson, 2006). In contrast, overexpression of HAS-3 suppressed growth (Bharadwaj et al., 2007). Two alternatives explain this dichotomy: HA synthesis is dose responsive, with enhanced tumorigenesis at low synthesis amounts and inhibited tumorigenesis at high amounts (Itano et al., 2004); and hyaluronidase co-expression with HA synthases is required for maximum tumorigenesis.

Overexpression of HYAL-1 in prostate tumor cells enhanced rate of proliferation (Simpson, 2006) and promoted cell cycle progression (Lin and Stern, 2001). However, tumorigenic potential either subcutaneously (Simpson, 2006) or in the prostate (Kovar et al., 2006) was not altered. Inhibition of HYAL-1 expression showed significant impairment of subcutaneous tumorigenesis and reduced proliferation through cell cycle arrest (Lokeshwar et al., 2005), consistent with a role for HYAL-1 as a tumor promoter. HA production by the tumor stroma correlates with HYAL-1 levels in tumor cells, suggesting crosstalk between the tumor and the tumor-associated stroma (Lokeshwar et al., 2005). Relative to cells expressing either HAS-2 or HYAL-1 alone, which showed modest but not significant increases in tumorigenesis, cells co-expressing HAS-2 and HYAL-1 gave rise to tumors of several-fold increased size (Simpson, 2006). Synergy between HA and its processing enzymes supports an active role for crosstalk between the HA metabolic pathways in tumor progression (Fig. 16.2).

In addition to its effects on transformation and proliferation of prostate cells, HA polymers diminish ECM adhesion and enhance motility of tumor cells. Excess production and retention of HA at the surface of prostate tumor cells by overexpression of HAS-3 was found to reduce adhesion to ECM proteins by downregulation of specific cell surface integrin receptors (Bharadwaj et al., 2007), thereby increasing migration potential. Time lapse videomicroscopy revealed that treatment of tumor cells with prostate fibroblast conditioned media, which caused formation of HA matrices on the tumor cells, also increased their motility (Ricciardelli et al., 2007).

VII. HYALURONAN AND INDIVIDUAL CANCERS

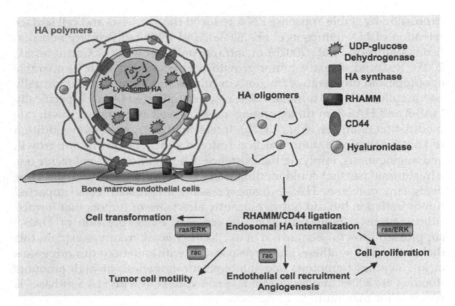

FIGURE 16.2   Model for HA involvement in tumor cell biology. The prostate stromal and tumor-associated HA matrices found in aggressively progressing tumors are maintained by a complex interplay of enzymes and receptors. Production of hairlike HA polymers occurs at the cell surface from membrane embedded HA synthases. High level HA synthesis places demands on intracellular UDP-glucose dehydrogenase to provide HA precursors in excess of the cell's energy needs. HA polymers are retained at the cell surface by transmembrane and extracellular HA binding proteins. Degradation of HA by ROS or hyaluronidases generates a decrement of lower molecular weight fragments of HA, which may be internalized through the action of specific cell surface receptors such as CD44 and RHAMM, and act in autocrine or paracrine fashion on both tumor cells and associated stromal cells. Increased cellular motility among endothelial cells within the prostate stroma in response to HA fragments may directly regulate angiogenesis. Response of prostate epithelial cells to HA fragments versus HA polymers determines cellular transformation, proliferation, motility, and ultimately may be required for sustained tumor growth and metastasis.

HA was most densely deposited at the trailing edge of the polarized motile cells, consistent with a potential role in de-adhesion. A related observation is that levels of the extracellular HA-binding proteoglycan, versican, rise as a function of prostate cancer severity (Ricciardelli et al., 1998) and are independently prognostic. Versican/HA complexes have been implicated in the mediation of prostate stromal cell motility. In fact, TGFβ1 secreted by tumor cells induces stromal fibroblast secretion of versican (Sakko et al., 2001). Thus, the combined accumulation of versican and HA matrix is required for prostate tumor cell remodeling of the ECM and acquisition of intrinsic motility (Ricciardelli et al., 2007).

Complexity of response to HA is increased by the involvement of HA receptors and binding proteins. In addition to versican and RHAMM, the cell surface HA receptor CD44 has been directly linked to prostate cancer progression. CD44-mediated responses are very complex, as it has been shown to function as a tumor suppressor, but activated signaling through CD44 promotes proliferation, motility, and metastasis. It has been proposed (Herrlich et al., 2000), that the activity of CD44 as a tumor suppressor is mediated by the local equilibrium of HA polymers and oligomers, whereas its signaling activities are the result of CD44 having the alternative HA independent function of growth factor presentation to relevant surface receptors. Interestingly, prostate cancer stem cells, recently implicated in prostate cancer progression (Maitland and Collins, 2005), were isolated and characterized by their capacity for self renewal, expressing among other markers, CD44 (Collins et al., 2005). Several splice variants of CD44 have been reported in prostate carcinomas (Liu, 1994), but there are conflicting reports on its relation to carcinoma stage. For example, CD44 standard isoform is reported to decrease with increasing cancer grade (De Marzo et al., 1998; Noordzij et al., 1997) due to hyper-methylation of the promoter (Kito et al., 2001), and loss of this form is an independent predictor of clinical recurrence (Aaltomaa et al., 2002; Vis et al., 2000). However, levels of CD44 variant isoforms, CD44v3 and CD44v6, have been shown to increase with Gleason sum and T-stage (Aaltomaa et al., 2002). Expression of CD44v3, but not CD44v6, indepen-dently predicts PSA biochemical recurrence. Among cultured prostate tumor cell lines, poorly aggressive androgen sensitive LNCaP cells do not express CD44, whereas metastatic androgen independent cell lines express standard and variant isoforms of CD44 (Liu, 1994; Welsh et al., 1995).

Certainly one important function of CD44 is to mediate the effects of HA polymers and oligosaccharides on cell signaling and response. In particular, CD44 is required for uptake and turnover of HA polymers (Culty et al., 1992; Harada and Takahashi, 2007). Opposite responses to polymers and oligo-mers are thought to result from multivalent clustering of receptors bound to polymers that transduces intracellular signals differently from the mono-valent unclustered CD44 monomer. CD44 is localized to lipid rafts in which HYAL-2 is associated, possibly through simultaneous binding of both to the same HA strand, and activated by CD44-mediated acidification through a Na+/H+ exchange pump (Bourguignon et al., 2004). CD44-mediated HA polymer adhesion is at least partially responsible for stabilization of cell-ECM contacts through interaction of its C-terminal tail with cytoskeletal proteins (Legg et al., 2002; Welsh et al., 1995; Zhu and Bourguignon, 1998), disruption of which can occur by HA-oligomeric activation of merlin, an inhibitor of cytoskeletal complex formation (Morrison et al., 2001). CD44 ligation with HA polymers can induce multidrug resistance (Misra et al., 2003; 2005) and disruption of signaling by oligomeric HA induces apoptosis (Ghatak et al., 2002), inhibits anchorage-independent growth and reduces *in vivo* tumorigenesis (Zeng et al., 1998).

VII. HYALURONAN AND INDIVIDUAL CANCERS

A final point to emphasize is that HA has been specifically implicated in the progression of prostate tumors from hormone dependence to independence following androgen ablation therapy. Rats treated long term with androgens developed neoplastic prostate tissue with elevated stromal HA (Li et al., 2001). Castration, on the other hand, reduced levels of HA in the rat prostate, and these were stimulated subsequently up to 10-fold by androgen administration (Terry and Clark, 1996). In fact, HA may directly influence the transition from androgen dependence to independence. In human tissue samples, elevation of the HA receptor, RHAMM, correlated with aggressive, hormone resistant cancer and metastasis (Lin et al., 2008). Stimulation of androgen dependent prostate tumor cells concurrently with androgen and HA increased proliferation *in vitro* and tumorigenesis *in vivo*, in an androgen receptor and RHAMM dependent fashion. In addition, siRNA knock-down of Rho kinase (ROK) could reverse HA-RHAMM malignant properties of androgen-insensitive prostate tumor cell lines (Lin et al., 2007), so antagonism of this signaling pathway may have potential therapeutic value for hormone resistant cancer patients.

ANGIOGENESIS AND LYMPHANGIOGENESIS

HA oligomers, specifically of 4–25 disaccharides, have been shown to stimulate angiogenesis (West et al., 1985), despite the antiproliferative effect of larger HA polymers on endothelial cells that suppresses angiogenesis (Rooney et al., 1995; West and Kumar, 1989). Furthermore, the antiangiogenic effect of HA polymers on endothelial cells is irreversible once engaged except in the sustained presence of antagonistic HA oligosaccharides (Deed et al., 1997). Thus, this normal function of HA has the potential to allow tumor cells to directly signal their own vascular development. Several reports have implicated HA in prostate tumor angiogenesis. For example, as mentioned above, inhibition of HA polymer synthesis suppressed growth of prostate tumors (McCarthy et al., 2005; Simpson et al., 2002b) but vascular density within the resultant tumors was also ≈80% diminished. Seemingly contrary to this finding, excess deposition of HA has been shown to suppress angiogenesis of prostate tumors. This suggests HA is required for angiogenesis but modification or further metabolism of the polymeric form is critical for the angiogenic response. In support of this hypothesis, angiogenic HA fragments (10–15 disaccharide units) have been detected in high-grade prostate cancer tissues (Lokeshwar et al., 1997; 2001) and the knock-down of HYAL-1 also dramatically impairs angiogenesis (Lokeshwar et al., 2005). Interestingly, HA production in relatively low quantities has a potently stimulatory effect on angiogenesis in prostate tumors (Simpson, 2006), which is consistent with motility experiments described above that show low HA concentrations stimulate, while high levels inhibit, migration.

Tumor lymphangiogenesis, growth of new lymphatic vessels within a developing tumor, has also been implicated in progression of prostate cancer. LYVE-1, a member of the link module domain containing family of HA receptors, like CD44, is primarily expressed on lymphatic endothelial cells and responsible for HA adhesion of lymphocytes, as well as mediation of HA clearance through the lymphatic system during inflammation (Johnson et al., 2007; Prevo et al., 2001). Expression of LYVE-1 is used as a marker for lymphangiogenesis and lymphatic content of human tumors, and has been used specifically to show that lymph vessels are destroyed in the transition from benign to malignant prostate cancer, rather than created (Trojan et al., 2004). Further studies showed a correlation between VEGF-C and lymph node metastasis, suggesting lymph vessel permeability, though not lymph vessel density, may be the key factor underlying lymph node metastasis (Trojan et al., 2006). Surprisingly, LYVE-1 null mice also showed no phenotype with respect to normal lymphatic development, and formed tumors comparably to wild-type mice, suggesting potential redundancy in lymphatic endothelial receptors (Gale et al., 2007). Mouse subcutaneous and orthotopic tumorigenesis and metastasis studies have given rise to conflicting results. In one study, VEGF-C stimulated lymphangiogenesis was strongly correlated with lymph node metastasis following surgical excision of subcutaneously implanted prostate tumors (Brakenhielm et al., 2007), while another group found lymphangiogenesis was not required for lymph node dissemination of orthotopically implanted tumor cells and that lymphatic vessel counts actually decreased (Wong et al., 2005). In either case, it appears that HA binding and turnover by LYVE-1 in lymphatic vessels is not required for metastasis.

METASTASIS AND BONE TURNOVER

HA deposition is observed in the stroma of cancer patients in early stages, but its levels appear to be highest in advanced cancer and remote metastases (Lokeshwar et al., 1996; 2001). In addition, high levels of HA (Lipponen et al., 2001) and/or HYAL-1 correlated with more invasive cancer (Madan et al., 1999), and metastatic prostate cancer lesions were found to have higher HYAL-1 levels than the high-grade primary tumor (Lokeshwar et al., 1996; 2001).

Prostate tumor cells that produce endogenous large quantities of HA polymer are more metastatic to lymph nodes when injected intra-prostatically in mice, and this can be increased by overexpressing HYAL-1 (Patel et al., 2002). In low HA-producing prostate tumor cells, normally not metastatic in a six-week time course, HYAL-1 overexpression induced rapid metastatic dissemination to lymph nodes following orthotopic implantation (Kovar et al., 2006). Stable antisense inhibition of HA synthesis abrogated ≈90% of spontaneous lymph node metastasis from orthotopic injections of prostate tumor cells (McCarthy et al., 2005). HA on the tumor cell surface may be functioning in metastasis at the level of facilitated arrest/adhesion to endothelial surfaces in remote tissues and also signaling of cell survival.

Predilection of metastatic prostate tumor cells for bone has been postulated to result from a combination of anatomic and molecular factors (Cher, 2001). Bone endothelium is the first constrained microvasculature encountered by cells that intravasate from the prostate, so physical arrest may be a significant component. Receptor-ligand interactions may then be engaged, signaling transendothelial migration. Colonization requires growth factors, new blood vessels, and bone tissue remodeling. Bone metastasis in prostate cancer patients is both osteoblastic, characterized by bone growth, and osteolytic, associated with bone breakdown. Prostate tumor cells have been termed osteomimetic because in coculture with bone derived cells, they acquire an osteoblast-like gene expression profile, and this ability to alter in response to bone microenvironment may underlie their tendency to colonize bone tissue (Chung et al., 2005; Koeneman et al., 1999).

Recent studies have shown that specific factors such as RANK, a surface receptor on osteoclasts, and RANKL, its cognate ligand expressed on osteoblasts, can be upregulated in the presence of tumor cells (Corey et al., 2002). Subsequent increased presentation of RANKL by tumor cells and osteoblasts triggers differentiation of osteoclasts, the mediators of bone breakdown and resorption, thereby creating physical space to be occupied by the tumor cells. Importantly, HA signaling through CD44 has been shown to increase the expression of RANKL on the surface of bone-derived stromal cells (Cao et al., 2005), and low molecular weight HA polymers specifically stimulated osteoclast differentiation (Ariyoshi et al., 2005), suggesting tumor-borne HA may be a significant contributing factor in osteomimicry and bone remodeling (Prince, 2004).

In addition to its effects on cell differentiation, there is a strong rationale for HA-mediated adhesion in metastatic potential of tumor cells. HA is a well-recognized substrate for circulating cell adhesion, as originally determined by its ability to arrest lymphocytes in high endothelial venules, and confirmed by *in vitro* adhesion assays (Aruffo et al., 1990; Miyake et al., 1990). Retention of high molecular weight pericellular HA by prostate tumor cell lines correlates with their metastatic potential in mice and promotes adhesion specifically to bone marrow endothelial cells *in vitro* (Simpson et al., 2001). This was proposed to be one mechanism underlying the skeletal metastatic frequency of clinical prostate cancer, and other bone-homing tumors have been shown to use CD44–HA interactions to orchestrate transmigration of the bone endothelium (Okada et al., 1999). Increased HA synthesis and pericellular retention by non-adherent prostate tumor cells rendered them adherent to bone marrow-derived endothelial cells and inhibition of HA synthesis in adherent cells by stable antisense RNA expression diminished their adhesion potential (Simpson et al., 2002a).

CD44 is the proposed receptor mediating HA-facilitated arrest on microvascular endothelial cells, since anti-CD44 antibody treatment blocks adhesion, while RHAMM is essential for migration *in vitro* and angiogenesis *in vivo* (Savani et al., 2001). The involvement of CD44 in HA-mediated adhesion of prostate and breast tumor cells to bone marrow-derived endothelial cells has specifically been demonstrated (Draffin et al., 2004), which highlights the potential capacity of CD44 to function as a bone "metastasis receptor" for HA-bearing tumor cells. However, as mentioned above, clinical studies do not consistently demonstrate CD44 association with metastasis to the bone or any other location.

In summary, HA and its metabolic enzymes and receptors clearly play a significant role in multiple aspects of the progression of prostate cancer. While HA is not intrinsically carcinogenic, its broad functions in cellular transformation, proliferation, adhesion, migration, inflammation, and angiogenesis are coordinately timed and executed in response to other genetic and epigenetic alterations that precede full-blown cancer. Thus, the sequence of responses initiated by elevated HA production are ultimately suited to potentiate tumor growth and metastatic spread. Since its signaling functions may additionally give rise to multi drug resistance, it is critical to understand the interplay of HA metabolism and cellular recognition from a therapeutic standpoint. Furthermore, the identification of aggressive prostate cancer stem cells that express CD44, in conjunction with the production of HA and its oligomers in the hostile environments that promote stem cell activation, highlight the immediate importance of defining these complex relationships.

References

Aaltomaa, S., Lipponen, P., Tammi, R., et al. (2002). Strong stromal hyaluronan expression is associated with PSA recurrence in local prostate cancer. *Urol Int* **69**, 266–272.

Al-Assaf, S., Navaratnam, S., Parsons, B. J., and Phillips, G. O. (2006). Chain scission of hyaluronan by carbonate and dichloride radical anions: potential reactive oxidative species in inflammation? *Free Radic Biol Med* **40**, 2018–2027.

Ariyoshi, W., Takahashi, T., Kanno, T., et al. (2005). Mechanisms involved in enhancement of osteoclast formation and function by low molecular weight hyaluronic acid. *J Biol Chem* **280**, 18967–18972.

Aruffo, A., Stamenkovic, I., Melnick, M., Underhill, C. B., and Seed, B. (1990). CD44 is the principal cell surface receptor for hyaluronate. *Cell* **61**, 1303–1313.

Bharadwaj, A. G., Rector, K., and Simpson, M. A. (2007). Inducible hyaluronan production reveals differential effects on prostate tumor cell growth and tumor angiogenesis. *J Biol Chem* **282**, 20561–20572.

Bourguignon, L. Y., Singleton, P. A., Diedrich, F., Stern, R., and Gilad, E. (2004). CD44 interaction with Na+–H+ exchanger (NHE1) creates acidic microenvironments leading to hyaluronidase-2 and cathepsin B activation and breast tumor cell invasion. *J Biol Chem* **279**, 26991–27007.

Brakenhielm, E., Burton, J. B., Johnson, M., et al. (2007). Modulating metastasis by a lymph-angiogenic switch in prostate cancer. *Int J Cancer* **121**, 2153–2161.

Camenisch, T. D., Spicer, A. P., Brehm-Gibson, T., et al. (2000). Disruption of hyaluronan synthase-2 abrogates normal cardiac morphogenesis and hyaluronan-mediated trans-formation of epithelium to mesenchyme [see comments]. *J Clin Invest* **106**, 349–360.

Cao, J. J., Singleton, P. A., Majumdar, S., et al. (2005). Hyaluronan increases RANKL expression in bone marrow stromal cells through CD44. *J Bone Miner Res* **20**, 30–40.

Cher, M. L. (2001). Mechanisms governing bone metastasis in prostate cancer. *Curr Opin Urol* **11** 438–483.

Chung, L. W., Baseman, A., Assikis, V., and Zhau, H. E. (2005). Molecular insights into prostate cancer progression: the missing link of tumor microenvironment. *J Urol* **173**, 10–20.

Collins, A. T., Berry, P. A., Hyde, C., Stower, M. J., and Maitland, N. J. (2005). Prospective identification of tumorigenic prostate cancer stem cells. *Cancer Res* **65**, 10946–10951.

Corey, E., Quinn, J. E., Bladou, F., et al. (2002). Establishment and characterization of osseous prostate cancer models: intra-tibial injection of human prostate cancer cells. *Prostate* **52**, 20–33.

Csoka, A. B., Frost, G. I., and Stern, R. (2001). The six hyaluronidase-like genes in the human and mouse genomes. *Matrix Biol* **20**, 499–508.

Culty, M., Nguyen, H. A., and Underhill, C. B. (1992). The hyaluronan receptor (CD44) participates in the uptake and degradation of hyaluronan. *J Cell Biol* **116**, 1055–1062.

De Klerk, D. P. (1983). The glycosaminoglycans of normal and hyperplastic prostate. *Prostate* **4**, 73–81.

De Klerk, D. P., Lee, D. V., and Human, H. J. (1984). Glycosaminoglycans of human prostatic cancer. *J Urol* **131**, 1008–1012.

De Marzo, A. M., Bradshaw, C., Sauvageot, J., Epstein, J. I., and Miller, G. J. (1998). CD44 and CD44v6 downregulation in clinical prostatic carcinoma: relation to Gleason grade and cytoarchitecture. *Prostate* **34**, 162–168.

De Marzo, A. M., Platz, E. A., Sutcliffe, S., et al. (2007). Inflammation in prostate carcino-genesis. *Nat Rev Cancer* **7**, 256–269.

Deed, R., Rooney, P., Kumar, P., et al. (1997). Early-response gene signalling is induced by angiogenic oligosaccharides of hyaluronan in endothelial cells. Inhibition by non-angio-genic, high-molecular-weight hyaluronan. *Int J Cancer* **71**, 251–256.

Degenhardt, K., Mathew, R., Beaudoin, B., et al. (2006). Autophagy promotes tumor cell survival and restricts necrosis, inflammation, and tumorigenesis. *Cancer Cell* **10**, 51–64.

Delpech, B., Girard, N., Bertrand, P., et al. (1997). Hyaluronan: fundamental principles and applications in cancer. *J Intern Med* **242**, 41–48.

Draffin, J. E., McFarlane, S., Hill, A., Johnston, P. G., and Waugh, D. J. (2004). CD44 poten-tiates the adherence of metastatic prostate and breast cancer cells to bone marrow endothelial cells. *Cancer Res* **64**, 5702–5711.

Ekici, S., Cerwinka, W. H., Duncan, R., et al. (2004). Comparison of the prognostic potential of hyaluronic acid, hyaluronidase (HYAL-1), CD44v6 and microvessel density for prostate cancer. *Int J Cancer* **112**, 121–129.

Enegd, B., King, J. A., Stylli, S., et al. (2002). Overexpression of hyaluronan synthase-2 reduces the tumorigenic potential of glioma cells lacking hyaluronidase activity. *Neuro-surgery* **50**, 1311–1318.

Evanko, S. P., Angello, J. C., and Wight, T. N. (1999). Formation of hyaluronan- and versican-rich pericellular matrix is required for proliferation and migration of vascular smooth muscle cells. *Arterioscler Thromb Vasc Biol* **19**, 1004–1013.

Fraser, J. R., Laurent, T. C., and Laurent, U. B. (1997). Hyaluronan: its nature, distribution, functions and turnover. *J Intern Med* **242**, 27–33.

Fujimoto, T., Hata, J., Yokoyama, S., and Mitomi, T. (1989). A study of the extracellular matrix protein as the migration pathway of neural crest cells in the gut: analysis in human embryos with special reference to the pathogenesis of Hirschsprung's disease. *J Pediatr Surg* **24**, 550–556.

Gakunga, P., Frost, G., Shuster, S., et al. (1997). Hyaluronan is a prerequisite for ductal branching morphogenesis. *Development* **124**, 3987–3997.

Gale, N. W., Prevo, R., Espinosa, J., et al. (2007). Normal lymphatic development and function in mice deficient for the lymphatic hyaluronan receptor LYVE-1. *Mol Cell Biol* **27**, 595–604.

Ghatak, S., Misra, S., and Toole, B. P. (2002). Hyaluronan oligosaccharides inhibit anchorage-independent growth of tumor cells by suppressing the phosphoinositide 3-kinase/Akt cell survival pathway. *J Biol Chem* **277**, 38013–38020.

Ghatak, S., Misra, S., and Toole, B. P. (2005). Hyaluronan constitutively regulates ErbB2 phosphorylation and signaling complex formation in carcinoma cells. *J Biol Chem* **280**, 8875–8883.

Goldstraw, M. A., Fitzpatrick, J. M., and Kirby, R. S. (2007). What is the role of inflammation in the pathogenesis of prostate cancer? *BJU Int* **99**, 966–968.

Gudmundsson, J., Sulem, P., Manolescu, A., et al. (2007). Genome-wide association study identifies a second prostate cancer susceptibility variant at 8q24. *Nat Genet* **39**, 631–637.

Haiman, C. A., Patterson, N., Freedman, M. L., et al. (2007). Multiple regions within 8q24 independently affect risk for prostate cancer. *Nat Genet* **39**, 638–644.

Harada, H. and Takahashi, M. (2007). CD44-dependent intracellular and extracellular catabolism of hyaluronic acid by hyaluronidase-1 and -2. *J Biol Chem* **282**, 5597–5607.

Harris, A. L. (2002). Hypoxia – a key regulatory factor in tumour growth. *Nat Rev Cancer* **2**, 38–47.

Herrlich, P., Morrison, H., Sleeman, J., et al. (2000). CD44 acts both as a growth- and inva-siveness-promoting molecule and as a tumor-suppressing cofactor. *Ann NY Acad Sci* **910**, 106–118. discussion 118–120.

Itano, N., Sawai, T., Atsumi, F., et al. (2004). Selective expression and functional character-istics of three mammalian hyaluronan synthases in oncogenic malignant transformation. *J Biol Chem* **279**, 18679–18687.

Jemal, A., Siegel, R., Ward, E., et al. (2006). Cancer statistics, 2006. *CA Cancer J Clin* **56**, 106–130.

Jiang, D., Liang, J., and Noble, P. W. (2007). Hyaluronan in tissue injury and repair. *Annu Rev Cell Dev Biol* **23**, 435–461.

Jin, S. and White, E. (2007). Role of autophagy in cancer: management of metabolic stress. *Autophagy* **3**, 28–31.

Johnson, L. A., Prevo, R., Clasper, S., and Jackson, D. G. (2007). Inflammation-induced uptake and degradation of the lymphatic endothelial hyaluronan receptor LYVE-1. *J Biol Chem* **282**, 33671–33680.

Kito, H., Suzuki, H., Ichikawa, T., et al. (2001). Hypermethylation of the CD44 gene is associated with progression and metastasis of human prostate cancer. *Prostate* **49**, 110–115.

Koeneman, K. S., Yeung, F., and Chung, L. W. (1999). Osteomimetic properties of prostate cancer cells: a hypothesis supporting the predilection of prostate cancer metastasis and growth in the bone environment. *Prostate* **39**, 246–261.

Kovar, J. L., Johnson, M. A., Volcheck, W. M., Chen, J., and Simpson, M. A. (2006). Hyal-uronidase expression induces prostate tumor metastasis in an orthotopic mouse model. *Am J Pathol* **169**, 1415–1426.

VII. HYALURONAN AND INDIVIDUAL CANCERS

Kuroda, K., Utani, A., Hamasaki, Y., and Shinkai, H. (2001). Up-regulation of putative hyaluronan synthase mRNA by basic fibroblast growth factor and insulin-like growth factor-1 in human skin fibroblasts. *J Dermatol Sci* **26**, 156–160.

Lapointe, J. and Labrie, C. (1999). Identification and cloning of a novel androgen-responsive gene, uridine diphosphoglucose dehydrogenase, in human breast cancer cells. *Endocrinology* **140**, 4486–4493.

Laurent, T. C., Laurent, U. B., and Fraser, J. R. (1996). The structure and function of hyaluronan: An overview. *Immunol Cell Biol* **74**, A1–A7.

Legg, J. W., Lewis, C. A., Parsons, M., Ng, T., and Isacke, C. M. (2002). A novel PKC-regulated mechanism controls CD44 ezrin association and directional cell motility. *Nat Cell Biol* **4**, 399–407.

Li, S. C., Chen, G. F., Chan, P. S., et al. (2001). Altered expression of extracellular matrix and proteinases in Noble rat prostate gland after long-term treatment with sex steroids. *Prostate* **49**, 58–71.

Lin, G. and Stern, R. (2001). Plasma hyaluronidase (HYAL-1) promotes tumor cell cycling. *Cancer Lett* **163**, 95–101.

Lin, S. L., Chang, D., Chiang, A., and Ying, S. Y. (2008). Androgen receptor regulates CD168 expression and signaling in prostate cancer. *Carcinogenesis* **29**, 282–290.

Lin, S. L., Chang, D., and Ying, S. Y. (2007). Hyaluronan stimulates transformation of androgen–independent prostate cancer. *Carcinogenesis* **28**, 310–320.

Lipponen, P., Aaltomaa, S., Tammi, R., et al. (2001). High stromal hyaluronan level is associated with poor differentiation and metastasis in prostate cancer. *Eur J Cancer* **37**, 849–856.

Liu, A. Y. (1994). Expression of CD44 in prostate cancer cells. *Cancer Lett* **76**, 63–69.

Liu, L. and Simon, M. C. (2004). Regulation of transcription and translation by hypoxia. *Cancer Biol Ther* **3** 427–492.

Lokeshwar, V. B., Cerwinka, W. H., Isoyama, T., and Lokeshwar, B. L. (2005). HYAL-1 hyaluronidase in prostate cancer: a tumor promoter and suppressor. *Cancer Res* **65**, 7782–7789.

Lokeshwar, V. B., Lokeshwar, B. L., Pham, H. T., and Block, N. L. (1996). Association of elevated levels of hyaluronidase, a matrix-degrading enzyme, with prostate cancer progression. *Cancer Res* **56**, 651–657.

Lokeshwar, V. B., Obek, C., Soloway, M. S., and Block, N. L. (1997). Tumor-associated hyaluronic acid: a new sensitive and specific urine marker for bladder cancer [published erratum appears in *Cancer Res* 1998;58(14):3191]. *Cancer Res* **57**, 773–777.

Lokeshwar, V. B., Rubinowicz, D., Schroeder, G. L., et al. (2001). Stromal and epithelial expression of tumor markers hyaluronic acid and HYAL-1 hyaluronidase in prostate cancer. *J Biol Chem* **276**, 11922–11932.

Madan, A. K., Pang, Y., Wilkiemeyer, M. B., Yu, D., and Beech, D. J. (1999). Increased hyaluronidase expression in more aggressive prostate adenocarcinoma. *Oncol Rep* **6**, 1431–1433.

Maitland, N. J. and Collins, A. (2005). A tumour stem cell hypothesis for the origins of prostate cancer. *BJU Int* **96**, 1219–1223.

Mast, B. A., Haynes, J. H., Krummel, T. M., Diegelmann, R. F., and Cohen, I. K. (1992). In vivo degradation of fetal wound hyaluronic acid results in increased fibroplasia, collagen deposition, and neovascularization. *Plast Reconstr Surg* **89**, 503–509.

McCarthy, J. B., Turley, E. A., Wilson, C. M., et al. (2005). Hyaluronan biosynthesis in prostate carcinoma growth and metastasis. In: Balasz, E.A. and Hascall V.C., (eds.). Hyaluronan: Structure, Metabolism, Biological Activities, Therapeutic Applications. Chapter 4, Hyaluronan and tumors. (Matrix Biology Institute, New Jersey).

Misra, S., Ghatak, S., and Toole, B. P. (2005). Regulation of MDR1 expression and drug resistance by a positive feedback loop involving hyaluronan, phosphoinositide 3-kinase, and ErbB2. *J Biol Chem* **280**, 20310–20315.

Misra, S., Ghatak, S., Zoltan-Jones, A., and Toole, B. P. (2003). Regulation of multidrug resistance in cancer cells by hyaluronan. *J Biol Chem* **278**, 25285–25288.

Misra, S., Toole, B. P., and Ghatak, S. (2006). Hyaluronan constitutively regulates activation of multiple receptor tyrosine kinases in epithelial and carcinoma cells. *J Biol Chem* **281**, 34936–34941.

Miyake, K., Underhill, C. B., Lesley, J., and Kincade, P. W. (1990). Hyaluronate can function as a cell adhesion molecule and CD44 participates in hyaluronate recognition. *J Exp Med* **172**, 69–75.

Mohamadzadeh, M., DeGrendele, H., Arizpe, H., Estess, P., and Siegelman, M. (1998). Proinflammatory stimuli regulate endothelial hyaluronan expression and CD44/HA-dependent primary adhesion. *J Clin Invest* **101**, 97–108.

Morrison, H., Sherman, L. S., Legg, J., et al. (2001). The NF2 tumor suppressor gene product, merlin, mediates contact inhibition of growth through interactions with CD44. *Genes Dev* **15**, 968–980.

Movsas, B., Chapman, J. D., Horwitz, E. M., et al. (1999). Hypoxic regions exist in human prostate carcinoma. *Urology* **53**, 11–18.

Noordzij, M. A., van Steenbrugge, G. J., Verkaik, N. S., et al. (1997). The prognostic value of CD44 isoforms in prostate cancer patients treated by radical prostatectomy. *Clin Cancer Res* **3**, 805–815.

Okada, T., Hawley, R. G., Kodaka, M., and Okuno, H. (1999). Significance of VLA-4-VCAM-1 interaction and CD44 for transendothelial invasion in a bone marrow metastatic myeloma model. *Clin Exp Metastasis* **17**, 623–629.

Patel, S., Turner, P. R., Stubberfield, C., et al. (2002). Hyaluronidase gene profiling and role of HYAL-1 overexpression in an orthotopic model of prostate cancer. *Int J Cancer* **97**, 416–424.

Pienta, K. J. and Smith, D. C. (2005). Advances in prostate cancer chemotherapy: a new era begins. *CA Cancer J Clin* **55**, 300–318. quiz 323–325.

Posey, J. T., Soloway, M. S., Ekici, S., et al. (2003). Evaluation of the prognostic potential of hyaluronic acid and hyaluronidase (HYAL-1) for prostate cancer. *Cancer Res* **63**, 2638–2644.

Prevo, R., Banerji, S., Ferguson, D. J., Clasper, S., and Jackson, D. G. (2001). Mouse LYVE-1 is an endocytic receptor for hyaluronan in lymphatic endothelium. *J Biol Chem* **276**, 19420–19430.

Prince, C. W. (2004). Roles of hyaluronan in bone resorption. *BMC Musculoskelet Disord* **5**, 12.

Pullen, M., Thomas, K., Wu, H., and Nambi, P. (2001). Stimulation of hyaluronan synthetase by platelet-derived growth factor bb in human prostate smooth muscle cells. *Pharmacology* **62**, 103–106.

Ricciardelli, C., Mayne, K., Sykes, P. J., et al. (1998). Elevated levels of versican but not decorin predict disease progression in early-stage prostate cancer. *Clin Cancer Res* **4**, 963–971.

Ricciardelli, C., Russell, D. L., Ween, M. P., et al. (2007). Formation of hyaluronan- and versican-rich pericellular matrix by prostate cancer cells promotes cell motility. *J Biol Chem* **282**, 10814–10825.

Rooney, P., Kumar, S., Ponting, J., and Wang, M. (1995). The role of hyaluronan in tumour neovascularization (review). *Int J Cancer* **60**, 632–636.

Sakko, A. J., Ricciardelli, C., Mayne, K., et al. (2001). Versican accumulation in human prostatic fibroblast cultures is enhanced by prostate cancer cell-derived transforming growth factor beta1. *Cancer Res* **61**, 926–930.

Savani, R. C., Cao, G., Pooler, P. M., et al. (2001). Differential involvement of the hyaluronan (HA) receptors CD44 and receptor for HA-mediated motility in endothelial cell function and angiogenesis. *J Biol Chem* **276**, 36770–36778.

Shyjan, A. M., Heldin, P., Butcher, E. C., Yoshino, T., and Briskin, M. J. (1996). Functional cloning of the cDNA for a human hyaluronan synthase. *J Biol Chem* **271**, 23395–23399.

VII. HYALURONAN AND INDIVIDUAL CANCERS

Simpson, M. A. (2006). Concurrent expression of hyaluronan biosynthetic and processing enzymes promotes growth and vascularization of prostate tumors in mice. *Am J Pathol* **169**, 247–257.

Simpson, M. A., Reiland, J., Burger, S. R., et al. (2001). Hyaluronan synthase elevation in metastatic prostate carcinoma cells correlates with hyaluronan surface retention, a prerequisite for rapid adhesion to bone marrow endothelial cells. *J Biol Chem* **276**, 17949–17957.

Simpson, M. A., Wilson, C. M., Furcht, L. T., et al. (2002a). Manipulation of hyaluronan synthase expression in prostate adenocarcinoma cells alters pericellular matrix retention and adhesion to bone marrow endothelial cells. *J Biol Chem* **277**, 10050–10057.

Simpson, M. A., Wilson, C. M., and McCarthy, J. B. (2002b). Inhibition of prostate tumor cell hyaluronan synthesis impairs subcutaneous growth and vascularization in immuno-compromised mice. *Am J Pathol* **161**, 849–857.

Spicer, A. P., Kaback, L. A., Smith, T. J., and Seldin, M. F. (1998). Molecular cloning and characterization of the human and mouse UDP-glucose dehydrogenase genes. *J Biol Chem* **273**, 25117–25124.

Spicer, A. P. and McDonald, J. A. (1998). Characterization and molecular evolution of a vertebrate hyaluronan synthase gene family. *J Biol Chem* **273**, 1923–1932.

Stern, R. (2005). Hyaluronan metabolism: a major paradox in cancer biology. *Pathol Biol (Paris)* **53**, 372–382.

Stern, R., Asari, A. A., and Sugahara, K. N. (2006). Hyaluronan fragments: an information-rich system. *Eur J Cell Biol* **85**, 699–715.

Terry, D. E. and Clark, A. F. (1996). Glycosaminoglycans in the three lobes of the rat prostate following castration and testosterone treatment. *Biochem Cell Biol* **74**, 653–658.

Toole, B. P. (1997). Hyaluronan in morphogenesis. *J Intern Med* **242**, 35–40.

Toole, B. P. (2002). Hyaluronan promotes the malignant phenotype. *Glycobiology* **12**, 37R–42R.

Toole, B. P. (2004). Hyaluronan: from extracellular glue to pericellular cue. *Nat Rev Cancer* **4**, 528–539.

Toole, B. P., Zoltan-Jones, A., Misra, S., and Ghatak, S. (2005). Hyaluronan: a critical component of epithelial–mesenchymal and epithelial–carcinoma transitions. *Cells Tissues Organs* **179**, 66–72.

Trojan, L., Michel, M. S., Rensch, F., et al. (2004). Lymph and blood vessel architecture in benign and malignant prostatic tissue: lack of lymphangiogenesis in prostate carcinoma assessed with novel lymphatic marker lymphatic vessel endothelial hyaluronan receptor (LYVE-1). *J Urol* **172**, 103–107.

Trojan, L., Rensch, F., Voss, M., et al. (2006). The role of the lymphatic system and its specific growth factor, vascular endothelial growth factor C, for lymphogenic metastasis in prostate cancer. *BJU Int* **98**, 903–906.

Tsuchiya, N., Kondo, Y., Takahashi, A., et al. (2002). Mapping and gene expression profile of the minimally overrepresented 8q24 region in prostate cancer. *Am J Pathol* **160**, 1799–1806.

Vis, A. N., Noordzij, M. A., Fitoz, K., et al. (2000). Prognostic value of cell cycle proteins p27(kip1) and MIB-1, and the cell adhesion protein CD44s in surgically treated patients with prostate cancer. *J Urol* **164**, 2156–2161.

Wagenlehner, F. M., Elkahwaji, J. E., Algaba, F., et al. (2007). The role of inflammation and infection in the pathogenesis of prostate carcinoma. *BJU Int* **100**, 733–737.

Watanabe, K. and Yamaguchi, Y. (1996). Molecular identification of a putative human hyaluronan synthase. *J Biol Chem* **271**, 22945–22948.

Weigel, P. H., Hascall, V. C., and Tammi, M. (1997). Hyaluronan synthases. *J Biol Chem* **272**, 13997–14000.

Welsh, C. F., Zhu, D., and Bourguignon, L. Y. (1995). Interaction of CD44 variant isoforms with hyaluronic acid and the cytoskeleton in human prostate cancer cells. *J Cell Physiol* **164**, 605–612.

Wernicke, M., Pineiro, L. C., Caramutti, D., et al. (2003). Breast cancer stromal myxoid changes are associated with tumor invasion and metastasis: a central role for hyaluronan. *Mod Pathol* **16**, 99–107.

West, D. C., Hampson, I. N., Arnold, F., and Kumar, S. (1985). Angiogenesis induced by degradation products of hyaluronic acid. *Science* **228**, 1324–1326.

West, D. C. and Kumar, S. (1989). The effect of hyaluronate and its oligosaccharides on endothelial cell proliferation and monolayer integrity. *Exp Cell Res* **183**, 179–196.

Wong, S. Y., Haack, H., Crowley, D., et al. (2005). Tumor-secreted vascular endothelial growth factor-C is necessary for prostate cancer lymphangiogenesis, but lymphangiogenesis is unnecessary for lymph node metastasis. *Cancer Res* **65**, 9789–9798.

Wouters, B. G., Weppler, S. A., Koritzinsky, M., et al. (2002). Hypoxia as a target for combined modality treatments. *Eur J Cancer* **38**, 240–257.

Yeager, M., Orr, N., Hayes, R. B., et al. (2007). Genome-wide association study of prostate cancer identifies a second risk locus at 8q24. *Nat Genet* **39**, 645–649.

Zeng, C., Toole, B. P., Kinney, S. D., Kuo, J. W., and Stamenkovic, I. (1998). Inhibition of tumor growth in vivo by hyaluronan oligomers. *Int J Cancer* **77**, 396–401.

Zhu, D. and Bourguignon, L. Y. (1998). The ankyrin-binding domain of CD44s is involved in regulating hyaluronic acid-mediated functions and prostate tumor cell transformation. *Cell Motil Cytoskeleton* **39**, 209–222.

VII. HYALURONAN AND INDIVIDUAL CANCERS

CHAPTER

17

Role of Hyaluronan and CD44 in Melanoma Progression

Carl Gebhardt, Marco Averbeck, Ulf Anderegg, and Jan C. Simon

HYALURONAN AND MELANOMA

Hyaluronan (hyaluronic acid, HA) is a polysaccharide found ubiquitously in the cutaneous extracellular matrix. It is synthesized as an unbranched polymer of repeating disaccharides of glucuronic acid and N-acetylglucosamin: $[-\beta(1,4)\text{-GlcUA-}\beta(1,3)\text{-GlcNAc-}]_n$ (for review see Toole, 2004)]. There is growing evidence that HA is involved in the course of melanoma: for example Dietrich and co-workers show that high expression levels of CD44 (the principal cell surface receptor for HA) correlate with poor prognosis of melanoma patients (Dietrich et al., 1997). Another study demonstrates a correlation between serum HA levels of

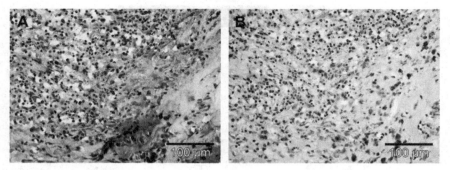

FIGURE 17.1   HA-deposition in the peritumoral stroma of melanoma metastasis. HA can be detected in high amounts in the stroma surrounding a melanoma metastasis (primary tumor: nodular malignant melanoma, Breslow thickness 4 mm) (**A**). HA was stained with the HA binding protein and visualized with amplex red, sections were counterstained with hematoxylin. As demonstrated by the strong red staining, high amounts of HA are synthesized by the melanoma cells (arrows) and the surrounding stroma tissue (arrowhead). Specificity was ensured by staining after incubation with hyaluronidase (**B**). These sections show no reactivity. (See Page 10 in Color Section at the back of the book).

melanoma patients and disease stage (Burchardt et al., 2003), showing high serum HA levels in patients with advanced stage disease. Furthermore, in histological sections of advanced melanoma, high levels of HA surrounding the tumor can be observed (Fig. 17.1). The contradiction to a report in which poor prognosis of melanoma patients was associated with reduced CD44 and HA expression in melanoma stage I patients (Karjalainen et al., 2000) is most likely due to several factors: (1) the different disease stages investigated in these studies. (2) Differences between *in vitro* and *in vivo* settings. For example, cutaneous melanomas evolve in basal epidermis, surrounded by keratinocytes with high levels of CD44 and an environment rich in HA. Thus, results on CD44 and HA metabolism in melanoma cell lines may not be comparable to the clinical behavior of human melanoma. (3) The same group demonstrate in a recent publication, that C8161 melanoma cells express different levels of CD44, depending on their depth of invasion into collagen lattices (Edward et al., 2005). This could mean that either a sub-population of CD44-positive cells in the original cell population has the ability to invade the lattices, or the process of invasion itself induces CD44 expression on the cells.

But importantly, direct CD44/HA interaction, i.e. the binding of HA to CD44, directly regulates tumor development (Bartolazzi et al., 1994), implicating a functional connection between the observed high expression levels of HA and CD44.

In addition, accumulating evidence shows the concomitant increase of HA in the course of melanoma. In mice, HA on the cell surface correlates with the metastatic potential of MM cells (Zhang et al., 1995) and in highly metastatic melanoma cell lines, HA synthesis is increased (Goebeler et al., 1996). Furthermore, inhibition of HA synthesis (Kudo et al., 2004; Yoshihara et al., 2005) and blocking HA by different means (Mummert et al., 2003; Xu et al., 2003) reduce tumorigenicity of melanoma cells.

INTERACTION OF HYALURONAN AND MELANOMA CELLS

The Hyaluronan Receptors

The interaction between HA and cells is mediated by different cell surface receptors, among which CD44 is the most prominent one on melanoma cells (reviewed in Ponta et al., 2003), although other HA receptors, such as RHAMM (receptor for hyaluronic acid mediated motility; reviewed in Cheung et al., 1999), HARE (hyaluronan receptor for endocytosis; Zhou et al., 2000) and LYVE1 (lymphatic vessel endothelial hyaluronan 1; Banerji et al., 1999; Prevo et al., 2001) have been identified. For example RHAMM is found to be expressed on melanoma cells and its expression increases during melanoma progression (Ahrens et al., 2001a). On the other hand we find RHAMM to be expressed exclusively within the cytoplasm of melanomas, thus rendering it unlikely that RHAMM mediates interaction of melanomas with extracellular HA (Ahrens et al., 2001a). Furthermore, CD44 interacts with a wide variety of different proteins, including osteopontin, collagen, fibronectin, growth factors, cytokines, and chemokines, as well as metalloproteinases (reviewed in Ponta et al., 2003).

CD44 Influences MM Proliferation

CD44 is a transcript of a single gene containing 20 exons. Since 10 of these exons are regulated by alternative splicing, multiple isoforms of CD44 exist. In addition to alternative splicing, CD44 is subject to different degrees of posttranslational modifications, such as glycosylation, sulfation, and phosphorylation, resulting in CD44 isoforms ranging from 85 to 200 kDa (reviewed in Ponta et al., 2003). The expression of standard CD44 and variant isoforms of CD44 has been discussed as prognostic marker in the course of MM (Dietrich et al., 1997; Manten-Horst et al., 1995). Taking together the expression studies and the findings that CD44 is the principal mediator of HA induced melanoma cell proliferation, one can conclude that the CD44–HA interaction promotes melanoma development (Ahrens et al., 2001a). Consequently, Ahrens and co-workers report, that (1) the expression of CD44 is increased during melanoma progression; (2) CD44 is the principal HA surface receptor on melanoma cells; and (3) HA induced increase of melanoma proliferation is mainly dependent on CD44–HA

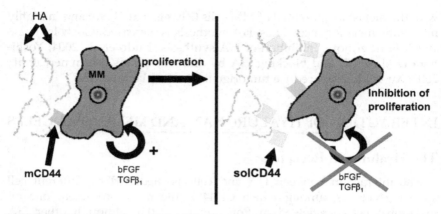

FIGURE 17.2 HA induces MM proliferation via membrane bound CD44; solCD44 is able to block this effect. HA binds to membrane CD44 (mCD44) and induces MM proliferation. This is mediated partly by autocrine secreted bFGF and TGFβ₁. Shedded/soluble CD44 (solCD44) blocks this autocrine proliferative loop and inhibits MM cell proliferation. (See Page 11 in Color Section at the back of the book).

interactions. The observed increase of the proliferative capacity of melanoma cells after HA binding is partly due to a stimulation of autocrine, secreted TGF-β1 and bFGF, thus showing for the first time a direct impact of HA on growth factor release in MM (Fig. 17.2).

In addition to the already mentioned post-translational modifications and the alternative splicing, the extracellular portion of CD44 can be cleaved by proteases from the cell surface thus generating a soluble form of CD44 (solCD44) (Ahrens et al., 2001b; Friedl et al., 1997; Okamoto et al., 2001). More than one protease seems to be involved in the shedding of CD44 from the cell surface, among which MT1-MMP (MMP14) is the best studied occurring in pancreatic tumor cell lines. Which protease is involved in CD44 cleavage in melanomas is currently under investigation (Kajita et al., 2001; Nakamura et al., 2004). The released CD44 is able to affect cellular behavior in several ways. (1) solCD44 is able to block the HA induced increase in proliferation of MM (Fig. 17.2) and importantly the HA binding ability of solCD44 is crucial for this effect, since solCD44 with mutations in the HA binding domain is not able to block the HA induced proliferation of MM (Ahrens et al., 2001b) and mammary carcinoma cells (Peterson et al., 2000). (2) CD44 cleavage regulates cell migration (Goebeler et al., 1996; Kajita et al., 2001), since enhanced shedding of CD44 by MT1-MMP induces cell migration in pancreatic tumor cell lines (Kajita et al., 2001) and highly aggressive MM shed significant amounts of CD44 compared to MM with lower tumorigenicity (Goebeler et al., 1996).

Interestingly, cytokines like oncostatin M and TGFβ are able to attenuate the shedding process itself and the subsequent capacity of HA binding by the soluble CD44 molecule (Cichy et al., 2005). These findings obtained with human lung squamous carcinoma cell line HTB58 might indicate that the local cytokine environment may also determine the effect exerted by the shed solCD44.

Hyaluronan Affects MM Migration via CD44

Since HA is the major component of the extracellular matrix of the skin, the question arises if HA influences migration and motility of MM, with possible consequences for local tumor progression and metastasis. Indeed several publications show an impact of HA on MM migration. It was demonstrated that migration of MM on HA coated surfaces is increased (Fig. 17.3) in cell lines expressing high amounts of CD44 (Thomas et al., 1993) and that MM cells with high metastatic potential synthesize high amounts of HA and CD44 (Goebeler et al., 1996). Blocking of CD44 by different means abolishes the observed effects of HA in both studies. Moreover, Goebeler and colleagues demonstrate that the HA binding epitope of CD44 is responsible for the observed changes in MM migration and metastasis. Furthermore, in mice the HA cell surface levels correlate with the metastatic potential of cells (Zhang et al., 1995) and transfection of MM cells with cDNAs encoding the HA synthesizing enzymes HAS-1 and HAS-2 induces melanoma cell motility as well (Ichikawa et al., 1999).

Similarly Sugahara et al. report that the small HA degradation products (HA-oligomers, sHA) are able to induce tumor cell migration (Sugahara et al., 2003). Again, the CD44 molecule seems to be involved in this process: the sHA fragments are able to induce CD44 shedding and increase cell motility that coincides with rearrangement of F-Actin and CD44 in the cells (Murai et al., 2004; Sugahara et al., 2003). Recently this group could demonstrate hyaluronidase activity in a pancreatic carcinoma cell line generating active sHA fragments. Obviously, the tumor cells can stimulate

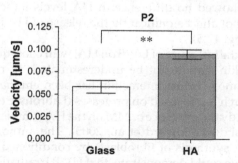

FIGURE 17.3 HA increases melanoma cell motility. Random motility of P2 melanoma cells is displayed on uncoated glass coverslips (clear bars) and HA coated coverslips (grey bars). Values are displayed as means ± SEM. The significance of differences was determined by t-test analysis (*$P < 0.05$; **$P < 0.01$).

their own CD44 shedding and migration in a vicious cycle driven by sHA fragments (Sugahara et al., 2006). These mechanistic findings underline the facts of increased hyaluronidase activities found in many aggressive tumors (Bertrand et al., 1997; Lokeshwar et al., 1997; Pham et al., 1997). This suggests that tumor spread and metastasis is enabled by hyaluronidase-generated sHA. These fragments decrease the CD44 anchoring of the cells by induced shedding and facilitate detachment and cell motility (Stern et al., 2006).

HYALURONAN METABOLISM, MELANOMA, AND UV-B

The HA metabolism and its role in cancer is complex and as of yet, not fully understood (Stern, 2005). There are numerous enzymes involved in HA metabolism (three HA synthases termed HAS-1-3 and three validated hyaluronidases in the skin, HYAL-1-3) which are susceptible to different steps of regulation (Stern, 2003; 2004; 2005). Several recent publications suggest a prominent role of HA in the progression of melanoma (Ichikawa et al., 1999; Mummert et al., 2003; Yoshihara et al., 2005), but less is known about the expression of different enzymes regulating HA anabolism and catabolism in MM. Recently, a link between UV-B and HA metabolism was observed (Sudel et al., 2004). This finding, together with the known importance of HA in MM progression and the fact that UV-B is critically involved in the development of MM (Berking et al., 2004; De Fabo et al., 2004; Jhappan et al., 2003) prompted us to question if all enzymes of HA metabolism are expressed in MM and if they are affected by UV-B. Indeed, the three HA synthases and the three hyaluronidases are expressed by melanoma cell lines and are, at least in the case of HAS-1, HAS-2, HYAL-1, and HYAL-3, subject to regulation by UV-B irradiation (Fig. 17.4). However, the subsequent analysis of HA content in the supernatants of irradiated cells showed no difference in HA levels indicating that UV-B irradiation does not affect significantly the release of HA into extracellular spaces *in vitro* (Fig. 17.5).

Nevertheless, the influence of UV-B on HA, with consequences for tumor progression in skin should not be underestimated. In recent years, the importance of tumor–stroma interaction has been emphasized. Although most of the research has focused on proteases (Labrousse et al., 2004; Pasco et al., 2004), recently the influence of MM on the HA synthesizing capacity of fibroblasts was shown (Edward et al., 2005). The authors report an up-regulation of HA synthesis of fibroblasts by conditioned supernatants of MM. Moreover, we could demonstrate, that UV-B irradiation influences the enzymes of the HA metabolism in HaCat keratinocytes and fibroblasts *in vitro* with an up-regulation of HAS-2 and HAS-3 in HaCats and to a lower extend in fibroblasts 24 hours after irradiation with 30 mJ/cm^2 (Averbeck

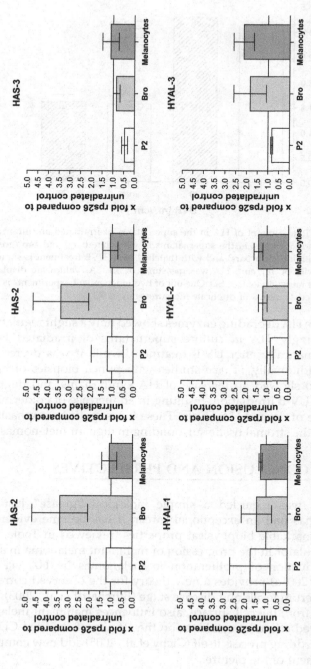

FIGURE 17.4 UVB influences the expression of HA synthases and hyaluronidases. Quantitative rt-PCR the three HA synthesizing enzymes (HAS-1-3) and the three HA degrading enzymes (HYAL-1-3). Values were normalized to the housekeeping gene (rps26) of un-irradiated controls. RNA of irradiated cells was collected 24 h after UVB treatment with 30 mJ/cm². Values are displayed as means of three independent experiments ± SEM.

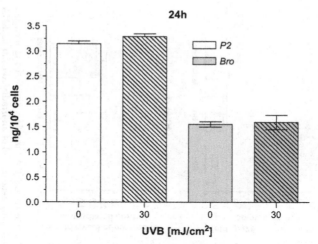

FIGURE 17.5   Total amount of HA in the supernatants of irradiated and un-irradiated cells. The total amount of HA in the supernatants of P2 (white bars) and Bro (grey bars) melanoma cells without (clear bars) and with (hatched bars) UVB treatment is shown 24 h after irradiation with 30 mJ/cm^2. HA was measured by ELISA. Values are displayed as means of duplicate measurements ± SD. One out of two independent experiments is shown. Values are displayed as means of duplicate measurements ± SD.

et al., 2007). The HA degrading enzymes showed only a slight regulation by UV-B. Interestingly, HA in culture supernatants of irradiated HaCats increased dramatically after UV-B treatment, while it was decreased in fibroblasts. Additionally, *in vivo* studies with punch biopsies of healthy volunteers also showed an increase of HA synthesizing and degrading enzymes after UV-B irradiation, resulting in an increase of epidermal HA and a decrease of dermal HA levels. These investigations emphasize the importance of the stromal tissue surrounding malignant melanomas.

CONCLUSION AND PERSPECTIVES

HA, which was regarded a simple biological "grease" just some decades ago, has had an exceptional career. It has become evident that besides it's remarkable biophysical properties (reviewed in Toole, 2004), HA is a co-regulator in the progression of malignant melanoma in several ways. The induction of proliferation in melanomas by HA via CD44 (Ahrens et al., 2001a) provides a new theory for the observed correlation between HA serum levels and disease stage (Burchardt et al., 2003). CD44 which is shed from the cell surface, also influences malignant melanoma, but the observed effects of cytokines on the affinity of shedded CD44 for HA and the shedding process itself (Cichy et al., 2005) add new complexity to this component of the picture.

Regarding the influence of HA on melanoma cell motility, a recent publication showed that the pro-migratory effects of HA are also performed by HA fragments, and that tumor cells are able by themselves to generate these fragments (Sugahara et al., 2003; Sugahara et al., 2006), resulting in autocrine motility induction, detachment, and spread (Stern et al., 2006). It appears that even if a part of the puzzle is unraveled, new investigations bring further insight into known effects, but also add new complexity to the total picture.

Regarding the influence of UV-B on the HA metabolism of MM, our investigations show no increase in the net amount of HA released into the extracellular space, but we demonstrate a regulation of HA metabolizing enzymes at the mRNA level. For skin, further studies are needed to investigate the quality of the HA following UV-B exposure. If UV-B leads to an increase in low molecular weight HA, a tumor modulating effect can be expected without a change in the total amount of HA, since the assays used to detect hyaluronan bind also short chains of HA. Another possibility is that cytokines secreted by MM stimulate HA production of tumor surrounding stroma cells, or modulate the enzymes of HA metabolism (Edward et al., 2005).

The more complex the picture becomes, the more difficult are the questions to be answered. Future experiments will focus more on the interplay of MM, its surrounding stroma and HA, including HA metabolism and the quality of the accumulated HA. Not only is HA metabolism not fully understood (Stern, 2003; 2005), but also the regulation of the HA receptors. Especially CD44, with its complex regulation (alternative splicing, post-translational modifications, and shedding) remains an important topic in the struggle to fully understand the role of HA in MM and tumor progression. Further insight into this network will hopefully enable us to find possible new targets for therapeutic strategies that will inhibit melanoma growth and metastasis.

ACKNOWLEDGMENT

This work was supported by a grant from the Deutsche Forschungsgemeinschaft (DFG) SFB 610 TP B7.

References

Ahrens, T., Assmann, V., Fieber, C., et al. (2001a). CD44 is the principal mediator of hyaluronic-acid-induced melanoma cell proliferation. *J Invest Dermatol* 116, 93–101.

Ahrens, T., Sleeman, J. P., Schempp, C. M., et al. (2001b). Soluble CD44 inhibits melanoma tumor growth by blocking cell surface CD44 binding to hyaluronic acid. *Oncogene* 20, 3399–3408.

Averbeck, M., Gebhardt, C. A., Voigt, S., et al. (2007). Differential regulation of hyaluronan metabolism in the epidermal and dermal compartments of human skin by UVB irradiation. *J Invest Dermatol* **127**, 687–697.

Banerji, S., Ni, J., Wang, S. X., Clasper, S., et al. (1999). LYVE-1, a new homologue of the CD44 glycoprotein, is a lymph-specific receptor for hyaluronan. *J Cell Biol* **144**, 789–801.

Bartolazzi, A., Peach, R., Aruffo, A. and Stamenkovic, I. (1994). Interaction between CD44 and hyaluronate is directly implicated in the regulation of tumor development. *J Exp Med* **180**, 53–66.

Berking, C., Takemoto, R., Satyamoorthy, K., et al. (2004). Induction of melanoma phenotypes in human skin by growth factors and ultraviolet B. *Cancer Res* **64**, 807–811.

Bertrand, P., Girard, N., Duval, C., et al. (1997). Increased hyaluronidase levels in breast tumor metastases. *Int J Cancer* **73**, 327–331.

Burchardt, E. R., Hein, R. and Bosserhoff, A. K. (2003). Laminin, hyaluronan, tenascin-C and type VI collagen levels in sera from patients with malignant melanoma. *Clin Exp Dermatol* **28**, 515–520.

Cheung, W. F., Cruz, T. F., and Turley, E. A. (1999). Receptor for hyaluronan-mediated motility (RHAMM), a hyaladherin that regulates cell responses to growth factors. *Biochem Soc Trans* **27**, 135–142.

Cichy, J., Kulig, P., and Pure, E. (2005). Regulation of the release and function of tumor cell-derived soluble CD44. *Biochim Biophys Acta* **1745**, 59–64.

De Fabo, E. C., Noonan, F. P., Fears, T., and Merlino, G. (2004). Ultraviolet B but not ultraviolet A radiation initiates melanoma. *Cancer Res* **64**, 6372–6376.

Dietrich, A., Tanczos, E., Vanscheidt, W., Schopf, E. and Simon, J. C. (1997). High CD44 surface expression on primary tumours of malignant melanoma correlates with increased metastatic risk and reduced survival. *Eur J Cancer* **33**, 926–930.

Edward, M., Gillan, C., Micha, D., and Tammi, R. H. (2005). Tumour regulation of fibroblast hyaluronan expression: a mechanism to facilitate tumour growth and invasion. *Carcinogenesis* **26**, 1215–1223.

Friedl, P., Maaser, K., Klein, C. E., et al. (1997). Migration of highly aggressive MV3 melanoma cells in 3-dimensional collagen lattices results in local matrix reorganization and shedding of alpha2 and beta1 integrins and CD44. *Cancer Res* **57**, 2061–2070.

Goebeler, M., Kaufmann, D., Brocker, E. B., and Klein, C. E. (1996). Migration of highly aggressive melanoma cells on hyaluronic acid is associated with functional changes, increased turnover and shedding of CD44 receptors. *J Cell Sci* **109** (Pt 7), 1957–1964.

Ichikawa, T., Itano, N., Sawai, T., et al. (1999). Increased synthesis of hyaluronate enhances motility of human melanoma cells. *J Invest Dermatol* **113**, 935–939.

Jhappan, C., Noonan, F. P., and Merlino, G. (2003). Ultraviolet radiation and cutaneous malignant melanoma. *Oncogene* **22**, 3099–3112.

Kajita, M., Itoh, Y., Chiba, T., et al. (2001). Membrane-type 1 matrix metalloproteinase cleaves CD44 and promotes cell migration. *J Cell Biol* **153**, 893–904.

Karjalainen, J. M., Tammi, R. H., Tammi, M. I., et al. (2000). Reduced level of CD44 and hyaluronan associated with unfavorable prognosis in clinical stage I cutaneous melanoma. *Am J Pathol* **157**, 957–965.

Kudo, D., Kon, A., Yoshihara, S., et al. (2004). Effect of a hyaluronan synthase suppressor, 4-methylumbelliferone, on B16F-10 melanoma cell adhesion and locomotion. *Biochem Biophys Res Commun* **321**, 783–787.

Labrousse, A. L., Ntayi, C., Hornebeck, W., and Bernard, P. (2004). Stromal reaction in cutaneous melanoma. *Crit Rev Oncol Hematol* **49**, 269–275.

Lokeshwar, V. B., Obek, C., Soloway, M. S., and Block, N. L. (1997). Tumor-associated hyaluronic acid: a new sensitive and specific urine marker for bladder cancer. *Cancer Res* **57**, 773–777.

Manten-Horst, E., Danen, E. H., Smit, L., et al. (1995). Expression of CD44 splice variants in human cutaneous melanoma and melanoma cell lines is related to tumor progression and metastatic potential. *Int J Cancer* **64**, 182–188.

Mummert, M. E., Mummert, D. I., Ellinger, L., and Takashima, A. (2003). Functional roles of hyaluronan in B16-F10 melanoma growth and experimental metastasis in mice. *Mol Cancer Ther* **2**, 295–300.

Murai, T., Miyazaki, Y., Nishinakamura, H., et al. (2004). Engagement of CD44 promotes Rac activation and CD44 cleavage during tumor cell migration. *J Biol Chem* **279**, 4541–4550.

Nakamura, H., Suenaga, N., Taniwaki, K., et al. (2004). Constitutive and induced CD44 shedding by ADAM-like proteases and membrane-type 1 matrix metalloproteinase. *Cancer Res* **64**, 876–882.

Okamoto, I., Kawano, Y., Murakami, D., et al. (2001). Proteolytic release of CD44 intracellular domain and its role in the CD44 signaling pathway. *J Cell Biol* **155**, 755–762.

Pasco, S., Ramont, L., Maquart, F. X., and Monboisse, J. C. (2004). Control of melanoma progression by various matrikines from basement membrane macromolecules. *Crit Rev Oncol Hematol* **49**, 221–233.

Peterson, R. M., Yu, Q., Stamenkovic, I., and Toole, B. P. (2000). Perturbation of hyaluronan interactions by soluble CD44 inhibits growth of murine mammary carcinoma cells in ascites. *Am J Pathol* **156**, 2159–2167.

Pham, H. T., Block, N. L., and Lokeshwar, V. B. (1997). Tumor-derived hyaluronidase: a diagnostic urine marker for high-grade bladder cancer. *Cancer Res* **57**, 778–783.

Ponta, H., Sherman, L., and Herrlich, P. A. (2003). CD44: from adhesion molecules to signalling regulators. *Nat Rev Mol Cell Biol* **4**, 33–45.

Prevo, R., Banerji, S., Ferguson, D. J., Clasper, S., and Jackson, D. G. (2001). Mouse LYVE-1 is an endocytic receptor for hyaluronan in lymphatic endothelium. *J Biol Chem* **276**, 19420–19430.

Stern, R. (2003). Devising a pathway for hyaluronan catabolism: are we there yet? *Glycobiology* **13**, 105R–115R.

Stern, R. (2004). Hyaluronan catabolism: a new metabolic pathway. *Eur J Cell Biol* **83**, 317–325.

Stern, R. (2005). Hyaluronan metabolism: a major paradox in cancer biology. *Pathol Biol (Paris)* **53**, 372–382.

Stern, R., Asari, A. A., and Sugahara, K. N. (2006). Hyaluronan fragments: an information-rich system. *Eur J Cell Biol* **85**, 699–715.

Sudel, K. M., Venzke, K., Mielke, H., et al. (2004). Novel aspects of intrinsic and extrinsic aging of human skin: beneficial effects of soy extract. *Photochem Photobiol.* 2005 May–Jun; **81(3)**: 581–587.

Sugahara, K. N., Hirata, T., Hayasaka, H., et al. (2006). Tumor cells enhance their own CD44 cleavage and motility by generating hyaluronan fragments. *J Biol Chem* **281**, 5861–5868.

Sugahara, K. N., Murai, T., Nishinakamura, H., et al. (2003). Hyaluronan oligosaccharides induce CD44 cleavage and promote cell migration in CD44-expressing tumor cells. *J Biol Chem* **278**, 32259–32265.

Thomas, L., Etoh, T., Stamenkovic, I., Mihm, M. C., Jr., and Byers, H. R. (1993). Migration of human melanoma cells on hyaluronate is related to CD44 expression. *J Invest Dermatol* **100**, 115–120.

Toole, B. P. (2004). Hyaluronan: from extracellular glue to pericellular cue. *Nat Rev Cancer* **4**, 528–539.

Xu, X. M., Chen, Y., Chen, J., et al. (2003). A peptide with three hyaluronan binding motifs inhibits tumor growth and induces apoptosis. *Cancer Res* **63**, 5685–5690.

Yoshihara, S., Kon, A., Kudo, D., et al. (2005). A hyaluronan synthase suppressor, 4-methylumbelliferone, inhibits liver metastasis of melanoma cells. *FEBS Lett* **579**, 2722–2726.

Zhang, L., Underhill, C. B., and Chen, L. (1995). Hyaluronan on the surface of tumor cells is correlated with metastatic behavior. *Cancer Res* **55**, 428–433.

Zhou, B., Weigel, J. A., Fauss, L., and Weigel, P. H. (2000). Identification of the hyaluronan receptor for endocytosis (HARE). *J Biol Chem* **275**, 37733–37741.

VII. HYALURONAN AND INDIVIDUAL CANCERS

CHAPTER

18

Role of Hyaluronan Metabolism in the Initiation, Invasion, and Metastasis of Breast Cancer

Tracey J. Brown and Natalie K. Thomas

THE TUMOR MICROENVIRONMENT OF BREAST CANCER

The evolutionary process of breast cancer involves progression through defined pathological and clinical stages, initiating with ductal hyperproliferation, subsequent development into *in situ* and invasive carcinomas, and finally metastatic disease (Allred and Fuqua, 2001; Burstei et al., 2004). In carcinogenesis, the importance of multiple genetic changes in epithelial cell oncogenes and tumor suppressor genes is well recognized (Vogelstein, 2004) but more recent focus on the role of host stroma has emphasized that efficient tumorigenesis requires an interaction between epithelial cells and the microenvironment. In pre-cancerous tissue, epigenetic changes can alter and activate the stroma, thereby providing the aberrant microenvironment necessary for tumor cell proliferation (Bhowmick and Moses, 2005). The activated stroma consists of altered extracellular matrix (ECM), various non-transformed cells (e.g. fibroblasts, myofibroblasts, leukocytes, and myoepithelial and endothelial cells) and growth factors that play a role in the initiation and progression of breast cancer (Bhowmick and Moses, 2005; Hu and Polyak, 2008). Reciprocal communication between epithelial cells and the tumor microenvironment enables transmittance of signals to the host stroma resulting in ECM remodeling adjacent to cancer cells (Howlett and Bissell, 1993; Elenbaas et al., 2001). The tumor-stroma pathology is similar to the embryonic state which is rich in glycosaminoglycans (Burstein et al., 2004). This pathology promotes an environment suited to cellular proliferation, angiogenesis, and stimulation of tissue-degrading proteases (Bissell et al., 2002), all cellular processes essential to carcinogenesis. This chapter will focus on the intratumoral metabolism of the ECM glycosaminoglycan, hyaluronan (HA) and more specifically the role that its anabolic and catabolic products play in the initiation, progression, and invasion of breast cancer.

INTRATUMORAL HYALURONAN LOCALIZATION DURING BREAST CANCER TUMORIGENESIS AND PROLIFERATION

Hyaluronan is a major constituent of the ECM where its linear structure, comprising of repeating disaccharide units of (β1-3) D-glucuronate-(β1-4) N-acetyl-D-glucosamine, and its large molecular weight (reaching up to 10 mega Daltons) provides unique physiochemical properties enabling it to bind up to 1000 times its weight in water (Fraser et al., 1997). Even at very low concentrations, high molecular weight (MW) HA forms inter- and intra-molecular associations resulting in a viscous milieu, making it an ideal biological scaffold to provide structural integrity in tissue, maintenance of hydration homeostasis (Toole, 2004), and the sequestration of growth factors, cytokines, and nutrients essential for cellular proliferation (Day and de la Motte, 2005). The multi-factorial functions of HA within the tumor microenvironment are regulated by its interaction with a variety of HA binding proteins such as versican and TSG-6, which promote the formation of both an HA-rich pericellular and extracellular matrix (Evanko et al., 2007). Through interaction with receptors, CD44, RHAMM, and

LYVE-1, the HA-containing ECM is able to be retained by cells, consequently reinforcing the structural integrity of the tumor while ensuring the maintenance of biologically relevant concentrations. The interface of HA and its receptors play a crucial role in cancer where the interaction has been implicated in the intracellular signaling cascades associated with tumor growth (Entwhistle et al., 1996), tumor cell adhesion (Toole and Hascall, 2002), neovascularization (Rooney et al., 1995), and metastasis (Toole, 2002). Refer to chapter 15 for an in-depth review of the localization of HA within tumors.

Relevant information on the clinical significance of HA in breast cancer is still emerging. In studies investigating clinical breast carcinoma samples, immunohistochemistry demonstrated that the expression of intra-tumoral HA was up-regulated in the surrounding stroma enabling it to be used as an independent prognostic factor for patient survival (Auvinen et al., 2000). The presence of high levels of HA in stromal myxoid changes in breast cancer was strongly associated with high tumor grade, tumor emboli with lymph node involvement, and increased mortality (Wernicke et al., 2003). Sonographic examination of invasive breast cancer showed that the tumor shape correlated with that of the HA ECM (Vignal et al., 2002). Furthermore, serum HA levels were significantly elevated in women with metastatic breast cancer when compared to non-metastatic carcinoma and benign breast diseases (Delpech et al., 1990). Contrary to the increase in extracellular HA, cytosolic HA has been studied in breast cancer where a strong correlation exists between cancer progression and reduced HA levels (Ruibal et al., 2000; 2001a; 2001b) suggesting that in progressive breast cancer the primary HA requirements are within the tumor microenvironment, where it can act as an integral component of the ECM and associated metabolic pathways.

METABOLISM OF HYALURONAN IN BREAST CANCER – DYSREGULATION OF A FINELY TUNED PROCESS

In healthy tissue the catabolic and anabolic rates of HA are tightly regulated resulting in precise concentrations which are indicative of the location and the specific function of the HA (Fraser et al., 1997). In general, medium and high MW HA (>400 kDa) is representative of the non-diseased state, but when processes such as inflammation occur this prompts the generation and accumulation of depolymerized or degraded HA, a scenario evident in cancer progression and dissemination (Stern, 2005). Due to the multi-factorial role of HA within the tumor microenvironment, changes in its concentration and molecular weight can exert significant epigenetic effects on tumor cells and the associated stroma. In breast cancer, the invasive phenotype can be correlated with the differential expression of specific isoforms of hyaluronan's catabolic and anabolic enzymes, thereby highlighting the functional importance of the metabolic end-products.

Comparative Expression and Isoform Functionality of the Hyaluronan Synthases

The hyaluronan synthases (HAS) are integral plasma membrane glycosyltransferases that are predicted to be topologically and structurally homologous, with clusters of transmembrane domains and a large intracellular region that constitutes approximately half of the overall protein. Three isoenzymes (HAS-1, HAS-2, and HAS-3) share 55–71% amino acid sequence identity, where the majority of HAS sequence consensus is in the large cytoplasmic loop which is predicted to contain the glycosyltransferase activity and substrate binding sites (Itano and Kimata, 2002). Hyaluronan biosynthesis is thought to occur by alternate addition of UDP-glucuronic acid and UDP-N-acetyl glucosamine substrates to the reducing end of the elongating chain, allowing unconstrained polymerization and extrusion into the extracellular matrix (Prehm, 1983; 2006). The mechanism of export of HA into the ECM is yet to be fully elucidated but it is thought that HAS may interact with activity-related phospholipids to form a pore which associates and dissociates from the growing chain as it is synthesized (Weigel et al., 2006). Additionally, it has been proposed that HA may be exported via an ABC transport system (Ouskova et al., 2004; Prehm and Schumacher, 2004).

The HA synthases appear to be separately and differentially regulated and expressed (Jacobson et al., 2000) where responses to stimuli such as TGF-β and IL-1β (Oguchi and Ishiguro, 2004) or mechanical stretching (Mascarenhas et al., 2004) can result in transcriptional up-regulation of the genes. As changes in HA production do not always coincide with alterations in prevalence of HAS transcripts (Recklies et al., 2001) modulation of the synthetic rate could also occur through regulation of post-translational modifications such as activation by phosphorylation (Goentzel et al., 2006), altered availability of synthetic substrates (Spicer et al., 1998), or cytokine-mediated changes in protein turnover or synthetic rate. Characterization of enzymatic properties of the HAS isoforms has revealed notable differences in enzyme kinetics and molecular size of their end product (Itano et al., 1999a). The HA elongation rate is inherently faster in HAS-1 and -2 when compared to HAS-3; these isoforms are also associated with the synthetic products of high MW HA ($2 \times 10^5 – 2 \times 10^6$) while HAS-3 is associated with production of lower molecular weight HA ($1 \times 10^5 – 1 \times 10^6$) (Itano et al., 1999a). The synthetic products of the HAS isoforms have generally been determined in transformed cells or under experimental conditions where simultaneous degradation had not been inhibited. One study investigated the MW produced by breast cancer cell lines after the inhibition of concurrent degradation (Udabage et al., 2005a). It was demonstrated that all HAS isoforms produced a 10 mega Dalton product and that the simultaneous degradation generated the variety of MWs detected and reported by other researchers. Functionally, the different roles of HAS isoforms are reiterated by their varied catalytic rates, substrate affinities and size distributions of the synthetic products.

HAS-2 is a Prognostic Indicator of Breast Cancer Tumorigenesis and Metastasis

To date, all synthase isoforms have been demonstrated to elicit functions within breast cancer initiation and progression, but the bulk of the experimental data highly implicates HAS-2 as the strongest prognostic factor in the evolution of breast cancer.

Breast cancer initiation appears to require HAS-1 and/or HAS-2 expression, as oncogenic malignant transformation of fibroblasts up-regulated both HAS-1 and -2, while tumorigenicity was reduced after HAS-2 inhibition in v-Ha-ras transformed cell lines (Itano et al., 2004). However, rigid control of expression appears to be required to facilitate tumor initiation, as HAS overexpression can reduce tumorigenic potential (Itano et al., 2004; Bharadwaj et al., 2007; Enegd et al., 2002), thereby indicating a narrow functional range in the optimal HAS expression and HA concentrations required to promote growth. The role of HAS-2 in spontaneous breast cancer was investigated using a transgenic murine model that purposely overexpressed HAS-2 with the objective of simulating the hyperproduction of HA found in breast cancer (Koyama et al., 2007). Forced expression of HAS-2 within spontaneous mammary tumors contributed to the formation of an HA-rich ECM, which promoted microvessel infiltration and an intratumoural stroma (Koyama et al., 2007). Malignant transformation in breast cancer is evident through the matrix remodeling conducted by stromal myofibroblasts and stiffening of the stroma (Kass et al., 2007) which is mediated by the production of matrix components and the release of angiogenic factors such as SDF-1/CXCL12 (Orimo et al., 2005). HAS-2 and HA production induce increased expression of pro-angiogenic factors in breast cancer stromal cells (Koyama et al., 2007) readying the microenvironment for progression and survival of the developing tumor mass.

During *adenocarcinoma progression*, the breast cancer epithelial cells lose their stable, polarized, non-malignant properties and transdifferentiate into fibroblastic, migratory cells acquiring mesenchymal characteristics, a process known as the epithelial–mesenchymal transition (EMT). EMT is characterized by epithelial cell loss of cell junctional proteins and cytoskeletal elements, increased expression of mesenchymal cadherins and vimentin rich intermediate filaments, and acquisition of motility observed in actin cytoskeletal rearrangement (Thiery and Sleeman, 2006). Transition can manifest in resistance to anoikis (Frisch, 2001) and enhanced survival (Thiery and Sleeman, 2006) where it can occur in up to 18% of breast carcinomas *in vivo* (Dandachi et al., 2001; Jones et al., 2001; Sørlie et al., 2001). The crucial role that HA synthesis plays in EMT and subsequently cancer progression was demonstrated when breast cancer cells infected with an HAS-2 adenovirus acquired several mesenchymal characteristics, including upregulation of vimentin, dispersion of cytokeratin, and loss of organized adhesion proteins at intercellular boundaries (Zoltan-Jones et al., 2003). In addition, increased HA production induced several other tumor cell survival mechanisms such as the capacity for anchorage-independent growth, resistance to apoptosis, stimulated invasiveness and gelatinase production, increased phosphoinositol 3-kinase (PI3 kinase)/ Akt pathway activity, and recruitment of stromal cells (Koyama et al., 2007; Zoltan-Jones et al., 2003). Interestingly, EMT has recently been identified to induce expression of stem cell markers (Rodriguez-Pinilla et al., 2007) suggesting that HA may indirectly promote formation of a sub-population of tumorigenic stem cells. This notion is supported by the role of HA in the hemopoietic stem cell niche (Haylock and Nilsson, 2006) and the identification of a $CD44^+/CD24^{-/low}$ tumorigenic stem cell subset within a heterogeneous breast cancer population (Al-Hajj et al., 2003) which indicated that HA–CD44 interactions may be of intrinsic importance in cancer stem cell biology. The role of HA in regulation of ErbB2 activation and downstream activation of the anti-apoptotic phosphoinositide 3-kinase/Akt pathway and its implications in tumour growth, progression, and chemoresistance are reviewed elsewhere in this series.

The correlation between HAS-2 and breast cancer progression was further established when Udabage et al. characterized HAS expression and HA production in 10 breast cancer cell lines and demonstrated a correlation between HAS-2 expression and invasive potential (Udabage et al., 2005a). Low levels of HAS-3 were detected in all cell lines, however, a glycocalyx could only be visualized in exponentially growing cells which predominantly expressed HAS-2 (Udabage et al., 2005a). On reaching a plateau in proliferation, all cell lines produced detectable amounts of cell-associated HA (Udabage et al., 2005a) demonstrating that after development of the critical tumor mass, retention of a cancer cell glycocalyx was not HAS isoform-specific. Characterization of the size distribution of the synthesized HA highlighted that both HAS-2 and HAS-3 extruded high MW HA and that invasive cell lines rapidly depolymerized this into fragments ranging from 10–40 kDa (Udabage et al., 2005a), which would feasibly be capable of initiating tumor neovascularization *in vivo* (Rooney et al., 1993; Sattar et al., 1994; West et al., 1985).

To date, the localization of HAS isoform expression within cancerous tissues has not been established due to the lack of widely available and reliable antibodies, therefore limiting the immunodetection of the native proteins (Weigel and DeAngelis, 2007). The essential requirement for high MW HA during invasion suggests that HAS-2 would be significantly up-regulated at the invasion front of the tumor which would account in part, for the increased HA concentration found at the stromal-tumor interface (Bertrand et al., 1992). The functional importance of HAS-2-mediated enhancement of HA production in breast cancer was identified by anti-sense inhibition of HAS-2 in a highly invasive breast cancer cell line (Udabage et al., 2005b). Inhibition of HAS-2 reduced proliferative potential and induced transient cell cycle arrest in 79% of the cell population *in vitro*, and completely abrogated tumor initiation and progression following subcutaneous mammary fat pad inoculation into nude mice (Fig. 18.1A) (Udabage et al., 2005b). These observations support a recent study which indicated that RNA interference mediated silencing of HAS-2 expression in an invasive breast cancer cell line reduced HA production leading to a less aggressive phenotype of reduced proliferative and migratory potential (Li et al., 2007).

HAS-2 has also been implicated *in breast cancer metastasis*. Injection of antisense-inhibited HAS-2 breast cancer cells in an intracardiac metastasis and localized intradermal mammary pad model demonstrated that HAS-2 expression was essential for breast cancer metastasis because the inhibition of HAS-2 significantly reduced organ metastasis (Fig. 18.1B & C) and conveyed longer survival times (Fig. 18.1D) (Udabage et al., 2005b). It has been reported that breast cancer cell lines secrete soluble factors which mediate up-regulation of HAS-1 and HAS-2 in CD44/RHAMM expressing osteoblasts (Bose and Masellis, 2005). Breast cancer cells interact with osteoblasts and release products which activate osteoclast formation and bone resorption (Bendre et al., 2002; Dickson and Lippman, 1995; Guise et al., 1996; Pederson et al., 1999), inhibit differentiation of mature osteo-blasts, or induce apoptosis in differentiated (Mastro et al., 2004; Mercer et al., 2004; Thomas et al., 1999). It is conceivable that osteoblast HA-receptors may anchor HA producing breast cancer cells to the malignant site, resulting in paracrine stimulation of HA production and chemotactic migration of additional breast cancer cells. While breast cancer cells exhibit a predilection for metastasis to bone, the absence of osteolytic lesions in the previously described HAS-2 inhibition study (Udabage et al., 2005b) suggests that the absence of HA production may inhibit anchoring of breast cancer cells, limiting stimulation of osteoblast HA synthesis and subsequent bone resorption. The extracellular phosphoglycoprotein, osteopontin, may also influence regulation of HAS-2 expression in breast cancer as osteopontin overexpression dramatically increased HAS-2 tran-scripts and HA production (Cook et al., 2006). Osteopontin is associated with the progression and metastasis of several cancers, and can bind cell surface receptors including HA-receptor CD44, to instigate signaling

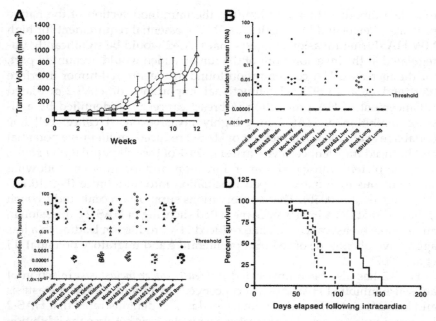

FIGURE 18.1   HAS-2 expression is an essential component of breast cancer tumorige-nicity and metastasis. The MDA-MB 231 invasive breast cancer cell line was stably trans-fected with an anti-sense HAS-2. Parental, mock, and ASHAS-2 transfectants were inoculated into the mammary fat pad of nude mice.

A, Primary tumor growth was followed over a 12-week period with tumor progression recorded twice weekly. The results graphed represent the average tumor volume (mm³) ± SE, where $n = 9$ to 13.

B, Alu PCR was used to detect the metastasis of human cancer cells to the brain, kidney, liver, and lung of the mouse. Results are expressed as the percentage of human tumor DNA in mouse soft organs ($n = 8$ per group).

C, Alu PCR was used to determine the extent of soft organ metastasis after intracardiac inoculation of the parental, mock, and ASHAS-2 cells lines. Results are expressed as the percentage of human tumor DNA in mouse soft organs ($n = 9$ per group). No metastasis to these organs could be detected where animals had been inoculated with MDA-MB-231 ASHAS-2-transfected cells.

D, The survival rate and number of days elapsed after intracardiac inoculations. There were no differences in the animal survival rate ($P = 0.0840$) between the parental and mock-transfected mice. Survival curve for ASHAS-2 was significantly different ($P < 0.0001$) from the both control groups. MDA-MB-231 ASHAS-2 transfectants (solid); parental cell line (short dash); mock transfectants (long dash). (This figure was reproduced from Udabage, L., et al. (2005). *Cancer Res* **65**, 6139–6150.)

cascades promoting migration, invasion, cell survival, and angiogenesis (Heldermon et al., 2001; Senger et al., 1996; Tuck et al., 2000).

Whilst there is a significant body of work in breast cancer indicating that HAS-2 overproduces HA promoting tumorigenesis and invasion, HAS-1 transfection can restore the metastatic potential of highly metastatic mammary carcinoma cells which have been rendered non-invasive by treatment with a chemical mutagen that abrogates HA production (Itano et al., 1999b). Therefore, it appears that although HAS-2 is preferentially expressed in breast cancer, the HA produced by HAS-1 is capable of eliciting similar functional effects. Moreover, the dynamic balance of HA production and HA degradation and the profoundly different signalling capacity of native and degraded HA ensures that anabolism and catabolism of HA are of significant importance in carcinogenesis.

Hyaluronan Degradation – An Essential Component of Breast Cancer Progression

The degradation of high MW HA within tissue occurs through the concerted effort of non-enzymic reactive oxygen species (ROS) (Knudson et al., 2002; Karihtala et al., 2007) and a group of multi-isoform enzymes known as the hyaluronidases (HAase) (Stern, 2003). The extensive volumetric domain of endogenous HA necessitates that it must be processed into more manageable fragments by ROS or hyaluronidase cleavage before cellular catabolism can occur (Knudson et al., 2002). Following the preliminary depolymerization, the HA follows two fates: (i) passive diffusion into the lymphatics and subsequently bloodstream where it binds to the lymph node LYVE-1 receptor (Prevo et al., 2001) or the liver endothelial cell receptor which rapidly (plasma half life of 2–5 min) clears and metabolizes the HA to monosaccharides (Fraser et al., 1984); or (ii) in the tissues, HA is internalized via CD 44 receptor-mediated endocytosis and degraded.

The degradation of HA occurs in cells by a series of sequential enzymatic reactions which involve several HAase isoforms, namely, HYAL-1 and -2 and PH-20 (refer to chapter 13 for a comprehensive review of hyaluronidases in cancer biology). In brief, the hyaluronidases associated with HA cellular metabolism are HYAL-1, known as serum hyaluronidase, which is thought to be an acid-active (pH 4.0–4.2) lysosomal enzyme (Afify et al., 1993; Frost et al., 1997) that is able to utilize HA of any size as a substrate, and which predominantly generates tetrasaccharides. HYAL-2 (Lepperdinger et al., 1998; 2001) is also proposed to be an acid-active enzyme and is linked to plasma membranes by a glycosylphosphatidylinositol anchor (Rai et al., 2001). This hyaluronidase has unique substrate specificity, cleaving high molecular mass HA polymers to intermediate size fragments of ~20 kDa, or about 50 disaccharide units. A third HAase, PH-20 is considered a neutral HAase as it does have activity at pH 7, but it can demonstrate activity at pH 4.0 to 8.0 (Cherr et al., 1999). PH-20 is thought to be the testicular HAase, but there are a number of reports of PH-20 in cancer cells (Liu et al., 1996; Madan et al., 1999), and particularly in breast cancer where it is thought to promote tumorigenesis by accelerating the release of FGF-2 from tumor cells, decomposing HA into small fragments, and promoting angiogenesis (Gao and Underhill, 2002; Wang et al., 2004).

In tissues that are not involved in the plasma or tissue clearance of HA, the primary metabolic receptor for HA is CD44 (Culty and Underhill, 1992) more particularly in breast cancer, where CD44 has been definitively identified as the receptor responsible for the binding and internalization of HA (Culty et al., 1994). Initiation of HA depolymerization and catabolism is thought to occur within a caveolin-rich lipid raft which contains CD44, HA synthase, and HYAL-2 (Stern, 2005). The HA is anchored to the cell surface via the cooperation of the GPI-anchored HYAL-2 and CD44 (Rai et al., 2001). Before metabolism occurs the high MW HA is initially cleaved by hyaluronidases (Stern, 2003) or ROS which generates intermediate molecules suitable for internalization within unique acid endocytic vesicles (Tammi et al., 2001) and further degradation by HYAL-2, resulting in 20-kDa HA fragments. The fragments are internalized, transported to endosomes and then to lysosomes where they are further digested by HYAL-1 to generate tetrasaccharides oligosaccharides (Rai et al., 2001) and the final catabolic end-products are monosaccharides generated through the concerted activity of β-exoglycosidases, β-glucuronidase, and β-N-acetyl glucosaminidase. As previously highlighted, oligosaccharide by-products of HA degradation are pivotal in cancer progression and invasion. For brevity, refer to Table 18.1 which correlates HA chain length with functionality within the malignant phenotype implicating the diversity of function of HA catabolic products.

Due to the lack of verification of the exact nature of HA turnover and the accumulation of HA degradation products within tumors *in vivo*, Brown et al. (manuscript in preparation) intravenously injected 825 kDa [^{3}H] HA into mice bearing CD44-positive, human breast cancer xenografts. The tumor cells and ECM were separated via trypsin digestion and the MW of the intra- and intercellular HA was determined using size exclusion chromatography. These data demonstrated that within a 72 h time-frame, the extracellular half-life of the 2 ± 0.4 μg of HA (amount distributed per gram of tumor) was 11 h. Within the initial 4 h, the high MW HA persisted in the tumor ECM primarily as 825 kDa after which 5% of the radioactivity presented as a heterogeneous HA species ranging from 1.2 kDa to 125 kDa, with the most prevalent molecular weight being 20 kDa; while the remaining metabolic products were [^{3}H]acetate and [^{3}H]water. Only minor concentrations (82–220 fg/cell) of tetrasaccharides were identified within the cells; no 20 kDa HA was observed, indicating that the intracellular metabolism of HA is a rapid event. These data suggest that a broad range pH-active such as PH-20 is active within the ECM and HYAL-2 is the predominant ECM hyaluronidase. In addition there was no prolonged accumulation of high concentrations of low MW HA fragments.

TABLE 18.1 The Molecular Weight of Hyaluronan Catabolic Products Promote Specific Functionality Within the Initiation, Progression, and Invasion of Breast Cancer

Oligosaccharide length	Function in cancer	Reference
Tumorigenic potential		
8	Abrogation of hyaluronan-rich stroma	Hosono et al., 2007
Progression		
1000	Induction of inflammatory chemokines required as a stimulus for HA synthesis	Noble, 2002
10–40	Promotion of tumor cell migration	Sugahara et al., 2003
	Induction of CD44 cleavage	Sugahara et al., 2003; 2006
8–32	Stimulation of angiogenesis	Sattar et al., 1994; West et al., 1985; Slevin et al., 1998; 2002; 2004; Lokeshwar, 2000
	Stimulation of tumor neovascularization	Rooney et al., 1995
12	Endothelial cell differentiation	Takahashi et al., 2005
12	Inhibition of anchorage-independent growth	Ghatak et al., 2002
Invasion		
~170	Activation of MAP kinase facilitating invasion	Kobayashi et al., 2002
4–6	Transcription of metalloproteinases	Fieber et al., 2004

The functional diversity of high MW HA and its lower MW degradation products in breast cancer is well established (see Stern et al., 2005 for review) but there is a lack of experimental evidence correlating HA catabolism with breast cancer phenotype. Quantitative identification of HAase genes and the functional characterization of the gene products in breast cancer cell lines found that invasive tumor cells primarily expressed HYAL-2 where the co-expressed HYAL-1 had a 5- to 10-fold lower transcript number. The turnover of HA was significantly higher in aggressive cell lines, but the differentiating factor was the ability of growth arrested cells to continue metabolizing HA, contrary to the less invasive phenotype which ceased turnover after reaching plateau growth (Udabage et al., 1999a). These findings were consistent with the proposed function of the

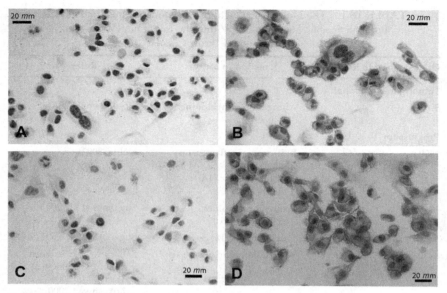

FIGURE 18.2   In breast cancer the expression and function of HAS-2, HYAL-2 and CD44 are intra-dependently regulated. The MDA-MB 231 invasive breast cancer cell line was stably transfected with an anti-sense HAS-2-pCl-Neo construct and mock control (pCl-neo vector without insert) using LipofectAMINE. One month prior to commencing the experiments, transfected cells were selected using G418 antibiotic and stable cell lines were established by harvesting and pooling of antibiotic-resistant colonies. The effect of HAS-2 inhibition on the expression of CD44 was determined using immunohistochemical detection of HAS-2 and CD44. Subconfluent stable transfectants of MDA-MD-231 expressing antisense mRNA to HAS-2 (A) and parental MDA-MB-231 (B) were reacted with an HAS-2-specific antibody. Cell surface reactivity to CD44H was also tested in subconfluent stable transfectants containing ASHAS-2 (C) and parental MDA-MB-231 (D) using an antihuman CD44 monoclonal antibody. (This figure was reproduced from Udabage, L., et al. (2005). *Cancer Res* **65**, 6139–6150.) (See Page 11 in Color Section at the back of the book).

catabolic products of HA within tumors (Stern, 2005). The co-dependency of CD44 and HYAL-2 in HA turnover in breast cancer was confirmed when the inhibition of HAS-2 concomitantly decreased CD44 and HYAL-2 expression (Fig. 18.2) which resulted in elevated concentrations of 10 mega Dalton HA (Fülop et al., 1997). The inability of the cancer cells to degrade the HA totally ablated tumor initiation, an observation potentially explained by the proposed anti-angiogenic effects of high MW HA (West and Kumar, 1989) emphasizing that when a tumor reaches a critical volume, it is necessary to establish neovascularization via the degradation of HA into small angiogenic fragments (<10 kDa). See Chapters 12 and 13 which further explore the role of HAases in other cancers and in the regulation of CD44 expression.

THE CO-ORDINATED FUNCTION AND IMPORTANCE OF HA METABOLISM IN BREAST CANCER

A comprehensive understanding of the role of HA metabolism and the functional diversity of the resultant polymers is yet to be fully elucidated. The genetic manipulation of the expression of HAS and HAases, and the use of specific inhibitors for these enzymes has made it possible to form a collective understanding of the role of HA metabolism in tumorigenesis, progression, and invasion (Fig. 18.3). It is posed that, during malignant transformation the acquisition of multiple genetic mutations in the epithelial cells (Vogelstein, 2004) result in altered rates of apoptosis and proliferation which is evidenced by an accumulation of pericellular and extracellular high MW HA, primarily synthesized by breast cancer HAS-2. The increase in HA concentration contributes to (i) the formation of a hydrated and volumetrically expanded versican/HA matrix which supports cancer–host interactions via the unhindered penetration of stromal cells and diffusion of nutrients (Folkman, 2002; Koyama et al., 2007); and (ii) the inactivation of tumor macrophages ensuring the immunological suppression of a potential attack on the tumor cells (Kuang et al., 2007). The next step in breast cancer progression eventuates when the increased synthesis of HA represses E-cadherin transcription which induces epithelial–mesenchymal transition (Haylock and Nilsson, 2006). In concert with ECM degrading enzymes, the tumor cells detach and migrate into the underlying stroma where they contribute to the formation of intratumoral stroma (Zoltan-Jones et al., 2003). The specialized stroma and tumor cells simultaneously produce, and then rapidly degrade the newly synthesized HA and the degradation products in turn promote angio-genesis (West et al., 1985; West and Kumar, 1989; Liu et al., 1996), increased inflammatory cell and fibroblast recruitment to the specialized stroma (Ronnov-Jessen et al., 1996; Tlsty and Hein, 2001), and enhanced expres-sion of matrix metalloproteinases, MMP-9 and MMP-13 (Fieber et al., 2004). The inflammatory cells appear to have dual roles: (i) secretion of cytokines, growth factors, and chemokines necessary for stimulation of HA synthesis; and (ii) production of ROS required for the initiation of extracellular HA degradation and the generation of HA fragments which drive the endothelial cell proliferation and neovascularization essential for tumor progression. In addition, it is thought that circulating bone marrow-derived endothelial cells are recruited to the area of neovascularization where they form new blood vessels (Gottfried et al., 2008). Finally, in preparation for cancer cell dissemination to distant sites, the development of the peritumoral lymphatic system is initiated by the secretion of lymphangiogenic activators (VEGF-C and -D) by stromal and tumor cells (Von Marschall et al., 2005).

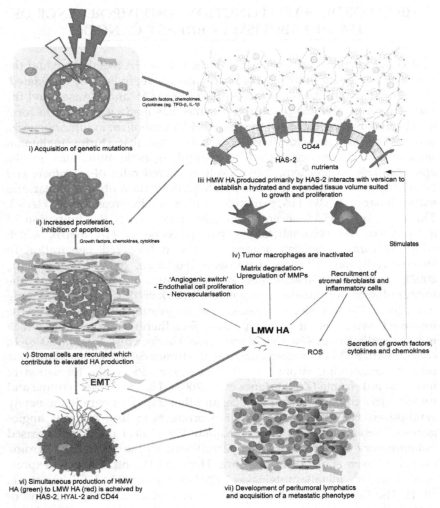

FIGURE 18.3 Dynamic hyaluronan metabolism contributes to the pathogenesis of breast cancer. Multiple genetic mutations (i) can lead to epithelial cells being at a proliferative advantage where apoptosis is inhibited (ii); this is manifested in an accumulation of pericellular and extracellular high MW HA primarily synthesized by HAS-2 (iii). HA within the pericellular and extracellular matrix can bind to the HA synthase (red), cell surface receptors such as CD44 (blue), and interact with other matrix components including versican, creating a hydrated and expanded tissue volume which facilitates the diffusion of nutrients and promotes growth (iii). Hyaluronan also inactivates tumor macrophages ensuring the immunological suppression of potential attack on the tumor cells (iv). Increased HA synthesis can stimulate recruitment of stromal cells which contribute to microenvironmental changes (v), and induction of epithelial–mesenchymal transition (EMT) which facilitates acquisition of a more invasive, metastatic phenotype. In concert with ECM degrading enzymes, EMT can induce tumor cell can detachment and migration into the underlying stroma where they contribute to the formation of intratumoral stroma. The synergistic

◀ actions of HAS-2 (red), CD44 (green), and HYAL-2 (yellow) ensure simultaneous production and degradation of HA (vi). Low molecular weight HA (red) can promote angiogenesis, increased inflammatory cell and fibroblast recruitment and enhanced expression of matrix metalloproteinases. Inflammatory cells can in turn secrete cytokines, growth factors, chemokines necessary for ongoing stimulation of HA synthases, and produce reactive oxygen species (ROS) which degrade HA. These events contribute to acquisition of a metastatic phenotype, which is accompanied by development of the peritumoral lymphatic system (vii). (See Page 12 in Color Section at the back of the book).

CONCLUSIONS

The dynamic balance between HA synthesis and degradation within the tumor microenvironment plays an integral role in the complex, multistep process of carcinogenesis. Substantial preclinical work has elegantly elucidated the potential participation of both the synthetic and degradative enzymes in the metabolic processes of breast cancer. Tumor initiation, progression and maintenance appear to be highly dependent on the accumulation of high MW HA within the breast cancer stroma, where it provides a hydrated growth matrix for tumor cells, promotes tumor survival by prevention of apoptosis, camouflages cancer cells from cytotoxic attack by host immunocompetent cells and ultimately stimulates invasion. In addition, after the epithelial–mesenchymal transition of cancer epithelial cells, the high MW HA acts as the substrate for further degradation into biologically active, low MW fragments that have the primary function of initiating cell signaling cascades during ECM remodeling. The specific signaling events induce neovascularization, lymphangiogenesis, and ECM degradation via the enhanced expression of matrix metalloproteinases. By no means is HA and its receptors the only instigators of these processes, to date numerous cellular ligands and molecules have been shown to participate in the process of HA metabolism and carcinogenesis of the breast, but a complete understanding of the significance with translation into clinical benefit requires substantially more work and elucidation.

References

Afify, A. M., Stern, M., Guntenhoener, M., and Stern, R. (1993). Purification and characterization of human serum hyaluronidase. *Arch Biochem Biophys* 305, 434–441.

Al-Hajj, M., Wicha, M. S., Benito-Hernandez, A., Morrison, S. J., and Clarke, M. F. (2003). Prospective identification of tumorigenic breast cancer cells. *Proc Natl Acad Sci USA* 100, 6890–6899.

Allred, D. C. and Fuqua, S. A. (2001). Histological and biological evolution of human premalignant breast disease. *Endocr Relat Cancer* 8, 47–61.

Auvinen, P., Tammi, R., Parkkinen, J., et al. (2000). Hyaluronan in peritumoral stroma and malignant cells associates with breast cancer spreading and predicts survival. *Am J Pathol* 156, 529–536.

Bendre, M. S., Gaddy-Kurten, D., Mon-Foote, T., et al. (2002). Expression of interleukin 8 and not parathyroid hormone-related protein by human breast cancer cells correlates with bone metastasis. *in vivo. Cancer Res* 62, 5571–5579.

Bertrand, P., Girard, N., Delpech, B., et al. (1992). Hyaluronan (hyaluronic acid) and hyaluronectin in the extracellular matrix of human breast carcinomas: comparison between invasive and non-invasive areas. *Int J Cancer* **52**, 1–6.

Bharadwaj, A. G., Rector, K., and Simpson, M. A. (2007). Inducible hyaluronan production reveals differential effects on prostate tumor cell growth and tumor angiogenesis. *J Biol Chem* **282**, 20561–20572.

Bhowmick, N. A. and Moses, H. L. (2005). Tumor–stroma interactions. *Curr Opin Genet Dev* **15**, 97–101.

Bissell, M. J., Radisky, D. C., Rizki, A., Weaver, V. M., and Petersen, O. W. (2002). The organizing principle: microenvironmental influences in the normal and malignant breast. *Differentiation* **70**, 537–546.

Bose, N. and Masellis, A. (2005). Secretory products of breast cancer cells upregulate hyaluronan production in a human osteoblast cell line. *Clin Exp Met* **22**, 629–642.

Burstein, H. J., Polyak, K., Wong, J. S., Lester, S. C., and Kaelin, C. M. (2004). Ductal carcinoma *in situ* of the breast. *N Engl J Med* **350**, 1430–1441.

Cherr, A. I. Y., Li, M. W., and Overstreet, J. W. (1999). Hyaluronic acid and the cumulus extracellular matrix induce increases in intracellular calcium in macaque sperm via the plasma membrane protein PH-20. *Zygote* **7**, 211–222.

Cook, A. C., Chambers, A. F., Turley, E. A., and Tuck, A. B. (2006). Osteopontin induction of hyaluronan synthase 2 expression promotes breast cancer malignancy. *J Biol Chem* **281**, 24381–24389.

Culty, M. and Underhill, C. B. (1992). The hyaluronan receptor (CD44) participates in the uptake and degradation of hyaluronan. *J Cell Biol* **116**, 55–62.

Culty, M., Thompson, E. W., and Underhill, C. B. (1994). Binding and degradation of hyaluronan by human breast cancer cell lines expressing different forms of CD44: correlation with invasive potential. *J Cell Physiol* **160**, 275–286.

Dandachi, N., Hauser-Kronberger, C., Moré, E., et al. (2001). Co-expression of tenascin-C and vimentin in human breast cancer cells indicates phenotypic transdifferentiation during tumour progression: correlation with histopathological parameters, hormone receptors, and oncoproteins. *J Pathol* **193**, 181–189.

Day, A. J. and de la Motte, C. A. (2005). Hyaluronan cross-linking: a protective mechanism in inflammation? *Trends Immunol* **26**, 637–643.

Delpech, B., Chevallier, B., Reinhardt, N., et al. (1990). Serum hyaluronan (hyaluronic acid) in breast cancer patients. *Int J Cancer* **46**, 388–390.

Dickson, R. B. and Lippman, M. E. (1995). Growth factors in breast cancer. *Endocr Rev* **16**, 559–589.

Elenbaas, B., Spirio, L., Koerner, F., et al. (2001). Human breast cancer cells generated by oncogenic transformation of primary mammary epithelial cells. *Genes Dev* **15**, 50–65.

Enegd, B., King, J. A., Stylli, S., et al. (2002). Overexpression of hyaluronan synthase-2 reduces the tumorigenic potential of glioma cells lacking hyaluronidase activity. *Neurosurgery* **50**, 1311–1318.

Entwistle, J., Hall, C. L., and Turley, E. A. (1996). HA receptors: regulators of signalling to the cytoskeleton. *J Cell Biochem* **61**, 569–577.

Evanko, S. P., Tammi, M. I., Tammi, R. H., and Wight, T. N. (2007). Hyaluronan-dependent pericellular matrix. *Adv Drug Deliv Rev* **59**, 1351–1365.

Fieber, C., Baumann, P., Vallon, R., et al. (2004). Hyaluronan-oligosaccharide-induced transcription of metalloproteases. *J Cell Sci* **117**, 359–367.

Folkman, J. (2002). Role of angiogenesis in tumor growth and metastasis. *Semin Oncol* **29**, 15–18.

Fraser, J. R., Laurent, T. C., and Laurent, U. B. (1997). Hyaluronan: its nature, distribution, functions and turnover. *J Intern Med* **242**, 27–33.

Fraser, J. R., Laurent, T. C., Engström-Laurent, A., and Laurent, U. G. (1984). Elimination of hyaluronic acid from the blood stream in the human. *Clin Exp Pharmacol Physiol* **11**, 17–25.

Frisch, S. M. (2001). Tumor suppression activity of adenovirus E1a protein: anoikis and the epithelial phenotype. *Adv Cancer Res* **80**, 39–49.

Frost, G. I., Csoka, T. B., Wong, T., and Stern, R. (1997). Purification, cloning, and expression of human plasma hyaluronidase. *Biochem Biophys Res Commun* **236**, 10–15.

Gao, F. Z. L. and Underhill, C. B. (2002). Promotion of growth of human breast cancer cells MDA231 by human sperm membrane-bound hyaluronidase: an experimental study. *Zhonghua Yi Xue Za Zhi* **82**, 207–210.

Ghatak, S., Misra, S., and Toole, B. P. (2002). Hyaluronan oligosaccharides inhibit anchorage-independent growth of tumor cells by suppressing the phosphoinositide 3-kinase/Akt cell survival pathway. *J Biol Chem* **277**, 38013–38020.

Goentzel, B. J., Weigel, P. H., and Steinberg, R. A. (2006). Recombinant human hyaluronan synthase 3 is phosphorylated in mammalian cells. *Biochem J* **396**, 347–354.

Gottfried, E., Kreutz, M., Haffner, S., et al. (2008). Differentiation of human tumour-associated dendritic cells into endothelial-like cells: an alternative pathway of tumour angiogenesis. *Scand J Immunol* **65**, 329–335.

Guise, T. A., Yin, J. J., Taylor, S. D., et al. (1996). Evidence for a causal role of parathyroid hormone-related protein in the pathogenesis of human breast cancer-mediated osteolysis. *J Clin Invest* **98**, 1544–1549.

Haylock, D. N. and Nilsson, S. K. (2006). The role of hyaluronic acid in hemopoietic stem cell biology. *Regen Med* **1**, 437–445.

Heldermon, C., DeAngelis, P. L., and Weigel, P. H. (2001). Topological organization of the hyaluronan synthase from *Streptococcus pyogenes*. *J Biol Chem* **276**, 2037–2046.

Hosono, K., Nishida, Y., Knudson, W., et al. (2007). Hyaluronan oligosaccharides inhibit tumorigenicity of osteosarcoma cell lines MG-63 and LM-8 *in vitro* and *in vivo* via perturbation of hyaluronan-rich pericellular matrix of the cells. *Am J Pathol* **171**, 274–286.

Howlett, A. R. and Bissell, M. J. (1993). The influence of tissue microenvironment (stroma and extracellular matrix) on the development and function of mammary epithelium. *Epithelial Cell Biol* **2**, 79–89.

Hu, M. and Polyak, K. (2008). Microenvironmental regulation of cancer development. *Curr Opin Genet Dev* **18**, 27–34.

Itano, N., Sawai, T., Atsumi, F., et al. (2004). Selective expression and functional characteristics of three mammalian hyaluronan synthases in oncogenic malignant transformation. *J Biol Chem* **279**, 18679–18687.

Itano, N. and Kimata, K. (2002). Mammalian hyaluronan synthases. *IUBMB Life* **54**, 195–199.

Itano, N., Sawai, T., Yoshida, M., et al. (1999a). Three isoforms of mammalian hyaluronan synthases have distinct enzymatic properties. *J Biol Chem* **274**, 25085–25092.

Itano, N., Sawai, T., Miyaishi, O. and Kimata, K. (1999b). Relationship between hyaluronan production and metastatic potential of mouse mammary carcinoma cells. *Cancer Res* **59**, 2499–2504.

Jacobson, A., Brinck, J., Briskin, M. J., Spicer, A. P., and Heldin, P. (2000). Expression of human hyaluronan synthases in response to external stimuli. *Biochem J* **348**, 29–35.

Jones, C., Nonni, A. V., Fulford, L., et al. (2001). CGH analysis of ductal carcinoma of the breast with basaloid/myoepithelial cell differentiation. *Br J Cancer* **85**, 422–427.

Karihtala, P., Soini, Y., Auvinen, P., et al. (2007). Hyaluronan in breast cancer: correlations with nitric oxide synthases and tyrosine nitrosylation. *J Histochem Cytochem* **55**, 1191–1198.

Kass, L., Erler, J. T., Dembo, M., and Weaver, V. M. (2007). Mammary epithelial cell: influence of extracellular matrix composition and organization during development and tumorigenesis. *Int J Biochem Cell Biol* **39**, 1987–1994.

Knudson, W., Chow, G., and Knudson, C. B. (2002). CD44-mediated uptake and degradation of hyaluronan. *Matrix Biol* **21**, 15–23.

Kobayashi, H., Suzuki, M., Kanayama, N., et al. (2002). CD44 stimulation by fragmented hyaluronic acid induces upregulation of urokinase-type plasminogen activator and its receptor and subsequently facilitates invasion of human chondrosarcoma cells. *Int J Cancer* **102**, 378–379.

Koyama, H., Hibi, T., Isogai, Z., et al. (2007). Hyperproduction of hyaluronan in neu-induced mammary tumor accelerates angiogenesis through stromal cell recruitment: possible involvement of versican/PG-M. *Am J Pathol* **170**, 1086–1099.

Kuang, D. M., Chen, N., Cheng, J., Zhuang, S. M., and Zheng, L. (2007). Tumor-derived hyaluronan induces formation of immunosuppressive macrophages through transient early activation of monocytes. *Blood* **110**, 587–595.

Lepperdinger, G., Mullegger, J., and Kreil, G. (2001). HYAL-2 – less active, but more versatile? *Matrix Biol* **20**, 509–514.

Lepperdinger, G., Strobl, B., and Kreil, G. (1998). HYAL-2, a human gene expressed in many cells, encodes a lysosomal hyaluronidase with a novel type of specificity. *J Biol Chem* **273**, 22466–22470.

Li, Y., Li, L., Brown, T. J., and Heldin, P. (2007). Silencing of hyaluronan synthase 2 suppresses the malignant phenotype of invasive breast cancer cells. *Int J Cancer* **120**, 2557–2567.

Liu, D. P. E., Diaconu, E., Guo, K., et al. (1996). Expression of hyaluronidase by tumor cells induces angiogenesis *in vivo*. *Proc Natl Acad Sci USA* **93**, 7832–7837.

Lokeshwar, V. B. and Selzer, M. G. (2000). Differences in hyaluronic acid-mediated functions and signaling in arterial, microvessel, and vein-derived human endothelial cells. *J Biol Chem* **275**, 27641–27649.

Madan, A. K., Dhurandhar, K. Y. N., Cullinane, C., Pang, Y., and Beech, D. J. (1999). Association of hyaluronidase and breast adenocarcinoma invasiveness. *Oncol Rep* **6**, 607–609.

Mascarenhas, M. M., Day, R. M., Ochoa, C. D., et al. (2004). Low molecular weight hyaluronan from stretched lung enhances interleukin-8 expression. *Am J Respir Cell Mol Biol* **30**, 51–60.

Mastro, A. M., Gay, C. V., Welch, D. R., et al. (2004). Breast cancer cells induce osteoblast apoptosis: a possible contributor to bone degradation. *J Cell Biochem* **91**, 265–276.

Mercer, R. R., Miyasaka, C., and Mastro, A. M. (2004). Metastatic breast cancer cells suppress osteoblast adhesion and differentiation. *Clin Exp Met* **21**, 427–435.

Noble, P. W. (2002). Hyaluronan and its catabolic products in tissue injury and repair. *Matrix Biol* **21**, 25–29.

Oguchi, T. and Ishiguro, N. (2004). Differential stimulation of three forms of hyaluronan synthase by TGF-beta, IL-1beta, and TNF-alpha. *Connect Tissue Res* **45**, 197–205.

Orimo, A., Gupta, P. B., Sgroi, D. C., et al. (2005). Stromal fibroblasts present in invasive human breast carcinomas promote tumor growth and angiogenesis through elevated SDF-1/CXCL12 secretion. *Cell* **121**, 335–348.

Ouskova, G., Spellerberg, B., and Prehm, P. (2004). Hyaluronan release from *Streptococcus pyogenes*: export by an ABC transporter. *Glycobiology* **14**, 931–938.

Pederson, L., Winding, B., Foged, N. T., Spelsberg, T. C., and Oursler, M. J. (1999). Identification of breast cancer cell line-derived paracrine factors that stimulate osteoclast activity. *Cancer Res* **59**, 5849–5855.

Prehm, P. (1983). Synthesis of hyaluronate in differentiated teratocarcinoma cells. Mechanism of chain growth. *Biochem J* **211**, 191–198.

Prehm, P. (2006). Biosynthesis of hyaluronan: direction of chain elongation. *Biochem J* **398**, 469–473.

Prehm, P. and Schumacher, U. (2004). Inhibition of hyaluronan export from human fibroblasts by inhibitors of multidrug resistance transporters. *Biochem Pharmacol* **68**, 1401–1410.

Prevo, R., Banerji, S., Ferguson, D. J., Clasper, S., and Jackson, D. G. (2001). Mouse LYVE-1 is an endocytic receptor for hyaluronan in lymphatic endothelium. *J Biol Chem* **276**, 19420–19430.

Rai, S. K., Vigdorovich, V., Danilkovitch-Miagkova, A., Lerman, M. I., and Miller, A. D. (2001). Candidate tumor suppressor HYAL-2 is a glycosylphosphatidylinositol (GPI)-anchored cell-surface receptor for jaagsiekte sheep retrovirus, the envelope protein of which mediates oncogenic transformation. *Proc Natl Acad Sci USA* **98**, 4443–4448.

Recklies, A. D., White, C., Melching, L., and Roughley, P. J. (2001). Differential regulation and expression of hyaluronan synthases in human articular chondrocytes, synovial cells and osteosarcoma cells. *Biochem J* **354**, 17–24.

Rodríguez-Pinilla, S. M., Rodríguez-Gil, Y., Moreno-Bueno, G., et al. (2007). Sporadic invasive breast carcinomas with medullary features display a basal-like phenotype: an immuno-histochemical and gene amplification study. *Am J Surg Pathol* **31**, 501–508.

Ronnov-Jessen, L., Petersen, O. W., and Bissell, M. J. (1996). .Cellular changes involved in conversion of normal to malignant breast, importance of the stromal reaction. *Physiol Rev* **76**, 69–125.

Rooney, P., Kumar, S., Ponting, J., and Wang, M. (1995). The role of hyaluronan in tumour neovascularization (review). *Int J Cancer* **60**, 632–636.

Rooney, P., Wang, M., Kumar, P., and Kumar, S. (1993). Angiogenic oligosaccharides of hyaluronan enhance the production of collagens by endothelial cells. *J Cell Sci* **105**, 213–218.

Ruibal, A., Arias, J., Del Río, M. C., Resino, C., and Tejerina, A. (2001a). Infiltrating ductal carcinomas of the breast in women over 60 years of age. Association with higher cellular proliferation and lower PS2, cell surface and cytosolic hyaluronic acid concentrations. *Rev Esp Med Nucl* **20**, 525–529.

Ruibal, A., Arias, J. I., Carmen del Río, M., et al. (2000). The cytosolic hyaluronic acid level defines several clinico-biological properties of CD44v5-positive infiltrating ductal carcinoma of the breast. *Med Clin (Barc)* **115**, 201–207.

Ruibal, A., Arias, J. I., Del Río, M. C., et al. (2001b). Histological grade in breast cancer: association with clinical and biological features in a series of 229 patients. *Int J Biol Markers* **16**, 56–61.

Sattar, A., Rooney, P., and Kumar, S. (1994). Application of angiogenic oligosaccharides of hyaluronan increases blood vessel numbers in rat skin. *J Invest Dermatol* **103**, 576–579.

Senger, D. R., Ledbetter, S. R., Claffey, K. P., et al. (1996). Stimulation of endothelial cell migration by vascular permeability factor/vascular endothelial growth factor through cooperative mechanisms involving the alphavbeta3 integrin, osteopontin, and thrombin. *Am J Pathol* **149**, 293–305.

Slevin, M., Krupinski, J., Kumar, S., and Gaffney, J. (1998). Angiogenic oligosaccharides of hyaluronan induce protein tyrosine kinase activity in endothelial cells and activate a cytoplasmic signal transduction pathway resulting in proliferation. *Lab Invest* **78**, 987–1003.

Slevin, M., Kumar, S., and Gaffney, J. (2002). Angiogenic oligosaccharides of hyaluronan induce multiple signaling pathways affecting vascular endothelial cell mitogenic and wound healing responses. *J Biol Chem* **277**, 41046–41059.

Slevin, M., West, D., Kumar, P., Rooney, P., and Kumar, S. (2004). Hyaluronan, angiogenesis and malignant disease. *Int J Cancer* **109**, 793–794.

Sørlie, T., Perou, C. M., Tibshirani, R., et al. (2001). Gene expression patterns of breast carcinomas distinguish tumor subclasses with clinical implications. *Proc Natl Acad Sci USA* **98**, 10869–10874.

Spicer, A. P., Kaback, L. A., Smith, T. J., and Seldin, M. F. (1998). Molecular cloning and characterization of the human and mouse UDP-glucose dehydrogenase genes. *J Biol Chem* **273**, 25117–25124.

Stern, R. (2003). Devising a pathway for hyaluronan catabolism: are we there yet? *Glycobiology* **13**, 105R–115R.

Stern, R. (2005). Hyaluronan metabolism: a major paradox in cancer biology. *Pathol Biol (Paris)* **53**, 372–382.

Stern, R., Asari, A. A., and Sugahara, K. N. (2006). Hyaluronan fragments: an information-rich system. *Eur J Cell Biol* **85**, 699–715.

Sugahara, K. N., Hirata, T., Hayasaka, H., et al. (2006). Tumor cells enhance their own CD44 cleavage and motility by generating hyaluronan fragments. *J Biol Chem* **281**, 5861–5868.

VII. HYALURONAN AND INDIVIDUAL CANCERS

Sugahara, K. N., Murai, T., Nishinakamura, H., et al. (2003). Hyaluronan oligosaccharides induce CD44 cleavage and promote cell migration in CD44-expressing tumor cells. *J Biol Chem* **278**, 32259–32265.

Takahashi, Y., Li, L., Kamiryo, M., et al. (2005). Hyaluronan fragments induce endothelial cell differentiation in a CD44- and CXCL1/GRO1-dependent manner. *J Biol Chem* **280**, 24195–241204.

Tammi, R., Pienimaki, D. K., MacCallum, M., et al. (2001). Hyaluronan enters keratinocytes by a novel endocytic route for catabolism. *J Biol Chem* **276**, 35111–35122.

Thiery, J. P. and Sleeman, J. P. (2006). Complex networks orchestrate epithelial–mesenchymal transitions. *Nat Rev Mol Cell Biol* **7**, 131–142.

Thomas, R. J., Guise, T. A., Yin, J. J., et al. (1999). Breast cancer cells interact with osteoblasts to support osteoclast formation. *Endocrinology* **140**, 4451–4458.

Tlsty, T. D. and Hein, P. W. (2001). Know thy neighbor: stromal cells can contribute oncogenic signals. *Curr Opin Genet Dev* **11**, 54–59.

Toole, B. P. (2002). Hyaluronan promotes the malignant phenotype. *Glycobiology* **12**, 37R–42R.

Toole, B. P. (2004). Hyaluronan: from extracellular glue to pericellular cue. *Nat Rev Cancer* **4**, 528–539.

Toole, B. P. and Hascall, V. C. (2002). Hyaluronan and tumor growth. *Am J Pathol* **161**, 745–747.

Tuck, A. B., Elliott, B. E., Hota, C., Tremblay, E., and Chambers, A. F. (2000). Osteopontin-induced, integrin-dependent migration of human mammary epithelial cells involves activation of the hepatocyte growth factor receptor (Met). *J Cell Biochem* **78**, 465–475.

Udabage, L., Brownlee, G.R., Nilsson, S.K. and Brown, T.J. (2005a). The over-expression of HAS-2, HYAL-2 and CD44 is implicated in the invasiveness of breast cancer. *Exp Cell Res* **310**, 205–217.

Udabage, L., Brownlee, G.R., Waltham, M., et al. (2005b). Antisense-mediated suppression of hyaluronan synthase 2 inhibits the tumorigenesis and progression of breast cancer. *Cancer Res* **65**, 6139–6150.

Vignal, P., Meslet, M. R., Roméo, J. M., and Feuilhade, F. (2002). Sonographic morphology of infiltrating breast carcinoma: relationship with the shape of the hyaluronan extracellular matrix. *J Ultrasound Med* **21**, 532–538.

Vogelstein, B. K. K. (2004). Cancer genes and the pathways they control. *Nature Med* **10**, 789–799.

Von Marschall, Z., Scholz, A., Stacker, S. A., et al. (2005). Vascular endothelial growth factor-D induces lymphangiogenesis and lymphatic metastasis in models of ductal pancreatic cancer. *Int J Oncol* **27**, 669–679.

Wang, L. P., Ning, H. Y., Yang, S. M., et al. (2004). Expression of PH-20 in primary and metastatic breast cancer and its pathological significance. *Zhonghua Bing Li Xue Za Zhi* **33**, 320–323.

Weigel, P. H. and DeAngelis, P. L. (2007). Hyaluronan synthases: a decade-plus of novel glycosyltransferases. *J Biol Chem* **282**, 36777–36781.

Weigel, P. H., Kyossev, Z., and Torres, L. C. (2006). Phospholipid dependence and liposome reconstitution of purified hyaluronan synthase. *J Biol Chem* **281**, 36542–36551.

Wernicke, M., Piñeiro, L. C., Caramutti, D., et al. (2003). Breast cancer stromal myxoid changes are associated with tumor invasion and metastasis: a central role for hyaluronan. *Mod Pathol* **16**, 99–107.

West, D. C. and Kumar, S. (1989). The effect of hyaluronate and its oligosaccharides on endothelial cell proliferation and monolayer integrity. *Exp Cell Res* **183**, 179–196.

West, D. C., Hampson, I. N., Arnold, F., and Kumar, S. (1985). Angiogenesis induced by degradation products of hyaluronic acid. *Science* **228**, 1324–1326.

Zoltan-Jones, A., Huang, L., Ghatak, S., and Toole, B. P. (2003). Elevated hyaluronan production induces mesenchymal and transformed properties in epithelial cells. *J Biol Chem* **278**, 45801–45810.

CLINICAL USES OF HYALURONAN-RELATED BIOMATERIALS AS ANTI-CANCER AGENTS

CHAPTER

19

Clinical Use of Hyaluronidase in Combination Cancer Chemotherapy: A Historic Perspective

Gerhard Baumgartner and Gerhard Hamilton

INTRODUCTION

Testicular hyaluronidase was first described in 1928 as "spreading factor" by Duran-Reynals and further characterized in 1940 by Chain and Duthie as mycolytic enzyme that mediated increased uptake of substances into tissues (Duran-Reynals, 1929; Chain et al., 1939; 1940). Clinical application of this enzyme started in 1952, when Breu demonstrated that hyaluronidase helps to distribute dyes and other substances in tissues and increases the permeability of dermis and connective tissue by degradation of hyaluronan (Breu, 1952). Atkinson pioneered the use of hyaluronidase in ophthalmology using the enzyme in combination with epinephrine and procaine in 1949 (Atkinson, 1949). The effects of hyaluronidase enables this enzyme to be used therapeutically to increase the speed of absorption and to diminish discomfort due to subcutaneous or intramuscular injection of fluid, to promote resorption of excess fluids and extravasated blood in tissues, and to increase the effectiveness of local anesthesia. Hyaluronidase has been used to reduce the extent of tissue damage following extravasation of parental nutrition solution, electrolyte infusions, antibiotics, aminophyline, and 'mannitol' (Farr et al., 1997; Frost, 2007). Hyaluronidase was widely used in many fields, i.e. in orthopedics, surgery, ophthalmology, internal medicine, oncology, dermatology, gynecology, etc (Kluza and Moritz, 1985). The application of this enzyme in tumor therapy in combination with chemotherapeutic agents has been studied for more than two decades. However, many aspects of hyaluronidase usage in anticancer regimens continues to be experimental (Baumgartner et al., 1998).

The term hyaluronidase was introduced to denote enzymes which degrade hyaluronan, although these enzymes are able to cleave other glycosaminoglycans (Stern, 2006). There are three main groups of enzymes with the same specificity, however, with different reaction mechanisms. The one available for clinical pilot experiments belongs to the class of testicular-type hyaluronidases (hyaluronate 4-glycanohydrolase, EC 3.2.1.35), like Permease®, which randomly cleaves β-N-acetyl-hexosamine $(1\rightarrow4)$ glycosidic bonds in hyaluronan, chondroitin, and chondroitin sulfates. The reaction generates even-numbered oligosaccharides, with mainly tetrasaccharides as the smallest fragments. In the human body, hyaluronidase is found in many organs (testes, spleen, skin, eye, liver, kidney, uterus, and placenta), body liquids (tears, blood, urine, sperm and others), and in tumors (Farr et al., 1997). Generally, hyaluronidases are extracted from bovine and sheep testes or from bacteria, respectively. Highly purified preparations, e.g. hyaluronidase from bovine testes, achieve an activity of 40,000–50,000 IU/mg. The most frequent contaminating enzymes in less pure preparations are proteases and glucuronidases. Hyaluronidase is rapidly eliminated from plasma: upon intravenous (i.v.) injection of 5000 IU hyaluronidase per kg body weight, no activity can be demonstrated after 45 min (serum half-time 2.1 ± 0.2 min; Oettl et al., 2003). The short half-life of the enzyme in plasma cannot be accounted for by excretion in urine and bile. Tissue distribution measurements reveal that the major site of uptake is the liver (Muckenschnabel et al., 1998). All hyaluronidases, including tumor hyaluronidases, are most active at an acid pH 3.7–4.0. Testicular hyaluronidase has a broad pH-optimum and at pH 5.0 has approximately 70% of the activity at pH 4.0. When measured by viscosity reduction or by turbidity methods, the enzyme is still active at neutral pH (Oettl et al., 2003).

CLINICAL APPLICATION OF HYALURONIDASE IN PERIPHERAL CIRCULATORY DISORDER

A review of the various applications of hyaluronidase in benign diseases was published by Farr and Menzel (Farr et al., 1997). One peculiar clinical application, which may be of relevance for the use of the enzyme as an adjunct in chemotherapy, is described by Baumgartner et al., namely for the experimental therapy of peripheral circulatory disorders (Baumgartner et al., 1998). In a patient with gangrene of the toes, the problem was successfully treated with 7500 IU hyaluronidase intravenously, in close agreement with published observations (Elder et al., 1980). Following this clinical observation, 16 patients were treated with 200,000 IU hyaluronidase intra-arterially (12/16 patients with diabetes) resulting in complete recovery of three patients, improvement in eight patients and progression of disease in three patients, leading to amputation. To confirm these results, a randomized trial comparing 200,000 IU hyaluronidase intra-arterially (nine patients) with placebo (nine patients) was performed. In the hyaluronidase group a statistically significant reduction of ulceration was achieved compared to the placebo group. Since parameters of local circulation were not improved in large vessels, the enzyme seems to improve transport of oxygen and nutrients in the small capillaries, a property which could similarly assist delivering chemotherapeutic drugs to solid tumors.

CLINICAL APPLICATIONS OF HYALURONIDASE IN MALIGNANT DISEASE

Expression of hyaluronidase was detected in various tumors and at the time was regarded as an important factor supporting tumor growth and dissemination. This led to an interest in inhibitors of this enzyme (Kiriluk et al., 1950; Rosenthal, 1952; Cameron et al., 1979). Paradoxically, both hyaluronan and hyaluronidases can correlate with cancer progression. Increased levels of hyaluronan on the surface of tumor cells are an indicator of poor outcome (Stern, 2005). Such information would indicate that cancer progression is inhibited by hyaluronidase. Yet progression of certain cancers also correlates with levels of hyaluronidase activity (Lokeshwar et al., 1996). Although hyaluronidase became an established preparation for the management of the extravasation of cytotoxic drugs, its use in treatment of tumors was considered contra-productive in regard to promotion of metastasis (Herp et al., 1968, Bertelli, 1995).

One observation of tumor response accidentally made during treatment of extravasation of a chemotherapeutic infusion in 1982 stimulated interest here in Vienna in the possibility that the enzyme hyaluronidase may improve the transport of cytotoxic drugs to and within tumor tissue by degradation of the extracellular matrix (Baumgartner and Baumgartner, 1985; Baumgartner, 1987). A myeloma patient treated with a polypeptide complex of L-phenylalanine mustard (Peptichemio®, Peptichemio AG, Switzerland; drug now discontinued) who had received several prior administrations without response, suffered from an extensive extravasation. A large dose (approximately 54,000 IU) of hyaluronidase was applied in order to resolve the extravasation. The combination of the cytotoxic drug with the enzyme however led to rapid reduction in bone pain from the bone manifestations of myeloma. Since this surprising observation pointed to hyaluronidase as crucial additive, a more systematic investigation of the enzyme as an adjunct to chemotherapy was initiated and reported for the first series of patients by Baumgartner and Baumgartner in 1985. Forty-six patients (17 myelomas, 11 malignant lymphomas, eight mammary carcinomas, seven head and neck carcinomas, two gastrointestinal carcinomas, and one ovarian carcinoma) that became resistant to their specific type of chemotherapy were treated again with the same regimen combined with bovine testicular hyaluronidase (7500 IU Permease®, Sanabo, Vienna, Austria), either intravenously, intramuscularly one hour prior to or intraperitoneally in combination with cytotoxic chemotherapy. Two cases of local irritation at the site of injection and one case of reversible anaphylactoid reaction were described. Responses reported in patients were as follows (complete response – CR; partial response – PR): myeloma – CR 2/9, subjective improvement – 7/9; non-Hodgkin-lymphoma (NHL) – CR 2/5, PR 2/5, and breast cancer – PR 2/4. Complete regression of ascites was achieved in most cases of intraperitoneal therapy in later studies (Baumgartner et al., 1998). The effectiveness of Permease® was ascribed to a reversal of the resistance of tumor cells or an alteration of pharmacokinetic parameters of cytotoxic agents, respectively.

HYALURONIDASE PREPARATIONS IN ADDITIONAL CLINICAL STUDIES IN VIENNA

Since the pilot study described above had yielded promising results, further attempts were made to improve and expand clinical applications of hyaluronidase by using more enriched preparations with higher specific activity. Bovine hyaluronidase had been available in various formulations for therapeutic use for over 40 years. These preparations were registered during different times and varied widely with regard to purification standards and dosage per vial. The two preparations of lyophilized bovine testicular hyaluronidase (Stettbacher, 1953; Aderhold, 1954) that were used for clinical application by our group at the III Department of Internal Medicine, Hanusch Hospital Vienna, V Department of Internal Medicine, Hietzing Hospital Vienna, and the Ludwig Boltzmann Institute for Clinical Oncology and Photodynamic Therapy were Permease® 750 IU/vial (available from 1982, Biochemie, Vienna, no longer available) and the further purified Neopermease® 200,000 IU/vial (available from 1984; discontinued 1998, Sanobo/Sandoz, Austria). The third available hyaluronidase preparation for clinical use at that time was Hylase® Dessau 1500 IU/vial (Impfstoffe, Dessau, Germany; now Riemser, Germany). Results of Oettl et al. demonstrated that Neopermease® and Hylase® Dessau are hyaluronidase preparations with nearly the same enzymatic properties (Oettl et al., 2003), though the molecular sizes and other properties of the enzymes appeared to differ. Although the existence of isoenzymes could not be definitely ruled out, the pattern of proteins appeared to depend primarily on different proteolytic cleavage reactions during the isolation procedures. Within the limits of experimental error, both pharmaceutical bovine testicular hyaluronidase preparations are characterized by identical pH dependencies. Enzyme activity was observed over a broad range of up to pH 8, with a distinct maximum at pH 3.5. For intramuscular application, 7500 IU Permease® (10 vials containing 750 IU each, dissolved in 5 ml isotonic NaCl) were given 1 h prior to chemotherapy, for intravenous application, 7500–22,500 IU Permease® (infusion time 15 min) or 200,000 IU Neopermease® (infusion time 15 min–24 h) in 100 and 500 ml isotonic NaCl, respectively. For intravesical therapy, 200,000 IU Neopermease® in 20 ml isotonic NaCl were instilled in combination with 20 mg mitomycin C following transurethral resection of bladder cancer.

EXTENDED STUDIES OF THE USE OF HYALURONIDASE AS ADJUNCTS IN CANCER CHEMOTHERAPY

During the years that followed, a total of 260 patients were treated with combinations of hyaluronidase and chemotherapeutic drugs at our institutions and in cooperation with other departments and hospitals. These included cases of brain tumors, head and neck cancers, bladder cancer, myeloma, lymphoma, breast cancer, non-small cell and small cell lung cancer, colon cancer, renal cell cancer, glioma, and others (Baumgartner, 1987). Sixty-nine patients received low-dose hyaluronidase intramuscularly, 11 patients low-dose intravenously, and 180 patients high-dose (20,000 IU) intravenously or intravesically.

To obtain more definite information concerning the contribution of hyaluronidase to the therapeutic effect of chemotherapy, the unchanged treatment cycle was repeated with addition of hyaluronidase in 103 patients, who had been resistant to various types of chemotherapy. The other patients were treated with a combination therapy from the beginning. Cytotoxic drugs included alkylating compounds, doxorubicin, Vinca alkaloids, cisplatin, etoposide, 5-fluorouracil, and methotrexate. At that time no alternative chemotherapy regimens were available for those patients. In a significant number of these patients, new responses could be achieved by the addition of hyaluronidase. Regression or remission could be maintained for varying but limited periods: myeloma patients (5 CR + PR/23), Morbus Hodgkin patients (5 CR + PR/9), NHL high-grade patients (9 CR + PR/14), mammary carcinoma patients (7 CR + PR; 4 stable disease/14) and colorectal carcinoma patients (5 PR + 7 stable disease/16). In 17 out of the 23 myeloma patients bone pain was reduced and mobility improved for a prolonged period.

In 27 patients with squamous cell carcinomas of the head and neck region, hyaluronidase was added to cytostatic chemotherapy (bleomycin/cisplatin/methotrexate/5-fluorouracil), in some from the beginning of treatment, and in others, after the development of chemoresistance. We administered either vials containing 750 IU of Permease® in a dosage of 7500 IU or 22,500 IU or a preparation of Neopermease® containing 200,000 IU. Hyaluronidase was well tolerated and reversible allergic reactions were observed in only two patients. Overall CR 14/27, PR 5/27 and stable disease 3/27 was achieved. After giving hyaluronidase to chemoresistant patients, CR 8/16, PR 3/16 and stable disease 3/16 were achieved. The course of disease in chemoresistant cases makes it very likely that hyaluronidase improved the outcome in such patients. Disease-free survival times have been extremely long. For example, in resistant patients with squamous cell head and neck carcinomas, complete remissions (median of 37 months) were achieved by the addition of hyaluronidase (Baumgartner and Neumann, 1987).

Hyaluronidase in combination with chemotherapeutics from the beginning was also tested in patients bearing brain tumors, either primary cerebral tumors or secondary tumor metastases. Furthermore this pilot study included high-grade astrocytomas and NHL, including HIV-positive patients. Although hyaluronidase is not able to cross the blood–brain barrier under normal conditions, it may pass a leaky barrier that appears to be prevalent in brain tumor patients (Vick et al., 1997; Baumgartner et al., 1987).

In six patients with mammary carcinoma and cerebral metastases, chemotherapy according to either the ACO-scheme (adriamycin/cyclophosphamide/vincristine) or MTX/DDP (methotrexate/cisplatin) was combined with hyaluronidase without radiation therapy. There was one

CR (brain and lung metastases), two PR, two that remained stable, and one case in which there was progression of disease. These remissions were maintained for a median of nine months. Eight patients with high-grade cerebral lymphomas (seven primary and one secondary) were treated with MTX/DDP/hyaluronidase without radiation therapy. Three CR and three PR were achieved (Baumgartner et al., 1987b). CR could be maintained for 11, nine and one month, respectively. In three further HIV-positive patients with primary cerebral lymphoma, one CR was achieved with CHOP (cyclophosphamide/doxorubicin/vincristine/prednisone)/hyaluronidase lasting for 6 months without radiation therapy.

In 1986, a 75-year-old female patient with inoperable glioblastoma stage IV in poor general condition was treated. Cytotoxic chemotherapy consisted of 60 mg lomustine in a single dose, 150 mg methotrexate, and 2 mg vincristine at 3-week intervals and additionally 200,000 IU hyaluronidase were injected intravenously prior to cytostatic application. Without any additive radiotherapy, a complete remission lasting for 9 months was obtained, proven by using computer-assisted tomography. This patient died from the tumor after 12 months.

Based on this observation, 39 patients with high-grade astrocytomas, including primary tumors and relapses, were treated with poly-chemotherapy and hyaluronidase in a second pilot study. In six patients who suffered from primary tumors, after partial resection, three partial remissions could be achieved. The same was observed in two of 12 patients with recurrent astrocytoma. The longest disease-free interval achieved by chemotherapy with added hyaluronidase was nine months. In the next randomized study involving 43 patients with high-grade astrocytoma, in the randomized arm A without hyaluronidase the median survival was 8.7 months, while in arm B with hyaluronidase the median survival averaged 9.5 months, indicating no significant differences. The median survival time achieved in this study (9.2 months) correlated well with results by others in grade IV astrocytomas with chemotherapy plus radiation therapy, however, quality of life was improved in our own study, due to partial responses and a lower requirement for surgical interventions for reducing intracranial pressure (Baumgartner et al., 1998).

Intravesical instillation of mitomycin C can significantly reduce the rate of relapse after radical resection of superficial bladder cancer (Bolenz et al., 2006). In our group, a randomized study using mitomycin C with and without hyaluronidase applied by intravesical instillation after radical resection of superficial bladder cancer was performed. The aim of this trial was to investigate whether hyaluronidase can further reduce the rate of relapse. Of major concern was the possibility that this enzyme would lead to an enhanced release of mitomycin C into the blood stream, resulting in increased systemic toxicity. However, intravesical instillation of 20 mg mitomycin C with and without 200,000 IU hyaluronidase and measurement of peak plasma concentrations ruled out any differences in resorption of the cytotoxic drug between these two patient groups (Maier and Baumgartner, 1988).

Following this pharmacokinetic study, 56 patients were randomized for mitomycin C with and without hyaluronidase treatment. The recurrence rate after a median observation time of 21.1 months was nine of the 28 patients (32%) who received 20 mg mitomycin C as monotherapy, in contrast to the two of 28 patients (7%) who received 20 mg mitomycin C together with 200,000 IU hyaluronidase intravesically. The local toxicity of mitomycin C was not enhanced by hyaluronidase (Maier and Baumgartner, 1988). After a median observation time of 50.4 months, the rate of relapse with mitomycin C single agent instillation was 59% compared to 27% after mitomycin C plus hyaluronidase (Maier and Baumgartner, 1989). It was concluded therefore that addition of hyaluronidase enhances the local effect of mitomycin C in the intravesical neoadjuvant treatment of bladder cancer.

Based on the results of this study, a group of 43 patients undergoing transurethral resection of Ta–T1 tumors and mitomycin C/hyaluronidase combination treatment was analyzed retrospectively after a mean observation period of 48.5 months. During two years of neoadjuvant therapy, tumor recurrence was seen in six patients (13.9%). Of the 37 patients who remained disease-free under treatment, five (13.5%) exhibited tumors later during a mean observation period of 24.5 months after treatment. These values were significantly lower than those obtained previously from a group of 63 patients treated with mitomycin C alone (mean observation period 50.4 months), with recurrence rates of 33.3% during and 26.2% after neoadjuvant chemotherapy (Hoebarth et al., 1989).

SIDE EFFECTS OBSERVED DURING CLINICAL AND EXPERIMENTAL USES OF HYALURONIDASE

Previously, Hylase® Dessau had been used clinically in large doses intravenously for the therapy of Bechterew's disease (Bellmann et al., 1972). The immunogenicity, formation of humoral antibodies, and frequency of anaphylactic reactions were studied in experimental animals in response to application of hyaluronidase (Storch et al., 1978). Intravenous and intramuscular administrations of 150 to 75,000 IU of Hylase® Dessau were followed by production of IgG antibodies against the enzyme. The formation of antibodies was reported to occur more extensively after intramuscular application. One third of the animals showed anaphylactic responses. 26% of patients developed antibodies after application of Hylase® Dessau. No anaphylactic reactions were observed in 17 patients with antibodies when intravenous application of hyaluronidase was continued.

Our first clinical studies were designed as phase I studies, mainly to obtain information concerning the tolerability of the hyaluronidases, the widely differing dosages (7500 or 200,000 IU) and the significance of different application routes. The large difference in the hyaluronidase content per vial in the two available preparations (Permease® and Neopermease®) was the main reason for the large variation in dosage schemes.

In general, hyaluronidase itself was very well tolerated when applied by the intramuscular or intravenous route with 20 out of 229 patients exhibiting reversible allergic reactions (Baumgartner, 1987). When given by the intravesical route, no side effects occurred. Local allergic reactions at the injection site were observed in three of 72 patients treated intramuscularly, with one patient showing systemic symptoms, such as nausea, vomiting and circulatory problems, which is typical of the allergic reactions when hyaluronidase is given intravenously. In all other systemically treated patients, the rate of allergic reactions was 10%, with the exception of astrocytomas where the rate was 20%, but always reversible and easily manageable by corticoids and calcium (4/10 patients received 22,500 IU and 13/149 patients 200,000 IU). This higher rate was observed in astrocytomas even though all these patients received corticoids as symptomatic prophylaxis against cerebral edema.

In 38 patients, measurements of the antibody level against hyaluronidase were performed. In 12 of these patients, antibody levels were already elevated before therapy. In further 18 patients, the increase in antibody levels was manifest until 6 weeks after the initiation of therapy (nine patients: 10–100 fold; one allergic reaction; Baumgartner, 1987). Five out of 16 pediatric patients with CNS tumors treated with hyaluronidase (Neopermease®) in addition to chemotherapy developed symptoms of immediate type allergic reactions (Szépfalusi et al., 1997). In sera from these five patients binding of specific IgE antibodies to proteins of the hyaluronidase preparation were found. No specific IgE binding was detected either in the sera of atopic patients, or in the control group. Further application of hyaluronidase is not indicated in cases of an allergic reaction that may be increased by impurities in these preparations.

APPLICATION OF HYALURONIDASE IN EXPERIMENTAL ANIMAL MODELS OF CANCER

Following studies concerning the use of hyaluronidase as an adjunct to chemotherapy for different tumor entities with promising results, investigations in regard to the pharmacokinetics and toxicity of the bovine enzyme were initiated in experimental animal models. Pharmacokinetic studies for hyaluronidase were performed by the group of Professor Schoenenberger at the Institute of Pharmacy, University of Regensburg

(Germany) in 1990 and have been continued by his successor, Professor Buschauer, systematically characterizing the distribution and effects of this enzyme *in vitro* and *in vivo*.

A first trial using a murine mammary carcinoma animal model showed that the efficacy of doxorubicin was enhanced by application of hyaluronidase in the vicinity of the tumor (Beckenlehner et al., 1992). The antitumor activity of doxorubicin was enhanced *in vitro* and *in vivo* by high doses of bovine hyaluronidase that had no cytotoxic activities themselves in different cell lines and in a mammary cancer tumor model (100,000 IU/kg).

A second trial tested the subcutaneous application of vinblastine close to a transplanted melanoma with and without hyaluronidase. Vinblastine as single agent therapy as well as with hyaluronidase alone were not effective in this resistant xenotransplant model, while the combination of hyaluronidase and vinblastine achieved curative results using one melanoma cell line and delayed tumor growth in three other melanoma lines (Spruss et al., 1995). A further trial in a malignant melanoma xenotransplant model investigated the impact of intraperitoneal hyaluronidase on the accumulation of melphalan in various organs and the tumor. It could be shown that accumulation of melphalan was increased by a factor of 5–32 after intraperitoneal application of hyaluronidase. This result indicates that hyaluronidase is able to cross the blood vessel barrier and reach the tumor, even when applied at some distance away from the tumor (Muckenschnabel et al., 1996). A further conclusion was that the extremely short time of decay of hyaluronidase in plasma, which amounts to a few minutes, is not due to the rapid inactivation but to a very efficient distribution to all tissues and also to a larger extent to tumor tissue. In summary, hyaluronidase alone even in very high concentrations revealed no antitumor activity, however, in combination with chemotherapeutic drugs significantly increased permeation and cytotoxicity (Muckenschnabel et al., 1998).

OVERVIEW AND ADDITIONAL OBSERVATIONS

Hyaluronidase has been in clinical use involving chemotherapy for the treatment of the extravasation, in order to prevent tissue necrosis. A key event in alerting us to the possible additional uses of hyaluronidase was the case of one myeloma patient who received a large dose of hyaluronidase locally to resolve extravasation of a highly toxic melphalan-containing drug. This local application of hyaluronidase not only resolved the marked extravasation, however, in addition, rapidly reduced pain from the bone manifestations of myeloma, a phenomenon not observed with the chemotherapeutic drug alone during previous administrations (Baumgartner and Baumgartner, 1985). Based on this observation, a series of patients were treated with cytotoxic drug combinations including hyaluronidase, most frequently after progression of disease following state-of-the-art regimens, but with the last cycle of chemotherapy repeated with addition of hyaluronidase. For diverse tumor entities a significant number of complete or partial responses were observed or patient status was changed to stable disease.

Modified treatments were most successful in squamous cell head and neck carcinoma, breast cancer, and myeloma. Responses were documented in astrocytomas, resulting in improved quality of life due to lower frequency of required surgical care, although survival was not prolonged. Under normal conditions hyaluronidase is not able to cross the blood–brain barrier, however, this limitation can be overcome in patients with brain tumors, many of whom appeared to have compromised blood–brain barriers (Vick et al., 1997). This was additionally demonstrated by an improved accumulation of the label sodium borocaptate for boron neutron capture therapy in astrocytomas following intravenous injection of hyaluronidase (Haselsberger et al., 1996). Two trials using hyaluronidase as an adjunct to intravesical mitomycin C therapy for bladder cancer, one of the studies randomized and one using a historical patient control group, pointed to a significant reduction in the number of recurrences. The levels of mitomycin C in the systemic circulation were not increased by hyaluronidase, and in all patients treated with hyaluronidase signs of increased metastasis were not observed.

Several findings of other groups supported our own observation of a significant role of the added hyaluronidase in the improved outcome of patients undergoing combination chemotherapy (Possinger, 1988; Thiruvengadam and Moran, 1995). Hyaluronidase enhanced the therapeutic effect of vinblastine in intralesional treatment of Kaposi's sarcoma (Smith et al., 1997). A pilot study using hyaluronidase in combination with chemotherapy and radiation therapy was performed in inoperable head and neck tumors (Klocker et al., 1995).

Another pilot study tested the combination in cases of mesotheliomas, where in consideration of the excessive production of hyaluronic acid, the concept of hyaluronidase seemed particularly promising (Jones et al., 1995). Thirty-eight advanced cases with mesothelioma were treated with good results, namely a 44% response rate (Israel, 1992; Breau et al., 1993). Thus hyaluronidase was shown to be effective also as adjuvant to systemic palliative therapy, although most of these results have up to now, not been confirmed by randomized trials.

The hyaluronidase preparations available for clinical application more than 20 years ago were of limited purity, containing mixtures of hyaluronidase, fragmented proteins and other enzymatic activities and of limited potency (activities of 1500 or 7500 IU), except the 200,000 IU package size. These first pilot results gained with these widely differing dosages did not indicate a clear-cut correlation of hyaluronidase dosage to therapeutic and side effects. Finally, those clinical studies had to be terminated due to the withdrawal of the hyaluronidase preparations because of safety concerns by the pharmaceutical companies. New preparations containing human recombinant hyaluronidase may solve the problems associated with the use of bovine protein and of enzymatic contaminations. Experiments using recombinant hyaluronidase will help to define the possible role of proteolytic and glycolytic contaminations in bovine testicular hyaluronidase preparations.

Important support for the concept of hyaluronidase as adjuvants for chemotherapy have come from multiple pharmacokinetic experiments in animal models. Application of high doses of hyaluronidase improved the effects of cytotoxic drugs in a breast cancer and a melanoma xenograft animal model significantly, not only when applied locally but most importantly, also when applied intraperitoneally, far from the tumor site (Muckenschnabel et al., 1996; 1998). Hyaluronidase alone had no effect *in vivo*, however, in combination with melphalan markedly increased the concentration of the drug in tumor tissue. Furthermore, it was reported that treatment with hyaluronidase blocked lymph node invasion by tumor cells in an animal model of T cell lymphoma (Zahalka et al., 1995).

CONCLUDING REMARKS AND SUMMARY

In conclusion, hyaluronidase as an adjunct to chemotherapy increases the response of tumors to chemotherapeutic drugs in experimental animal models and in the clinical trials described above. The mechanisms responsible for the synergism of hyaluronidase with cytostatics *in vivo* are not clear. The most simple explanation is a breakdown of the physical barrier shielding tumor cells against drugs (Gately et al., 1984; Kohno et al., 1994; Jones et al., 1995). Solid tumors are partially protected against cytotoxic drugs by a limited physical access due to irregular vascularization and blood supply and dense stromal areas that impede the diffusion of chemotherapeutics, in addition to mechanisms of resistance at the cellular level. Transient disintegration of the intercellular structures, such as the degradation of hyaluronan by hyaluronidase, would be expected to facilitate the transport of drugs to tumor cells and thereby increase their therapeutic efficacy, as demonstrated for liver metastases of colorectal cancer for cisplatin (Civalleri et al., 1996). Interestingly, many human solid tumors contain high concentrations of hyaluronan, which correlate with high hyaluronan serum levels of the patients (Toole, 2004).

The interstitial pressure, which is known to be elevated in tumor tissues, can be decreased by hyaluronidase as recently shown (Eikenes et al., 2005). Intratumoral injection of hyaluronidase (1500 IU) one hour prior to i.v.

injection of liposomal doxorubicin increased tumor uptake four-fold. Additonally, in human osteosarcoma xenografts grown in mice, the closure of vessels was prevented and perfusion improved in response to hyaluronidase or collagenase (Tufto et al., 2007).

A further resistance mechanism of solid tumors is the vascularization deficit resulting in hypoxia, which might also be modulated by hyaluronidase. Degradation of hyaluronan by hyaluronidase yields a heterogeneous mixture of oligosaccharides and hyaluronan fragments of different sizes, to which certain biological functions have been ascribed, e.g. induction of irreversible phenotypic and functional maturation of dendritic cells during inflammation, induction of angiogenesis by stimulation of proliferation, and migration of vascular endothelial cells via multiple signaling pathways (Toole, 2004). These fragments may be responsible for effects of hyaluronidase distant from a local injection site, since the enzyme may become rapidly inactivated in the peripheral circulation. Jain et al. has demonstrated that transient normalization of tumor vascularization creates a "window of opportunity" for chemotherapeutic intervention (Jain, 2005). At the tumor cell level, the group of Kerbel has shown an adhesion-dependent multicellular tissue chemoresistance which has been demonstrated to be sensitive to hyaluronidase (St. Croix et al., 1996).

The role of hyaluronan has undergone re-evaluation from a passive component of connective tissues controlling viscosity and permeability, to an active role in cell survival and modulation of the malignant state (Toole, 2004; Girish et al., 2007). This glycosaminoglycan can bind to specific cell surface receptors of tumor cells, such as CD44 and RHAMM (receptor of hyaluronic acid-mediated motility) providing a cell coat as well as for intracellular signaling. Hyaluronan strongly promotes anchorage-independent growth by supplying survival signals. Hyaluronan oligomers are both high and low affinity ligands, and together with soluble CD44 or RHAMM disrupt this positive regulatory mechanism via hyaluronan receptors. Oligosaccharides generated by hyaluronidase may act synergistically with increases in permeability of the cellular matrix to improve the cytotoxic effects of drugs. Various combinations of all of these mechanisms may be the basis for the clinical results described above.

In summary, hyaluronidase seems to constitute a valuable adjunct to chemotherapy. When administered locally for bladder cancer and mesothelioma, the protective layer of cells may be removed and the physical barriers of the tumors compromised. The shutdown of intratumoral vessels and blood supply, increased interstitial pressure, and impeded permeability may be reduced. In systemic therapy, indirect effects of the enzyme, for example release of oligosaccharides and hyaluronan fragments may play important roles in addition to the direct effects. Otherwise, the synergistic effects of systemic chemotherapy with small amounts of locally administered hyaluronidase, for example in myeloma patients with bone manifestations, may be difficult to explain. Since expression of hyaluronan and hyaluronidase seems to have different effects in different tumors, hyaluronidase in combination with chemotherapy may be helpful only in specific tumor entities, however, in a wide dose range. The clinical studies performed more than 20 years ago need to be repeated and evaluated using human recombinant hyaluronidase, in order to exclude side activities of impurities of the bovine preparations. Additionally, investigations into the mechanisms by which hyaluronidase improves the efficacy of specific anti-cancer drugs should be conducted.

References

Aderhold, K. (2003). Possibilities of using hyalase, a new hyaluronidase preparation in surgery. *Zentralbl Chir* **79** (8), 1954.

Atkinson, W. S. (1949). Use of hyaluronidase with local anesthesia in ophthalmology; preliminary report. *Arch Ophthalmol* **42**, 628–633.

Baumgartner, G. and Baumgartner, M. (1985). Results of a pilot study of hyaluronidase as an adjunct to cytostatic therapy in malignant diseases. *Wien Klin Wochenschr* **97**, 148–153.

Baumgartner, G. (1987). Hyaluronidase in der Therapie maligner Erkrankungen. Beilage Wr. Klin. Wochenschr 99, supplement.

Baumgartner, G. and Neumann, H. (1987). Hyaluronidase in der zytostatischen Therapie von HNO-Tumoren. *Laryng Rhinol Otol* **66**, 195–199.

Baumgartner, G., Gomar-Hoess, C., Sakr, L., Ulsperger, E., and Wogritsch, C. (1998). The impact of extracellular matrix on the chemoresistance of solid tumors–experimental and clinical results of hyaluronidase as additive to cytostatic chemotherapy. *Cancer Lett* **131**, 85–99.

Baumgartner, G., Horaczek, A., Grunert, P., Kitz, K., and Wunsch, M. (1987). Hyaluronidase als Zusatz zur zytostatischen Chemotherapie bei Glioblastomen. *Onkologie* **10**, 100–103.

Baumgartner, G., Fortelny, A., Zaenker, K. S., and Kroczek, R. (1988). Phase I study in chemoresistant loco-regional malignant disease with hyaluronidase. *Reg Cancer Treat* **1**, 55–58.

Beckenlehner, K., Bannke, S., Spruss, T., et al. (1992). Hyaluronidase enhances the activity of adriamycin in breast cancer models in vitro and in vivo. *J Cancer Res Clin Oncol* **118**, 591–596.

Bellmann, H., Kothe, W., Fleischmann, H., et al. (1972). Therapy of Bechterew's disease using large doses of intravenously administered hyalase "Dessau. *Dtsch Gesundheitsw* **27**, 2391–2395.

Bertelli, G. (1995). Prevention and management of extravasation of cytotoxic drugs. *Drug Saf* **12**, 245–255.

Bolenz, C., Cao, Y., Arancibia, M. F., et al. (2006). Intravesical mitomycin C for superficial transitional cell carcinoma. *Expert Rev Anticancer Ther* **6**, 1273–1282.

Breau, J. L., Boaziz, C., Morère, J. F., Sadoun, D., and Israel, L. (1993). Chemotherapy with cisplatin, adriamycin, bleomycin and mitomycin C, combined with systemic and intrapleural hyaluronidase in stage II and III mesothelioma. *Eur Respir Rev* **3**, 223–225.

Breu, W. (1952). Hyaluronidase. *Wien Klin Wschr* **64**, 435–437.

Cameron, E., Pauling, L., and Leibovitz, B. (1979). Ascorbic acid and cancer, a review. *Cancer Res* **39**, 663–681.

Chain, E. and Duthie, E. S. (1939). A mucolytic enzyme in testis extracts. *Nature* **144**, 977–978.

Chain, E. and Duthie, E. S. (1940). Identity of hyaluronidase and spreading factor. *Br J Exp Pathol* 21, 324–338.

Civalleri, D., Esposito, M., De Cian, F., et al. (1996). Effect of hyaluronidase on the pharmacokinetics of free and total platinum species after intra-arterial cisplatin in refractory patients with colorectal liver metastases. *Clin Drug Invest* 122, 94–104.

Duran-Reynals, F. (1929). The effect of extracts of certain organs from normal and immunized animals on the infecting power of the vaccine virus. *J Exp Med* 50, 327–339.

Elder, J. B., Raftary, A. C., and Cope, V. (1980). Intraarterial hyaluronidase in severe peripheral arterial disease Lancet., 648–649.

Eikenes, L., Tari, M., Tufto, I., et al. (2005). Hyaluronidase induces a transcapillary pressure gradient and improves the distribution and uptake of liposomal doxorubicin (Caelyx®) in human osteosarcoma xenografts. *Br J Cancer* 93, 81–88.

Farr, C., Menzel, J., Seeberger, J., and Schweigle, B. (1997). Clinical pharmacology and possible applications of hyaluronidase with reference to Hylase "Dessau". *Wien Med Wochenschr* 147, 347–355.

Frost, G. I. (2007). Recombinant human hyaluronidase (rHuPH20): an enabling platform for subcutaneous drug and fluid administration. *Expert Opin Drug Deliv* 4, 427–440.

Gately, C. L., Muul, L. M., Greenwood, M. A., et al. (1984). In vitro studies on the cell-mediated immune response to human brain tumors. II. Leukocyte induced coats of glycosaminoglycans increase the resistance of glioma cells to cellular immune attack. *J Immunol* 133, 3387–3395.

Girish, K. S. and Kemparaju, K. (2007). The magic glue hyaluronan and its eraser hyaluronidase, a biological overview. *Life Sci* 80, 1921–1943.

Haaselsberger, K., Radner, H., and Pendl, G. (1996). Na2B12H11SH (BSH) in combination with systemic hyaluronidase: a promising concept for boron neutron capture therapy for glioblastoma. *Neurosurgery* 39, 321–326.

Herp, A., DeFilippi, J., and Fabianek, J. (1968). The effect of serum hyaluronidase on acidic polysaccharides and its activity in cancer. *Biochim Biophys Acta* 158, 150–153.

Hoebarth, K., Maier, U., and Marberger, M. (1992). Topical chemoprophylaxis of superficial bladder cancer with mitomycin C and adjuvant hyaluronidase. *Eur Urol* 21, 206–210.

Israel, L. (1992). Hyaluronidase in addition to chemotherapy in the management of mesothelioma. *Proc Annu Meet Am Soc Clin Oncol* 11, A1209.

Jain, R. K. (2005). Normalization of tumor vasculature: an emerging concept in anti-angiogenic therapy. *Science* 307, 58–62.

Jones, L. M., Gardner, M. J., Catterall, J. N., and Turner, G. A. (1995). Hyaluronic acid secreted by mesothelial cells: a natural barrier to ovarian cancer cell adhesion. *Clin Exp Metastasis* 13, 373–380.

Kiriluk, L. B., Kremen, A. J., and Glick, D. (1950). Mucolytic enzyme systems XII. Hyaluronidase in human and animal tumors, and further studies on the serum hyaluronidase inhibitor in human cancer. *J Natl Cancer Inst* 10, 993–1000.

Klocker, J., Sabitzer, H., Raunik, W., Wieser, S., and Schumer, J. (1995). Combined application of cisplatin, vindesine, hyaluronidase and radiation for treatment of advanced squamous cell carcinoma of the head and neck. *Am J Clin Oncol* 185, 425–428.

Kluza, H. P. and Moritz, A. J. (1985). Hyaluronidase. Neue Aspekte der klinischen Anwendung. *Muench Med Wochenschr* 21, 561–562.

Kohno, N., Ohnuma, T., and Truog, P. (1994). Effects of hyaluronidase on doxorubicin penetration into squamous carcinoma multicellular tumor spheroids and its cell lethality. *J Cancer Res Clin Oncol* 120, 293–297.

Lokeshwar, V. B., Lokeshwar, B. L., Pham, H. T., and Block, N. L. (1996). Association of elevated levels of hyaluronidase, a matrix-degrading enzyme, with prostate cancer progression. *Cancer Res* 56, 651–657.

VIII. HYALURONAN-RELATED BIOMATERIAL

Maier, U. and Baumgartner, G. (1988). Mitomycin C plasma levels after intravesical instillation with or without hyaluronidase. *J Urol* **195**, 845.

Maier, U. and Baumgartner, G. (1989). Metaphylactic effect of mitomycin C with and without hyaluronidase after transurethral resection of bladder cancer, randomized trial. *Urology* **141**, 529–530.

Muckenschnabel, I., Bernhardt, G., Spruss, T., and Buschauer, A. (1996). Hyaluronidase pretreatment produces selective melphalan enrichment in malignant melanoma implanted in nude mice. *Cancer Chemother Pharmacol* **38**, 88–94.

Muckenschnabel, I., Bernhardt, G., Spruss, T., and Buschauer, A. (1998). Pharmacokinetics and tissue distribution of bovine testicular hyaluronidase and vinblastine in mice, an attempt to optimize the mode of adjuvant hyaluronidase administration in cancer chemotherapy. *Cancer Lett* **131**, 71–84.

Oettl, M., Hoechstetter, J., Asen, I., Bernhardt, G., and Buschauer, A. (2003). Comparative characterization of bovine testicular hyaluronidase and a hyaluronate lyase from *Streptococcus agalactiae* in pharmaceutical preparations. *Eur J Pharm Sci* **18**, 267–277.

Possinger, K. (1988). Additive Hyaluronidase-Behandlung bei Patienten mit metastasiertem kolorektalen Karzinom. *Wr Klin Wochenschr 100/24*, 13–14.

Rosenthal, L. (1952). Hyaluronidase as a factor in tumor metastasis. *Zentralbl Chir* **77**, 1885–1886.

Smith, K. J., Skelton, H. G., Turiansky, G., and Wagner, K. F. (1997). Hyaluronidase enhances the therapeutic effect of vinblastine in intralesional treatment of Kaposi's sarcoma. *J Am Acad Dermatol* **36**, 239–342.

Spruss, T., Bernhardt, G., Schoenenberger, H., and Schiess, W. (1995). Hyaluronidase significantly enhances the efficacy of regional vinblastine chemotherapy of malignant melanoma. *J Cancer Res Clin Oncol* **121**, 193–202.

St. Croix, B., Rak, J. W., Kapitain, S., et al. (1996). Reversal by hyaluronidase of adhesion-dependent multicellular drug resistance in mammary carcinoma cells. *J Natl Cancer Inst* **88**, 1285–1296.

Stern, R. (2005). Hyaluronan metabolism: a major paradox in cancer biology. *Pathol Biol (Paris)* **53**, 372–382.

Stern, R. and Jedrzejas, M. J. (2006). Hyaluronidases: their genomics, structures, and mechanisms of action. *Chem Rev* **106**, 818–839.

Stettbacher, A. (1953). Clinical studies on the use of hyaluronidase (Permease Cilag). *Helv Med Acta* **20**, 206–210.

Storch, H., Dellas, T., and Bellmann, H. (1978). Animal experimental studies on immunogenicity, humoral response and danger of anaphylaxis in parenteral administration of hyaluronidase. *Z Exp Chir* **11**, 128–133.

Szépfalusi, Z., Nentwich, I., Dobner, M., Pillwein, K., and Urbanek, R. (1997). IgE-mediated allergic reaction to hyaluronidase in paediatric oncological patients. *Eur J Pediatr* **156**, 199–203.

Thiruvengadam, R. and Moran, E. (1995). Hyaluronidase in chemoresistant malignancies. *Proc Annu Meet Am Soc Clin Oncol* **14**, A1385.

Toole, B. P. (2004). Hyaluronan, from extracellular glue to pericellular cue. *Nat Rev Cancer* **4**, 528–539.

Tufto, I., Hansen, R., Byberg, D., et al. (2007). The effect of collagenase and hyaluronidase on transient perfusion in human osteosarcoma xenografts grown orthotopically and in dorsal skinfold chambers. *Anticancer Res* **27**, 1475–1481.

Vick, N. A., Shondekor, J. D., and Bigner, D. D. (1997). Chemotherapy of brain tumors. The blood brain barrier is not a factor. *Arch Neurol* **34**, 523–526.

Zahalka, M. A., Okon, E., Gosslar, U., Holzmann, B., and Naor, D. (1995). Lymph node (but not spleen) invasion by murine lymphoma is both CD44- and hyaluronate-dependent. *J Immunol* **154**, 5345–5355.

CHAPTER

20

Exploiting the Hyaluronan–CD44 Interaction for Cancer Therapy

Virginia M. Platt and Francis C. Szoka, Jr.

EXPLOITING THE HYALURONAN–CD44 INTERACTION FOR CANCER THERAPY

Hyaluronan (HA) mediates the connection between a cell and its local environment by acting as a structural component of the extracellular matrix and by signaling through cell surface receptors. CD44, the main HA receptor, regulates cell motility, growth, and survival during tissue growth and maintenance (McKee et al., 1996). Malfunctions in tissue development caused by aberrant interactions between CD44 and HA (CD44–HA) contribute to cancer growth and progression (Naor et al., 2002). In many types of cancer, CD44 expression differs substantially from the expression normally seen in healthy tissue; CD44 expression is up-regulated or CD44 is alternately spliced to produce non-native variants (Naor et al., 2002). Signaling pathways downstream of CD44 activation, when deregulated, lead to tumor growth, progression, and metastasis (Bourguignon et al., 2000; 2003; 2007). The composition of the extracellular matrix is also directly modulated during progression of some cancers; HA levels within the tumor extracellular matrix change (Toole, 2004). Because of the intimate relationship between the CD44–HA system and cell survival and growth, it is an increasingly investigated area for applications to anti-cancer chemotherapeutics.

This chapter will review the potential ways in which the CD44–HA interaction can be exploited in cancer therapy. HA is a high-molecular weight saccharide signaling molecule; it can be used as a drug (Alaniz et al., 2006; Hosono et al., 2007; Gilg et al., 2008), a drug carrier (Luo and Prestwich, 1999; Luo et al., 2000; 2002), or a CD44-targeting ligand (Eliaz and Szoka, 2001; Eliaz et al., 2004a; 2004b; Peer and Margalit, 2004a; 2004b).

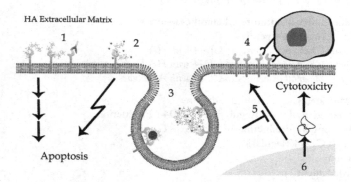

FIGURE 20.1   Schematic of CD44 Mediated Anti-Cancer Strategies (1) HA or anti-CD44 antibody induced matrix detachment which may lead to anti-growth signaling cascades, (2) HA induced cell sensitization to chemotherapeutic agents, (3) CD44 mediated HA drug carrier endocytosis, (4) CD44 targeted immune response (5) Genetic downregulation of CD44 production and (6) Production of cytotoxic proteins via CD44 splice-linked gene regulation within the nucleus.

This chapter will also discuss methods that reduce CD44 surface presentation (Subramaniam et al., 2007), interfere with the interaction of CD44 and the extracellular matrix (Peterson et al., 2000; Ahrens et al., 2001; Song et al., 2004), target CD44 (Tijink et al., 2006; Rupp et al., 2007), or activate CD44-mediated cell destruction (Harada et al., 2001; 2002; 2004; Dall et al., 2005).

The six distinct approaches to anti-cancer therapy which exploit the CD44–HA relationship are illustrated in Fig. 20.1: (1) Interference with CD44–HA interactions causing physical matrix disruption or modulation of the CD44 anti-growth signaling pathway; (2) HA-induced sensitization to chemotherapeutics; (3) Targeting of drugs to CD44; (4) Generation of an immune response to CD44; (5) Genetic down regulation of CD44 production; and (6) Expression of cytotoxic proteins by CD44 regulatory pathways.

INTERFERENCE WITH CD44-HA INTERACTIONS

Interference with CD44–HA Interactions by Soluble CD44

CD44, interacting with extracellular matrix HA, stabilizes and supports CD44 expressing cells. Disrupting association of tumor cells with the extracellular matrix can reduce the tumor burden by causing anti-growth signaling or by physically interfering with tumor expansion into surrounding tissue. Soluble HA binding proteins, anti-CD44 antibodies and exogenously added HA have the potential to reduce cancer progression by influencing CD44–matrix interactions through these mechanisms. A summary of studies which explore the effect of CD44–HA interactions on cancer progression is given in Table 20.1.

Soluble HA binding receptors, either expressed via stable transfection or added as immunoglobulin fusion constructs, efficiently decreased tumor growth (Sy et al., 1992; Bartolazzi et al., 1994; Peterson et al., 2000; Ahrens et al., 2001). The therapeutic effects of the soluble binding proteins were correlated with HA binding. Mutant proteins which could not bind HA had no anti-tumor benefits, therefore the disruption of CD44–HA interactions is implicated in the mechanism by which these treatments act to decrease tumor growth (Peterson et al., 2000; Ahrens et al., 2001). In murine models of lymphoma (Sy et al., 1992) and melanoma (Bartolazzi et al., 1994), CD44–immunoglobin fusion proteins inhibited tumor invasion and decreased cancer progression. Cancerous cells transfected to overexpress soluble CD44 exhibited significantly decreased HA binding *in vitro* and reduced tumor progression *in vivo* (Peterson et al., 2000; Ahrens et al., 2001). Mammary carcinoma cells that stably overexpressed soluble CD44 formed a much smaller primary tumor which failed to attach to the

TABLE 20.1A   *In vitro* Interference with CD44–HA Interactions

Acting agent	Cell lines	Effect	Controls	Reference
Soluble CD44				
Soluble CD44	TA3/St (transfected murine mammary carcinoma)	Inhibited HA binding and internalization	Untransfected cells	Yu, 1997
Soluble CD44	TA3/St (transfected murine mammary carcinoma)	Inhibited anchorage-independent growth	Non-HA binding soluble CD44	Peterson, 2000
Soluble CD44	MV3 (human melanoma)	Inhibited HA binding, inhibited proliferation	Non-HA binding soluble CD44	Ahrens, 2001
Antibodies				
Mouse anti-CD44 monoclonal antibody (GKW.A3)	SMMU-2 (human melanoma)	Inhibited HA binding	HMW-HA	Guo, 1994
Mouse anti-CD44H monoclonal antibody (2C5)	IPNT-H (human pilocytic astrocytoma) IPSB-18 (human anaplastic astrocytoma) GO-G-CCM (human anaplastic astrocytoma) GO-G-UVW (human anaplastic astrocytoma)	Reduced matrigel invasion	Control antibodies	Merzak, 1994; Koochekpour, 1995
Mouse anti-CD44 monoclonal antibody (F10-44-2)	G-CCM (human anaplastic astrocytoma)	Reduced HA-matrigel invasion	HA-free matrigel	Radotra, 1997
Rat anti-CD44 monoclonal antibody	G-26 (mouse oligodendroglioma)	Reduced matrigel invasion	Untreated cells	Wiranowska, 1998

Mouse anti-CD44 monoclonal antibody	9L (rat gliosarcoma)	Inhibited ECM binding	Isotype immunoglobulins	Gunia, 1999
Mouse anti-CD44 antibodies (H90 and A3D8)	AML (human acute myeloid leukemia)	Induced differentiation	Control antibodies, HMW-HA, retinoic acid	Charrad, 1999
Mouse anti-CD44 monoclonal antibody	C6 (rat glioblastoma multiforme)	Inhibited ECM binding	Control antibodies	Breyer, 2000
Mouse anti-CD44 antibody (A3D8)	HL60 (human myeloblast leukemia) NB-4 (human promyelocytic leukemia)	Inhibited proliferation, inhibited drug-induced apoptosis	Control antibodies	Allouche, 2000
Mouse anti-CD44 antibodies (H90 and A3D8)	KG1a (human myeloblast leukemia) HL60 (human myeloblast leukemia) NB4 (human promyelocytic leukemia) THP-1 (human monocytic leukemia)	Inhibited proliferation, induced differentiation	Control antibodies, retinoic acid	Charrad, 2002
Anti-CD44 monoclonal antibody (HI44a)	AML (human acute myeloid leukemia)	Induced differentiation, induced apoptosis	Untreated cells	Song, 2004

(continued)

Table 20.1A (continued)

Acting agent	Cell lines	Effect	Controls	Reference
Free hyaluronan				
HA-O(3–12 disaccharides)	B16F10 (murine melanoma)	Inhibited proliferation	Chondroitin sulfate, HMW-HA	Zeng, 1998
HA-O(3–10 disaccharides)	TA3/St (murine mammary carcinoma)	Inhibited anchorage-independent growth, induced apoptosis	Chitin, chondroitin sulfate, HMW-HA, anti-CD44 antibodies	Ghatak, 2002
	HCT-116 (human colon carcinoma)			
	LX-1 (human lung carcinoma)			
HA-O(3–10 disaccharides)	C6 (rat glioblastoma multiforme)	Inhibited invasion, inhibited anchorage-independent growth	Chitin, glucuronic acid, N-acetyl glucosamine, HMW-HA, anti-CD44 antibodies	Ward, 2003
	A172 (human glioblastoma)			
	U87 (human glioblastoma)			
HA-O(2–7 disaccharides)	LBLa (invasive murine T-lymphoma)	Increased apoptosis	HMW-HA, LMW-HA, anti-CD44 antibodies	Alaniz, 2006
	LBLc (murine T-lymphoma)			
HA-O(2–6 disaccharides)	MG-63 (human osteoblast osteosarcoma)	Inhibited growth, increased apoptosis, disrupted matrix attachment, inhibited matrigel invasion	HMW-HA, HA4, anti-CD44 antibodies	Hosono, 2007
	LM-8 (murine osteosarcoma)			

TABLE 20.1B *In vivo* Interference with CD44–HA Interactions

Acting agent	Tumor model	Injection method	Effect	Reference
Soluble CD44				
Soluble CD44–Rg fusion protein	CD44N2.1 (transfected human Burkitt lymphoma) in BALB/c mice via intravenous injection	Co-injection with tumor implantation	Reduced tumor formation	Sy, 1992
Soluble CD44–Rg fusion protein	B16F10 (murine melanoma) in SCID mice via subcutaneous injection at the pump site	Subcutaneous pump	Reduced tumor formation	Bartolazzi, 1994
Soluble CD44	TA3/St (transfected murine mammary carcinoma) in A/jax mice via tail vein injection		Prevented tumor formation	Yu, 1997
Soluble CD44	TA3/St (transfected murine mammary carcinoma) in A/jax mice via intraperitoneal injection		Prevented peritoneal wall invasion	Peterson, 2000
Soluble CD44	1F6 (transfected human melanoma) in MF1 nude mice via subcutaneous flank injection		Prevented primary tumor growth	Ahrens, 2001

(continued)

VIII. HYALURONAN-RELATED BIOMATERIAL

Table 20.1B (*continued*)

Acting agent	Tumor model	Injection method	Effect	Reference
Antibodies				
Mouse anti-CD44 monoclonal antibody (1.1ASML)	AS-14 (CD44v transfected rat metastatic pancreatic adenocarcinoma) in BDX rats via subcutaneous foot-pad injection	Intravenous with tumor implantation	Prevented distant metastasis, increased survival time	Seiter, 1993
Mouse anti-CD44 monoclonal antibody (GKW.A3)	SMMU-2 (human melanoma) in SCID mice via subcutaneous injection	Co-injection with tumor implantation	Prevented primary tumor formation	Guo, 1994
		Intravenous	Prevented primary tumor formation if treated with or shortly after tumor cell injection	
		Intravenous	Prevented metastasis formation if given 7 days after tumor cell injection	
Rat anti-CD44 monoclonal antibody (IIM 7.8.1)	LB (T-cell lymphoma) in BALB/c mice via subcutaneous flank injection	Subcutaneous	Prevented lymph node metastasis	Zahalka, 1995
Mouse anti-CD44 monoclonal antibody	9L (rat gliosarcoma) in athymic nude rats via intracerebral injection	Intracerebral	Reduce tumor size	Gunia, 1999
Mouse anti-CD44 monoclonal antibody	C6 (rat glioblastoma multiforme) in athymic nude rats via intracerebral injection	Intracerebral	Reduce tumor size	Breyer, 2000

Bispecific anti-CD44 × anti-Id antibody	38C-13 (murine B-cell lymphoma) in C3H/eB mice via subcutaneous injection.	Intraperitoneal	Reduced lymph node, bone marrow and spleen metastasis, increased survival time	Avin, 2004
Mouse anti-CD44 antibody (H90)	AML (human acute myeloid leukemia) in sublethally irradiated NOD-SCID mice via intravenous or intrafemoral injection.	Intraperitoneal	Reduced leukemic repopulation	Jin, 2006
Mouse anti-CD44 antibody (IM 7)	CML (murine bone marrow transduced chronic myeloid leukemia) in lethally irradiated B6X129 F2 mice via intravenous injection	Co-injection with tumor implantation	Reduced induction of CML-like leukemia	Krause, 2006

(continued)

VIII. HYALURONAN-RELATED BIOMATERIAL

Table 20.1B (*continued*)

Acting agent	Tumor model	Injection method	Effect	Reference
Free hyaluronan				
HA-O(3–12 disaccharides)	B16F10 (murine melanoma) in BALB/c mice via subcutaneous retroscapular injection	Subcutaneous pump	Reduced primary tumor growth	Zeng, 1998
HA-O(3–10 disaccharides)	LX-1 (human lung carcinoma) in BALB/c mice via subcutaneous dorsal injection	Subcutaneous pump	Reduced primary tumor growth	Ghatak, 2002
	Tas3/St (murine mammary carcinoma) in A/jax mice via subcutaneous dorsal injection			
HA-O(2–6 disaccharides)	LM-8 (murine osteosarcoma) in C3H/He mice via subcutaneous dorsal flank injection	Daily intratumoral	Prevented distant metastasis, slightly reduced primary tumor growth	Hosono, 2007
HA-O(3–10 disaccharides)	C6 (rat glioblastoma multiforme) in Sprague–Dawley rats via intraspinal injection	Intratumoral	Inhibited primary tumor growth, increased apoptosis	Gilg, 2008
HMW-HA	Hs578T (human breast carcinoma) in CD1 mice via orthotopic mammary fat pad injection	Intratumoral	Regressed primary tumor growth in [1/3] of animals	Herrera-Gayol, 2002

VIII. HYALURONAN-RELATED BIOMATERIAL

peritoneal wall. When implanted in mice, these transfected cells caused fewer fatalities than either untransfected cells or cells that expressed mutant soluble CD44 (Peterson et al., 2000). This decreased primary tumor growth was linked, at least partially, to increased apoptosis (Yu et al, 1997). Primary malignant melanoma tumors which were comprised of cells transfected to stably overexpress soluble CD44 grew more slowly than tumors comprised of untransfected cells or cells transfected to express the non-HA binding mutant soluble CD44 (Ahrens et al., 2001).

Interference with CD44–HA Interactions by Antibodies

Metastasis formation may be treated with therapies other than those used to decrease primary tumor growth. In several studies involving the treatment of tumors with anti-CD44 antibodies, metastatic progression could be efficiently blocked, but such treatment had no effect on the primary tumor growth. Substantial decreases in metastasis were seen in a murine model of human melanoma (Guo et al., 1994), a rat pancreatic adenocarcinoma model (Seiter et al., 1993), and a murine model of T-cell lymphoma (Zahalka et al., 1995).

The timing of treatment was critical as antibodies were most efficacious when given either with tumor implantation or after tumor implantation but before metastasis. In a spontaneous, highly metastatic rat pancreatic adenocarcinoma model, treatment with anti-CD44 variant antibodies caused only transient relief if given after metastasis formation. When treatment was given before metastasis occurred, substantial decreases in colonization were observed (Seiter et al., 1993). Following intravenous treatment of a human melanoma mouse model there was no decrease in primary tumor growth, but a substantial decrease in metastasis. Although primary tumor growth could be inhibited if antibody treatment began shortly after cell implantation, primary tumors could not be treated once they were formed. Intravenous antibody treatment was only effective at preventing metastasis at this point (Guo et al., 1994). In a murine T-cell lymphoma model, lymph node invasion was prevented when treated by subcutaneous anti-CD44 antibodies soon after tumor cell implantation (Zahalka et al., 1995).

An anti-metastatic effect was also observed with a bispecific antibody which recognized both CD44 and a cell surface marker of the malignant B lymphocyte. For therapeutic efficacy, this antibody required both antigens to be present on the surface of the target cell. Adhesion was largely achieved by the high affinity of the antibody for the malignant B lymphocyte marker, so less binding to non-malignant lymphocytes occurred. Once associated with malignant cells the anti-CD44 antibody portion could cause anti-CD44 mediated metastasis blockage. The low affinity of the anti-CD44 portion prevented off-target toxicities to CD44 expressing cells elsewhere in the body. *In vivo* treatment using the bispecific antibody resulted in significantly decreased lymphoma cell dissemination to the lymph nodes, spleen, and bone marrow (Avin et al., 2004).

In vitro studies of several human glioma cell lines suggested that anti-CD44 antibodies prevented invasion by significantly decreasing adhesion (Merzak et al., 1994; Koochekpour et al., 1995; Radotra and McCormick, 1997; Wiranowska et al., 1998). *In vivo*, tumor size could be decreased 80% or more in rats with gliosarcoma and glioblastoma multiforme brain tumors by treating the tumors with anti-CD44 antibodies. *In vitro* studies of both of these cell types showed these antibodies caused CD44-specific, dose-dependent cell detachment (Gunia et al., 1999; Breyer et al., 2000) without significant cytotoxicity (Breyer et al., 2000). This suggested the antibodies were functional *in vivo* partly due to their ability to physically limit matrix association.

Other anti-CD44 antibodies were found to be effective largely due to their ability to induce anti-growth signaling pathways. Anti-CD44 antibodies promoted differentiation and apoptosis in human acute myeloid leukemia (AML) (Charrad et al., 1999; Song et al., 2004). Two anti-CD44 antibodies induced cell differentiation in a manner similar to that of high molecular weight HA (HMW-HA) and oligosaccharide HA (HA-O). However, a third antibody that recognized a different domain of CD44 did not induce differentiation, suggesting that the mechanism required a ligand-like association with CD44 (Charrad et al., 1999). Decreased proliferation and differentiation was seen, to varying degrees, in several human myeloid cell line models of AML subtypes. Very immature AML subtype cells could be induced to differentiate by a combination of anti-CD44 antibodies and retinoic acid, a drug used for treatment of only one subtype of AML (Charrad et al., 2002). One of the antibodies that induced differentiation also interfered with drug-induced apoptosis for several common chemotherapeutics in two myeloid cell lines. The antibody that was not effective at inducing differentiation caused no reduction of drug effectiveness, so the mechanism of differentiation may be antagonistic to the pathway sensitized for drug induced apoptosis (Allouche et al., 2000).

Intraperitoneal administration of antibodies in a human AML transplanted mouse model increased differentiation, caused loss of homing capacity of leukemic stem cells and reduced leukemic repopulation, possibly due to decreased AML engraftment (Jin et al., 2006). Anti-CD44 antibodies also decreased the presentation of chronic myeloid leukemia (CML) in lethally irradiated mice injected with CML-generating bone marrow. Although blockage of engraftment did not reach significance, antibody pretreatment of bone marrow resulted in a moderate increase in survival time. This suggested the antibody treatment may be successful for multiple types of CD44 presenting leukemia (Krause et al., 2006).

Interference with CD44–HA Interactions by HA

The nature of HA as a matrix component and a signaling molecule allows it to act as a pro-apoptotic agent (Ghatak et al., 2002; Alaniz et al., 2006; Hosono et al., 2007; Gilg et al., 2008). In lung, mammary and colon carcinoma cell lines, anti-growth effects were seen with HA-O but not with HMW-HA, chitin, or chondroitin sulfate (chitin and chondroitin are saccharides with similar structure to HMW-HA) (Ghatak et al., 2002). T-lymphoma cells underwent apoptosis when exposed to HA-O but not when exposed to HMW-HA (Alaniz et al., 2006). Anti-tumor effects were traced, in part, to the decrease in anti-apoptotic signaling (Ghatak et al., 2002; Alaniz et al., 2006; Gilg et al., 2008).

Both HMW-HA and HA-O are effective as *in vivo* therapeutic agents for the treatment of cancer. Daily intratumoral injection of HMW-HA in a murine xenograft breast cancer model resulted in complete tumor regression in 25% of animals studied (Herrera-Gayol and Jothy, 2002). HA-O, injected subcutaneously via a slow infusion pump, inhibited tumor growth *in vivo* in a murine melanoma model (Zeng et al., 1998), a murine osteosarcoma model, and a murine lung carcinoma model (Ghatak et al., 2002). Ghatak and colleagues delivered HA oligomers via an osmotic pump, prior to implantation, in subcutaneously injected lung and mammary carcinoma murine tumor models. In both models, treatment with HA-O reduced the weight of the resultant primary tumor compared to untreated controls (Ghatak et al., 2002). Intratumoral injection of octasaccharide HA, but not HMW-HA or tetrasaccharide HA, inhibited primary tumor growth and significantly decreased lung metastasis in a murine osteosarcoma model (Hosono et al., 2007). In a rat spinal cord glioma model, HA-O injected near the tumor site suppressed growth by down regulating Akt-related anti-apoptotic signaling (Gilg et al., 2008).

Exogenously added HA may also decrease cell growth by limiting matrix interactions independent of cell growth and survival (Ward et al., 2003). Addition of exogenous HA-O, but not chitin oligosaccharides, prevented glioma cell invasion through a reconstituted basement membrane which contained HMW-HA. Invasion was also inhibited by both anti-CD44 antibodies and overexpressed soluble CD44, suggesting the anti-invasive effect of HA-O was largely mediated through interruption of matrix HA–CD44 interactions (Ward et al., 2003). In this case, HA-O did not induce apoptosis or substantially influence cell migration in the absence of HMW-HA, potentially because this type of motility is CD44 independent (Ward et al., 2003).

The viability of HA as a stand alone anti-cancer treatment is unclear, as results from a similar study showed HMW-HA increased metastatic potential when co-delivered during cell implantation in a colorectal mouse model (Tan et al., 2001). Furthermore, the consequences of introducing free HA of any size into the body at concentrations high enough to exhibit anti-tumor effects may result in side effects. These side effects may be based on HA's interaction with endogenous CD44 on other cell types, such as leukocytes within the epithelium of the skin (Mackay et al., 1994).

In summary, multiple types of therapeutics can disrupt the CD44–HA interaction, resulting in decreased tumor burden by physically shielding the tumors from the surrounding milieu, which prevents their spread, or actively engaging CD44 to induce anti-growth and apoptotic signaling.

HA-INDUCED SENSITIZATION TO CHEMOTHERAPEUTICS

The signaling pathways induced by HA–CD44 association may cause cells to become more sensitive to co-delivered chemotherapeutic agents (Misra et al., 2003; Cordo Russo et al., 2008). In drug resistant human mammary carcinoma cells, HA-O suppressed two kinase pathways: MAP kinase and phosphoinositide 3-kinase. Suppression of these pathways sensitized the cells to several commonly used chemotherapeutics: doxorubicin, taxol, vincristin, methotrexate, and BCNU (Misra et al., 2003). Efflux of doxorubicin from drug-resistant lymphoma cells was inhibited by treatment with HA-O (Cordo Russo et al., 2008). HA modulation of the signaling may lead to potentiated anti-cancer therapies, as in the case of HMW-HA co-administered with doxorubicin, but this augmentation has yet to be clinically established (Rosenthal et al., 2005).

Conflicting evidence suggests the administration of free HA may also be antagonistic in certain HA-drug combinations. In T-cells, HA oligomers of approximately 18 disaccharides up regulated P-glycoprotein, resulting in increased dexamethasone efflux (Tsujimura et al., 2006). Head and neck squamous carcinoma cells exposed to HMW-HA were less sensitive to three commonly used anti-cancer therapeutics: cisplatin, methotrexate and Adriamycin, in a manner that could be blocked by anti-CD44 antibodies (Wang and Bourguignon, 2006a; Wang and Bourguignon, 2006b). This suggests sensitization may be specific to the drug, HA length or cell type.

TARGETING CHEMOTHERAPEUTICS

Targeting Chemotherapeutics to CD44 with HA

Deregulation of CD44 maintenance in cancers often leads to its over-expression on the surface of tumor cells (Naor et al., 2002). Targeting cytotoxic drugs to CD44 can localize therapies to areas where CD44 is highly expressed. This targeting is advantageous for successful anti-tumor activity and can limit off-target effects. The area of HA targeting therapies has recently been reviewed (Platt and Szoka, 2008; Yadav et al., 2008).

CD44-associated HA can be endocytosed (Luo and Prestwich, 1999; Luo et al., 2000) carrying conjugated cargo, such as drugs or drug carriers, into the cell. Cellular uptake increases therapeutic efficiency of HA conjugated drugs (Coradini et al., 2004a, b; Eliaz et al., 2004a). The polymeric nature of HA yields multiple functional groups for chemical conjugation. Attaching drugs to HA benefits anti-cancer therapies by increasing drug cytotoxicity to CD44 overexpressing cancer cells while decreasing toxicity to healthy cells.

Carboranes, used in boron neutron capture therapy, must be localized to the tumor at high enough concentrations to deliver a therapeutic dose of thermal neutrons. Carboranes conjugated to the side-chains of HMW-HA accumulated to a high concentration within human bladder carcinoma, colorectal adenocarcinoma, and ovarian adenocarcinoma cell lines due to CD44 specific uptake mechanism (Di Meo et al., 2007; 2008). So far, this approach has yet to be evaluated in animal models of cancer.

A number of cytotoxic anti-cancer drugs including paclitaxel (Luo and Prestwich, 1999; Luo et al., 2000; Auzenne et al., 2007), doxorubicin (Luo et al., 2002), sodium butyrate (Coradini et al., 1999; 2004a,b), mitomycin c (Akima et al., 1996), and epirubicin (Akima et al., 1996) had increased CD44 specific internalization and cell cytotoxicity when conjugated to HA.

HA–paclitaxel increased cytotoxic delivery to CD44 overexpressing cancer cell lines including breast, colon, and ovarian carcinoma (Luo and Prestwich, 1999; Luo et al., 2000). HA-conjugated paclitaxel was taken up in a dose-dependent manner that could be blocked by competition with anti-CD44 antibodies or HMW-HA but not chondroitin sulfate (Luo et al., 2000), a sulfated polymeric sugar composed of the same two saccharides as HA.

HA–drug conjugates treat cancer *in vivo*. Studies performed by Akima and coworkers showed potent anti-metastatic effects of HA–mitomycin c injected subcutaneously in a murine model of Lewis lung adenocarcinoma (Akima et al., 1996). HA–paclitaxel conjugates injected locally to the tumor increased the mean survival time in a human ovarian carcinoma xenograft mouse model (Auzenne et al., 2007). Sodium butyrate–HA conjugates injected intratumorally or subcutaneously reduced primary tumor growth and lung metastasis in a murine Lewis lung carcinoma model (Coradini et al., 2004a, b) and a murine melanoma model (Coradini et al., 2004b). However, HMW-HA conjugates injected intravenously exhibited a short half-life due to HA specific clearance mechanisms. Conjugates collected in the liver and spleen (Coradini et al., 2004b).

Targeting Chemotherapeutics to CD44 with HA Drug Carriers

Incorporating the targeting properties of HA into large carriers may improve CD44 expressing cell-specific uptake, prolong circulation time and limit drug access to many organs that would normally be damaged by exposure to chemotherapeutic agents. Liposomes, drug carriers comprised of a phospholipid bilayer and an aqueous core, coated with HMW-HA (Peer and Margalit, 2004a, b) or HA-O (Eliaz et al., 2004a, b) showed specific uptake into CD44 expressing cells. HA coated liposomes can be created in two ways: conjugating HMW-HA to the surface via multiple site attachment or conjugating HA-O to individual lipids via a single connection at HA's reducing end. Either of these liposomes improved *in vitro* uptake of doxorubicin (Eliaz and Szoka, 2001; Eliaz et al., 2004a, b; Peer and Margalit, 2004a) in a manner that was shown to increase the potency of doxorubicin (Eliaz et al., 2004a) and mitomycin c (Peer and Margalit, 2004b) in CD44 expressing cells.

HMW-HA liposomes released mitomycin c more slowly than uncoated liposomes (Peer and Margalit, 2000; 2004b). *In vivo* HMW-HA bearing liposomes delivering either doxorubicin or mitomycin c increased the mean survival time in murine models of intraperitoneal ascites tumors, solid colon or pancreatic carcinoma tumors and metastatic melanoma (Peer and Margalit, 2004a, b). Mice with foot-pad colon carcinoma tumors treated with three doses of HMW-HA liposomes, encapsulating mitomycin c or doxorubicin, experienced long-term survival (Peer and Margalit, 2004a, b). Mice with metastatic lung melanomas also experienced long term survival when treated with doxorubicin containing HMW-HA liposomes (Peer and Margalit, 2004a).

HA targeted liposomes showed specific uptake in CD44 expressing cells. These carriers also have *in vivo* characteristics that improved tumor toxicity, such as prolonged circulation time, increased tumor accumulation, and sustained release parameters (Peer and Margalit, 2000; 2004a,b). Combining the tumor targeting characteristics of HA with the pharmacokinetic benefits of drug carriers appeared to efficiently treat tumors.

Nanoparticle systems can be constructed from ionic polymers, such as HMW-HA. HA nanoparticles were developed to improve the delivery of cisplatin. The physical solubility limitation of cisplatin was overcome by creating an ionic HA–cisplatin complex (Jeong et al., 2007). HMW-HA was partially degraded to low molecular weight HA (LMW-HA). An ionic interaction between cisplatin and HA formed nanoparticles of approximately 100–200 nm. The nanoparticles could be degraded by hyaluronidase to release cisplatin. The group did not investigate CD44 mediated cell uptake or test the anti-tumor activity of these compounds (Jeong et al., 2007).

Targeting Chemotherapeutics to CD44 with Antibodies

Anti-CD44 antibodies conjugated to radiolabels and cytotoxic drugs successfully target to CD44 (Heider et al., 1996). Antibody conjugates are the most clinically advanced CD44 targeting therapy. Murine anti-CD44v6 monoclonal antibody conjugates were given to patients with squamous cell carcinoma of the head and neck. Patients receiving radiolabeled anti-CD44 antibodies experienced low anti-mouse immune responses. The conjugate accumulated preferentially in tumors and had a half-life of approximately 35 hours (Stroomer et al., 2000). Anti-CD44 antibodies conjugated to an anti-cancer drug were also well tolerated. No immune response was observed and the antibodies circulated for approximately 4 days (Sauter et al., 2007).

Humanized monoclonal antibodies against CD44v6 stabilized disease in refractory head and neck cancers when coupled to the radioisotope rhenium-186 (Verel et al., 2002; Borjesson et al., 2003; Colnot et al., 2003). Although the antibody–drug conjugate showed favorable disease stabilization in several patients with squamous cell carcinoma of the head and neck (Tijink et al., 2006), it showed unfavorable tumor to non-tumor ratios in patients with CD44 variant up regulated breast cancer (Koppe et al., 2004; Rupp et al., 2007). These trials were stopped due to fatal off-target toxicities within the skin. These effects show the importance of restricting the targeted carrier from healthy sites in the body that express moderate levels of CD44, such as the skin (Tijink et al., 2006).

CD44 MEDIATED CANCER IMMUNOTHERAPY

Although the interaction of CD44 and HA affords opportunities for delivery of therapeutic anti-cancer drugs, invoking an immune response against CD44 variants expressed on tumors could eliminate the need for systemic delivery of cytotoxic drugs and may result in a longer lasting effect.

Most cancer vaccines target tumors through stimulation of cytotoxic T-lymphocytes that recognize tumor specific antigens. One method to induce this type of T-cell response is to expose the T-cells to dendritic cells primed with antibody coated tumor cells (Pilon-Thomas et al., 2006). Dendritic cells exposed to CD44 expressing B16 melanoma cells coated with anti-CD44 antibodies were injected intravenously on three occasions to vaccinate mice. Sixty percent of mice immunized with these pulsed dendritic cells did not develop primary tumors after challenge with subcutaneous injection of CD44 positive metastatic melanoma cells (Pilon-Thomas et al., 2006). The number of metastatic lung tumors derived from intravenously injected metastatic melanoma cells also decreased upon treatment with dendritic cells. The dendritic cells were injected one day after the metastatic melanoma cells (Pilon-Thomas et al., 2006). Most of the control mice, that were either untreated or treated with unpulsed dendritic cells, formed some distant metastases in the gastrointestinal track or brain

(Pilon-Thomas et al., 2006). All mice injected with pulsed dendritic cells remained free of distant metastases.

Cytotoxic T-cells can be designed to recognize CD44 by genetically engineering a response into the T-cells (Hekele et al., 1996; Dall et al., 1997; 2005). This modification gives them the ability to lyse CD44 variant expressing cells. Murine cytotoxic T-cells transduced with a fusion gene, a single chain anti-CD44 variant antibody fused to a T-cell receptor signal transduction domain, gained the ability to recognize and lyse CD44v6-expressing pancreatic cells. Daily intravenous injection of 3×10^7 T-cells into mice bearing established subcutaneous rat pancreatic CD44v6-expressing carcinoma tumors significantly decreased tumor growth (Hekele et al., 1996). Daily intratumoral injection of 1×10^6 cytotoxic T-cells and the cytokine signaling molecule IL-2 resulted in consistent growth inhibition of established CD44v7/8 expressing cervical carcinoma tumors in a murine model (Dall et al., 2005). These cytotoxic T-cells were engineered to express a fusion construct of a single chain anti-CD44v7/8 antibody and a T-cell receptor signal transduction domain (Dall et al., 2005).

A bispecific antibody was developed that crosslinked a macrophage complement receptor and CD44. Macrophages are effector immune cells and, when associated with cancer cells, lyse them. An anti-CD44v6 anti-CR3 bispecific antibody increased macrophage–tumor cell interaction in a CD44v6 specific manner. However, when used *in vivo* this type of antibody was less effective than a control antibody which recognized only CD44v6. The bispecific antibody increased animal mortality when injected intraperitoneally in a rat adenocarcinoma model (Somasundaram et al., 1996). Macrophage mediated immune recruitment has yet to show robust anti-tumor activity.

Each of these immunotherapy methods successfully initiated a response from the host immune system directly against the tumor. Although these systems eliminate the need for cytotoxic drugs, their complexity and the fact that the treatment must be optimized on a patient to patient basis may limit their clinical usage.

GENE THERAPY BY DOWN REGULATION OF CD44 PRODUCTION

Genetic therapies which decrease the concentration of CD44 on the surface of the cell may also prevent metastatic spread of primary tumors (Harada et al., 2001; Subramaniam et al., 2007). CD44 expression was nearly completely down regulated by stably transfecting human colon adenocarcinoma cells with plasmids encoding antisense CD44 cDNA. These cells had a significantly reduced ability to bind HA *in vitro* (Harada et al., 2001). When mice were injected with transfected cells, tumors developed more slowly than when they were injected with non-transfected cells. Intrasplenically injected cells developed no metastatic liver colonies and fewer peritoneal nodules (Harada et al., 2001).

Stable transfection of human metastatic colorectal carcinoma cells, with plasmids for CD44 variant specific siRNA, down regulated CD44v6 but not CD44 (Reeder et al., 1998). This transfection did not affect cell proliferation or HA binding ability *in vitro*. When transfected colorectal cells were allowed to grow as subcutaneous flank tumors, no difference in primary tumor growth was observed compared to untransfected cells (Reeder et al., 1998). However, the metastatic ability of the siRNA producing cells was drastically reduced. When the cells were injected into the spleen of nude mice these mice developed no metastatic liver tumors (Reeder et al., 1998).

In another study of human metastatic colon carcinoma, cells stably transfected to express siRNA against the transmembrane domain of CD44 showed decreased adhesion and decreased cell survival *in vitro*. Mice injected subcutaneously with siRNA transfected cells formed smaller tumors than mice injected with control carcinoma cells that expressed normal levels of CD44 (Subramaniam et al., 2007).

Tumors established from untransfected colon carcinoma cells were treated with direct injection of a polyethylenimine complex containing siRNA CD44 plasmid DNA. Mice were given twelve bi-weekly intratumoral injections beginning one week after tumor cell implantation. These mice experienced decreased tumor growth and one mouse had complete tumor regression; the study was continued into the seventh week. Long-term survival was not investigated (Subramaniam et al., 2007).

In each of the studies, stable transfection of cells was required to produce substantially decreased metastatic spread. Repeated intratumoral injection of plasmid DNA encoding siRNA provided some suppression of tumor growth (Subramaniam et al., 2007), but clinical siRNA therapy is not currently available. If simple robust delivery systems can be developed, the use of siRNA for therapeutic anti-cancer therapies may become a clinical option (Szoka, 2008).

GENE THERAPY BY PRODUCTION OF CYTOTOXIC GENES THROUGH CD44 REGULATION

Genetic deregulation of the CD44 system may promote cancer progression. Aspects of this deregulation, specifically up regulation of CD44 variants, can be exploited in anti-cancer therapeutics without attempting to alter CD44 levels. This is possible because cytotoxic genes activated by a CD44 variant-specific mRNA splicing mechanism produce cytotoxic protein in cells that express the CD44 variant isotypes. Asman and colleagues cloned a prodrug converting enzyme (cytosine deaminase) under the control of the CD44v8/v9 splice domain (Asman et al., 1995). This recombinant CD44-cytosine deaminase fusion protein actively converted prodrug to drug (Asman et al., 1995). A similar approach placed alkaline phosphatase under the control of several different CD44 variant splice domains. Genes under the control of a v8/v9 domain were spliced in both CD44 and CD44 variant cells. Those genes controlled by a v9/v10 domain were only accurately spliced in CD44 variant expressing cells (Hayes et al., 2002; 2004). Prostate adenocarcinoma cells express CD44 variants 8, 9, and 10. When transfected with the v9/v10 alkaline phosphatase plasmid, the cells were sensitized to the prodrug etoposide phosphate (Hayes et al., 2002). Cells expressing CD44 but not the variant form were sensitive to etoposide, but not to the prodrug (Hayes et al., 2002).

These studies suggest that deregulation of the CD44 pathway at a genetic level can be usurped to express cytotoxic treatments at the tumor site. This may yield a treatment localized to areas of CD44 variant up-regulation. However, some variants are endogenously expressed and may give rise to off-target effects. If efficient and robust gene carriers can be developed, this toxicity may be avoided by delivering the genes specifically to tumor sites.

CONCLUSION

This chapter describes six approaches which use the CD44–HA interaction to treat cancer in animal models and clinical settings. There are numerous opportunities for highly specific cancer therapy using drug conjugates, carriers, or antibodies that target drugs to CD44 or act to disrupt the interaction between CD44 and HA. Many of these methods can decrease tumor burden in animal models but have yet to show significant clinical utility. Further advances in the use of HA as a carrier or targeting molecule, antibodies as active agents, and methods of CD44 inactivation will have to be made before these treatments can be seen as clinically viable options.

Drugs conjugated to anti-CD44 variant antibodies are currently the most clinically explored therapy. However, these antibodies have severe off-target toxicities. With the current advances in siRNA delivery, downregulation of CD44 overexpression presents an attractive future avenue. The most obvious therapy for near-future clinical relevance is delivery of currently available cancer drugs via targeted carriers. These carriers may utilize HA or antibody targeting while still retaining long circulation and tumor specific accumulation. Furthermore, particles may be specifically endocytosed by CD44 specific mechanisms, increasing drug potency. The many different approaches available for exploiting the CD44–HA interaction for anti-cancer treatments present opportunities for discoveries to develop and optimism that improved cancer therapies are on the horizon.

ACKNOWLEDGMENT

This work was partially supported by NIH RO1 CA 107268.

References

Ahrens, T., Sleeman, J. P., Schempp, C. M., et al. (2001). Soluble CD44 inhibits melanoma tumor growth by blocking cell surface CD44 binding to hyaluronic acid. *Oncogene* **20**, 3399–3408.

Akima, K., Ito, H., Iwata, Y., et al. (1996). Evaluation of antitumor activities of hyaluronate binding antitumor drugs: synthesis, characterization and antitumor activity. *J Drug Target* **4**, 1–8.

Alaniz, L., Garcia, M. G., Gallo-Rodriguez, C., et al. (2006). Hyaluronan oligosaccharides induce cell death through PI3-K/Akt pathway independently of NF-kappaB transcription factor. *Glycobiology* **16**, 359–367.

Allouche, M., Charrad, R. S., Bettaieb, A., et al. (2000). Ligation of the CD44 adhesion molecule inhibits drug-induced apoptosis in human myeloid leukemia cells. *Blood* **96**, 1187–1190.

Asman, D. C., Dirks, J. F., Ge, L., et al. (1995). Gene therapeutic approach to primary and metastatic brain tumors: I. CD44 variant pre-RNA alternative splicing as a CEPT control element. *J Neurooncol* **26**, 243–250.

Auzenne, E., Ghosh, S. C., Khodadadian, M., et al. (2007). Hyaluronic acid–paclitaxel: antitumor efficacy against CD44(+) human ovarian carcinoma xenografts. *Neoplasia* **9**, 479–486.

Avin, E., Haimovich, J., and Hollander, N. (2004). Anti-idiotype • anti-CD44 bispecific antibodies inhibit invasion of lymphoid organs by B cell lymphoma. *J Immunol* **173**, 4736–4743.

Bartolazzi, A., Peach, R., Aruffo, A., and Stamenkovic, I. (1994). Interaction between CD44 and hyaluronate is directly implicated in the regulation of tumor development. *J Exp Med* **180**, 53–66.

Borjesson, P. K., Postema, E. J., Roos, J. C., et al. (2003). Phase I therapy study with (186) re-labeled humanized monoclonal antibody BIWA 4 (Bivatuzumab) in patients with head and neck squamous cell carcinoma. *Clin Cancer Res* **9**, 3961S–3972S.

Bourguignon, L. Y., Peyrollier, K., Gilad, E., and Brightman, A. (2007). Hyaluronan–CD44 interaction with neural Wiskott–Aldrich syndrome protein (N-WASP) promotes actin polymerization and ErbB2 activation leading to beta-catenin nuclear translocation, transcriptional up-regulation, and cell migration in ovarian tumor cells. *J Biol Chem* **282**, 1265–1280.

Bourguignon, L. Y., Singleton, P. A., Zhu, H., and Diedrich, F. (2003). Hyaluronan-mediated CD44 interaction with RhoGEF and rho kinase promotes Grb2-associated binder-1 phosphorylation and phosphatidylinositol 3-kinase signaling leading to cytokine (macrophage-colony stimulating factor) production and breast tumor progression. *J Biol Chem* **278**, 29420–29434.

Bourguignon, L. Y., Zhu, H., Shao, L., and Chen, Y. W. (2000). CD44 interaction with tiam1 promotes rac1 signaling and hyaluronic acid-mediated breast tumor cell migration. *J Biol Chem* **275**, 1829–1838.

Breyer, R., Hussein, S., Radu, D. L., et al. (2000). Disruption of intracerebral progression of C6 rat glioblastoma by in vivo treatment with anti-CD44 monoclonal antibody. *J Neurosurg* **92**, 140–149.

Charrad, R. S., Gadhoum, Z., Qi, J., et al. (2002). Effects of anti-CD44 monoclonal antibodies on differentiation and apoptosis of human myeloid leukemia cell lines. *Blood* **99**, 290–299.

Charrad, R. S., Li, Y., Delpech, B., et al. (1999). Ligation of the CD44 adhesion molecule reverses blockage of differentiation in human acute myeloid leukemia. *Nat Med* **5**, 669–676.

Colnot, D. R., Roos, J. C., de Bree, R., et al. (2003). Safety, biodistribution, pharmacokinetics, and immunogenicity of 99mTc-labeled humanized monoclonal antibody BIWA 4 (Bivatuzumab) in patients with squamous cell carcinoma of the head and neck. *Cancer Immunol Immunother* **52**, 576–582.

Coradini, D., Pellizzaro, C., Abolafio, G., et al. (2004a). Hyaluronic-acid butyric esters as promising antineoplastic agents in human lung carcinoma: a preclinical study. *Invest New Drugs* **22**, 207-217.

Coradini, D., Zorzet, S., Rossin, R., et al. (2004b). Inhibition of hepatocellular carcinomas in vitro and hepatic metastases in vivo in mice by the histone deacetylase inhibitor HA–BUT. *Clin Cancer Res* **10**, 4822-4830.

Coradini, D., Pellizzaro, C., Miglierini, G., Daidone, M. G., and Perbellini, A. (1999). Hyaluronic acid as drug delivery for sodium butyrate: improvement of the anti–proliferative activity on a breast–cancer cell line. *Int J Cancer* **81**, 411–416.

Cordo Russo, R. I., Garcia, M. G., Alaniz, L., et al. (2008). Hyaluronan oligosaccharides sensitize lymphoma resistant cell lines to vincristine by modulating P-glycoprotein activity and PI3K/Akt pathway. *Int J Cancer* **122**, 1012–1018.

Dall, P., Herrmann, I., Durst, B., et al. (2005). In vivo cervical cancer growth inhibition by genetically engineered cytotoxic T cells. *Cancer Immunol Immunother* **54**, 51–60.

Dall, P., Hekele, A., Beckmann, M. W., et al. (1997). Efficient lysis of CD44v7/8-presenting target cells by genetically engineered cytotoxic T-lymphocytes – a model for immunogene therapy of cervical cancer. *Gynecol Oncol* **66**, 209–216.

Di Meo, C., Panza, L., Campo, F., et al. (2008). Novel types of carborane-carrier hyaluronan derivatives via "click chemistry." *Macromol Biosci.* **8**, 670–681.

Di Meo, C., Panza, L., Capitani, D., et al. (2007). Hyaluronan as carrier of carboranes for tumor targeting in boron neutron capture therapy. *Biomacromolecules* **8**, 552–559.

Eliaz, R. E., Nir, S., Marty, C., and Szoka, F. C., Jr. (2004a). Determination and modeling of kinetics of cancer cell killing by doxorubicin and doxorubicin encapsulated in targeted liposomes. *Cancer Res* **64**, 711–718.

Eliaz, R. E., Nir, S. and Szoka, F. C., Jr. (2004b). Interactions of hyaluronan-targeted liposomes with cultured cells: modeling of binding and endocytosis. *Methods Enzymol* **387**, 16–33.

Eliaz, R. E. and Szoka, F. C., Jr. (2001). Liposome-encapsulated doxorubicin targeted to CD44: a strategy to kill CD44-overexpressing tumor cells. *Cancer Res* **61**, 2592–2601.

Ghatak, S., Misra, S., and Toole, B. P. (2002). Hyaluronan oligosaccharides inhibit anchorage-independent growth of tumor cells by suppressing the phosphoinositide 3-kinase/Akt cell survival pathway. *J Biol Chem* **277**, 38013–38020.

Gilg, A. G., Tye, S. L., Tolliver, L. B., et al. (2008). Targeting hyaluronan interactions in malignant gliomas and their drug-resistant multipotent progenitors. *Clin Cancer Res* **14**, 1804–1813.

Gunia, S., Hussein, S., Radu, D. L., et al. (1999). CD44s – targeted treatment with monoclonal antibody blocks intracerebral invasion and growth of 9L gliosarcoma. *Clin Exp Metastasis* **17**, 221–230.

Guo, Y., Ma, J., Wang, J., Che, X., et al. (1994). Inhibition of human melanoma growth and metastasis in vivo by anti-CD44 monoclonal antibody. *Cancer Res* **54**, 1561–1565.

Harada, N., Mizoi, T., Kinouchi, M., et al. (2001). Introduction of antisense CD44S CDNA down-regulates expression of overall CD44 isoforms and inhibits tumor growth and metastasis in highly metastatic colon carcinoma cells. *Int J Cancer* **91**, 67–75.

Hayes, G. M., Dougherty, S. T., Davis, P. D., and Dougherty, G. J. (2004). Molecular mechanisms regulating the tumor-targeting potential of splice-activated gene expression. *Cancer Gene Ther* **11**, 797–807.

Hayes, G. M., Carpenito, C., Davis, P. D., et al. (2002). Alternative splicing as a novel of means of regulating the expression of therapeutic genes. *Cancer Gene Ther* **9**, 133–141.

Heider, K. H., Sproll, M., Susani, S., et al. (1996). Characterization of a high-affinity monoclonal antibody specific for CD44v6 as candidate for immunotherapy of squamous cell carcinomas. *Cancer Immunol Immunother* **43**, 245–253.

Hekele, A., Dall, P., Moritz, D., et al. (1996). Growth retardation of tumors by adoptive transfer of cytotoxic t lymphocytes reprogrammed by CD44v6-specific scFv:zeta-chimera. *Int J Cancer* **68**, 232–238.

Herrera-Gayol, A. and Jothy, S. (2002). Effect of hyaluronan on xenotransplanted breast cancer. *Exp Mol Pathol* **72**, 179–185.

Hosono, K., Nishida, Y., Knudson, W., et al. (2007). Hyaluronan oligosaccharides inhibit tumorigenicity of osteosarcoma cell lines MG-63 and LM-8 in vitro and in vivo via perturbation of hyaluronan-rich pericellular matrix of the cells. *Am J Pathol* **171**, 274–286.

Jeong, Y. I., Kim, S. T., Jin, S. G., et al. (2007). Cisplatin-incorporated hyaluronic acid nanoparticles based on ion-complex formation. *J Pharm Sci* **97**, 1268–1276.

Jin, L., Hope, K. J., Zhai, Q., Smadja-Joffe, F., and Dick, J. E. (2006). Targeting of CD44 eradicates human acute myeloid leukemic stem cells. *Nat Med* **12**, 1167–1174.

Koochekpour, S., Pilkington, G. J., and Merzak, A. (1995). Hyaluronic acid/CD44H interaction induces cell detachment and stimulates migration and invasion of human glioma cells in vitro. *Int J Cancer* **63**, 450–454.

Koppe, M., Schaijk, F., Roos, J., et al. (2004). Safety, pharmacokinetics, immunogenicity, and biodistribution of (186) Re-labeled humanized monoclonal antibody BIWA 4 (Bivatuzumab) in patients with early-stage breast cancer. *Cancer Biother Radiopharm* **19**, 720–729.

Krause, D. S., Lazarides, K., von Andrian, U. H., and Van Etten, R. A. (2006). Requirement for CD44 in homing and engraftment of BCR-ABL-expressing leukemic stem cells. *Nat Med* **12**, 1175–1180.

Luo, Y., Bernshaw, N. J., Lu, Z. R., Kopecek, J., and Prestwich, G. D. (2002). Targeted delivery of doxorubicin by HPMA copolymer–hyaluronan bioconjugates. *Pharm Res* **19**, 396–402.

Luo, Y., Ziebell, M. R., and Prestwich, G. D. (2000). A hyaluronic acid–taxol antitumor bioconjugate targeted to cancer cells. *Biomacromolecules* **1**, 208–218.

Luo, Y. and Prestwich, G. D. (1999). Synthesis and selective cytotoxicity of a hyaluronic acid–antitumor bioconjugate. *Bioconjug Chem* **10**, 755–763.

Mackay, C. R., Terpe, H. J., Stauder, R., et al. (1994). Expression and modulation of CD44 variant isoforms in humans. *J Cell Biol* **124**, 71–82.

McKee, C. M., Penno, M. B., Cowman, M., et al. (1996). Hyaluronan (HA) fragments induce chemokine gene expression in alveolar macrophages. The role of HA size and CD44. *J Clin Invest* **98**, 2403–2413.

Merzak, A., Koocheckpour, S., and Pilkington, G. J. (1994). CD44 mediates human glioma cell adhesion and invasion in vitro. *Cancer Res* **54**, 3988–3992.

Misra, S., Ghatak, S., Zoltan-Jones, A., and Toole, B. P. (2003). Regulation of multidrug resistance in cancer cells by hyaluronan. *J Biol Chem* **278**, 25285–25288.

Naor, D., Nedvetzki, S., Golan, I., Melnik, L., and Faitelson, Y. (2002). CD44 in cancer. *Crit Rev Clin Lab Sci* **39**, 527–579.

Peer, D. and Margalit, R. (2004a). Tumor-targeted hyaluronan nanoliposomes increase the antitumor activity of liposomal doxorubicin in syngeneic and human xenograft mouse tumor models. *Neoplasia* **6**, 343–353.

VIII. HYALURONAN-RELATED BIOMATERIAL

Peer, D. and Margalit, R. (2004b). Loading mitomycin C inside long circulating hyaluronan targeted nano-liposomes increases its antitumor activity in three mice tumor models. *Int J Cancer* **108**, 780–789.

Peer, D. and Margalit, R. (2000). Physicochemical evaluation of a stability-driven approach to drug entrapment in regular and in surface-modified liposomes. *Arch Biochem Biophys* **383**, 185–190.

Peterson, R. M., Yu, Q., Stamenkovic, I., and Toole, B. P. (2000). Perturbation of hyaluronan interactions by soluble CD44 inhibits growth of murine mammary carcinoma cells in ascites. *Am J Pathol* **156**, 2159–2167.

Pilon-Thomas, S., Verhaegen, M., Kuhn, L., Riker, A., and Mule, J. J. (2006). Induction of anti-tumor immunity by vaccination with dendritic cells pulsed with anti-CD44 IgG opson-ized tumor cells. *Cancer Immunol Immunother* **55**, 1238–1246.

Platt, V. M. and Szoka, F. C., Jr. (2008). Anticancer therapeutics: targeting macromolecules and nanocarriers to hyaluronan or CD44, a hyaluronan receptor. *Mol Pharm* **5**, 474–486.

Radotra, B. and McCormick, D. (1997). Glioma invasion in vitro is mediated by CD44–hyaluronan interactions. *J Pathol* **181**, 434–438.

Reeder, J. A., Gotley, D. C., Walsh, M. D., Fawcett, J., and Antalis, T. M. (1998). Expression of antisense CD44 variant 6 inhibits colorectal tumor metastasis and tumor growth in a wound environment. *Cancer Res* **58**, 3719–3726.

Rosenthal, M. A., Gibbs, P., Brown, T. J., et al. (2005). Phase I and pharmacokinetic evaluation of intravenous hyaluronic acid in combination with doxorubicin or 5-fluorouracil. *Chemotherapy* **51**, 132–141.

Rupp, U., Schoendorf-Holland, E., Eichbaum, M., et al. (2007). Safety and pharmacokinetics of Bivatuzumab mertansine in patients with CD44v6-positive metastatic breast cancer: final results of a Phase I study. *Anticancer Drugs* **18**, 477–485.

Sauter, A., Kloft, C., Gronau, S., et al. (2007). Pharmacokinetics, immunogenicity and safety of Bivatuzumab mertansine, a novel CD44v6-targeting immunoconjugate, in patients with squamous cell carcinoma of the head and neck. *Int J Oncol* **30**, 927–935.

Seiter, S., Arch, R., Reber, S., et al. (1993). Prevention of tumor metastasis formation by anti-variant CD44. *J Exp Med* **177**, 443–455.

Somasundaram, C., Arch, R., Matzku, S., and Zoller, M. (1996). Development of a bispecific F(Ab')2 conjugate against the complement receptor CR3 of macrophages and a variant CD44 antigen of rat pancreatic adenocarcinoma for redirecting macrophage-mediated tumor cytotoxicity. *Cancer Immunol Immunother* **42**, 343–350.

Song, G., Liao, X., Zhou, L., et al. (2004). HI44a, an anti-CD44 monoclonal antibody, induces differentiation and apoptosis of human acute myeloid leukemia cells. *Leuk Res* **28**, 1089–1096.

Stroomer, J. W., Roos, J. C., Sproll, M., et al. (2000). Safety and biodistribution of 99mTech-netium-labeled anti-CD44v6 monoclonal antibody BIWA 1 in head and neck cancer patients. *Clin Cancer Res* **6**, 3046–3055.

Subramaniam, V., Vincent, I. R., Gilakjan, M., and Jothy, S. (2007). Suppression of human colon cancer tumors in nude mice by siRNA CD44 gene therapy. *Exp Mol Pathol* **83**, 332–340.

Sy, M. S., Guo, Y. J., and Stamenkovic, I. (1992). Inhibition of tumor growth in vivo with a soluble CD44-immunoglobulin fusion protein. *J Exp Med* **176**, 623–627.

Szoka, F. (2008). Molecular biology. The art of assembly. *Science* **319**, 578–579.

Tan, B., Wang, J. H., Wu, Q. D., Kirwan, W. O., and Redmond, H. P. (2001). Sodium hy-aluronate enhances colorectal tumour cell metastatic potential in vitro and in vivo. *Br J Surg* **88**, 246–250.

Tijink, B. M., Buter, J., de Bree, R., et al. (2006). A Phase I dose escalation study with anti-CD44v6 Bivatuzumab mertansine in patients with incurable squamous cell carcinoma of the head and neck or esophagus. *Clin Cancer Res* **12**, 6064–6072.

Toole, B. P. (2004). Hyaluronan: from extracellular glue to pericellular cue. *Nat Rev Cancer* **4**, 528–539.

Tsujimura, S., Saito, K., Kohno, K., and Tanaka, Y. (2006). Fragmented hyaluronan induces transcriptional up-regulation of the multidrug resistance-1 gene in CD4+ T Cells. *J Biol Chem* **281**, 38089–38097.

Verel, I., Heider, K. H., Siegmund, M., et al. (2002). Tumor targeting properties of monoclonal antibodies with different affinity for target antigen CD44V6 in nude mice bearing head-and-neck cancer xenografts. *Int J Cancer* **99**, 396–402.

Wang, S. J. and Bourguignon, L. Y. (2006a). Hyaluronan–CD44 promotes phospholipase C-mediated Ca2+ signaling and cisplatin resistance in head and neck cancer. *Arch Otolaryngol Head Neck Surg* **132**, 19–24.

Wang, S. J. and Bourguignon, L. Y. (2006b). Hyaluronan and the interaction between CD44 and epidermal growth factor receptor in oncogenic signaling and chemotherapy resistance in head and neck cancer. *Arch Otolaryngol Head Neck Surg* **132**, 771–778.

Ward, J. A., Huang, L., Guo, H., Ghatak, S., and Toole, B. P. (2003). Perturbation of hyaluronan interactions inhibits malignant properties of glioma cells. *Am J Pathol* **162**, 1403–1409.

Wiranowska, M., Tresser, N., and Saporta, S. (1998). The effect of interferon and anti-CD44 antibody on mouse glioma invasiveness in vitro. *Anticancer Res* **18**, 3331–3338.

Yadav, A. K., Mishra, P., and Agrawal, G. P. (2008). An insight on hyaluronic acid in drug targeting and drug delivery. *J Drug Target* **16**, 91–107.

Yu, Q., Toole, B. P., and Stamenkovic, I. (1997). Induction of apoptosis of metastatic mammary carcinoma cells in vivo by disruption of tumor cell surface CD44 function. *J Exp Med* **186**, 1985–1996.

Zahalka, M. A., Okon, E., Gosslar, U., Holzmann, B., and Naor, D. (1995). Lymph node (but not spleen) invasion by murine lymphoma is both CD44- and hyaluronate-dependent. *J Immunol* **154**, 5345–5355.

Zeng, C., Toole, B. P., Kinney, S. D., Kuo, J. W., and Stamenkovic, I. (1998). Inhibition of tumor growth in vivo by hyaluronan oligomers. *Int J Cancer* **77**, 396–401.

A NEW PERSPECTIVE

CHAPTER

21

Hyaluronidase-2 and Its Role as a Cell-Entry Receptor for Sheep Retroviruses That Cause Contagious Respiratory Tract Cancers

A. Dusty Miller

OUTLINE

ONCOGENIC SHEEP RETROVIRUSES

Jaagsiekte sheep retrovirus (JSRV) causes pulmonary adenocarcinoma (also called sheep pulmonary adenomatosis or jaagsiekte) in sheep and goats (Palmarini et al., 1999). JSRV-induced tumors arise from epithelial cells in the lower airway, and tumor cells express markers of type II alveolar and/or bronchiolar epithelial cells (Palmarini et al., 1995). Two strains of a closely related retrovirus called enzootic nasal tumor virus (ENTV) have been cloned from sheep (ENTV-1) (Cousens et al., 1999) and goats (ENTV-2) (Ortin et al., 2003) that share ~95% overall amino acid similarity with JSRV. ENTV can be found in the nasal fluid of animals with intranasal tumors, which eventually progress and cause severe cranial deformations and respiratory blockage, resulting in death (Vitellozzi et al., 1993). JSRV and ENTV can increase production of lung and nasal fluid and can spread by aerosolization of virus present in these secretions. JSRV and ENTV are present in many countries worldwide and have a significant economic and animal health impact. In addition, the disease induced by JSRV exhibits histological features similar to those of many human pulmonary adenocarcinomas, and study of adenocarcinoma induced by JSRV and ENTV may provide insights into the etiology of human lung cancer. While ENTV and JSRV do not appear to cause lung cancer in humans having occupational exposure to these viruses, this possibility has not been definitively excluded.

Until recently, oncogenic retroviruses were divided into those that cause cancer with long latency and do so by insertional activation of host oncogenes, and acutely transforming retroviruses that rapidly induce cancer as a result of virus acquisition and expression of host cell oncogenes. For example, Moloney murine leukemia virus induces leukemia over weeks to months by insertional activation of host cell oncogenes such as *lck* and *myc*, while the acutely-transforming Harvey murine sarcoma virus carries a mutant cellular *ras* oncogene and can acutely transform cells in culture and in animals (Rosenberg and Joli-coeur, 1997). JSRV and ENTV are examples of a small but growing new class of retroviruses that are acutely transforming and induce cancer as a direct result of expression of viral genes that show no relation to host cell genes. In the case of JSRV, cancer induction can occur in as little as 10 days in newborn sheep (Sharp et al., 1983), showing that it is acutely transforming, yet JSRV does not contain sequences related to mammalian genes. JSRV and ENTV are simple retroviruses (Fig. 21.1) that carry genes required for viral replication and that lack accessory genes typical of complex retroviruses or cell-derived genes typical of most acutely-transforming retroviruses. Analysis of the transforming activity of JSRV and ENTV in cell culture has revealed that the *env* genes of these viruses are necessary and sufficient to induce trans-formation (Dirks et al., 2002; Maeda et al., 2001; Rai et al., 2001). The primary role of Env in viral replication is to promote virus entry into cells following binding to specific cell-surface receptors, and it seemed likely that these receptors might play a key role in transformation as well.

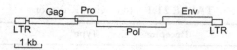

FIGURE 21.1 Genetic structure of JSRV and ENTV. The structure of the integrated DNA form of the retroviruses is shown. Long terminal repeat (LTR) sequences that function to initiate and terminate mRNA transcription are shown flanking the protein coding regions. Protein coding regions are: Gag, virion core polypeptide; Pro, protease; Pol, reverse transcriptase (polymerase) and integrase; Env, viral coat protein (envelope) required for cell entry. The reading frames of the protein coding regions are indicated by the elevation of the boxes, for example, the Pro and Env coding regions are in the same reading frame but Pol and Gag are in different reading frames. kb, distance in kilobases.

IDENTIFICATION OF HYAL-2 AS THE CELL-ENTRY RECEPTOR FOR JSRV AND ENTV

Retrovirus entry into cells depends on the presence of specific proteins that bind the viral Env protein and help trigger conformational changes in Env that lead to fusion of the virus and cell membranes and entry of the virus core into the cell. A wide variety of proteins have been found to serve as receptors for different retroviruses, based primarily on their ability to promote virus entry after expression in cells that are not naturally permissive for virus entry (Table 21.1). In most cases, a single protein suffices to render otherwise non-permissive cells susceptible to virus entry. Typically, these proteins promote virus binding, and may also promote virus fusion with the cell membrane. For other viruses (for example, HIV) there are distinct binding and fusion receptors that are required for virus entry. Retrovirus receptors are key determinants of the species and cell types that a retrovirus can infect, and thus are primary determinants of the host range and the type of disease induced by the virus.

To identify the cell-entry receptor for JSRV we used a retroviral vector that encodes human placental alkaline phosphatase (AP) and that was packaged into virions bearing the JSRV Env protein on the virion surface (Rai et al., 2000). In early experiments we found that this vector could transfer and express (transduce) the AP marker protein gene to sheep and human cells, but not to rodent cells, including those from mice, rats, and hamsters. This allowed us to develop a genetic screen to identify the human gene that when expressed in rodent cells would allow vector transduction. As target cells we used a set of 80 hamster cell lines carrying different fragments of DNA that had been produced by fusing hamster cells with irradiated human cells. This allowed us to identify the chromosomal location of the receptor within a few hundred kilobase pairs of DNA (Rai et al., 2000). We were lucky to find that this region had been cloned as a set of overlapping cosmid clones, and it was relatively straightforward to identify the gene encoding the receptor by testing

TABLE 21.1 Retrovirus Receptors

Retrovirus	Receptor	Type[a]	Function
Human immunodeficiency virus, simian immunodeficiency virus	CD4 and CXCR4, CCR5, others	TM1 TM7	Immune function G protein-coupled chemokine receptors
Feline immunodeficiency virus	CD134 and CXCR4	TM1 TM7	Immune function G protein-coupled chemokine receptor
Human T-cell leukemia virus	GLUT-1	TM12	Glucose transport
Ecotropic murine leukemia virus	CAT-1 (SLC7A1)	TM14	Basic amino acid transport
Gibbon ape leukemia virus, amphotropic murine leukemia virus, 10A1 murine leukemia virus, feline leukemia virus type B, woolly monkey virus	Pit1 (SLC20A1) or Pit2 (SLC20A2)	TM10-13 TM10-13	Phosphate transport Phosphate transport
RD114, type D simian retroviruses, baboon endogenous virus, human endogenous retrovirus type W	RDR (SLC1A5) or RDR2 (SLC1A4)	TM9-10 TM9-10	Neutral amino acid transport Glutamate and neutral amino acid transport
Xenotropic and polytropic murine leukemia viruses	XPR1	TM8	G protein-coupled signaling? Transport?
Feline leukemia virus type A	Thtr1	TM12	Thiamine transport
Feline leukemia virus type C	Flvcr	TM12	Heme export
Feline leukemia virus type T	Felix and Pit1 (SLC20A1)	soluble TM10-13	Env-like protein Phosphate transport
Pig endogenous retrovirus type A	Par-1 (GPR172A) or Par-2 (GPR172B)	TM10-11 TM10-11	G protein-coupled receptor G protein-coupled receptor
M813 murine leukemia virus	Smit1 (SLC5A3)	TM14	*myo*-inositol transport
Avian leukosis virus type A	Tva	TM1	LDL receptor-like protein
Avian leukosis virus types B, D, E	Tvb	TM1	Fas/TNFR-like receptor

Table 21.1 (*continued*)

Retrovirus	Receptor	Type[a]	Function
Avian leukosis virus type C	Tvc	TM1	Butyrophilin-like (immunoglobulin superfamily)
Avian leukosis virus type J	NHE1 (SLC9A1)	TM12	Na+/H+ antiporter
Mouse mammary tumor virus	Tfr1	TM1	Transferrin receptor
Jaagsiekte sheep retrovirus, enzootic nasal tumor virus	HYAL-2	GPI-anchored	Hyaluronidase (weak)

[a]TM indicates a transmembrane protein and the number after TM indicates the number of times the protein is predicted to span the membrane. GPI-anchored indicates a glycosylphosphatidylinositol-anchored membrane protein.

hamster cells for JSRV vector susceptibility following transfection of individual cosmids into the cells (Rai et al., 2001). This genetic analysis indicated that only one gene served as a receptor for JSRV, but to reinforce this conclusion, we tested all of the human hyaluronidase family members for receptor activity, and found no activity associated with human HYAL-1, HYAL-3, HYAL-4, or Spam1 (Rai et al., 2001). These data indicate that HYAL-2 is the only protein in the human genome that functions as a JSRV receptor.

We also tested hyaluronidase family members from other species for receptor function to determine if receptor function correlates with the ability of the JSRV vector to transduce cells from different species (Fig. 21.2). Indeed, mouse HYAL-2 functioned poorly as a receptor for JSRV vector cell-entry when expressed in mouse or hamster cells, consistent with the inability of the JSRV vector to transduce mouse cells (Rai et al., 2001). In contrast, sheep HYAL-2 functioned well as a receptor for JSRV vector cell-entry when expressed in mouse or hamster cells, consistent with the high susceptibility of sheep cells to JSRV vector transduction (Dirks et al., 2002). An intermediate result was obtained for rat HYAL-2, where overexpression of rat HYAL-2 in mouse, hamster, or rat cells rendered the cells susceptible to JSRV vector transduction, but rat cells are normally resistant to vector transduction (Liu et al., 2003a). Additional experiments showed that JSRV Env binds rat HYAL-2 less well than it does human HYAL-2, supporting the interpretation that higher levels of rat HYAL-2 are required to promote efficient JSRV vector transduction than are normally expressed on rat cells (Liu et al., 2003a). In conclusion, these experiments show that expression of a functional HYAL-2 protein is the primary determinant of JSRV Env-mediated virion entry into cells.

Using a retroviral vector encoding AP packaged into virions bearing either the ENTV or JSRV Env proteins on the virion surface, we found that

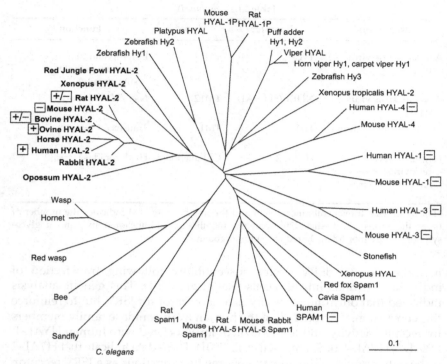

FIGURE 21.2 Receptor activity of HYAL-2 orthologs (bold) and paralogs. Proteins related to HYAL-2 obtained by BLAST search of GenBank are shown. Proteins that exhibit high receptor activity when expressed in cells that are normally non-permissive for JSRV and ENTV vector transduction are indicated by "+," those that exhibit moderate receptor activity are indicted by "+/–," and those that exhibit very low to no activity are indicated by "–" (Dirks et al., 2002; Duh et al., 2005; Rai et al., 2001; Van Hoeven and Miller, 2005).

ENTV Env promotes infection of a more restricted range of cell types than does the JSRV Env (Dirks et al., 2002). Given the similarity in ENTV and JSRV Env amino-acid sequences, we first tested whether ENTV Env might use HYAL-2 for cell entry as does JSRV Env. Indeed, both sheep and human HYAL-2 can serve as cell-entry receptors for virus bearing the ENTV Env. Interestingly, HYAL-2 expression was not sufficient to promote entry of the ENTV vector into all cells, indicating that other factors are important for ENTV Env-mediated entry into cells (Van Hoeven and Miller, 2005).

HYAL-2 LOCATION AND ENZYMATIC ACTIVITY

HYAL-2 was initially identified as a lysosomal hyaluronidase by addition of a green fluorescent protein (GFP) tag to the carboxy terminus of HYAL-2 and by showing that GFP fluorescence localized to lysosomes after expression of the hybrid protein in a rat glioma cell line (Lepperdinger et al., 1998). HYAL-2 exhibited low but detectable hyaluronidase activity with an acidic pH optimum in these experiments. However, later studies have conclusively shown that HYAL-2 is actually a glycosylphosphatidylinositol (GPI)-anchored cell-surface protein, consistent with its role as a cell-surface receptor for sheep retroviruses (Duh et al., 2005; Liu et al., 2003a; 2004; Miller et al., 2006; Rai et al., 2001; Van Hoeven and Miller, 2005). GPI-anchored proteins have an endoplasmic reticulum (ER) signal sequence at the amino terminus, which directs the proteins to the ER and is removed during protein translocation into the ER, and have hydrophobic carboxy termini that are replaced with a GPI anchor that tethers the proteins to the cell surface. The latter feature likely explains the original result indicating that HYAL-2 was a lysosomal protein — the GFP tag added to the carboxy end of HYAL-2 in those experiments was likely removed during GPI anchor addition and was sent to lysosomes for degradation, while the processed HYAL-2 was exported to the cell surface.

HYAL-2 exhibits very low hyaluronidase activity in comparison to HYAL-1 or Spam1 under all conditions analyzed to date. While the hyaluronidase activity of HYAL-2 could be detected in cells engineered to greatly over-express HYAL-2 by infection with a vaccinia virus vector encoding human HYAL-2 (Lepperdinger et al., 1998), we have had difficulty detecting HYAL-2 activity in cells transduced with a retroviral vector that encodes human HYAL-2 (Rai et al., 2001). Therefore we studied a soluble preparation of HYAL-2 made by inserting a stop codon into the human HYAL-2 cDNA at the position of the GPI-anchor cleavage site, and by expressing this truncated protein in insect cells using a baculovirus vector (Vigdorovich et al., 2005; 2007). The endoplasmic reticulum signal sequence is properly removed from the translated protein, and without the hydrophobic tail and GPI-anchor signal sequence, the protein is secreted from the cells. This secreted form of HYAL-2 (sHYAL-2) corresponds exactly to native HYAL-2 expressed on the cell surface except that it lacks the GPI anchor. The sHYAL-2 protein appears to be a properly folded monomeric protein by size-exclusion chromatography and is stable in solution at 4°C for months. Initially, the hyaluronidase activity of purified sHYAL-2 appeared to have a neutral pH optimum (Vigdorovich et al., 2005), but we later showed that this was due to the presence of a very small amount of a highly active baculoviral hyaluronan lyase (Vigdorovich et al., 2007), and that the hyaluronidase activity of sHYAL-2 actually has an acidic pH optimum consistent with previous analysis of HYAL-2 activity in cultured cells. This activity could be greatly reduced by mutation of amino acid residues that correspond to the active site residues common to the bee venom and Spam1 hyaluronidases, indicating that the active site for hyaluronan digestion in HYAL-2 is similar to those of other hyaluronidases (Vigdorovich et al., 2007). However, the hyaluronidase activity of sHYAL-2 is ~400-fold lower than that of Spam1 (Vigdorovich et al., 2007).

The availability of purified sHYAL-2 allowed us to further analyze the kinetics of hyaluronan degradation by HYAL-2 (Vigdorovich et al., 2005; 2007). Others have noted a 20 kDa intermediate of hyaluronan degradation that appeared to be uniquely associated with hyaluronan degradation by HYAL-2, and that this intermediate did not disappear even after prolonged incubation with HYAL-2 (Lepperdinger et al., 1998). We also find this intermediate following digestion of hyaluronan with sHYAL-2, but in contrast, this intermediate can be completely digested following prolonged incubation with sHYAL-2. Indeed, similar kinetics of hyaluronan digestion are observed for HYAL-1 and Spam1, with initial rapid digestion of hyaluronan to a 20 kDa intermediate followed by a 25-fold slower digestion of the 20 kDa form to smaller products. Thus digestion of hyaluronan by HYAL-1, HYAL-2 and Spam1 appears to follow similar biphasic kinetics involving a relatively stable 20 kDa intermediate corresponding to 50–60 disaccharide units.

Although purified HYAL-2 has low hyaluronidase activity, it is possible that other cellular proteins or cofactors might modulate HYAL-2 activity and/or be required for conversion of HYAL-2 into a more active enzyme. Indeed, a requirement for CD44 to promote acidification of the extracellular environment and activate the hyaluronidase activity of HYAL-2 has been described (Bourguignon et al., 2004; Harada and Takahashi, 2007). Given the long incubation times used for detection of hyaluronidase activity in these experiments, it still appears that HYAL-2 is a relatively weak hyaluronidase, although a direct comparison of HYAL-2 activities to a highly active hyaluronidase such as Spam1 was not performed. Perhaps in the local space adjacent to the cell only a small amount of hyaluronidase is required to mediate biologically relevant changes in hyaluronan properties and production of a highly active enzyme would be deleterious.

HYAL-2 ROLE IN SHEEP RETROVIRUS ONCOGENESIS?

Interaction of JSRV and ENTV Env proteins with human HYAL-2, location of the human HYAL-2 gene in the 3p21.3 lung cancer tumor suppressor locus, and the presumed role of HYAL-2 in metabolism of the extracellular matrix all pointed to a potential role of HYAL-2 in transformation by the sheep retrovirus Env proteins. Support for this hypothesis was provided by studies in the human bronchial epithelial cell line BEAS-2B (Danilkovitch-Miagkova et al., 2003). In these cells, HYAL-2 can bind to the RON receptor tyrosine kinase rendering it inactive. JSRV Env can transform the cells, and in cells expressing Env, Env associated with HYAL-2 and caused its degradation, releasing

RON from suppression by HYAL-2 and activating the Akt and mitogen-activated protein kinase oncogenic pathways. Most importantly, expression of a dominant-negative RON protein blocked Env transformation of the cells indicating that RON played a critical role in Env transformation.

On the other hand, JSRV and ENTV Env proteins cannot mediate virus entry into mouse cells and do not bind mouse HYAL-2 (Liu et al., 2003a), yet both Env proteins can transform cultured NIH 3T3 mouse fibroblasts (Liu et al., 2003a, b; Maeda et al., 2001). Furthermore, expression of either JSRV or ENTV Env in mouse lung can induce lung adenocarcinoma similar to that seen in sheep infected with replication-competent JSRV (Fig. 21.3) (Wootton et al., 2005; 2006a,b). These results indicate that mouse HYAL-2 plays no role in oncogenic transformation by either Env protein in mice. Whether HYAL-2 plays some role in sheep tumorigenesis is uncertain but an interaction of Env with HYAL-2 seems unlikely to be required based on the results in mice.

So how do the JSRV and ENTV Env proteins transform cells if not by interaction with HYAL-2? Several studies have shown that sequences in the cytoplasmic domain of the Env proteins are critical for transformation,

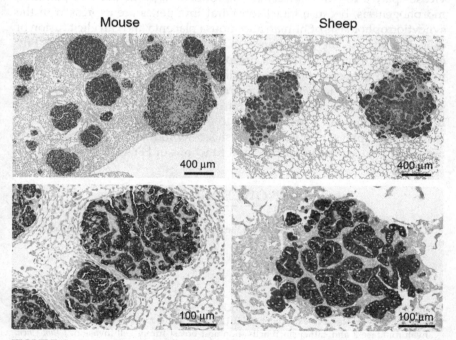

FIGURE 21.3 Lung tumors induced by Env expression in mice and JSRV infection in sheep. Fixed paraffin-embedded lung sections were stained for JSRV Env expression using a monoclonal antibody against Env which stains transformed lung cells in tumors. For more details see Wootton et al. (2006b).

and that oncogenic signaling occurs through the phosphoinositide 3-kinase (PI3K)/Akt and mitogen-activated protein kinase (MAPK) pathways (Liu and Miller, 2007). The mechanisms by which the Env proteins activate these pathways are currently unknown.

HYAL-2 ROLE IN SHEEP PLACENTAL MORPHOGENESIS

Perhaps one of the most interesting findings relating to the interaction of sheep retrovirus Env proteins with HYAL-2 is the role of Env proteins synthesized from endogenous sheep retroviruses and HYAL-2 in placental morphogenesis in sheep. Mammals carry many copies of retroviruses in their genomes. Sheep carry ~ 20 copies of endogenous retroviruses related to JSRV and ENTV, but the Env proteins synthesized from these viruses are either nonfunctional or contain mutations that render the Env proteins non-transforming. However, some of these Env proteins can still interact with HYAL-2, and in this case, can mediate fusion not between virions and cells but between cells. It has been hypothesized that endogenous retroviruses play a role in mammalian reproduction, particularly in placental morphogenesis, because intact retroviral Env genes are expressed in the syncytiotrophoblasts of human and mouse placenta and can elicit fusion of cells in culture. The importance of endogenous sheep retrovirus Env expression during pregnancy in sheep was confirmed by administration of morpholino antisense oligonucleotides to block Env expression *in utero*, which resulted in termination of pregnancy (Dunlap et al., 2006). This study dramatically confirms that retroviruses are not simply pathogens but can contribute in a positive way to mammalian evolution, and shows that HYAL-2 plays a critical role in sheep reproduction. Further work is necessary to decipher other potential functions of HYAL-2 in mammals.

ACKNOWLEDGMENTS

This work was supported by grants from the Fred Hutchinson Cancer Research Center and the National Institutes of Health.

References

Bourguignon, L. Y., Singleton, P. A., Diedrich, F., Stern, R., and Gilad, E. (2004). CD44 interaction with Na$^+$–H$^+$ exchanger (NHE1) creates acidic microenvironments leading to hyaluronidase-2 and cathepsin B activation and breast tumor cell invasion. *J Biol Chem* **279**, 26991–27007.

Cousens, C., Minguijon, E., Dalziel, R. G., et al. (1999). Complete sequence of enzootic nasal tumor virus, a retrovirus associated with transmissible intranasal tumors of sheep. *J Virol* **73**, 3986–3993.

Danilkovitch-Miagkova, A., Duh, F. M., Kuzmin, I., et al. (2003). Hyaluronidase 2 negatively regulates RON receptor tyrosine kinase and mediates transformation of epithelial cells by jaagsiekte sheep retrovirus. *Proc Natl Acad Sci USA* **100**, 4580–4585.

Dirks, C., Duh, F. M., Rai, S. K., Lerman, M. I. and Miller, A. D. (2002). Mechanism of cell entry and transformation by enzootic nasal tumor virus. *J Virol* **76**, 2141–2149.

Duh, F. M., Dirks, C., Lerman, M. I. and Miller, A. D. (2005). Amino acid residues that are important for HYAL-2 function as a receptor for jaagsiekte sheep retrovirus. *Retrovirology* **2**, 59.

Dunlap, K. A., Palmarini, M., Varela, M., et al. (2006). Endogenous retroviruses regulate peri-implantation placental growth and differentiation. *Proc Natl Acad Sci USA* **103**, 14390–14395.

Harada, H. and Takahashi, M. (2007). CD44-dependent intracellular and extracellular catabolism of hyaluronic acid by hyaluronidase-1 and -2. *J Biol Chem* **282**, 5597–5607.

Lepperdinger, G., Strobl, B., and Kreil, G. (1998). HYAL-2, a human gene expressed in many cells, encodes a lysosomal hyaluronidase with a novel type of specificity. *J Biol Chem* **273**, 22466–22470.

Liu, S. L., Duh, F. M., Lerman, M. I. and Miller, A. D. (2003a). Role of virus receptor HYAL-2 in oncogenic transformation of rodent fibroblasts by sheep beta retrovirus env proteins. *J Virol* **77**, 2850–2858.

Liu, S. L., Lerman, M. I. and Miller, A. D. (2003b). Putative phosphatidylinositol 3-kinase (PI3K) binding motifs in ovine beta retrovirus Env proteins are not essential for rodent fibroblast transformation and PI3K/Akt activation. *J Virol* **77**, 7924–7935.

Liu, S. L., Halbert, C. L., and Miller, A. D. (2004). Jaagsiekte sheep retrovirus envelope efficiently pseudotypes human immunodeficiency virus type 1-based lentiviral vectors. *J Virol* **78**, 2642–2647.

Liu, S. L. and Miller, A. D. (2007). Oncogenic transformation by the jaagsiekte sheep retrovirus envelope protein. *Oncogene* **26**, 789–801.

Maeda, N., Palmarini, M., Murgia, C. and Fan, H. (2001). Direct transformation of rodent fibroblasts by jaagsiekte sheep retrovirus DNA. *Proc. Natl. Acad. Sci. USA* **98**, 4449–4454.

Miller, A. D., Vigdorovich, V., Strong, R. K., Fernandes, R. J. and Lerman, M. I. (2006). Letter to the Editor: HYAL-2, where are you? *Osteoarthritis Cartilage* **14**, 1315–1317.

Ortin, A., Cousens, C., Minguijon, E., et al. (2003). Characterization of enzootic nasal tumour virus of goats: complete sequence and tissue distribution. *J Gen Virol* **84**, 2245–2252.

Palmarini, M., Dewar, P., De las Heras, M., et al. (1995). Epithelial tumour cells in the lungs of sheep with pulmonary adenomatosis are major sites of replication for Jaagsiekte retrovirus. *J Gen Virol* **76**, 2731–2737.

Palmarini, M., Sharp, J. M., De las Heras, M. and Fan, H. (1999). Jaagsiekte sheep retrovirus is necessary and sufficient to induce a contagious lung cancer in sheep. *J Virol* **73**, 6964–6972.

Rai, S. K., DeMartini, J. C., and Miller, A. D. (2000). Retrovirus vectors bearing jaagsiekte sheep retrovirus Env transduce human cells by using a new receptor localized to chromosome 3p21.3. *J Virol* **74**, 4698–4704.

Rai, S. K., Duh, F. M., Vigdorovich, V., et al. (2001). Candidate tumor suppressor HYAL-2 is a glycosylphosphatidylinositol (GPI)-anchored cell-surface receptor for jaagsiekte sheep retrovirus, the envelope protein of which mediates oncogenic transformation. *Proc Natl Acad Sci. USA* **98**, 4443–4448.

Rosenberg, N. and Jolicoeur, P. (1997). Retroviral pathogenesis: oncogenesis. In: Coffin, J. M., Hughes, S. H. and Varmus H.E., (Eds.). Retroviruses. Cold Spring Harbor: Cold Spring Harbor Laboratory Press, pp. 478–523.

Sharp, J. M., Angus, K. W., Gray, E. W. and Scott, F. M. (1983). Rapid transmission of sheep pulmonary adenomatosis (jaagsiekte) in young lambs. *Arch Virol* **78**, 89–95.

IX. A NEW PERSPECTIVE

Van Hoeven, N. S. and Miller, A. D. (2005). Improved enzootic nasal tumor virus pseudotype packaging cell lines reveal virus entry requirements in addition to the primary receptor HYAL-2. *J Virol* **79**, 87–94.

Vigdorovich, V., Miller, A. D. and Strong, R. K. (2007). Ability of hyaluronidase 2 to degrade extracellular hyaluronan is not required for its function as a receptor for jaagsiekte sheep retrovirus. *J Virol* **81**, 3124–3129.

Vigdorovich, V., Strong, R. K. and Miller, A. D. (2005). Expression and characterization of a soluble, active form of the jaagsiekte sheep retrovirus receptor, HYAL-2. *J Virol* **79**, 79–86.

Vitellozzi, G., Mughetti, L., Palmarini, M., et al. (1993). Enzootic intranasal tumour of goats in Italy. *J Vet Med A Physiol Pathol Clin Med* **40**, 459–468.

Wootton, S. K., Halbert, C. L., and Miller, A. D. (2005). Sheep retrovirus structural protein induces lung tumours. *Nature* **434**, 904–907.

Wootton, S.K., Halbert, C.L. and Miller, A.D. (2006a). Envelope proteins of jaagsiekte sheep retrovirus and enzootic nasal tumor virus induce similar bronchio-alveolar tumors in lungs of mice. *J Virol* **80**, 9322–9325.

Wootton, S.K., Metzger, M.J., Hudkins, K.L., et al. (2006b). Lung cancer induced in mice by the envelope protein of jaagsiekte sheep retrovirus (JSRV) closely resembles lung cancer in sheep infected with JSRV. *Retrovirology* **3**, 94.

Index

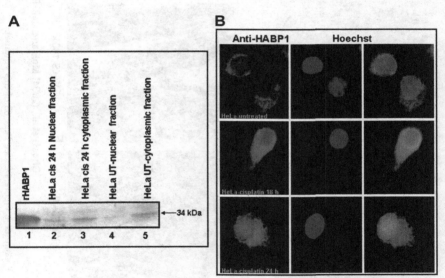

A

rHABP1

HeLa cis 24 h Nuclear fraction

HeLa cis 24 h cytoplasmic fraction

HeLa UT-nuclear fraction

HeLa UT-cytoplasmic fraction

←34 kDa

1   2   3   4   5

B

Anti-HABP1	Hoechst

HeLa-untreated

HeLa-cisplatin 18 h

HeLa-cisplatin 24 h

Taken from Kamal and Datta, Apoptosis, 2006, 11, 861-874

Chapter 4, Figure 4.1 (See Page 61 of this volume).

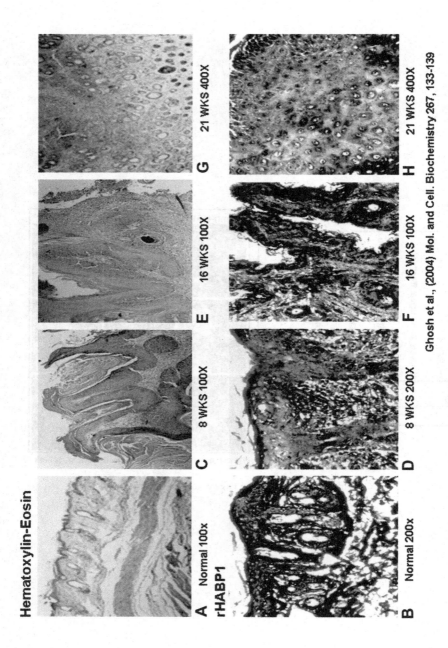

Hematoxylin-Eosin

rHABP1

A Normal 100x

B Normal 200x

C 8 WKS 100X

D 8 WKS 200X

E 16 WKS 100X

F 16 WKS 100X

G 21 WKS 400X

H 21 WKS 400X

Ghosh et al., (2004) Mol. and Cell. Biochemistry 267, 133-139

Chapter 4, Figure 4.2 (See Page 63 of this volume).

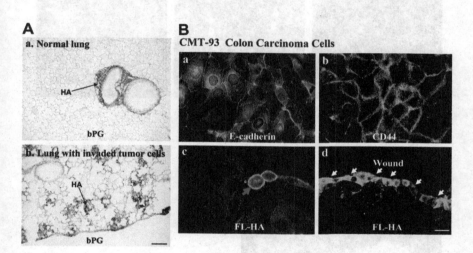

A

a. Normal lung

HA

bPG

b. Lung with invaded tumor cells

HA

bPG

B

CMT-93 Colon Carcinoma Cells

a   E-cadherin

b   CD44

c   FL-HA

d   Wound   FL-HA

Chapter 5, Figure 5.1 (See Page 72 of this volume).

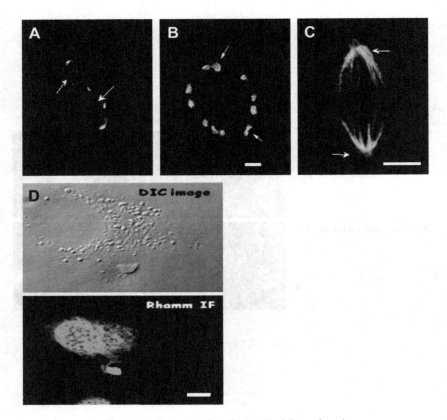

Chapter 9, Figure 9.6 (See Page 162 of this volume).

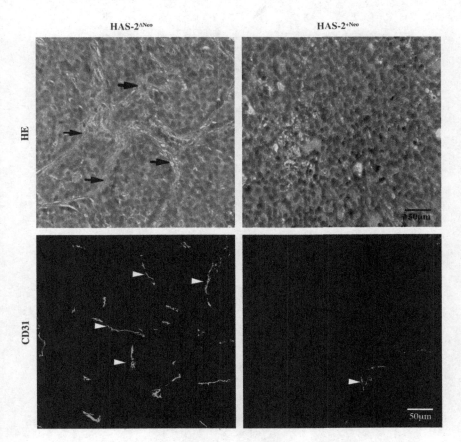

Chapter 10, Figure 10.2 (See Page 175 of this volume).

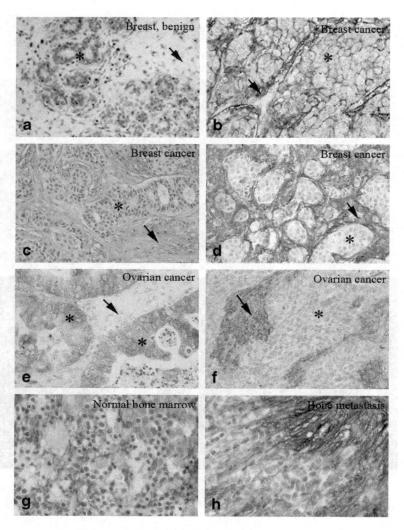

Chapter 14, Figure 14.1 (See Page 260 of this volume).

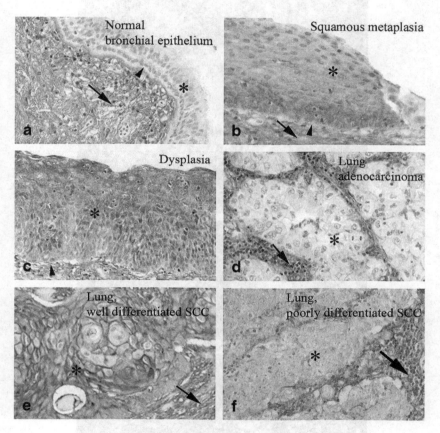

Chapter 14, Figure 14.2 (See Page 264 of this volume).

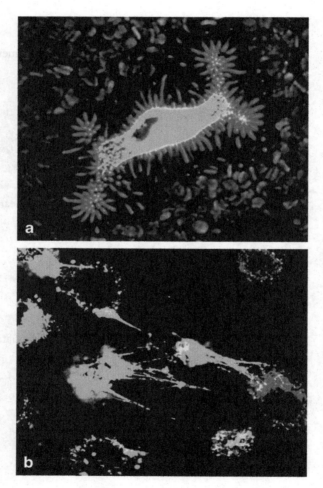

Chapter 14, Figure 14.4 (See Page 275 of this volume).

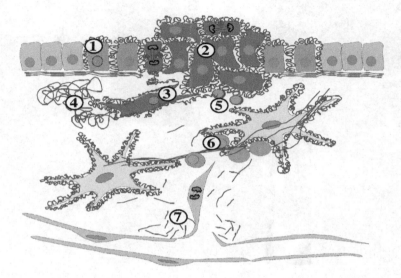

Chapter 14, Figure 14.5 (See Page 277 of this volume).

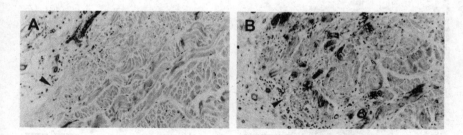

Chapter 15, Figure 15.4 (See Page 290 of this volume).

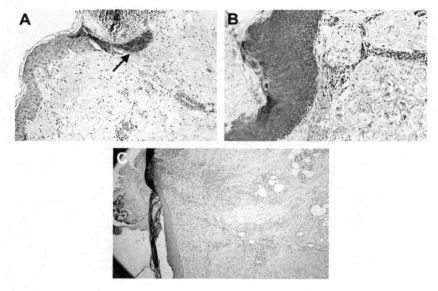

Chapter 15, Figure 15.5 (See Page 293 of this volume).

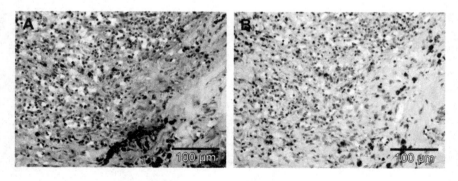

Chapter 17, Figure 17.1 (See Page 330 of this volume).

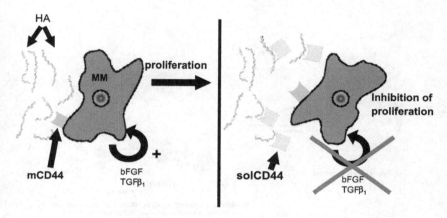

Chapter 17, Figure 17.2 (See Page 332 of this volume).

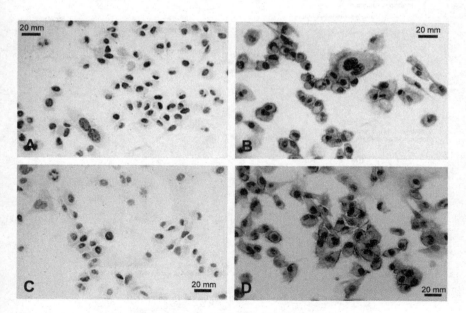

Chapter 18, Figure 18.2 (See Page 352 of this volume).

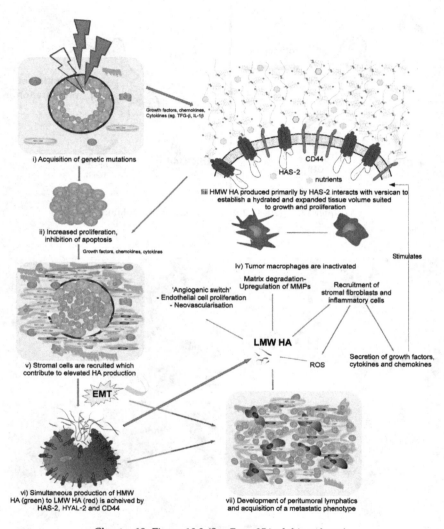

i) Acquisition of genetic mutations

ii) Increased proliferation, inhibition of apoptosis

Growth factors, chemokines, Cytokines (eg. TFG-β, IL-1β

CD44

HAS-2

nutrients

Iiii HMW HA produced primarily by HAS-2 interacts with versican to establish a hydrated and expanded tissue volume suited to growth and proliferation

Iv) Tumor macrophages are inactivated

Stimulates

Matrix degradation-
Upregulation of MMPs

Recruitment of stromal fibroblasts and inflammatory cells

'Angiogenic switch'
- Endothelial cell proliferation
- Neovascularisation

Growth factors, chemokines, cytokines

LMW HA

ROS

Secretion of growth factors, cytokines and chemokines

v) Stromal cells are recruited which contribute to elevated HA production

EMT

vi) Simultaneous production of HMW HA (green) to LMW HA (red) is acheived by HAS-2, HYAL-2 and CD44

vii) Development of peritumoral lymphatics and acquisition of a metastatic phenotype

Chapter 18, Figure 18.3 (See Page 354 of this volume).

Cognitive Approach to Natural Language Processing

Series Editor
Florence Sèdes

Cognitive Approach to Natural Language Processing

Edited by

Bernadette Sharp
Florence Sèdes
Wiesław Lubaszewski

First published 2017 in Great Britain and the United States by ISTE Press Ltd and Elsevier Ltd

ISTE Press Ltd
27-37 St George's Road
London SW19 4EU
UK

www.iste.co.uk

Elsevier Ltd
The Boulevard, Langford Lane
Kidlington, Oxford, OX5 1GB
UK

www.elsevier.com

Notices

Knowledge and best practice in this field are constantly changing. As new research and experience broaden our understanding, changes in research methods, professional practices, or medical treatment may become necessary.

Practitioners and researchers must always rely on their own experience and knowledge in evaluating and using any information, methods, compounds, or experiments described herein. In using such information or methods they should be mindful of their own safety and the safety of others, including parties for whom they have a professional responsibility.

To the fullest extent of the law, neither the Publisher nor the authors, contributors, or editors, assume any liability for any injury and/or damage to persons or property as a matter of products liability, negligence or otherwise, or from any use or operation of any methods, products, instructions, or ideas contained in the material herein.

For information on all our publications visit our website at http://store.elsevier.com/

British Library Cataloguing-in-Publication Data
A CIP record for this book is available from the British Library
Library of Congress Cataloging in Publication Data
A catalog record for this book is available from the Library of Congress
ISBN 978-1-78548-253-3

Printed and bound in the UK and US

Contents

Chapter 5. Hidden Structure and
Function in the Lexicon   91

Philippe VINCENT-LAMARRE, Mélanie LORD,
Alexandre BLONDIN-MASSÉ, Odile MARCOTTE,
Marcos LOPES and Stevan HARNAD

Chapter 6. Transductive Learning Games
for Word Sense Disambiguation   109

Rocco TRIPODI and Marcello PELILLO

Chapter 7. Use Your Mind and Learn to Write: The Problem of Producing Coherent Text

Michael ZOCK and Debela Tesfaye GEMECHU

Chapter 8. Stylistic Features Based on Sequential Rule Mining for Authorship Attribution

Mohamed Amine BOUKHALED and Jean-Gabriel GANASCIA

Chapter 9. A Parallel, Cognition-oriented Fundamental Frequency Estimation Algorithm 177

Ulrike GLAVITSCH

Chapter 10. Benchmarking n-grams, Topic Models and Recurrent Neural Networks by Cloze Completions, EEGs and Eye Movements 197

Markus J. HOFMANN, Chris BIEMANN and Steffen REMUS

Preface

This book is a special issue dedicated to exploring the relationship between natural language processing and cognitive science, and the contribution of computer science to these two fields. Poibeau and Vasishth [POI 16] noted that research interest in cognitive issues may have been given less attention because researchers from the cognitive science field are overwhelmed by the technical complexity of natural language processing; similarly, natural language processing researchers have not recognized the contribution of cognitive science to their work. We believe that the international workshops of Natural Language and Cognitive Science (NLPCS), launched in 2004, have provided a strong platform to support the consistent determination and diversity of new research projects which acknowledge the importance of interdisciplinary approaches and bring together computer scientists, cognitive and linguistic researchers to advance research in natural language processing.

This book consists of 10 chapters contributed by the researchers at the recent NLPCS workshops. In Chapter 1, Philippe Blache explains that the process of understanding language is theoretically very complex; it must be carried out in real time. This process requires many different sources of information. He argues that the global interpretation of a linguistic input is based on the grouping of elementary units called chunks which constitute the backbone of the "interpret whenever possible" principle which is responsible for delaying the understanding process until enough information becomes available. The following two chapters address the problem of human association. In Chapter 2, Korzycki, Gatkowska and Lubaszewski discuss an experiment based on 900 students who participated in a free word

xii   Cognitive Approach to Natural Language Processing

association test. They have compared the human association list with the association list retrieved from text using three algorithms: the Church and Hanks algorithm, the Latent Semantic Analysis and Latent Dirichlet Allocation. In Chapter 3, Lubaszewski, Gatkowska and Godny describe a procedure developed to investigate word associations in an experimentally built human association network. They argue that each association is based on the semantic relation between two meanings, which has its own direction and is independent from the direction of other associations. This procedure uses graph structures to produce a semantically consistent subgraph. In Chapter 4, Rapp investigates whether human language generation is governed by associations, and whether the next content word of an utterance can be considered as an association with the representations of the content words, already activated in the speaker's memory. He introduces the concept of the Reverse Association Task and discusses whether the stimulus can be predicted from the responses. He has collected human data based on the reverse association task, and compared them to the machine-generated results. In Chapter 5, Vincent-Lamarre and his colleagues have investigated how many words, and which ones, are required to define all the rest of the words in a dictionary. To this end, they have applied graph-theoretic analysis to the Wordsmyth suite of dictionaries. The results of their study have implications for the understanding of symbol grounding and the learning and mental representation of word meaning. They conclude that language users must have the vocabulary to understand the words in definitions to be able to learn and understand the meaning of words from verbal definitions. Chapter 6 focuses on word sense disambiguation. Tripodi and Pelillo have explored the evolutionary game theory approach to study word sense disambiguation. Each word to be disambiguated is represented as a player and each sense as a strategy. The algorithm has been tested on four datasets with different numbers of labeled words. It exploits relational and contextual information to infer the meaning of a target word. The experimental results demonstrate that this approach has outperformed conventional methods and requires a small amount of labeled points to outperform supervised systems. In Chapter 7, Zock and Tesfaye have focused on the challenging task of text production expressed in terms of four tasks: ideation, text structuring, expression and revision. They have focused on text structuring which involves the grouping (chunking), ordering and linking of messages. Their aim is to study which parts of text production can be automated, and whether the computer can build one or several topic trees based on a set of inputs provided by the user. Authorship attribution is the focus of study in

Chapter 8. Boukhaled and Ganascia have analyzed the effectiveness of using sequential rules of function words and Part-of-Speech (POS) tags as a style marker that does not rely on the bag-of-words assumption or on their raw frequencies. Their study has shown that the frequencies of function words and POS n-grams outperform the sequential rules. Fundamental frequency detection (F0), which plays an important role in human speech perception, is addressed in Chapter 9. Glavitsch has investigated whether F0 estimation, using the principles of human cognition, can perform equally well or better than state-of-the-art F0 detection algorithms. The proposed algorithm, which operates in the time domain, has achieved very low error rates and outperformed the state-of-the-art correlation-based method RAPT in this respect, using limited resources in terms of memory and computing power. In neurocognitive psychology, manually collected cloze completion probabilities (CCPs) are used to quantify the predictability of a word from sentence context in models of eye movement control. As these CCPs are based on samples of up to 100 participants, it is difficult to generalize a model across all novel stimuli. In Chapter 10, Hofmann, Biemann and Remus have proposed applying language models which can be benchmarked by item-level performance on datasets openly available in online databases. Previous neurocognitive approaches to word predictability from sentence context in electroencephalographic (EEG) and eye movement (EM) data relied on cloze completion probability (CCP) data. Their study has demonstrated that the syntactic and short-range semantic processes of n-gram language models and recurrent neural networks (RNN) can perform more or less equally well when directly accounting CCP, EEG and EM data. This may help generalize neurocognitive models to all possible novel word combinations.

Bibliography

[POI 16] POIBEAU T., VASISHTH S., "Introduction: Cognitive Issues in Natural Language Processing", *Traitement Automatique des Langues et Sciences Cognitives*, vol. 55, no. 3, pp. 7–19, 2016.

Bernadette SHARP
Florence SÈDES
Wiesław LUBASZEWSKI
March 2017

Delayed Interpretation, Shallow Processing and Constructions: the Basis of the "*Interpret Whenever Possible*" Principle

We propose in this chapter to investigate the "*interpret whenever possible*" principle that consists of delaying the processing mechanisms until enough information becomes available. This principle relies on the identification of elementary units called chunks, which are identified by means of basic features. These chunks are segments of the input to be processed. In some cases, depending on the accessibility of the information they bear, chunks can be linguistically structured elements. In other cases, they are simple segments. Chunks are stored in a buffer of the working memory and progressively grouped (on the basis of a cohesion measure) when possible, progressively identifying the different constructions of the input. The global interpretation of a linguistic input is then not based anymore on a word-by-word mechanism, but on the grouping of these constructions that constitute the backbone of the "*interpret whenever possible*" principle.

1.1. Introduction

From different perspectives, natural language processing, linguistics and psycholinguistics shed light on the way humans process language. However, this knowledge remains scattered: classical studies usually focus on language processing subtasks (e.g. lexical access) or modules (e.g. morphology,

---

Chapter written by Philippe BLACHE.

syntax), without being aggregated into a unified framework. It then remains very difficult to find a general model unifying the different sources of information into a unique architecture.

One of the problems lies in the fact that we still know only little about how the different dimensions of language (prosody, syntax, pragmatics, semantics, etc.) interact. Some linguistic theories exist, in particular within the context of *Construction Grammars* [FIL 88, GOL 03, BLA 16], that propose approaches making it possible to gather these dimensions and implement their relations. These frameworks rely on the notion of *construction*, which is a set of words linked by specific properties at any level (lexical, syntactic, prosodic, etc.) and with which a specific meaning, which is often non-transparent or accessible compositionally (e.g. idioms or multi-word expressions), can be associated. Interestingly, these theories also provide a framework for integrating multimodal information (verbal and non-verbal). Interpreting a construction (i.e. accessing to its associated meaning) results from the interaction of all the different dimensions. In this organization, processing a linguistic production is not a linear process, but uses mechanisms for a global recognition of the constructions. Contrarily to incremental architectures (see, e.g., [FER 02, RAY 09]), the syntactic, semantic and pragmatic processing is not done word-by-word, but more globally, on the basis of such constructions.

This conception of language processing requires a *synchronization* procedure for the alignment of all the different sources of information in order to identify a construction and access its meaning. In natural situations (e.g. conversations), the different input flows can be verbal (prosody, syntactic, pragmatics, etc.) and non-verbal (gestures, attitudes, emotions, context, etc.); they are not strictly temporally synchronized. It is then necessary to explain how information can be temporarily stored and its evaluation delayed until enough information becomes available. In this perspective, the input linguistic flow (being read or heard) is segmented into elements that can be of any form, partially or entirely recognized: segments of the audio flow, set of characters, and also, when possible, higher-level segments made of words or even clusters of words. In this chapter, we address these problems through several questions:

1) What is the nature of the delaying mechanism?

2) What is the nature of the basic units and how can they be identified?

3) How is the delaying mechanism implemented?

1.2. Delayed processing

Different types of delaying effects can occur during language processing. For example, at the brain level, it has been shown that language processing may be impacted by the presentation rate of the input. This phenomena has been investigated in [VAG 12], claiming that when the presentation rate increases and becomes faster than the processing speed, intelligibility can collapse. This is due to the fact that language network seems to work in a constant of time: cortical processing speed is shown by the authors to be tightly constrained and cannot be easily accelerated. As a result, when the presentation rate increases, the processing speed remaining constant, a blocking situation can suddenly occur. Concretely, this means that when the presentation rate is accelerated, and because the processing speed remains constant, part of the input stream has to be buffered. Experiments show that the rate can be accelerated to 40% before reaching a collapse of intelligibility. This situation occurs when the buffer becomes saturated and is revealed at the cortical level by the fact that the activation of the higher-order language areas (that are said to reflect intelligibility [FRI 10]) drops suddenly, showing that the input signal becomes unintelligible.

This model suggests that words can be processed immediately when presented at a slow rate, in which case the processing speed is that of the sensory system. However, when the rate increases and words are presented more rapidly, the processing speed limit is reached and words cannot be processed in real time anymore. In such a situation, words have to be stored in a buffer, from which they are retrieved in a *first-in-first-out* manner, when cognitive resources become available again. When the presentation rate is higher than the processing speed, the number of words to be stored increases. A lock occurs when the maximal capacity of the buffer is reached, entailing a collapse of intelligibility.

Besides this buffering mechanism, other cues indicate that the input is probably not processed linearly, word-by-word, but rather only from time to time. This conception means that even in normal cases (i.e. without any intelligibility issue), the interpretation is only done periodically, the basic units being stored before being processed. Several studies have investigated such a phenomenon. At the cortical level, the analysis of stimulus intensity fluctuation reveals the presence of specific activity (spectral peaks) after phrases and sentences [DIN 16]. The same type of effect can also be found in eye-movement during reading: longer fixations are observed when reading words that end a phrase or a sentence. This *wrap-up effect* [WAR 09], as well as the presence of different timescales at the cortical level described above, constitute cues in favor of a delaying mechanism in which basic elements are stored temporarily, and an integration operation is triggered when enough material becomes available for the interpretation.

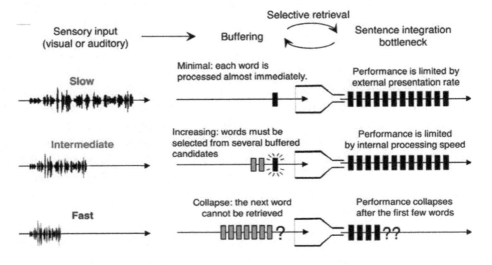

Figure 1.1. *Illustration of the bottleneck situation, when the presentation rate exceeds the processing speed (reproduced from [VAG 12])*

At the semantic level, other evidence also shows that language processing, or at least language interpretation, is not strictly incremental. Interesting experiments have been performed, which reveal that language comprehension can remain very superficial: [ROM 13] has shown that, in an idiomatic context, the access to the meaning of words can be completely switched off, replaced by a global access at the level of idiom. This effect has been shown at the cortical level: when introducing a semantic violation within an idiom, there is no difference between hard and soft semantic violations (which is not the case in a comparable non-idiomatic context); in some cases, processing a word does not mean integrating it into a structure. On the contrary, in this situation there is a simple shallow process of scanning the word, without doing any interpretation. The same type of observation has been made in reading studies: depending on the task (e.g. when very simple comprehension questions are expected), the reader may apply a superficial treatment [SWE 08]. This effect is revealed by the fact that ambiguous sentences are read faster, meaning that no resolution is done and the semantic representation remains underspecified. Such variation in the level of processing depends then on the context: when the pragmatic and semantic context carries enough information, it renders the complete processing mechanism useless, the interpretation being predictable. At the attentional level, this observation is confirmed in [AST 09], showing that the allocation of attentional resources to certain time windows depends on its predictability: minimal attention is allocated when information is predictable or, on the contrary, maximal attention is involved in case of mismatch with expectations. The same type of variation is observed when the listener adapts its perceptual strategy to the speakers, applying *perceptual accommodation* [MAG 07].

These observations are in line with the *good-enough theory* [FER 07] for which the interpretation of complex material is often considered to be shallow and incomplete. This model suggests that interpretation is only done from time to time, on the basis of a small number of adjacent words, and delaying the global interpretation until enough material becomes available. This framework and the evidence on which it relies also reinforce the idea that language processing is generally not linear and word-by-word. On the contrary, it can be very shallow and delayed when necessary.

1.3. Working memory

The delaying mechanism relies implicitly on a storage device which is implemented in the *short-term memory*, which is the basis of the cognitive system organization, by making it possible to temporarily store pieces of information of any nature. In general, it is considered that this memory is mainly devoted to storage. However, a specific short-term memory, called *working memory*, also allows for the manipulation of the information and a certain level of processing. It works as a buffer in which part of the information, which can be partially structured, is stored. Some models [BAD 86, BAD 00] propose an architecture in which the working memory plays the role of a supervisor, on top of different sensory-motor loops as well as an episodic buffer.

One important feature of the working memory (and short-term memory in general) is its limited capacity. In a famous paper, [MIL 56] evaluated this limit to a "magic" number of seven units. However, it has been observed that units to be stored in this memory are not necessarily atomic; they can also constitute groups that are considered as single units. For example, stored elements can be numbers, letters, words, or even sequences, showing that groups can be encoded as single units. In this case, the working memory does not directly store the set of elements, but more probably the set of pointers towards the location of the elements in another (lower) part of the short-term memory. These types of higher-level elements are called *chunks*, which basically consist, in the case of language, of a set of words.

Working memory occupies a central position in cognitive architectures such as ACT-R (*Adaptive Character of Thought-Rational*, see [AND 04]). In this model, short-term information (chunks) is stored in a set of buffers. The architecture, in the manner of that proposed by [BAD 86], is organized around a set of modules (manual control, visual perception, problem state, control state and declarative memory) coordinated by a supervising system (the production system). Each module is associated with a buffer that contains one chunk, defined as a unit containing a small amount of information. Moreover, in this organization, each buffer can contain only one unit of knowledge.

ACT-R has been applied to language processing, in which short-term buffers play the role of an interface between procedural and declarative memories (the different types of linguistic knowledge) [LEW 05, REI 11]. Buffers contain chunks (information units) that are represented as lists of attribute-value pairs. Chunks are stored in the memory, they form a unit and they can be directly accessible, as a whole. Their accessibility depends on a level of *activation*, making it possible to control their retrieval in the declarative memory. A chunk's activation consists of several parameters: latency since its last retrieval, weights of the elements in relation to the chunk as well as the strength of these relations. It can be integrated into the following formula, quantifying the activation A of a chunk i:

$$A_i = B_i + \sum_j W_j S_{ji}$$ [1.1]

In this formula, B represents the basic activation of the chunk (its frequency and the recency of its retrieval), W indicates the weights of the terms in relation to i and S the strength of the relations linking other terms to the chunks. It is then possible to associate a chunk with its level of activation. The interesting point is that chunk activation is partially dependent on the context: the strength of the relations with other elements has a consequence on the level of activation, controlling its probability as well as the speed of its retrieval.

This architecture implicitly contains the idea of delayed evaluation: the basic units are first identified and stored into different buffers, containing pieces of information that can be atomic or structured. Moreover, this proposal also gives indications on the type of the retrieval. The different buffers in which chunks are stored are not implemented as a stack, following a *first-in-first-out* retrieval mechanism. On the contrary, chunks can be retrieved in any order, with a preference given first to that with the higher activation value.

The ACT-R model and the activation notion give a more precise account of comprehension difficulties. In the previous section, we have seen that they can be the consequence of a buffer saturation (in computational terms, a *stack overflow*). Such difficulties are controlled thanks to the *decay of accessibility* of stored information [LEW 05]. This explanation is complementary with observations presented in the previous section: the activation level has a correlation with the processing speed. Chunks with a high activation will be retrieved rapidly, decreasing the number of buffered elements. When many chunks have a low activation, the processing speed decreases, resulting in a congestion of the buffers.

One important question in this architecture is the role of working memory in procedural operations, and more precisely the construction of the different elements to be stored. In some approaches, working memory plays a decisive role in terms of integration: basic elements (lexical units) are assembled into structured ones, as a function of their activation. In this organization, working memory becomes the site where linguistic analysis is done. This is what has been proposed, for example, in the *"Capacity theory of comprehension"* [JUS 92] for which working memory plays a double role of storage and processing. In this theory, elements of any level can be stored and accessed: words, phrases, thematic structures, pragmatic information, etc. It is, however, difficult to explain how such model can implement at the same time a delaying aspect (called *"wait-and-see"* by the authors) and an incremental comprehension system interpreting step-by-step. In their study on memory capacity, [VAG 12] propose a simpler view with a unique input buffer whose role is limited to storing words. In our approach, we adopt an intermediate position in which the buffer is limited to storage, but elements of different types can be stored, including partially structured ones such as chunks.

1.4. How to recognize chunks: the segmentation operations

The hypothesis of a delayed evaluation in language processing not only relies on a specific organization of the memory, but also requires a mechanism for the identification of the elements to be stored in the buffer. Two important questions are to be answered here: what is the nature of these elements, and how can they be identified. Our hypothesis relies on the idea that no deep and precise linguistic analysis is done at a first stage. If so, the question is to explain and describe the mechanisms, necessarily at a low level, for the identification of the stored elements.

These questions are more generally related to the general problem of segmentation. Given an input flow (e.g. connected speech), what types of element can be isolated and how? Some mechanisms, specific to the audio signal, are at work in speech segmentation. Many works addressing this question ([MAT 05], [GOY 10], [NEW 11], [END 10]) exhibit different cues, at different levels, that are used in particular (but not only) for word segmentation tasks, among which:

– *Prosodic level*: stress, duration and pitch information can be associated in some languages with specific positions in the word (e.g. initial or final), helping in detecting the word boundaries.

– *Allophonic level*: phonemes are variable and their realization can depend on their position within words.

– *Phonotactic level*: constraints on the ordering of the phonemes, which gives information about the likelihood that a given phoneme is adjacent to another one within and between words.

– *Statistical/distributional properties*: transitional probabilities between consecutive syllables.

Word segmentation results from the satisfaction of multiple constraints encoding different types of information, such as phonetic, phonological, lexical, prosodic, syntactic, semantic, etc. (see [MCQ 10]). However, most of these segmentation cues are at a low level and do not involve an actual lexical access. In this perspective, what is interesting is that some segmentation mechanisms are not dependent on the notion of word and then can also be used in other tasks than word segmentation. This is very important because the notion of word is not always relevant (because involving rather high-level features, including semantic ones). In many cases, other types of segmentations are used, without involving the notion of words, but staying at the identification of larger segments (e.g. prosodic units), without entering into a deep linguistic analysis.

At a higher level, [DEH 15] has proposed isolating five mechanisms making it possible to identify sequence knowledge:

– *Transition and timing knowledge*: when presenting a sequence of items (of any nature), at a certain pace, the transition between two items is anticipated thanks to the approximate timing of the next item.

– *Chunking*: contiguous items can be grouped into the same unit, thanks to the identification of certain regularities. A chunk is simply defined here in terms of a set of contiguous items that frequently co-occur and then can be encoded as a single unit.

– *Ordinal knowledge*: a recurrent linear order, independently of any timing, constitutes information for the identification of an element and its position.

– *Algebraic patterns*: when several items have an internal regular pattern, their identification can be done thanks to this information.

– *Nested tree structures generated by symbolic rules*: identification of a complex structure, gathering several items into a unique element (typically a phrase).

What is important in these sequence identification systems (at least the first four of them) is the fact that they apply to any type of information and rely on low-level mechanisms, based on the detection of regularities and when possible their frequency. When applied to language, these systems explain how syllables, patterns or groups can be identified directly. For example, algebraic patterns are specific to a certain construction such as in the following example, taken from a spoken language corpus: *"Monday, washing, Tuesday, ironing, Wednesday, rest"*. In this case, without any syntactic or high-level processing, and thanks to the regularity of the pattern /*date - action*/, it is possible to segment the three subsequences and group them into a unique general one. In this case, a very basic mechanism, *pattern identification*, offers the possibility to identify a construction (and access directly to its meaning).

When putting together the different mechanisms described in this section, we obtain a strong set of parameters that offer the possibility of segmenting the input into units. In some cases, when cues are converging enough, the segments can be words. In other cases, they are larger units. For example, long breaks (higher than 200ms) are a universal segmentation constraint in prosody: two such breaks identify the boundaries of a segment (that can correspond to a prosodic unit).

As a result, we can conclude that several basic mechanisms, which do not involve deep analysis, make it possible to segment the linguistic input, be it read or heard. Our hypothesis is that these segments are the basic units stored initially in the buffers. When possible, the stored units are words, but not necessarily. In the general case, they are sequences of characters or phonemes that can be retrieved later. This is what occurs when hearing a speaker without understanding: the audio segment is stored and accessed later when other sources of information (e.g. the context) become available and make it possible to refine the segmentation into words.

1.5. The delaying architecture

Following the different elements presented so far, we propose integrating the notion of delayed evaluation and chunking into the language processing organization. This architecture relies on the idea that the interpretation of a sentence (leading to its comprehension) is only done *whenever possible*, instead of word-by-word. The mechanism consists of accumulating enough information before any in-depth processing. Doing this means: first, the capacity to identify atomic units without making use of any deep parsing, and; second, to store these elements and retrieve them when necessary.

We do not address here the question of building an interpretation, but focus only on this preliminary phase of accumulating pieces of information. This organization relies on a two-stage process distinguishing between a first level of packaging and a second corresponding to a deeper analysis. Such a distinction recalls the well-known *"Sausage Machine"* [FRA 78] that distinguishes a first phase called the *Preliminary Phrase Packager (PPP)*, consisting of identifying the possible groups (or chunks) in a limited window made of 6 or 7 words. In this proposal, the groups correspond to phrases that can be incomplete. The second level is called the *Sentence Structure Supervisor (SSS)* and it groups the units produced in the *PPP* into larger structures. In this classical architecture, each level involves a certain type of parsing, relying on grammatical knowledge. Moreover, the interpretation is supposed to be done starting from the identification of the syntactic structure, in a classical compositional perspective.

Our proposal also relies on a two-stage organization:

1) segmenting and storing;

2) aggregating complex chunks.

However, this model does not have any *a priori* on the type of units to be built: they are not necessarily phrases, they can be simply made of unstructured segments of the input. Moreover, the second stage is not obligatory: the recognition of a construction, and the interpretation of the corresponding subpart of the input, can be done at the first level.

We detail in the following these two stages, on the basis of the more general *"interpretation whenever possible"* organization.

1.5.1. *Segment-and-store*

The first stage when processing a linguistic input (text or speech) is the segmentation into atomic chunks. Atomic means here that no structure is built, chunks being only segments of the input, identified thanks to low-level parameters. In other words, no precise analysis of the input is performed, the mechanism consisting of gathering all possible information available immediately. As a result, because the level of precision of the information can be very different, chunks can be of many different types and levels. Some of the segmentation mechanisms are indeed very general or even universal. For example, the definition of "*inter-pausal units*" relies on the identification of long breaks in the audio signal. The resulting chunk is a long sequence of phonemes without internal organization or sub-segmentation. In some (rare) cases, no other features than long breaks are available and the chunk remains large and stored as such. However, in most of the situations, more information is available, making it possible to identify finer chunks, and when possible words. Several such segmenting features exist, in particular:

– *Prosodic contours, stress*: pitch, breaks, duration and stress may indicate word boundaries.

– *Phonotactic constraints*: language-dependent constraints on the sequence of phonemes. The violation of such constraints may indicate boundaries.

– *Lexical frequency units*: in some cases, an entire unit can be highly predictable (typically very frequent words, named entities, etc.), making it possible to directly segment the input.

These features are subject to high variation and do not lead to a segmentation in all cases. When ambiguity is high, no finer segmentation is done at this stage. On the contrary, these low-level features can often lead to the possibility of segmenting into words. What is important is that these features correspond to information that can be directly assessed, independently of any other property or knowledge.

At this first stage, atomic chunks are stored into the buffers. We present in the following section the next step of this pre-processing phase, consisting of aggregating chunks.

1.5.2. *Aggregating by cohesion*

Constructions can be described as a set of interacting properties. This definition offers the possibility to conceive a measure based on the number of these properties and their weights, as proposed in [BLA 16]. At the syntactic level, the set of properties describing a construction corresponds to a graph in which nodes are words and edges represent the relations. The graph density then constitutes a first type of measure: a high density of the graph corresponds to a high number of properties, representing a certain type of cohesion between the words. Moreover, the quality of these relations can also be evaluated, some properties being more important than others (which are represented by their weighting). A high density of hard properties (i.e. with heavy weights) constitute a second type of information. Finally, some sentences can be non-canonical, bearing certain properties that are violated (e.g. in case of agreement or linear precedence violation). Taking into consideration the number of violated properties in comparison with the satisfied ones is the last type of indication we propose to use in the evaluation of the cohesion.

Our hypothesis is that a correlation exists between the cohesion measure, defined on the basis of these three types of information, and the identification of a construction. In other words, a construction corresponds to a set of words linked with a high number of properties, of heavy weights, with no or few violations.

The first parameter of the cohesion measure relies on the number of properties that are assessed for a given construction, in comparison with the possible properties in the grammar. The following graph illustrates the set of properties *in the grammar* describing the nominal construction[1]:

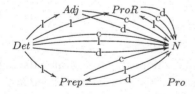

$$[1.2]$$

The number of possible relations in which a category is involved can be estimated by the number of incident relations of the corresponding vertex in the graph (called in graph theory the *vertex* degree). We then propose to define the degree of a category by this measure. In the previous graph, we have the following degrees: $deg_{[gram]}(N) = 9; deg_{[gram]}(ProR) = 2; deg_{[gram]}(Adj) = 1$.

---

1 The letters d, l, c stand respectively for *dependency, linearity* and *co-occurrence* properties.

During a parse (i.e. knowing the list of categories), the same type of evaluation can be applied to the constraint graph describing a construction, as in the following example:

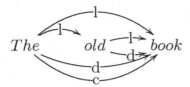

Each word is involved in a set of relations. The degree of a word is, similarly to the grammar, the set of incident edges of a word. In this example, we have: $deg_{[sent]}(N) = 5; deg_{[sent]}(Adj) = 1; deg_{[sent]}(Det) = 0$.

The first parameter of our estimation of the cohesion relies on a comparison of these two values: for a given word, we know from the grammar the number of properties in which it could theoretically be involved. We also know from the parsing of a given sentence how many of these properties are effectively assessed. We can then define a value, the *completeness ratio*, indicating the density of the category: the higher the number of relations in the grammar being verified, the higher the completeness value:

$$Comp(cat) = \frac{deg_{[sent]}(cat)}{deg_{[gram]}(cat)} \qquad [1.3]$$

Besides this completeness ratio, it is also interesting to examine the density of the constraint graph itself. In graph theory, this value is calculated as a ratio between the number of edges and the number of vertices. It is more precisely defined as follows (S is the constraint graph of a sentence, E the set of edges and V the set of vertices):

$$Dens(S) = \frac{|E|}{5 * |V|(|V| - 1)} \qquad [1.4]$$

In this formula, the numerator is the number of existing edges and the denominator is the total number of possible edges (each edge connecting two different vertices, multiplied by 5, the number of different types of properties). This value makes it possible to distinguish between *dense* and *sparse* graphs. In our hypothesis, a dense graph is correlated with a construction.

The last parameter taken into account is more qualitative and takes into account the weights of the properties. More precisely, we have seen that all properties can be either satisfied or violated. We define then a normalized satisfaction ratio as follows (where W^+ is the sum of the weights of the satisfied properties and W^- that of the violated ones):

$$Sat(S) = \frac{W^+ - W^-}{W^+ + W^-}$$ [1.5]

Finally, the cohesion value can be calculated as a function of the three previous parameters as follows (C being a construction and G_C its corresponding constraint graph):

$$Cohesion(C) = \sum_{i=1}^{|S|} Comp(w_i) * Dens(G_C) * Sat(G_C)$$ [1.6]

Note that the *density* and *satisfaction* parameters can be evaluated directly, without depending on the context and without needing to know the type of the construction. On the contrary, evaluating the *completeness* parameter requires knowing the construction in order to extract from the grammar all the possible properties that describe it. In a certain sense, the two first parameters are *basic*, in the same sense as described for properties, and can be assessed automatically.

The *cohesion* measure offers a new estimation of the notion of *activation*. Moreover, it also provides a way to directly identify constructions on the basis of simple properties. Finally, it constitutes an explicit basis for the implementation of the general parsing principle stipulating that constructions or chunks are set of words with a high density of relations of heavy weights. This definition corresponds to the *Maximize On-line Processing* principle [HAW 03], which stipulates that *"the human parser prefers to maximize the set of properties that are assignable to each item X as X is parsed. [...] The maximization difference between competing orders and structures will be a function of the number of properties that are misassigned or unassigned to X in a structure S, compared with the number in an alternative"*.

This principle offers a general background to our conception of language processing. Instead of building a syntactic structure serving as support of the comprehension of a sentence, the mechanisms consist of a succession of chunks, maximizing the cohesion function estimated starting from the available information. When the density of information (or the cohesion) reaches a certain threshold, the elements can be grouped into a unique chunk, stored in the working memory. When the threshold is not reached, the state of the buffer is not modified and a new element of the input stream is scanned. This general parsing mechanism offers the possibility to integrate different sources of information when they become available by delaying the evaluation, waiting until a certain threshold of cohesion can be identified. This constitutes a framework for implementing the basic processing of the good-enough theory: *interpret whenever possible*.

1.6. Conclusion

Understanding language is theoretically a very complex process, involving many different sources of information. Moreover, it has to be done in real time. Fortunately, in many cases, the understanding process can be facilitated thanks to different parameters: predictability of course, and also the fact that entire segments of the input can be processed directly. This is the case of most of the *constructions*, in which the meaning can be accessed directly, the construction being processed as a whole. At a lower level, it is also possible to identify subparts of the input (e.g. patterns, prosodic units, etc.) from which global information can be retrieved directly. Different observations show that low-level features usually make it possible to identify such global segments. The language processing architecture we propose in this chapter relies on this: instead of recognizing words and then trying to integrate them step-by-step into a syntactic structure to be interpreted, segments are first identified. These segments can be of any type: sequences of phonemes, words, group of words, etc. Their common feature is that they do not need any deep level information or process to be recognized.

Once the segments (called *chunks*) are identified, they are stored in a buffer, without any specific interpretation. In other words, the interpretation mechanism is *delayed* until enough information becomes available. When a new chunk is buffered, an evaluation of its *cohesion* with the existing ones in the buffer is done. When the cohesion between different chunks (that corresponds to the notion of activation in cognitive architectures) reaches a certain threshold, they are merged into a unique one, replacing them in the buffer, as a single unit. This mechanism makes it possible to progressively recognize constructions and directly access their meaning.

This organization, instead of a word-by-word incremental mechanism, implements the *"interpret whenever possible"* principle. It constitutes a framework for explaining all the different delaying and shallow processing mechanisms that have been observed.

1.7. Bibliography

[AND 04] ANDERSON J.R., BOTHELL D., BYRNE M.D. *et al.*, "An integrated theory of the mind", *Psychological Review*, vol. 111, no. 4, pp. 1036–1060, 2004.

[AST 09] ASTHEIMER L.B., SANDERS L.D., "Listeners modulate temporally selective attention during natural speech processing", *Biological Psychology*, vol. 80, no. 1, pp. 23–34, 2009.

[BAD 86] BADDELEY A., *Working Memory*, Clarendon Press, Oxford, 1986.

[BAD 00] BADDELEY A., "The episodic buffer: a new component of working memory?", *Trends in Cognitive Sciences*, vol. 4, no. 11, pp. 417–423, 2000.

[BLA 16] BLACHE P., "Representing syntax by means of properties: a formal framework for descriptive approaches", *Journal of Language Modelling*, vol. 4, no. 2, 2016.

[DEH 15] DEHAENE S., MEYNIEL F., WACONGNE C. *et al.*, "The neural representation of sequences: from transition probabilities to algebraic patterns and linguistic trees", *Neuron*, vol. 88, no. 1, 2015.

[DIN 16] DING N., MELLONI L., ZHANG H. *et al.*, "Cortical tracking of hierarchical linguistic structures in connected speech", *Nature Neuroscience*, vol. 19, no. 1, pp. 158–164, 2016.

[END 10] ENDRESS A.D., HAUSER M.D., "Word segmentation with universal prosodic cues", *Cognitive Psychology*, vol. 61, no. 2, pp. 177–199, 2010.

[FER 02] FERREIRA F., SWETS B., "How incremental is language production? Evidence from the production of utterances requiring the computation of arithmetic sums", *Journal of Memory and Language*, vol. 46, no. 1, pp. 57–84, 2002.

[FER 07] FERREIRA F., PATSON N.D., "The 'Good Enough' approach to language comprehension", *Language and Linguistics Compass*, vol. 1, no. 1, 2007.

[FIL 88] FILLMORE C.J., "The mechanisms of 'Construction Grammar'", *Proceedings of the Fourteenth Annual Meeting of the Berkeley Linguistics Society*, pp. 35–55, 1988.

[FRA 78] FRAZIER L., FODOR J.D., "The sausage machine: a new two-stage parsing model", *Cognition*, vol. 6, no. 4, pp. 291–325, 1978.

[FRI 10] FRIEDERICI A., KOTZ S., SCOTT S. *et al.*, "Disentangling syntax and intelligibility in auditory language comprehension", *Human Brain Mapping*, vol. 31, no. 3, pp. 448–457, 2010.

[GOL 03] GOLDBERG A.E., "Constructions: a new theoretical approach to language", *Trends in Cognitive Sciences*, vol. 7, no. 5, pp. 219–224, 2003.

[GOY 10] GOYET L., DE SCHONEN S., NAZZI T., "Words and syllables in fluent speech segmentation by French-learning infants: an ERP study", *Brain Research*, vol. 1332, no. C, pp. 75–89, 2010.

[HAW 03] HAWKINS J., "Efficiency and complexity in grammars: three general principles", in MOORE J., POLINSKY M. (eds), *The Nature of Explanation in Linguistic Theory*, CSLI Publications, 2003.

[JUS 92] JUST M.A., CARPENTER P.A., "A capacity theory of comprehension: individual differences in working memory", *Psychological Review*, vol. 99, no. 1, pp. 122–149, 1992.

[LEW 05] LEWIS R.L., VASISHTH S., "An activation-based model of sentence processing as skilled memory retrieval", *Cognitive Science*, vol. 29, pp. 375–419, 2005.

[MAG 07] MAGNUSON J., NUSBAUM H., "Acoustic differences, listener expectations, and the perceptual accommodation of talker variability", *Journal of Experimental Psychology: Human Perception and Performance*, vol. 33, no. 2, pp. 391–409, 2007.

[MAT 05] MATTYS S.L., WHITE L., MELHORN J.F., "Integration of multiple speech segmentation cues: a hierarchical framework", *Journal of Experimental Psychology*, vol. 134, no. 4, pp. 477–500, 2005.

[MCQ 10] MCQUEEN J.M., "Speech perception", in LAMBERTS K., GOLDSTONE R. (eds), *The Handbook of Cognition*, Sage, London, 2010.

[MIL 56] MILLER G., "The magical number seven, plus or minus two: some limits on our capacity for processing information", *Psychological Review*, vol. 63, no. 2, pp. 81–97, 1956.

[NEW 11] NEWMAN R.S., SAWUSCH J.R., WUNNENBERG T., "Cues and cue interactions in segmenting words in fluent speech", *Journal of Memory and Language*, vol. 64, no. 4, 2011.

[RAY 09] RAYNER K., CLIFTON C., "Language processing in reading and speech perception is fast and incremental: implications for event related potential research", *Biological Psychology*, vol. 80, no. 1, pp. 4–9, 2009.

[REI 11] REITTER D., KELLER F., MOORE J.D., "A computational cognitive model of syntactic priming", *Cognitive Science*, vol. 35, no. 4, pp. 587–637, 2011.

[ROM 13] ROMMERS J., DIJKSTRA T., BASTIAANSEN M., "Context-dependent semantic processing in the human brain: evidence from idiom comprehension", *Journal of Cognitive Neuroscience*, vol. 25, no. 5, pp. 762–776, 2013.

[SWE 08] SWETS B., DESMET T., CLIFTON C. *et al.*, "Underspecification of syntactic ambiguities: evidence from self-paced reading", *Memory and Cognition*, vol. 36, no. 1, pp. 201–216, 2008.

[VAG 12] VAGHARCHAKIAN L.G.D.-L., PALLIER C., DEHAENE S., "A temporal bottleneck in the language comprehension network", *Journal of Neuroscience*, vol. 32, no. 26, pp. 9089–9102, 2012.

[WAR 09] WARREN T., WHITE S.J., REICHLE E.D., "Investigating the causes of wrap-up effects: evidence from eye movements and E–Z reader", *Cognition*, vol. 111, no. 1, pp. 132–137, 2009.

Can the Human Association Norm Evaluate Machine-Made Association Lists?

This chapter presents a comparison of a word association norm created by a psycholinguistic experiment to association lists generated by algorithms operating on text corpora. We compare lists generated by the Church–Hanks algorithm and lists generated by the LSA algorithm. An argument is presented on how those automatically generated lists reflect semantic dependencies present in human association norm, and that future comparisons should take into account a deeper analysis of the human association mechanisms observed in the association list.

2.1. Introduction

For more than three decades, there has been a commonly shared belief that word occurrences retrieved from a large text collection may define the lexical meaning of a word. Although there are some suggestions that co-occurrences retrieved from texts [RAP 02, WET 05] reflect the text's contiguities, there also exist suggestions that algorithms, such as the LSA, are unable to distinguish between co-occurrences which are corpus-independent semantic dependencies (elements of a semantic prototype) and co-occurrences which are corpus-dependent factual dependencies [WAN 05, WAN 08]. We shall adopt the second view to show that existing statistical algorithms use mechanisms which improperly filter word co-occurrences retrieved from texts. To prove this supposition, we shall

Chapter written by Michał KORZYCKI, Izabela GATKOWSKA and Wiesław LUBASZEWSKI.

compare the human association list to the association list retrieved from a text by three different algorithms, i.e. the Church–Hanks [CHU 90] algorithm, the Latent Semantic Analysis (LSA) algorithm [DEE 90] and the Latent Dirichlet Allocation (LDA) algorithm [BLE 03].

LSA is a word/document matrix rank reduction algorithm, which extracts word co-occurrences from within a text. As a result, each word in the corpus is related to all co-occurring words and all texts in which it occurs. This makes a base for an associative text comparison. The applicability of the LSA algorithm is the subject of various types of research, which range from text content comparison [DEE 90] to the analysis of human association norm [ORT 12]. However, there is still little interest in studying the linguistic significance of machine-made associations.

It seems obvious that a comparison of the human association norm and machine-created association list should be the base of this study. And we can find some preliminary studies based on such a comparison: [WAN 05, WET 05, WAN 08], the results of which show that the problem needs further investigation. It is worth noting that all the types of research referred to used a stimulus–response association strength to make a comparison. The point is that, if we compare association strength computed for a particular stimulus–response pair in association norms for different languages, we can find that association strength differs, e.g. butter is the strongest (0.54) response to stimulus bread in the Edinburgh Associative Thesaurus (EAT), but in the Polish association norm described below the association chleb "bread"–masło "butter" is not the strongest (0.075). In addition, we can observe that association strength may not distinguish a semantic and non-semantic association, e.g. roof 0.04, Jack 0.02 and wall 0.01, which are responses to the stimulus "house" in EAT. Therefore, we decided to test machine-made association lists against human association norms excluding association strength. As a comparison, we use the norm made by Polish speakers during a free word association experiment [GAT 14], hereinafter referred to as the author's experiment. Because both LSA and LDA use the whole text to generate word associations, we also tested human associations against the association list generated by the Church–Hanks algorithm [CHU 90], which operates on a sentence-like text window. We also used three different text corpora.

2.2. Human semantic associations

2.2.1. *Word association test*

Rather early on, it was noted that words in the human mind are linked. American clinical psychologists G. Kent and A.J. Rosanoff [KEN 10] perceived the diagnostic usefulness of an analysis of the links between words. In 1910, the duo created and conducted a test of the free association of words. They conducted research on 1,000 people of varied educational backgrounds and professions, asking their research subjects to give the first word that came into their minds as the result of a stimulus word. The study included 100 stimulus words (principally nouns and adjectives). The Kent–Rosanoff list of words was translated into several languages, in which this experiment was repeated, thereby enabling comparative research to be carried out. Word association research was continued in [PAL 64], [POS 70], [KIS 73], [MOS 96], [NEL 98], and the repeatability of results allowed the number of research subjects to be reduced, while at the same time increasing the number of word stimuli to be employed, for example 500 kids and 1,000 mature research subjects and 200 words [PAL 64] or 100 research subjects and 8,400 words [KIS 73]. Research on the free association of words has also been conducted in Poland [KUR 67], the results of which are the basis for the experiment described below.

Computational linguistics also became involved in research on the free association of words, though at times these experiments did not employ the rigors used by psychologists when conducting experiments, for example, those that permitted the possibility of providing several responses to an individual stimulus word [SCH 12] or those that used word pairs as a stimulus [RAP 08].

There exist some algorithms which generate an association list on the basis of text corpora. However, automatically generated associations were rather reluctantly compared with the results of psycholinguistic experiments. The situation is changing; Rapp's results [RAP 02] were really encouraging.

Finally, association norms are useful for different tasks, for example information extraction [BOR 09] or dictionary expansion [SIN 04, BUD 06].

2.2.2. *The author's experiment*

Some 900 students of the Jagiellonian University and AGH University of Technology participated in a free word association test as described in this chapter. A Polish version of the Kent–Rosanoff list of stimulus words, which was previously used by I. Kurcz, was employed [KUR 67]. After an initial analysis, it was determined that we would employ as a stimulus, each word from the Kent–Rosanoff list, which grammatically speaking is a noun, as well as the five most frequent word associations for each of those nouns obtained in Kurcz's experiment [KUR 67]. If given associations appeared for various words, for example, *white* for *doctor, cheese, sheep*, that word as a stimulus appeared only once in our experiment. The resulting stimulus list contained 60 words from the Kent–Rosanoff list, in its Polish version, as well as 260 words representing those associations (responses) which most frequently appeared in Kurcz's research. It, therefore, is not an exact repetition of the experiment conducted 45 years ago.

The conditions of the experiment conducted, as well as the method of analyzing the results, have been modified. The experiment was conducted in a computer lab, with the aid of a computer system which was created specifically for the requirements of this experiment. This system presents a list of stimuli and then stores the associations in a data base. Instructions appeared on the computer screens of each participant, which in addition were read aloud by the person conducting the experiment. After the instructions were read, the experiment commenced, whereby a stimulus word appeared on the computer screen of each participant, who then wrote the first free association word which came to their mind – only one response was possible. Once the participant wrote down their association (or the time ran out for him to write down his association), the next stimulus word appeared on the screen, until the experiment was concluded. The number of stimulus words, as well as their order, was the same for all participants.

As a result, we obtained 260 association lists which consisted of more than 16,000 associated words. Association list derived from the experiment will be used to evaluate algorithm-derived association lists.

2.2.3. *Human association topology*

In this chapter, the associations coming from various sources are compared based on ranked word lists. This does not reflect, however, the complex structure of human associations. These can be represented as weighted graphs with specific words in nodes and associations in the vertices. This graph can be then sub-divided into subnets by starting from a specific stimulus (word) and cutting off the net at a certain distance from this central stimulus. Those subnets can be treated as representative for a specific meaning of a word. The strongest associations are always correlated with the fact that they are bi-directional. But if we look at the each pair of connected words to find what the connection means, we see that connections can differ in meaning, e.g. *home–mother* says that a home is a place specific to a mother in contrast to *home–roof* which says that roof is a part of a building. Having analyzed all the pairs we may find that some of them connect the stimulus word in the same way, e.g. *parents* and *family* connect the *home* on the same principle as *mother*, and *chimney* and *wall* are parts of a building along with the *roof*. This observation shows that the lexical meanings of the stimulus word are organizing subnets in an association network. We show two of them to illustrate the phenomenon. Figure 2.1 shows the sub-net for the meaning *dom* ("home", as a place for family) and Figure 2.2 *dom* ("home", as a building).

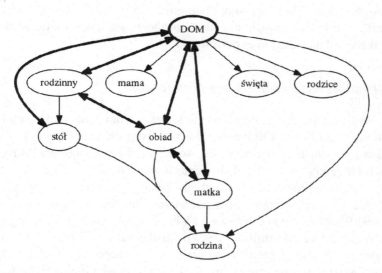

Figure 2.1. *Human association subnet for dom ("home", as a place for family)*

The graph in Figure 2.1 shows the relations between the words: *dom* and *rodzinny* (family; adjective), *stół* (table), *mama* (mum), *matka* (mother), *obiad* (dinner), *święta* (holidays), *rodzice* (parents) and *rodzina* (family).

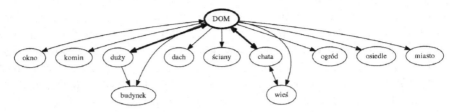

Figure 2.2. *Human association subnet for dom ("house", as a building)*

The graph in Figure 2.2 shows the relations between words: *dom* and *komin* (chimney), *big* (duży), *budynek* (building), *dach* (roof), *ściany* (walls), *chata* (cottage), *wieś* (village), *ogród* (garden), *osiedle* (estate) and *miasto* (city).

Clearly both subnets were identified manually and it is hard to believe that it would be possible to extract those subnets automatically by a use of an algorithm operating solely on the network [GAT 16]. Then, we shall treat all the associations to a particular stimulus as a list disregarding what association means. Then, we may distinguish semantically valid associations comparing the Polish association list with the English associations obtained in the free word association experiment.

2.2.4. *Human associations are comparable*

We shall compare a Polish list derived from our experiment to a semantically equivalent English list derived from the Edinburgh Associative Thesaurus. To illustrate the problem, we selected an ambiguous Polish word *dom*, which refers to the English words *home* and *house*. Those lists will present words associated with their basic stimulus, and ordered in accordance to their strength of association. Due to the varied number of responses (95 for *home* and *house* and 540 for *dom*), we will be using a more qualitative measure of similarity based on the rank of occurring words on them, rather than on a direct comparison of association strength. That list measure $LM_w(l_1, l_2)$, given two word lists l_1 and l_2 and a comparison window, which will be equivalent to the amount of words matching in l_1 and l_2 in a window of w words taken from the beginning of the lists.

In order to establish some basic expected levels of similarity, we will compare the list obtained in our experiment for the stimulus word *dom*, whose meaning covers both English words *home* and *house*. First, each Polish association word was carefully translated into English, and then the lists automatically looked for identical words. Because words may differ in rank on the compared lists, the table includes the window size needed to match a word on both lists.

dom	home	house
rodzinny (*adv. family*)	house	home
mieszkanie (*flat*)	family	garden
rodzina (*n.family*)	mother	door
spokój (*peace*)	away	boat
ciepło (*warmth*)	life	chimney
ogród (*garden*)	parents	roof
mój (*my*)	help	flat
bezpieczeństwo (*security*)	range	brick
mama (*mother*)	rest	building
pokój (*room*)	stead	bungalow

Table 2.1. *Top 10 elements of the experiment lists for dom (author's experiment) and the EAT lists for home and house*

The lists can be compared separately, but considering the ambiguity of *dom*, we can compare the list of association of *dom* with a list of interspersed (i.e. a list composed of the 1st word related to *home*, next to the 1st word associated with *house*, then the 2nd word related to *home*, etc.) associations of both *home* and *house* lists from the EAT.

w	home + house vs. *dom*	w	home vs. *dom*	w	*house* vs. *dom*
3	family	3	Family	3	Family
6	garden	9	Mother	6	Flat
9	mother	18	Cottage	6	Garden
12	roof	24	Garden	11	Roof
14	flat	26	Parents	14	Room
18	building	35	Peace	15	Building
19	chimney	41	Security	19	Chimney
26	parents			21	Cottage
30	room			30	Mother
32	brick			32	Brick
35	cottage			34	Security
64	security			40	Warm
65	peace			41	Warmth
74	warm				
75	warmth				

Table 2.2. *Comparison of the experiment list and the EAT lists. Matching words are shown for their corresponding window sizes w for the $LM_w(l_1,l_2)$ measure*

The original, i.e. used for comparison human association list, is a list of words associated with a stimulus word ordered by frequency of responses. Unfortunately, we cannot automatically distinguish words which enter into semantic relation to the stimulus word by frequency or by computed association strength, for example in the list associated to the word *table* a semantically unrelated *cloth* is substantially more frequent than *legs* and *leg*, which enter into "part of" relation to the *table* [PAL 64]. The described observation is language independent. The proposed method of comparison truncates from the resulting list language-specific semantic associations, e.g. *home – house* and *house – home* the most frequent on EAT as well as all non-semantic associations, e.g. *home – office* or *house – Jack*. Each resulting list consists of words, each of which is semantically related to a stimulus word. In other words, the comparison of the human association list will automatically extract a sub-list of semantic associations.

2.3. Algorithm efficiency comparison

2.3.1. *The corpora*

In order to compare the association lists with the LSA lists, we have prepared three distinct corpora to train the algorithm. The first consists of 51,574 press notes of the Polish Press Agency and contains over 2,900,000 words. That corpus represents a very broad description of reality, but can be somehow seen as restricted to only a more formal subset of the language. This corpus will be referred to as PAP.

The second corpus is a fragment of the National Corpus of Polish [PRZ 11] with 3,363 separate documents spanning over 860,000 words. That corpus is representative in the terms of the dictionary of the language; however, the texts occurring in it are relatively random, in the sense that they are not thematically grouped or following some deeper semantic structure. This corpus will be referred to as the NCP.

The last corpus is composed of 10 short stories and one novel *Lalka* (The Doll) by Bolesław Prus – a late 19th Century novelist using a modern version of Polish similar to the one used nowadays. The texts are split into 10,346 paragraphs of over 300,000 words. The rationale behind this corpus was to try to model some historically deeply rooted semantic associations with such basic notions as *dom*. This corpus will be referred to in as PRUS.

All corpora were lemmatized using a dictionary-based approach [KOR 12].

2.3.2. *LSA-sourced association lists*

Latent Semantic Analysis is a classical tool for automatically extracting similarities between documents, through dimensionality reduction. A term-document matrix is filled with weights corresponding to the importance of the term in the specific document (term-frequency/inverted document frequency in our case) and then is reduced via Singular Value Decomposition to a lower dimensional space called the concept space.

Formally, the term-document matrix X of dimension $n \times m$ (n terms and m documents) can be decomposed into U and V orthogonal matrices and $\sum$ a diagonal matrix through singular value decomposition:

$$X = U \sum V^T \tag{2.1}$$

This in turn can be represented through a rank k approximation of X in a smaller dimensionally space ($\sum$ becomes a $k \times k$ matrix). We used an arbitrary rank value of 150 in our experiment:

$$X_k = U_k \sum_k V_k^T \tag{2.2}$$

This representation is often used to compare documents in this new space, but as the problem is symmetrical it can be used to compare words. The U_k matrix of dimensions $n \times k$ represents the model of words in the new k-dimensional concept space. We can thus compare the relative similarity of each word by taking the cosine distance between their representations.

The LSA-sourced lists of associations are composed of the ordered list (by cosine distance) from the given word in a model build on each of the tree corpora as described above.

A crucial element in the application of Latent Semantic Analysis [LAN 08] is determining k, the number of concepts that are used to project the data to the reduced k-dimensional concept space. As this parameter is a characteristic of the corpus, and to some degree of the specific application, in this case it has been determined experimentally. For each corpus (PRUS, NCP and PAP), an LSA model has been built for a range of dimensions between 25 and 400 with an increment of 25. For each corpus, the dimension has been chosen as the one that gave the highest sum of matching words from 10 association lists in a window of 1,000 words. The final results, as presented in section 3.4, correspond to a dimension of 75 for PRUS and NCP and 300 for PAP. The calculations were made using the *gensim* topic modeling library.

2.3.3. *LDA-sourced lists*

Latent Dirichlet Allocation is a mechanism used for topic extraction [BLE 03]. It treats documents as probabilistic distribution sets of words or topics. These topics are not strongly defined – as they are identified on the basis of the likelihood of co-occurrences of words contained in them.

In order to obtain ranked lists of words associated with a given word w_n, we take the set of topics generated by LDA, and then for each word contained, we take the sum of the weight of each topic multiplied by the weight of given word w_n in this topic.

Formally, for N topics and w_{ji} denoting the weight of the word i in the topic j, the ranking weight for the word i is computed as follows:

$$w_i = \sum\nolimits_{j=1..N} w_{ij} * w_{nj} \qquad\qquad [2.3]$$

This representation allows us to create a ranked list of words associated with a given word w_n based on their probability of co-occurrence in the documents.

2.3.4. *Association ratio-based lists*

In order to evaluate the quality of the relatively advanced mechanism of Latent Semantic Analysis, we will compare its efficiency to the *association ratio* as presented in [CHU 90], with some minor changes related to the nature of the processed data. For two words x and y, their association ratio $f_w(x,y)$ will be defined as the number of times y follows or precedes x in a window of w words. The original association ratio was asymmetric, considering only words y following the parameter x. This approach will, however, fail in the case of texts that are written in languages with no strict word ordering in sentences (Polish in our case) where syntactic information is represented through rich inflection rather than through word ordering. We will use the same value for w as is in Church and Hanks [CHU 90] that suggested a value of 5. This measure can be seen as simplistic in comparison with LSA, but, as the results will show, is useful nonetheless.

2.3.5. *List comparison*

First, we have to compare the list obtained automatically from the three corpora for the word *dom* (*home/hose*) with the reference list, i.e. human association list obtained from human subjects in the author's experiment. The comparison will be presented in terms of $LM_w(l_1,l_2)$ for l_1 being the human association list and l_2 being the lists obtained through LSA/LDA similarities and the *association ratio* f_5 as described above. In the comparison we shall apply the three different window sizes to the reference list.

To begin, we shall compare the full human association list that is 151 words long to the lists generated by the algorithms described above. We will arbitrarily restrict the length of automatically generated lists to 1,000 words.

W	PRUS f_5	PAP f_5	NCP f_5	PRUS LSA	PAP LSA	NCP LSA	PRUS LDA	PAP LDA	NCP LDA
10									
25							1		
50	2	1	2				1		1
75	2	4	3	1		1	3		1
100	4	7	9	2		2	4	2	4
150	11	14	17	2	2	2	7	3	6
300	19	24	30	2	6	3	13	8	12
600	34	25	41	4	11	12	22	15	23
1,000	36	43	49	7	13	18	39	23	39

Table 2.3. *$LM_w(l_1,l_2)$ values for different w, for different l_2 from various list sources, l_1 being the human experiment result list*

This may seem excessive, as it contains also a random association of low interest to us – the lists obtained through EAT and the author's list comparison contain only 15 words.

Thus, we will restrict the human association list to only the first 75 words – that was also the length needed to obtain the combined list for *home* and *house* from the EAT.

W	PRUS f_s	PAP f_s	NCP f_s	PRUS LSA	PAP LSA	NCP LSA	PRUS LDA	PAP LDA	NCP LDA
10									
25							1		
50	2		2				1		1
75	2	4	3	1		1	3		1
100	3	5	8	2		1	3		3
150	8	9	10	2	1	1	5		3
300	11	15	21	2	5	1	10	3	8
600	21	23	30	4	7	5	14	8	15
1,000	22	28	33	5	9	6	23	14	22

Table 2.4. $LM_w(l_1,l_2)$ values for different w, for different l_2 from various list sources, l_1 being the human experiment result list restricted to 75 entries

As can be seen, automatically generated association lists match some part of the human association list only if we use a large window size. Second, we can observe that the Church–Hanks algorithm seems to generate a list that is more comparable to a human-derived list.

The shorter word list in the EAT (*house*) contains 42 words. The 40 words is the window size, which applied to the author's list, allow us to find all the elements common to the EAT *home/house* combined list and the author's experiment list for *dom*. Therefore, we shall use a 40-word window for comparison.

As we can see this window size seems to be optimal, because it reduces substantially – if compared to the full list – the non-semantic associations for both algorithms.

Finally, we have to test automatically generated lists against the combined human association list, i.e. the list which consists of words, which are present both in the author's list and the EAT lists, presented in Table 2.2.

W	PRUS f_5	PAP f_5	NCP f_5	PRUS LSA	PAP LSA	NCP LSA	PRUS LDA	PAP LDA	NCP LDA
10									
25							1		
50	2		2				1		1
75	2	4	3	1		1	2		1
100	3	5	7	1		1	2		3
150	7	9	9	1		1	4		3
300	8	9	17	1	4	1	8	1	6
600	15	16	22	2	6	5	10	3	11
1,000	16	20	22	3	6	6	16	6	15

Table 2.5. *LM$_w$(l$_1$,l$_2$) values for different w, for different l$_2$ from various list sources, l$_1$ being the human experiment result list restricted to first 40 entries*

W	PRUS f_5	PAP f_5	NCP f_5	PRUS LSA	PAP LSA	NCP LSA	PRUS LDA	PAP LDA	NCP LDA
10						1			
25						1	1		
50			1			1	1		1
75		1	3			1	3		1
100		3	3			2	3		3
150	3	4	5			2	5		3
300	4	8	5		1	2	9	2	8
600	8	12	9		2	3	12	7	13
1,000	10	12	12	2	2	3	16	7	14

Table 2.6. *LM$_w$(l$_1$,l$_2$) values for different w, for different l$_2$ from various list sources, l$_1$ being the human experiment result list restricted to words that are present in both the author's and the EAT experiment, see Table 2.2*

The results show a tendency similar to that observed during the test of human association list in full length. First, the window size influences the matching number. The second observation is also similar: the list generated by the Church–Hanks algorithm matches the human association list better – it matches 10 or 12 out of 15 words semantically related to the stimulus.

To learn more, we repeated a comparison over a wider range of words. We selected eight words: *chleb* (bread), *choroba* (disease), *światło* (light), *głowa* (head), *księżyc* (moon), *ptak* (beard), *woda* (water) and *żolnierz* (soldier). Then, we used the described method to obtain a combined list for the author's experiment and the EAT.

Word	W	PRUS f_5	PAP f_5	NCP f_5	PRUS LSA	PAP LSA	NCP LSA	PRUS LDA	PAP LDA	NCP LDA
Bread	25		1		1	1		1		
	100		4	2	1	1	1		2	1
	1,000	1	8	3		2	2	1	3	2
Disease	25		1					1		
	100	1	3	5				1	1	2
	1,000	1	9	8	1	7	2	2	8	8
Light	25	1	1			1		1	1	
	100	3	4	3	1	1		2	2	1
	1,000	3	5	3	4	5	2	5	4	2
Head	25	1		2		1	1		1	1
	100	1	2	4		1	1	1	1	1
	1,000	3	6	6	1	2	3	2	3	2
Moon	25		3	3	1		2	1		1
	100	3	4	5	1		3	1	1	2
	1,000	3	4	6	4	2	5	3	3	7
Bird	25	1	2	1	1		1	1		1
	100	2	4	2	1		2	1	2	3
	1,000	2	5	7	4	3	3	3	2	3
Water	25		1	2	1	1		1	1	1
	100		4	6	2	3	2	3	3	3
	1,000	4	8	10	3	5	6	5	6	5
Soldier	25	2	2	2	2	1	3	2	2	3
	100	2	5	5	2	6	3	2	7	4
	1,000	2	12	9	3	10	4	3	11	5

Table 2.7. $LM_w(l_1, l_2)$ values for different word stimuli, different w, for different l_2 from various list sources, l_1 being the human experiment result list restricted to entries in both the author's and the EAT experiment

The table below contains similar comparison, but without restricting the association list to words contained in both experiments.

Word	W	PRUS f_5	PAP f_5	NCP f_5	PRUS LSA	PAP LSA	NCP LSA	PRUS LDA	PAP LDA	NCP LDA
Bread	25	1	1	2		1	2		1	1
	100	2	5	6	1	2	5	1	3	4
	1,000	4	19	12	3	4	9	5	6	8
Disease	25		1	1		1			1	
	100	1	3	7		2		1	4	4
	1,000	3	13	14	1	13	8	2	14	7
Light	25	2	1	1	1	1		1	1	
	100	6	6	4	3	1		6	1	1
	1,000	11	15	9	10	9	3	10	9	4
Head	25	3	1	3		3	1		3	1
	100	6	6	7		5	1	1	5	2
	1,000	17	17	12	7	9	7	4	9	7
Moon	25	1	4	6	1		2	1		3
	100	5	5	11	1	1	4	2	1	5
	1,000	5	9	15	7	5	12	6	5	15
Bird	25	1	8	2	2		2	2		2
	100	3	9	5	3	2	2	4	2	3
	1,000	5	13	19	8	9	9	9	9	9
Water	25	1	2	3	1	1	1	1	1	1
	100	3	7	8	2	4	3	3	4	4
	1,000	9	20	21	10	9	15	14	9	17
Soldier	25	1	5	4	1	2	3	1	2	3
	100	2	11	9	4	7	6	6	8	7
	1,000	3	25	22	9	20	11	8	16	10

Table 2.8. $LM_w(l_1, l_2)$ values for different word stimuli, different w, for different l_2 from various list sources, l_1 being the unrestricted human experiment result list

As can be seen, the values in the columns corresponding to the f_5 algorithm are clearly better than the corresponding LSA values, regardless of the size of the human lists.

2.4. Conclusion

If we look at our results, we may find that in general they are comparable with the results of related research of Wandmacher [WAN 05] and [WAN 08]. Generally speaking, both the LSA and LDA algorithms generate an association list, which contains only a small fraction of the lexical relationships, which are present in the human association norm. Surprisingly, the Church–Hanks algorithm does much better, which suggests that the problem of how machine-made associations relate to the human association norm should be investigated more carefully. The first suggestion may be derived from [WET 05] – we have to learn more about the relationship between the human association norm and the text to look for a method more appropriate than a simple list comparison. We argue that if a human lexicographer uses contexts retrieved from text by the Church–Hanks algorithm to select those which define lexical meaning, then the list of associations generated by three compared algorithms should be filtered by a procedure that is able to assess the semantic relatedness of two co-occurring words, or we shall look for a new method of co-occurrences selection.

A second suggestion may be derived from an analysis of the human association list. It is well known that such a list consists of responses, which are semantically related to the stimulus, responses which reflect pragmatic dependencies and so-called "clang responses". But within this set of semantically related responses, we can find more frequent direct associations, i.e. such as those which follow a single semantic relation, e.g. "whole – part": *house – wall* and not so frequent indirect associations like: *mutton – wool* (baranina – rogi), which must be explained by a chain of semantic relations, in our example "source" relationship, i.e. *ram* is a source of *mutton* followed by "whole – part" relation, i.e. *horns* is a part of *ram* or the association: *mutton – wool* (baranina – wełna), explained by a "source" relation, i.e. *ram* is a source of *mutton*, followed by "whole – part" *fleece* is a part of *ram*, which is followed by "source" relation, i.e. *fleece* is a source of *wool*. These association chains suggest that some associations are based on a semantic network, which may form the paths explaining indirect associations. Then, it would be very interesting to recognize that human associations may form a network [KIS 73] and test the machine associating mechanism against the association network.

2.5. Bibliography

[BLE 03] BLEI D.M., NG A.Y., JORDAN M.I., "Latent Dirichlet allocation", *Journal of Machine Learning Research*, vol. 3, nos. 4–5, pp. 993–1022, 2003.

[BOR 09] BORGE-HOLTHOEFER J., ARENAS A., "Navigating word association norms to extract semantic information", *Proceedings of the 31st Annual Conference of the Cognitive Science Society*, Groningen, available at: http://csjarchive.cogsci. rpi.edu/Proceedings/2009/papers/621/paper621.pdf, pp. 1–6, 2009.

[BUD 06] BUDANITSKY A., HIRST G., "Evaluating wordnet-based measures of lexical semantic relatedness", *Computational Linguistics*, vol. 32, no. 1, pp. 13–47, 2006.

[CHU 90] CHURCH K.W., HANKS P., "Word association norms, mutual information, and lexicography", *Computational Linguistics*, vol. 16, no. 1, pp. 22–29, 1990.

[DEE 90] DEERWESTER S., DUMAIS S., FURNAS G. *et al.*, "Indexing by latent semantic analysis", *Journal of the American Society for Information Science*, vol. 41, no. 6, pp. 391–407, 1990.

[GAT 14] GATKOWSKA I., "Word associations as a linguistic data", in CHRUSZCZEWSKI P., RICKFORD J., BUCZEK K. *et al.* (eds), *Languages in Contact*, vol. 1, Wrocław, 2014.

[GAT 16] GATKOWSKA I., "Dom w empirycznych sieciach leksykalnych", *Etnolingwistyka*, vol. 28, pp. 117–135, 2016.

[KEN 10] KENT G., ROSANOFF A.J., "A study of association in insanity", *American Journal of Insanity*, vol. 67, pp. 317–390, 1910.

[KIS 73] KISS G.R., ARMSTRONG C., MILROY R. *et al.*, "An associative thesaurus of English and its computer analysis", in AITKEN A.J., BAILEY R.W., HAMILTON-SMITH N. (eds), *The Computer and Literary Studies*, Edinburgh University Press, Edinburgh, 1973.

[KOR 12] KORZYCKI M., "A dictionary based stemming mechanism for Polish", in SHARP B., ZOCK M. (eds), *Natural Language Processing and Cognitive Science 2012*, SciTePress, Wrocław, 2012.

[KUR 67] KURCZ I., "Polskie normy powszechności skojarzeń swobodnych na 100 słów z listy Kent-Rosanoffa", *Studia Psychologiczne*, vol. 8, pp.122–255, 1967.

[LAN 08] LANDAUER T.K., DUMAIS S.T., "Latent semantic analysis", *Scholarpedia*, vol. 3, no. 11, pp. 43–56, 2008.

[MOS 96] MOSS H., OLDER L., *Birkbeck Word Association Norms*, Psychology Press, 1996.

[NEL 98] NELSON D.L., MCEVOY C.L., SCHREIBER T.A, *The University of South Florida Word Association, Rhyme, and Word Fragment Norms*, 1998.

[ORT 12] ORTEGA-PACHECO D., ARIAS-TREJO N., BARRON MARTINEZ J.B., "Latent semantic analysis model as a representation of free-association word norms", *11th Mexican International Conference on Artificial Intelligence (MICAI 2012)*, Puebla, pp. 21–25, 2012.

[PAL 64] PALERMO D.S., JENKINS J.J., *Word Associations Norms: Grade School through College*, Minneapolis, 1964.

[POS 70] POSTMAN L.J., KEPPEL G., *Norms of Word Association*, Academic Press, 1970.

[PRZ 11] PRZEPIÓRKOWSKI A., BAŃKO M., GÓRSKI R, *et al.*, "National Corpus of Polish", *Proceedings of the 5th Language & Technology Conference: Human Language Technologies as a Challenge for Computer Science and Linguistics*, Poznań, pp. 259–263, 2011.

[RAP 02] RAPP R., "The computation of word associations: comparing syntagmatic and paradigmatic approaches", *Proceedings of the 19th International Conference on Computational Linguistics*, Taipei, vol. 1, pp. 1–7, 2002.

[RAP 08] RAPP R., "The computation of associative responses to multiword stimuli", *Proceedings of the Workshop on Cognitive Aspects of the Lexicon (COGALEX 2008)*, Manchester, pp. 102–109, 2008.

[SCH 12] SCHULTE IM WALDE S., BORGWALDT S., JAUCH R., "Association norms of German noun compounds", *Proceedings of the 8th International Conference on Language Resources and Evaluation*, Istanbul, available at: http://lrec.elra.info/proceedings/lrec2012/pdf/584_Paper.pdf, pp. 1–8, 2012.

[SIN 04] SINOPALNIKOVA A., SMRZ P., "Word association thesaurus as a resource for extending semantic networks", *Proceedings of the International Conference on Communications in Computing, CIC '04*, Las Vegas, pp. 267–273, 2004.

[WAN 05] WANDMACHER T., "How semantic is latent semantic analysis", *Proceedings of TALN/RECITAL 5*, available at: https://taln.limsi.fr/tome1/P62.pdf, pp. 1–10, 2005.

[WAN 08] WANDMACHER T., OVCHINNIKOVA E., ALEXANDROV T., "Does latent semantic analysis reflect human associations", *Proceedings of the ESSLLI Workshop on Distributional Lexical Semantics,* Hamburg, available at: http://www.wordspace.collocations.de/lib/exe/fetch.php/workshop:esslli:esslli_2008_lexicalsemantics.pdf, pp. 63–70, 2008.

[WET 05] WETTLER M., RAPP R., SEDLMEIER P., "Free word associations correspond to contiguities between words in text", *Journal of Quantitative Linguistics*, vol. 12, no. 2, pp. 111–122, 2005.

3

How a Word of a Text Selects the Related Words in a Human Association Network

According to tradition, experimentally obtained human associations are analyzed in themselves, without relation to other linguistic data. In rare cases, human associations are used as the norm to evaluate the performance of algorithms, which generate associations on the basis of text corpora. This chapter will describe a mechanical procedure to investigate how a word embedded in a text context may select associations in an experimentally built human association network. Each association produced in the experiment has a direction from stimulus to response. On the other hand, each association is based on the semantic relation between the two meanings, which has its own direction which is independent from the direction of associations. Therefore, we may treat the network as a directed or an undirected graph. The procedure described in this chapter uses both graph structures to produce a semantically consistent sub-graph. A comparison of the results shows that the procedure operates equally well on both graph structures. This procedure is able to distinguish those words in a text which enter into a direct semantic relationship with the stimulus word used in the experiment employed to create a network, and is able to separate those words of the text which enter into an indirect semantic relationship with the stimulus word.

3.1. Introduction

It is easy to observe that semantic information may occur in human communication, which is not lexically present in a sentence. This phenomenon does not affect the human understanding process, but the performance of a text processing algorithm may suffer from that. Consider, e.g. this exchange: *Auntie, I've got a terrier! – That's really nice, but you'll have to take care of the animal.* The connection between the two sentences

Chapter written by Wiesław LUBASZEWSKI, Izabela GATKOWSKA and Maciej GODNY.

in this exchange suggests that there is a link between *terrier* and *animal* in human memory. A lexical semanticist may explain this phenomenon by the properties of the *hyponymy* relation, which is a transitive one: the pairs, a terrier is a dog, and a dog is an animal, imply that a terrier is an animal [LYO 63, MUR 03]. We can even process this phenomenon automatically using a dictionary such as WordNet. However, there are frequent situations when we require a more complex reasoning to decode the information encoded in a text. Let us take an example: *The survivor regained his composure as he heard a distant barking.* It seems to be obvious that a human reader, who is a native speaker of English, can easily explain the reason behind the change of a survivor's mental state. This person may say for example: *a dog barks* and *a dog lives close (is subsidiary) to a man* and *a man may help the survivor*. However, we can find it impossible to perform such reasoning by an algorithm which uses manually built semantic dictionaries such as WordNet or even FrameNet at their present stage of development [RUP 10].

Then, we can find it reasonable to study the properties of an experimentally built natural dictionary, i.e. an association network that consists of words and naturally preferred semantic connections between them. There exists a reliable method to build such a network. The free word association test [KEN 10], in which the tested person responds with a word associated with a stimulus word provided by a researcher, would provide naturally preferred connections between the stimulus word and the response word. If we perform a multiphase word association test using responses obtained in the initial phase as the stimuli in the next phase of the test, we would create a rich lexical network in which words are linked with multiple links [KIS 73].

Returning to our example, we shall look-up the Edinburgh Associative Thesaurus, which is the first large lexical network built experimentally. We can find here 35 *dog* associations such as, among others:

dog – man, bark, country, pet, gun, collar, leash, lead, whistle

We see the word *dog* directly connected to *bark* and *man*, and that both words co-occur with *collar, lead, leash, pet, gun*, which are attributes of the *dog – man* proximity. Then, if we look at the *dog* associations in the Polish lexical network [GAT 14] built in the experiment, where the word *dog* was not included in the stimuli set, and it is associated only by responses, we can find the following associations:

man, sheep, protector, smoke – dog

As we see in the Polish network, dog is also linked to a man, while other dog-linked words suggest that a dog is working for a man.

Therefore, we may suppose that a study of the meaning connections in the lexical network built by the free association test would provide the data to explain how a word of a text connects in the dictionary and how (if possible) those connections may provide information, which is lexically missing in a text. There are also phenomena observed in the network which may strengthen this supposition. If we look closer at the dog associations, we may find that most of them are directly explicable, e.g. a dog is a pet, a dog has a collar or a dog is a protector. However, there are also on both lists associations which need reasoning to be explained – we call them indirect associations. For example, the association dog – gun in the English network may be explained by the reasoning based on the chain of directly explicable connections: dog is subsidiary to a man and man hunts and man uses a gun. We can find the same situation in Polish for the dog – smoke association: dog is subsidiary to a man and man causes fire and fire produces smoke. Once we identify an indirect association such as dog – gun in the network, we may look in the network for the path which has the dog as a start node and the gun as the end node. If the path is found, we have to assess the path to find if it explains the dog – gun connection. It has been observed that, if a network is rich enough, we may identify more distant associations and explaining paths, e.g. *mutton – horns*, explained by the path *mutton – ram – horns* or the association *mutton – wool*, explained by the path *mutton – ram – fleece – wool*, which were identified manually in the Polish network [GAT 13].

However, before we start looking for the explaining paths in the network, we have to develop a reliable mechanic procedure that takes a word of a text as an input and can find in the network the sub-net (sub-graph) which is optimally related to the word of a text – where optimal means a sub-net in which each node (word) is semantically related to a word of a text. This chapter describes such a procedure.

The procedure to be described was originally designed to operate on the association network treated as an undirected graph [LUB 15]. However, the evaluations of the semantic consistency of the sub-net extracted by the procedure were really encouraging and we therefore decided to expand the procedure to make it able to operate simultaneously on the network treated as a directed graph. The expansion is important because it adapts the procedure to the nature of the network – the network built in the free word association experiment is a directed graph; each connection between two nodes (words) in the network has a direction, always from the stimulus word to the response word. This expansion enables us to really assess a procedure. We shall compare how it operates, on both directed and undirected network structures.

3.2. The network

The network described in this chapter was built via a free word association experiment [GAT 14], in which two sets of stimuli were employed, each in a different phase of the experiment. In the first phase, 62 words taken from the Kent–Rosanoff list were tested as the primary stimuli. In the second phase, the five most frequent responses to each primary stimulus obtained in phase one were used as stimuli. To reduce the amount of manual labor required to evaluate the algorithm output, we used a reduced network, which is based on:

– 43 primary stimuli taken from the Polish version of the Kent–Rosanoff list;

– 126 secondary stimuli which were the three most frequent associations with each primary stimulus.

The average number of associations with a particular stimulus, produced by more than 900 subjects, is approximately 150. Therefore, the total number of stimulus–response pairs obtained for the 168 stimuli, as a result of the experiment, is equal to 25,200. Due to the fact that the analysis of the results produced by an algorithm would require manual work, we reduced the association set through the exclusion of each stimulus–response pair, where the response frequency was equal to 1. As a result, we obtained 6,342 stimulus–response pairs, where 2,169 pairs contain responses to the primary stimuli, i.e. primary associations, and 4,173 pairs that contain responses to the secondary stimuli, i.e. secondary associations. The resulting network consists of 3,185 nodes (words) and 6,155 connections between the nodes.

The experimentally built association network can be described on a graph, where the *graph* is defined as tuple (V, E), where V is the set of nodes (vertices) and E is the set of connections between the two nodes from V. The connection between the two nodes may have a *weight*. The experiment result is a list of triples: (S,A,C), where S is the stimulus, A is the association and C is the number of participants, who associated A with S. The C represents the association strength, which can be converted into a connection weight of C_w, counted as follows: $C_w = S_c/C$, where S_c is the total of all responses given to the stimulus S. Then, we may treat the association network as a *weighted* graph, which is a tuple (V, E, w), where w is the function that assigns a weight to every connection.

As each stimulus–association (response) pair has a direction, always from the stimulus to the response, we may consider an association network as a *directed* graph [KIS 73], which means that each connection between the two nodes $(v1, v2)$ has a direction, i.e. it starts in $v1$ and ends in $v2$ – this kind of connection is called an *arc*. On the other hand, if we recognize that the connection $(v1, v2)$ is a semantic relation between the meanings of the two words, then we have to recognize that the stimulus–response direction and the direction of semantic relations between the two meanings may differ. Let us consider the associations: *chair – leg* and *leg – chair*. In both cases, the associated meanings are connected by the same semantic relation, i.e. *meronymy* [MUR 03], which has a direction from a part, e.g. *leg* to a whole *chair*. The same phenomenon may be observed with respect to the *hyponymy* relation, which goes from a subordinate *terrier* to a superordinate *dog* meaning and the direction of the relation does not depend on the direction of the association *terrier – dog* or *dog – terrier*. Therefore, we can treat the association network as an *undirected* graph, which means that the connection between the two nodes $(v1, v2)$ has no direction, i.e. $(v1,v2) = (v2, v1)$.

The *path* in the graph is a sequence of nodes that are connected by edges or arcs. The path *length* is the number of nodes along the path. Path *weight* is the sum of the weights of the connections in the path. The *shortest* path between two nodes $(v1, v2)$ is the path where the path weight is smaller than the weight of the direct connection between $v1$ and $v2$.

3.3. The network extraction driven by a text-based stimulus

If both the network and the text are structures built from words, then we may look for an efficient algorithm that can identify in the text the stimulus word used in an experiment performed to build a network and a reasonable number of direct associations with this stimulus. Words identified in the text may serve as the starting point to extract a sub-graph from the network, which will contain as many associations as possible. The semantic relationship between the nodes of a returned sub-graph will be the subject of an evaluation.

In more technical language, the algorithm should take a graph (association network) and the subset of its nodes identified in a text (*extracting nodes*) as input. Then, the algorithm creates a sub-graph with all extracting nodes as an initial node set. After that, all the connections between the extracting nodes which exist in the network are added to the resulting sub-graph – these connections are said to be *direct*. Finally, every direct connection is checked in the network to determine whether it can be replaced with a shortest path, i.e. a path which has a path weight lower than the weight of the direct connection and a node number less than or equal to the predefined path length. If such a path is found, it is added to the sub-graph – which means adding all the path's nodes and connections. If we apply this procedure to each text of a large text collection, and if we merge the resulting text sub-graphs, we may evaluate the sub-graph created for a particular stimulus word.

3.3.1. Sub-graph extraction algorithm

The source graph G, extracting nodes EN and maximum number of intermediate nodes in path l are given. First, an empty sub-graph, SG, is created, and all extracting nodes, EN, are added to a set of nodes (vertices) V_{sg}. In the next set of steps, the ENP of all pairs between the nodes in the EN is created. For every pair in the ENP the algorithm checks if the connection between the paired nodes $v1$, $v2$ exists in G. If it does, this connection is added to the sub-graph SG set of connections E_{sg}. Then, the shortest path sp between $v1$ and $v2$ is checked in G. If the shortest path sp is found, i.e. the sp weight is lower than the weight of the direct connection $(v1, v2)$ and the number of the shortest path intermediate nodes is less than l ($length(sp) - 2$, (–2 because the start and end nodes are not intermediate), then the sp path is added to the sub-graph *SG* by adding its nodes and connections to the appropriate sets V_{sg} and E_{sg}. Finally, the sub-graph *SG* is returned.

NEA (G, EN, l)
Input: G – (V_g, E_g, w_g) – graph (set of nodes, set of connections, weighting function), EN - extracting nodes, l - maximal number of intermediate nodes in the path
Output: SG - (V_{sg}, E_{sg}, w_{sg}) - extracted sub-graph

```
1   Vsg ← Vg;
2   Esg ← Ø;
3   wsg ← wg;
4   ENP ← pairs(EN );
5   for each v1 , v2 ∈ ENP do
6          if conns(v1, v2) ∈ G then
7                  Esg ← Esg+ conns(v1, v2) ;
8                  sp ← shortest_path(v1, v2);
9                  if weight(sp) < wg(conns(v1, v2)) and length(sp) - 2 <= l then
10                         for each v ∈ nodes(sp) do
11                                 if not v ∈ Vsg then
12                                         Vsg ← Vsg + v;
13                                 end
14                         end
10                         for each e ∈ conns(sp) do
11                                 if not e ∈ Esg then
12                                         Esg ← Esg + e;
13                                 end
14                         end
17                 end
18         end
19  end
    return SG
```

It seems to be clear that the size of the sub-graph created by the algorithm depends on the number of extracting nodes given at the input. As the texts may differ in the number of primary associations with a particular stimulus which would serve as extracting nodes, there is a need for a procedure that controls the number of extracting nodes used by the network extracting algorithm.

3.3.2. *The control procedure*

The procedure controls the number of extracting nodes *EN* and the sub-graph *SG* size. In order for it to be used to build a sub-graph for a given stimulus, the text must contain a stimulus *S* and at least *dAn* direct associations with the stimulus. The *dAn* = 2 is chosen as a starting value for the extracting algorithm, which means that if the text has *dAn* < 2, the text is omitted. If the text has *dAn* ≥ 2, the text is used for sub-graph extraction. First, the stimulus and the *dAn* = 2 primary associations are passed as extracting nodes to the network extracting algorithm *NEA*. Then, the number of nodes in the returned sub-graph is counted. In the next step, the *dAn* is incremented by 1 and the new set of extracting nodes is passed to the *NEA*. The returned sub-graph size is evaluated, i.e. the number of nodes of the sub-graph based on *dAn* + 1 is multiplied by the sub-graph size control parameter *Ss*, which tells us which fraction of the base sub-graph, i.e. the sub-graph created for a start value of *dAn* = 2, must exist in the sub-graph created for *dAn* + 1. For example, *Ss* = 0.5 means that at least half of the nodes from the base sub-graph must remain in the sub-graph created after the incrementing of the *dAn*. If the newly created sub-graph does not match the condition set by *Ss*, the procedure stops and the sub-graph created in the previous step becomes the final for a particular text. If the newly created sub-graph matches the condition set by *Ss*, the *dAn* is incremented by 1, and a new sub-graph is created.

3.3.3. *The shortest path extraction*

Figures 3.1 and 3.2 represent a subset of the experimental network, treated respectively as directed and undirected graphs. Each graph consists of such nodes as: *chleb* "bread", *masło* "butter", *jedzenie* "food", *ser* "cheese", *mleko* "milk", *dobry* "good", *kanapka* "sandwich" and *żółty* "yellow" linked by connections produced by the free word association experiment.

Figure 3.1 represents the concept of normalizing the directed network, if a path shorter than the one directly linking two nodes can be found. "Shorter" in this case means that the sum of weights of the path connections is smaller than the weight of the direct connection. On this specific example, the dotted connection between the nodes is replaces the original black one. It is caused by the fact that the path *ser* →*jedzenie* → *chleb* → *masło* has a weight sum of 84, which is lower than the direct connection weight of 200 for the nodes *masło* → *ser*.

The same reasoning applies to the experimental net being represented by an undirected weighted graph (Figure 3.2).

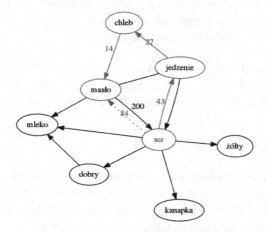

Figure 3.1. *Shortest path for a directed network. For a color version of this figure, see www.iste.co.uk/sharp/cognitive.zip*

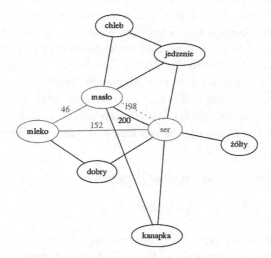

Figure 3.2. *Shortest path for an undirected network. For a color version of this figure, see www.iste.co.uk/sharp/cognitive.zip*

In the case of the undirected graph, we treat it just as a directed graph with symmetrical connections between nodes, i.e. $(v1, v2) = (v2, v1)$. We can see from Figure 3.2 that the connection *ser – masło* is replaced by the same path *ser –jedzenie – chleb – masło* as in the directed graph, and that another shortest path for the *ser – masło* connection is found, i.e. the path *masło – mleko – ser*, with a path weight of 198 which is smaller than 200, which is the weight of the *ser – masło* direct connection.

In both cases, the Dijkstra's classic shortest path algorithm has been applied. However, the sub-graph extracting algorithm *NEA* will reject any shortest path which does not meet the condition set by the *l* parameter.

3.3.4. *A corpus-based sub-graph*

First, a separate sub-graph for each primary stimulus was created for each text in the corpus. All sub-graphs were obtained with empirically adjusted parameters [HAR 14], such as: intermediate nodes in the path of $l = 3$ for the extracting algorithm, and direct associations with the stimulus minimum dAn = 2, with a sub-graph adjusting parameter of Ss = 0.5 for the control procedure. Then, the text-based sub-graphs obtained for a specific primary stimulus were merged into the corpus-based primary stimulus sub-graph, i.e. all sets of nodes and all sets of edges were merged, forming a multiple set union. Finally, the corpus-based primary stimulus sub-graph was trimmed, which means that each non-connected node was removed from the final sub-graph, and each open path (paths with which the end node had not connected) which had more than two edges between the stimulus and the end node was reduced to match the network-forming principle that a stimulus (A) produces an association (B), which then serves as a stimulus to produce an association (C). Afterwards the reduced path takes the form A – B – C.

3.4. Tests of the network extracting procedure

3.4.1. *The corpus to perform tests*

To test the original procedure, we employed the three stylistically and thematically different corpora, i.e. the PAP corpus which consists of 51,574 press releases of the Polish Press Agency, and contains over 2,900,000 words, the sub-corpus of the National Corpus of Polish with a size of 3,363 separate documents spanning over 860,000 words, and a literary text corpus which consists of 10 short stories and the novel *Lalka* (The Doll) written by the influential novelist, Bolesław Prus. All three corpora were lemmatized using a dictionary-based approach [KOR 12]. The procedure performed equally well on all three corpora. Then, we decided to perform the test described below on the largest corpus, which is that of the PAP.

3.4.2. *Evaluation of the extracted sub-graph*

To evaluate the quality of the extracted sub-graph, we shall use two separate evaluation criteria: first, to test the semantic consistency of the sub-graph and second, to test how the sub-graph matches the text collection.

3.4.2.1. *Semantic consistency of the sub-graph*

To perform the evaluation, each of the 6,342 stimulus–response pairs used to build the network were manually evaluated. The evaluation was necessary because of the observation that the free word association experiment may produce so-called clang associations, i.e. words that sound like a stimulus or rhyme with the stimulus, e.g. *house – mouse* and the idiom completion associations, e.g. *white – house*, which forms a multipart lexical unit and, therefore, does not reflect the meaning relation between stimulus and response [CLA 70]. We had expanded this observation treating all associations which introduce proper names, e.g. *river – Thames* and the not so frequent deictic association, e.g. *girl – me* as non-semantic.

The evaluation is as follows. If the stimulus is semantically related to the response as in *dom – ściana* (house – wall), the pair is marked as semantic, otherwise the pair is considered to be a non-semantic one, e.g. *góra – Tatry* (mountain – the proper name) or *dom – mój* (house – my).

Then, the sub-graph nodes were evaluated consecutively along a path in the following way. If the two connected nodes matched a stimulus–response pair marked as semantic, then the right node was marked as semantic – *Sn*. If the two connected nodes matched a non-semantic stimulus–response pair, then the right node was marked as non-semantic one – *nSn*. If the two connected nodes did not match any stimulus–response pair, then both nodes were marked as *nSn*, except the stimulus node which is in principle a semantic one. After the final pair of a path was evaluated, the evaluation of the connection of the start node (stimulus) and the end node of the path was evaluated to check the semantic consistency of the path. As a result, a non-semantic node nSn is considered to be any end node (association) that does not have a semantic relation to the start node (stimulus), even if it has a semantic relation with a preceding node, e.g. the path: *krzesło – stół – szwedzki* (chair – table – Swedish), where pairs *krzesło – stół* and *stół – szwedzki* enter into a semantic relation, but the stimulus *krzesło* "chair" does not enter into a semantic relationship with the association *szwedzki* "Swedish".

3.4.2.2. Matching the sub-graph and a text collection

To assess how the extracted sub-graph relates to a text collection, we have to match each text that contains a particular stimulus against the sub-graph extracted for this stimulus. Then, we have to count the network nodes (words) that were recognized both in the texts and in the sub-graph SnT. After that, we have to match the entire set of direct associations with a particular stimulus which is present in the network against the texts. This is performed in order to recognize network nodes (words) which are present in the network but were rejected by the algorithm, and therefore, are not present in the sub-graph TnS.

3.4.3. Directed and undirected sub-graph extraction: the comparison

Now, we can present the results for each primary stimulus, where the sub-graph for each primary stimulus word is evaluated. To compare directed and undirected sub-graphs extracted for each stimulus, we shall use all data obtained in the sub-graph evaluation process, i.e.:

– Sn: the number of nodes in the sub-graph created by the algorithm;

– nSn: the number of non-semantic nodes in the sub-graph recognized by a sub-graph evaluation;

– SnT: the number of network nodes (words) which were recognized both in the texts and in the sub-graph;

– TnS: the number of network nodes (words) which are present in the texts but were rejected by the algorithm, and therefore are not present in the sub-graph.

Before we start the evaluation per stimulus, we have to show the joint results of the evaluation of 43 stimuli. To perform this analysis, we must determine the total number of nodes in the network – Nn. Table 3.1 shows the joint result for all sub-graphs based on the PAP corpus.

Stimuli	Nn	Sn	nSn	SnT	TnS	Graph
43	3,185	898	65	710	38	undirected
43	3,185	878	64	788	64	directed

Table 3.1. *Joint evaluation of 43 stimuli*

If we look at Table 3.1 and compare the number of network nodes Nn and the sum of SnT (network nodes retrieved in the text to extract a sub-graph) and TnS (network nodes present in text but rejected by an algorithm), we can discern that only a fraction of the nodes (words) which are present in the network appear in the large text collection – 0.234 for the undirected network and 0.267 for the directed one. This score is substantially lower than the sub-graph node Sn-to-network node Nn ratio, which is 0.281 for an undirected network and 0.275 for a directed one. It can be said that these figures show the relation between the language dictionary (the network) and the use of a dictionary to produce texts. The nSn value (non-semantic nodes in the sub-graph) shows that the non-semantic nodes in sub-graphs are only 0.072 of the total sub-graph nodes, in both undirected and directed networks. This result shows both the semantic consistency of the empirically built association network, and the quality of the cautious method for building a sub-graph described in this chapter.

Finally, the difference in size of Sn, SnT and TnS may reflect the differences between the directed and undirected graph structures, which influence the use of the words of a text to extract a sub-graph. We shall provide a detailed analysis later on.

3.4.4. *Results per stimulus*

A more detailed evaluation of the results will be possible if we look at the results obtained for each particular primary stimulus. These results are shown in Table 3.2.

Stimulus	Directed network				Undirected network			
	Sn	nSn	SnT	TnS	Sn	nSn	SnT	TnS
baranina "mutton"	3	0	3	0	5	0	4	0
chata "cottage"	7	1	6	0	10	0	8	0
chleb "bread"	17	2	16	4	17	0	15	0
chłopiec "boy"	19	1	19	0	22	0	13	1
choroba "illness"	41	2	30	1	30	1	22	0
doktor "doctor"	19	2	17	2	24	0	13	0
dom "house/home"	51	1	51	4	35	1	32	0
dywan "carpet"	3	0	3	0	7	1	5	0
dziecko "child"	51	7	46	2	27	5	26	0
dziewczyna "girl"	14	2	12	0	9	0	7	0
głowa "head"	28	1	28	0	27	3	30	0
góra "mountain"	28	2	26	0	20	4	11	1
jedzenie "food"	27	4	23	1	22	2	13	0
kapusta "cabbage"	13	2	10	2	11	2	6	0
król "king"	14	1	14	2	15	0	13	1
krzesło "chair"	6	2	4	1	7	2	3	0
księżyc "moon"	15	0	15	4	18	0	15	0
lampa "lamp"	5	0	5	1	4	0	4	0
łóżko "bed"	11	2	7	2	7	1	7	1
mężczyzna "man"	33	2	33	6	61	5	38	0
miasto "city"	32	2	32	0	37	3	31	2
mięso "meat"	33	3	28	3	36	1	28	1
motyl "butterfly"	3	0	3	1	12	1	8	0
obawa "fear"	12	1	10	0	10	1	6	0
ocean "ocean"	9	0	8	0	17	1	16	0
okno "window"	22	5	16	1	25	3	16	0
orzeł "eagle"	15	1	15	0	14	1	10	0
owca "sheep"	15	2	11	4	19	2	19	3
owoc "fruit"	27	2	26	0	30	3	15	1
pająk "spider"	3	0	3	0	5	1	4	0
pamięć "memory"	24	1	20	2	22	0	32	3
podłoga "floor"	14	3	11	1	8	1	4	0
praca "work"	54	2	48	2	56	6	44	3
ptak "bird"	20	0	20	1	23	1	19	1
radość "joy"	19	1	16	1	25	0	26	9

ręka "hand"	33	5	29	0	10	1	12	1
rzeka "river"	24	1	19	2	20	0	19	2
ser "cheese"	8	0	7	1	10	0	8	0
sól "salt"	9	0	9	0	17	2	10	0
światło "light"	16	0	14	2	12	2	11	3
woda "water"	48	0	42	4	68	8	43	4
wódka "vodka"	10	1	10	4	10	0	10	0
żołnierz "soldier"	23	0	23	3	34	0	34	1

Table 3.2. *Evaluation for each primary stimulus word*

The joint evaluation has shown that the procedure which operates on an undirected network produces a slightly larger sub-graph. However, if we look at the differences per stimulus in Figure 3.3, which compare the sub-graph size of the directed network with the undirected one, we may find that any differences appear to be stimulus dependent. Figure 3.3 shows that *Sn* size rises simultaneously for both networks and only *Sn* for *dziecko* "child" (+24), *ręka* "hand" (+23), *dom* "home/house" (+16), *choroba* "illness" (+11), *żołnierz* "soldier" (−11), *woda* "water" (−20) and *mężczyzna* "man" (−28) may reflect a difference in network structures. We have to add that the listed words do not share substantial semantic features.

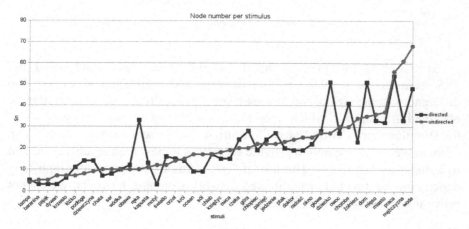

Figure 3.3. *Sub-graph node number per stimulus*

Having compared the sub-graph size, we can analyze the *nSn* – negative nodes in the sub-graph. This can be seen in Figure 3.4, which shows the *nSn*-to-*Sn* ratio per stimulus; the stimuli are ordered by the sub-graph size. We can see that only 17 out of 43 sub-graphs extracted from an undirected network do not contain a non-semantic node, while for a directed network, it is only 13. It is interesting to observe that only five stimuli words, i.e. *baranina* "mutton", *księżyc* "moon", *lampa* "lamp", *ser* "cheese" and *żołnierz* "soldier" share this property in both network structures. The differences in the *nSn*-to-*Sn* ratio seem to be network structure dependent.

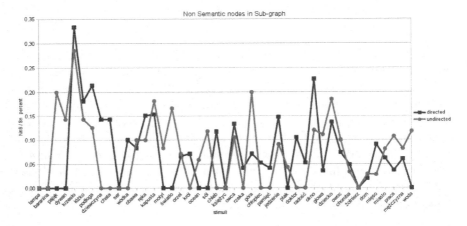

Figure 3.4. *Non-semantic nodes in the sub-graph*

At first glance, we may say that the stimuli status with respect to *SnT* and *TnS* seems to be similar for both directed and undirected networks. The differences in the size of the *SnT* (sub-graph nodes retrieved in text) may be observed in Table 3.2, which seem to be random and corpus dependent – e.g. the stimulus word *dywan* "carpet" occurred in only seven texts and only two of them were rich enough to provide extracting nodes (the stimulus word and two direct associations). The use of *SnT* words to create a sub-graph may depend on the directed or undirected network structure; however, we cannot prove this without separate research.

Finally, we have to analyze the *TnS*, i.e. associations which are present both in the network and texts but are not present in the sub-graph, as the algorithm had rejected them. First, we may observe that the algorithm operating on a directed network rejects more text occurring nodes, which we may correlate with the smaller number of sub-graph nodes for a directed network. The second observation is that there are only 10 stimuli which have the rejected text occurring nodes for both directed and undirected networks. It seems to be reasonable to look at these rejected network nodes. The full list of rejected nodes for all 10 stimuli is shown in Table 3.3. To save space, we shall use only the English translation of the rejected nodes.

Stimulus	*TnS* – undirected network	*TnS* – directed network
król "king"	law	kingdom, scepter
mięso "meat"	breakfast	pork, beef, cow
owca "sheep"	*horns, meat, baby sheep	meadow, mountain, skin, wool
pamięć "memory"	work, *mark, *will	limited, permanent
praca "work"	machine, *decision, hand	relax, difficult
radość "joy"	heartfelt, fear, to enjoy, disappointment, despair, big, disaster, worry, joyful	oda
rzeka "river"	*air, *earth	stream, *forest
światło "light"	lightness, reading, room	darkness, white
woda "water"	depth, salt, *sand, wave	thirst, *desert, drink, wet
żołnierz "soldier"	military	bravery, drill, *sea

Table 3.3. *Rejected nodes for 10 stimuli*

As we look at the nodes rejected by the algorithm operating on both networks, we find that all the words are semantically related to a stimulus, and for most of them, we can directly explain the connection between the stimulus and the association, e.g. *king* "makes/executes" a *law* for an undirected network and *king* "possesses" a *kingdom* and *scepter* is *king's* "attribute". However, some of those rejected nodes (marked by the asterisk) do not relate directly to the stimulus, e.g. *sheep – horns, water – desert*, and we can explain them by a sequence of direct connections, i.e. *sheep – ram – horns* and *water – thirst – search – desert*. That is to say that all words marked by an asterisk relate to a stimulus in the same way as indirect associations. Therefore, we can say that the method described in this chapter may help to identify indirect associations, which are present in the network. It is much easier to manually check a short list of nodes rejected by the algorithm, and then to manually check the entire network. Once the indirect associations are identified, we may rather easily construct an automatic procedure which would search for the paths explaining those indirect associations.

3.5. A brief discussion of the results and the related work

The proposed method for a text-driven extraction of an association network is simple and cautious on graph operations. The quality of sub-graphs extracted for stimuli words such as *pająk* "spider", *lampa* "lamp" and *dywan* "carpet", which occurred in a really small number of texts, seems to prove that the extracting algorithm does not depend on the number of texts used for network extracting. If it is true, the algorithm may serve as a reliable tool for extracting an association network on the basis of a single text, which may provide data to study how a particular direct association retrieved in a text may influence the sub-graph size and content. That is to say that we may observe how the sub-graph of lamp (Figure 3.5) may change if a text replaces the direct association *ulica* "street" with the direct association *krzesło* "chair", which has its own sub-graph created separately as shown in Figure 3.6.

The *lampa* sub-graph consists of directly associated nodes provided by the text *ulica* "street", *żarówka* "light bulb", *światło* "light" and the connection *żarówka – światło* added by the algorithm.

Figure 3.5. *Sub-graph lampa "Lamp". For a color version of this figure, see www.iste.co.uk/sharp/cognitive.zip*

The *krzesło* sub-graph consists of directly associated nodes: *stół* 'table', *dom* 'home', *stary* 'old' provided by the text and *obiad* 'dinner', *rodzinny* 'family' added by the text and *obiad* 'dinner', *rodzinny* 'family' added by the algorithm.

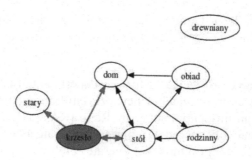

Figure 3.6. *Sub-graph krzesło "chair". For a color version of this figure, see www.iste.co.uk/sharp/cognitive.zip*

The study of a single text seems to be justified because a human reader comprehends just the text, not a text collection. It would be interesting to compare the graph of a text extracted word by word using our method with the graph of a text built solely on the basis of a text collection (e.g. [LOP 07, WU 11, AGG 13]). This should be the subject of further investigation.

The analysis of words which were rejected by the algorithm in the sub-graph extraction process suggests that the text-driven network extracting procedure may serve as a tool to provide the data which would locate the indirect association in a large network. This is a task which would be extremely hard to do manually. Having identified the indirect association, we may search the network automatically to find all paths that would explain those indirect associations. These explanatory paths may bring the new data to the study of the human association mechanism as analyzed by Clark [CLA 70].

However, if we look at an association network from the perspective of a computer program aimed to emulate human reasoning, we will find that an experimentally obtained connection between the two words does not provide explicit information as to what that connection means. However, it seems to be clear that only explicit information on the manner in which a *dog* relates to a *man* may serve as a basis for reasoning about a *survivor's* mental state in the introductory example. This means that we have to classify connections between words to make a network usable for computer programs to perform human-like reasoning. We shall not discuss the possible classification methods (e.g. [DED 08, GAT 15]), but we have to stress that the proper classification must recognize that the connection between two nodes in the lexical network reflects features that organize a particular type, e.g. *dog, flower*, or particular aggregate, e.g. *furniture, water* structures in the network [SOW 00].

Finally, we have to realize why an experimentally built association network does not entirely match a network built automatically from a text collection. Since an influential study by Rapp [RAP 02], an experimentally built association network served as a norm to evaluate associations generated by different statistical algorithms that operate solely on a text collection (e.g. [WAN 08, GAT 13, UHR 13]). We have to agree that text-generated associations reflect text contiguities [WET 05]. However, we may add that, contrary to text-derived associations, the associations obtained by the free word association experiment represent features that define a lexical meaning. If we compare the results of the *Wortschatz* algorithm operating on the Polish newspaper text collection [BIE 07], we can find that the word *dom* (home/house) is associated with many different verbs, e.g. *kupić* (buy), *uderzyć* (hit), *wybudować* (build), *spłonąć* (burn), *stoi* (standing), *wjechał* (struck), *zniszczyć* (destroy), and *mieć* (possess), which may be associated with many different objects, while, in the experimental network described in this paper, *dom* is associated with a single verb *mieszkać* "to dwell" and this particular verb is specific to *dom*, because *to dwell* defines the destination of the object and the place called *dom.*

3.6. Bibliography

[AGG 13] AGGARWAL C.C., ZHAO P., "Towards graphical models for text processing", *Knowledge and Information Systems*, vol. 36, no. 1, pp. 1–21, 2013.

[BIE 07] BIEMANN C., HEYRER G., QUASTHOFF U. *et al.*, "The Leipzig Corpora Collection – Monolingual corpora of standard size", *Proceedings of Corpus Linguistics 2007*, Birmingham, pp. 1–12, 2007.

[CLA 70] CLARK H.H., "Word associations and linguistic theory", in LYONS J. (ed.), *New Horizons in Linguistics*, Penguin Books, Harmondsworth, 1970.

[DED 08] DE DEYNE S., STORMS G., "Word associations: network and semantic properties", *Behavior Research Methods*, vol. 40, no. 1, pp. 213–231, 2008.

[GAT 13] GATKOWSKA I., KORZYCKI M., LUBASZEWSKI W., "Can human association norm evaluate latent semantic analysis", in SHARP B., ZOCK M. (eds), *Proceedings of the 10th International Workshop on Natural Language Processing and Cognitive Science 2013*, Marseille, 2013.

[GAT 14] GATKOWSKA I., "Word associations as a linguistic data", in CHRUSZCZEWSKI P., RICKFORD J., BUCZEK K. *et al.* (eds), *Languages in Contact*, vol. 1, Wrocław, 2014.

[GAT 15] GATKOWSKA I., "Empiryczna sieć powiązań leksykalnych", *Polonica*, vol. 35, pp. 155–178, 2015.

[HAR 14] HARĘZA M., Automatic text classification with use of empirical association network, Master Thesis, AGH, University of Science and Technology, Kraków, 2014.

[KEN 10] KENT G., ROSANOFF A.J., "A study of association in insanity", *American Journal of Insanity*, vol. 67, no. 2, pp. 317–390, 1910.

[KIS 73] KISS G.R., ARMSTRONG C., MILROY R. *et al.*, "An associative thesaurus of English and its computer analysis", in AITKEN A.J., BAILEY R.W., HAMILTON-SMITH N. (eds), *The Computer and Literary Studies*, Edinburgh University Press, 1973.

[KOR 12] KORZYCKI M., "A dictionary based stemming mechanism for Polish", in SHARP B., ZOCK M. (eds), *Proceedings of the 9th International Workshop on Natural Language Processing and Cognitive Science 2012*, Wrocław, 2012.

[LOP 07] LOPES A.A., PINHO R., PAULOVICH F.V. *et al.*, "Visual text mining using association rules", *Computers and Graphics*, vol. 31, pp. 316–326, 2007.

[LUB 15] LUBASZEWSKI W., GATKOWSKA I., HARĘZA M., "Human association network and text collection", in SHARP B., DELMONTE R. (eds), *Proceedings of the 12th International Workshop on Natural Language Processing and Cognitive Science 2015*, De Gruyter, Berlin, 2015.

[LYO 63] LYONS J., *Structural Semantics. An Analysis of Part of the Vocabulary of Plato*, Blackwell, Oxford, 1963.

[MUR 03] MURPHY M.L., *Semantic Relations and the Lexicon*, Cambridge University Press, Cambridge, 2003.

[RAP 02] RAPP R., "The computation of word associations: comparing syntagmatic and paradigmatic approaches", *Proceedings of the 19th International Conference on Computational Linguistics*, vol. 1, Taipei, pp. 1–7, 2002.

[RUP 10] RUPPENHOFER J., ELLSWORTH M., PETRUCK M.R.L. *et al.*, *FrameNet II: Extended Theory and Practice*, Berkeley University, 2010.

[SOW 00] SOWA J.F., *Knowledge Representation. Logical, Philosophical, and Computational Foundations*, vol. 13, Brooks/Cole, Pacific Grove, 2000.

[UHR 13] UHR P., KLAHOLD A., FATHI M., "Imitation of the human ability of word association", *International Journal of Soft Computing and Software Engineering*, vol. 3, no. 3, pp. 248–254, 2013.

[WAN 08] WANDMACHER T., OVCHINNIKOVA E., ALEXANDROV T., "Does latent semantic analysis reflect human associations", *Proceedings of the ESSLLI Workshop on Distributional Lexical Semantics*, Hamburg, 2008.

[WET 05] WETTLER M., RAPP R., SEDLMEIER P., "Free word associations correspond to contiguities between words in text", *Journal of Quantitative Linguistics*, vol. 12, no. 2, pp. 111–122, 2005.

[WU 11] WU J., XUAN Z., PAN D., "Enhancing text representation for classification tasks with semantic graph structures", *ICIC International*, vol. 7, no. 5(b), pp. 2689–2698, 2011.

The Reverse Association Task

Free word associations are the words human subjects spontaneously come up with upon presentation of a stimulus word. In experiments comprising thousands of test subjects, large collections of associative responses have been compiled. In previous publications, it was shown that these human associations can be resembled by statistically analyzing the co-occurrences of words in large text corpora. In this chapter, we consider the reverse question, namely whether the stimulus can be predicted from the responses. We call this reverse association task and present an algorithm for approaching it. We also collected human data on the reverse association task, and compared them with the machine-generated results.

4.1. Introduction

Word associations have always played an important role in psychological learning theory, and have been investigated not only in theory, but also in experimental work where, for example, such associations were collected from human subjects. Typically, the subjects are given questionnaires with lists of stimulus words, and were asked to write down for each stimulus word the spontaneous association which first came to mind. This led to collections of associations, the so-called association norms, as exemplified in Table 4.1. Among the best known association norms are the Edinburgh Associative Thesaurus (EAT) [KIS 73], the Minnesota Word Association Norms [JEN 70, PAL 64] and the University of South Florida Free

---

Chapter written by Reinhard RAPP.

Association Norms [NEL 98]. More recently, attempts have been made to use crowd sourcing methods for collecting associations in various languages (*Jeux de mots*[1] and *Word Association Study*[2]). In this way, researchers are able to collect much larger datasets than was previously possible.

Association theory, which can be traced back to Aristotle in ancient Greece, has often stated that our associations are governed by our experiences. For example, more than a century ago, William James [JAM 90] formulated this in his book, *The Principles of Psychology*, as follows:

> "Objects once experienced together tend to become associated in the imagination, so that when any one of them is thought of, the others are likely to be thought of also, in the same order of sequence or coexistence as before. This statement we may name the law of mental association by contiguity."

This citation is talking of *objects*, but the question arose whether for words the same principles might apply, and with the advent of corpus linguistics, it was possible to verify this experimentally by looking at the distribution of words in texts. Among the first to do so were [CHU 90], [SCH 89] and [WET 89].

Their underlying assumption was that in text corpora, strongly associated words often occur in close proximity. This is actually confirmed by corpus evidence: Figure 4.1 assigns to each stimulus word position 0, and displays the occurrence frequencies of its primary associative response (most frequent response as produced by the test persons) at relative distances between −50 and +50 words. However, to give a general picture and to abstract away from idiosyncrasies, the figure is not based on a single stimulus/response pair, but instead represents the average of 100 German stimulus/response pairs as used by Russell and Meseck [RUS 96]. The effect is in line with expectations: the closer we get to the stimulus word, the higher the chances that the primary associative response occurs. Only for distances of plus or minus one, there is an exception, but this is an artifact because content words are typically separated by function words, and among

---

1 http://www.jeuxdemots.org/jdm-accueil.php
2 https://www.smallworldofwords.org/en

our 100 primary responses there are no function words. In addition, test persons typically select content words only.

While such considerations are the basis underlying our work, in this chapter, the focus is on whether it is possible not only to compute the responses from the stimulus, but also to compute the stimulus from the responses. To the best of our knowledge, this has not been attempted before in a comparable (distributional semantics) framework, and therefore we are not aware of any directly related literature.

However, this task is somewhat related to the computation of associations when given several stimulus words simultaneously, which is sometimes referred to using the terms *multi-stimulus-* or *multiword associations* [RAP 08], or *remote association test* (RAT). A recent notable publication on the RAT, which gives pointers to other related works, is Smith *et al.* [SMI 13]. It applies this methodology on problems that require consideration of multiple constraints, such as choosing a job based on salary, location and work description. Another one is Griffiths *et al.* [GRI 07], which assumes that concept retrieval from memory can be facilitated by inferring the gist of a sentence, and using that gist to predict related concepts and disambiguate words. It implements this by using a topic model.

CIRCUS	FUNNY	NOSE
clown (24)	laugh (23)	face (16)
ring (10)	girl (11)	eyes (12)
elephant (6)	joke (8)	mouth (11)
tent (6)	laughter (6)	ear (10)
animals (5)	amusing (4)	eye (6)
top (5)	hilarious (4)	throat (4)
boy (4)	comic (3)	smell (3)
clowns (3)	ha ha (3)	bag (2)
horse (2)	ha-ha (3)	big (2)
horses (2)	sad (3)	handkerchief (2)

Table 4.1. *Top 10 sample associations for three stimulus words as taken from the Edinburgh Associative Thesaurus. The numbers of subjects responding with the respective word are given in brackets*

Our approach differs from this previous work in that it focuses on a related but different and particularly well-defined task. In our approach, we have eliminated all (for this particular task) unnecessary sophistication, such as *Latent Semantic Analysis* (which we used extensively in previous work) or *Topic Modeling*, resulting in a simple yet effective algorithm. For example, [GRI 07] reports 11.54% correctly predicted first associates. Rapp [RAP 08] presents a number of evaluations using various corpora and datasets, but with all results below 10%. The above-mentioned paper by Smith *et al.* [SMI 13] gives no such figures at all. In comparison, the best results presented here are at 54% (see section 4.3.3). It should be emphasized, however, that all comparisons have to be taken with caution, as there is no commonly used gold standard for this, and hence all authors used different test data, and different corpora. Note also that, in contrast to the related work, our focus is on the novel reverse association task, which gives us test data of unprecedented quality and quantity (as any word association norm can be used), but for which the previous test data is unsuitable as it relates to a somewhat different task.

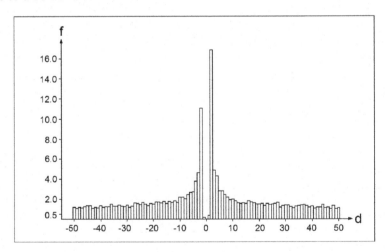

Figure 4.1. *Occurrence frequency f of a primary response at distance d from a stimulus word, averaged over 100 stimulus/response pairs [RAP 96]*

This paper, which is a substantially extended version of [RAP 13], is structured as follows: we first look at how we compute associations for single stimulus words. This lays the basis for the second part where we reverse our viewpoint and compute the stimulus word from its associations. In the third part, we want to find out how humans perform on the reverse association task. For this purpose, we conducted a reverse association experiment where human responses were collected. Finally, we compare the performances of man and machine.

4.2. Computing forward associations

4.2.1. *Procedure*

As discussed in the introduction, we assume that there is a relationship between word associations as collected from human subjects and word co-occurrences as observed in a corpus. As our source of human data, we use the *Edinburgh Associative Thesaurus* (EAT; [KIS 73, KIS 75]), which is the largest classical collection of its kind[3]. The EAT comprises about 100 associative responses as requested from British students for each of the 8,400 stimulus terms. As some of these stimulus terms are multiword units which we did not want to include here, we removed these from the thesaurus, such that 8,210 items remained.

To obtain the required co-occurrence counts, we aimed for a corpus which is as representative as possible for the language environment of the EAT's British test subjects. We therefore chose the *British National Corpus* (BNC), a 100-million-word corpus of written and spoken language, which was compiled with the intention of providing a balanced sample of British English [BUR 98]. For our purpose, it is also an advantage that the texts in the BNC are not very recent (from 1960 to 1993), thereby including the time period when the EAT data was collected (between June 1968 and May 1971).

Since function words were not considered important for our analysis of word semantics, we decided to remove them from the text to save memory requirements and processing time. This was done on the basis of a list of approximately 200 English function words. We also decided to lemmatize

---

3 An even larger, though possibly more noisy, association database has been collected via online gaming at www.wordassociation.org

the corpus using the lexicon of full forms provided by Karp *et al.* [KAR 92]. This not only improves the problem of data sparseness, but also significantly reduces the size of the co-occurrence matrix to be computed. Since most word forms are unambiguous concerning their possible lemmas, we only conducted a partial lemmatization that does not take the context of a word into account and thus leaves the relatively few words with several possible lemmas unchanged. For consistency reasons, we applied the same lemmatization procedure to the whole EAT. Note that, as the EAT contains only isolated words, in this case, a lemmatization procedure that takes the context of a word into account would not be possible.

For counting word co-occurrences, as in most other studies, a fixed window size is chosen and it is determined how often each pair of words occurs within a text window of this size. Choosing a window size usually means a trade-off between two parameters: specificity versus the sparse-data problem. The smaller the window, the more salient the associative relations between the words inside the window, but the more severe the problem of data sparseness. In our case, with ±2 words, the window size looks rather small. However, this can be justified since we have reduced the effects of data sparseness by using a large corpus and by lemmatizing the corpus. It should also be noted that a window size of ±2 applied after elimination of the function words is comparable to a window size of ±4 applied to the original texts (assuming that roughly every second word is a function word).

Based on the window size of ±2, we computed the co-occurrence matrix for the corpus. By storing it as a sparse matrix, it was feasible to include all of the approximately 375,000 lemmas occurring in the BNC.

Although word associations can be successfully computed based on raw word co-occurrence counts, the results can be improved when the observed co-occurrence-frequencies are transformed by some function that reduces the effects of absolute word frequency. As it is well established, we decided to use the log-likelihood ratio [DUN 93] as our association measure. It compares the observed co-occurrence counts with the expected co-occurrence counts, thus strengthening significant word pairs and weakening incidental word pairs. In the remainder of this paper, we refer to co-occurrence vectors and matrices that have been transformed this way as association vectors and matrices.

4.2.2. Results and evaluation

To compute the associations for a given stimulus word, we look at its association vector as computed in the way described above, and rank the words in the vocabulary according to association strength. Table 4.2 (two columns on the right) exemplifies the results for the stimulus word *cold*[4]. For comparison, the two columns on the left list the responses from the EAT, and words occurring in both lists are printed in bold. It can be seen that especially the test persons' most frequent responses are predicted rather well in the simulation: among the top eight experimental responses, six can be found among the computed responses.

Surprisingly, although the system solely relies on word co-occurrences, it predicts not only syntagmatic but also paradigmatic associations (e.g. not only *cold* → *ice* but also *cold* → *hot*; see [DE 96, RAP 02]).

We conducted a straightforward evaluation of the results. It is based on lemmatized versions of both the British National Corpus and, as this is the quasi-standard for evaluation in related works, the Kent and Rosanoff [KEN 10] subset of the Edinburgh Associative Thesaurus which comprises 100 words.

For 17% of the stimulus words, the system produced the primary associative response, which is the most frequent response as produced by the human subjects[5]. In comparison, the average participant in the Edinburgh Associative Thesaurus [KIS 73] produced 23.7% primary responses to these stimulus words. This means that the system performs reasonably but not quite as well as the test persons.

---

4 In order not to lose information, in contrast to all other results presented in this chapter, this table is based on an unlemmatized corpus and an unlemmatized association norm.

5 Wettler *et al.* [WET 05] report somewhat better results by additionally taking advantage of the observation that test persons typically answer with words from the mid-frequency range. As it is not clear how this affects the results when computing associations for several given words, we did not do so in the current paper.

Observed responses	# of subjects	Computed responses	# of subjects
hot	34	**water**	5
ice	10	**hot**	34
warm	7	weather	0
water	5	**wet**	3
freeze	3	blooded	0
wet	3	**ice**	10
feet	2	air	0
freezing	2	**winter**	2
nose	2	**freezing**	2
room	2	bitterly	0
sneeze	2	damp	0
sore	2	wind	0
winter	2	**warm**	7
arctic	1	felt	0
bad	1	**war**	1
beef	1	night	0
blanket	1	icy	0
blow	1	**heat**	1
cool	1	shivering	0
dark	1	cistern	0
drink	1	feel	0
flu	1	windy	0
flue	1	stone	0
frozen	1	morning	0
hay fever	1	shivered	0
head	1	eyes	0
heat	1	clammy	0
hell	1	sweat	0
ill	1	blood	0
north	1	shower	0
often	1	rain	0
shock	1	winds	0
shoulder	1	tap	0
snow	1	dry	0
store	1	**dark**	1
uncomfy	1	grey	0
war	1	hungry	0

Table 4.2. *Comparison between observed and computed associative responses to the stimulus word* cold *(matching words in bold; no lemmatization; capitalized words transferred to lower case)*

4.3. Computing reverse associations

4.3.1. *Problem*

Having seen that word associations with single stimulus words can be computed with a quality similar to that achieved by human subjects, let us now turn to the main question of this paper, namely whether it is also possible to reverse the task, i.e. to compute a stimulus word from its associations.

Let us look at an example: According to the Edinburgh Associative Thesaurus, the top three most frequent responses to *clown* are *circus* (produced by 26 out of 93, i.e. 28% of the test persons), *funny* (9% of the test persons) and *nose* (8% of the test persons). The question is now: given only the three words *circus*, *funny* and *nose*, is it possible to determine that their common stimulus word is *clown*? And if it is possible, what would be the quality of the results?

The above is an illustrative example, but, in other cases, it is often more difficult to guess the correct answer. To give a feeling for the difficulty of the task, let us provide a few more examples involving varying numbers of given words, with the solutions provided in Table 4.4:

apple, juice → ?

water, tub, clean → ?

grass, blue, red, yellow → ?

drink, gin, bottle, soda, Scotch → ?

4.3.2. *Procedure*

Our first idea on how to compute the stimulus given the responses was to look at the associations of the responses, and to determine their intersection. However, in preliminary experiments, we found out that this does not work well. The reason appears to be asymmetry in word association. But what do we mean by asymmetry in this context?

The co-occurrence counts that we extract from the corpus are symmetric, because whenever word A co-occurs with word B, word B also co-occurs with word A. Whether an association matrix computed from the co-occurrence matrix is also symmetric depends on the association measure used. However, even in the case of symmetric weights, associations can still be asymmetric. Let us illustrate this using Figure 4.2. This is the graphical equivalent of a symmetric association matrix[6]. As can be seen, the strongest association with *blue* is *black*. However, the opposite is not true: the strongest association with *black* is not *blue* as *black* has an even stronger association with *white*.

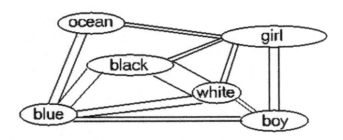

Figure 4.2. *Associative lexical network with symmetric weights*

To give an idea about the situation in the EAT: not considering multiword units, the EAT comprises 8,210 stimulus words and likewise 8,210 primary responses. However, there is not a complete overlap between these two vocabularies: only 7,387 words occur in both, which means that, only for these words, symmetry considerations are possible. Of these 7,387 cases, 63% of the responses were symmetric, and 37% were asymmetric. Table 4.3 shows some examples from the EAT for both types of associations.

---

6 In the asymmetric case, we would require two directed connections of (usually) different widths between each pair of nodes.

Symmetric Associations			Asymmetric Associations		
Stimulus	PR on Stimulus	PR on Response	Stimulus	PR on Stimulus	PR on Response
bed	sleep	bed	baby	boy	girl
black	white	black	bitter	sweet	sour
boy	girl	boy	comfort	chair	table
bread	butter	bread	cottage	house	home
butter	bread	butter	dream	sleep	bed
chair	table	chair	hand	foot	shoe
dark	light	dark	heavy	light	dark
girl	boy	girl	lamp	light	dark
hard	soft	hard	red	blue	sky
light	dark	light	sickness	health	wealth

Table 4.3. *Examples for symmetric and asymmetric associations (PR = primary response)*

Let us now return to our above example, namely *circus, nose, funny* → *clown*. Here, *circus* and *clown* are an example for the symmetric case. Both are each other's primary associative responses in the EAT, and therefore *circus* is the strongest association with *clown*, and likewise *clown* is the strongest association with *circus*. If this were always true, things would be straightforward. However, this is not the case. For example, *clown* is strongly associated with *nose*, but *nose* is not strongly associated with *clown*. In the EAT, among 97 test persons, given the stimulus word *nose*, none responded with *clown*. Likewise, given the word *funny*, among 98 test persons, again nobody answered with *clown*. Therefore, if we take the intersection of the associations with *circus, nose,* and *funny, clown* would be out. This is why an approach based on intersecting associations does not work well.

Instead, like in word sense disambiguation and like in multiword semantics, it appears that we have to take contextual information into account[7]. For example, in the context of *circus, nose* is clearly related to *clown*, but, in the context of *doctor*, it is not.

Such considerations resulted in the following approach: we utilize the observation that a stimulus word must have strong weights to all of its top associations, and that a strong association with only some of them does not suffice. Such a behavior can usually be put into practice by using a multiplication.

However, we do not multiply the association strengths, as the log-likelihood ratio has an inappropriate (exponential) value characteristic. This value characteristic has the effect that a weak association with one of the stimuli can easily be overcompensated by a strong association with another stimulus, which is not desirable. Instead of multiplying the association strengths, we therefore multiply their ranks. This improves the results considerably.

These considerations lead us to the following procedure (see [RAP 08]): given an association matrix of vocabulary V containing the log-likelihood ratios between all possible pairs of words, to compute the stimulus word causing the responses $a, b, c, ...$, the following steps are conducted:

1) For each word in V (by considering its association vector), look up the ranks of words $a, b, c, ...$ in its association vector, and compute the product of these ranks ("Product-of-Ranks algorithm").

2) Sort the words in V according to these products, with the sort order such that the lowest value obtains the top rank (i.e. conduct a reverse sort).

---

7 Further reflections on this may lead to the fundamental question whether asymmetry of word associations is the consequence of word ambiguity, or whether word ambiguity is the consequence of asymmetry of word associations.

Note that this procedure is somewhat time consuming as these computations are required for each word in a large vocabulary[8]. On the plus side, the procedure is in principle applicable to any number of given words, and with an increasing number of given words, there is only a slight increase in computational load.

A minor issue is the assignment of ranks to words that have identical log-likelihood scores, especially in the frequent case of zero co-occurrence counts. In such cases, the assignment of almost arbitrary ranks within such a group of words could adversely affect the results. We therefore suggest assigning corrected ranks, which are to be chosen as the average ranks of all words with identical log-likelihood scores.

In principle, the algorithm can also be used if there is only a single given word. However, this does not make much sense as the algorithm is computationally far more expensive than what we described in section 4.2, and the results are typically worse for the reason that ranks do not allow as fine-grained distinctions as do association strengths. For example, given the word *white*, the algorithm might find several words in the vocabulary where *white* is on rank 1 (e.g. *black* and *snow*). However, as (without further sophistication) no distinction is made between these, they will end up in arbitrary order, without taking into account that the top rank of *black* is more salient than that of *snow*[9].

On the other hand, if the number of given words becomes larger, depending on the application, it can be helpful to introduce a limit to the maximum rank, thereby reducing the effects of statistical variation, which is especially severe for the lower ranks. Note that for the current work, we used a rank limit of 10,000. However, the exact value is not critical because this usually has little impact if the focus is mainly on the top ranks, as is the case here.

---

8 Considerable time savings are possible by using an index of the non-zero co-occurrences.
9 In the EAT, 57 and 40 subjects, respectively, responded with *white* for these two stimulus words; another example is *lily*, where the primary associative response is also *white*, but is produced by only 19 subjects. In this case, the next frequent response, namely *flower*, is very close as it is produced by 17 subjects.

4.3.3. *Results and evaluation*

To give an impression of the results when applying the above algorithm to various responses from the EAT, Table 4.4 lists some results. For example, the EAT lists *apple* and *juice* as the top responses when given the stimulus word *fruit*, but our algorithm, when provided with *apple* and *juice*, computes that *orange* would be the best stimulus. This is not as expected, but also has some plausibility. The expected stimulus *fruit* at least shows up on the 8th position of the computed list of words.

For a quantitative evaluation, like for the forward associations, we consider only the Kent and Rosanoff [KEN 10] subset of the EAT[10]. We count in how many cases the expected word is ranked first in the list of computed words. This leads to conservative numbers as only exact matches are taken into account. For example, the last item in Table 4.4, where *whisky* instead of *whiskey* is on rank 1, would count as incorrect.

When predicting the stimuli from the associative responses, the question is how many of the responses should be taken into account, and in how far the quality of the results depends on the number of responses. To answer this question, we conducted the evaluation several times, each time with another number of given words (= EAT responses). There are three expectations:

– the more subjects have given a response, the more salient it is and the more helpful it should be for predicting the stimulus;

– responses given by only one or very few subjects might be arbitrary and therefore not helpful for predicting the stimulus;

– considering a larger number of salient responses should improve the results.

These expectations are confirmed by the results. Figure 4.3 shows the percentage of correctly computed stimuli depending on the number of top responses (from the EAT) that are taken into account. As can be seen, the quality of the results improves up to seven given words where it reaches 54% accuracy, and from then on degrades. This means that, on average, already the eighth response word is not helpful for determining the respective stimulus word.

---

10 It should be noted that the Kent and Rosanoff [KEN 10] subset typically leads to relatively high accuracies, as it mostly comprises familiar words with high corpus frequencies.

Top 2 responses from EAT: apple (1,385) juice (1,613)
Stimulus word from EAT: **fruit** (3,978)
Computed Stimuli: orange (2,333), grape (273), lemon (1,019), lime (612), pineapple (220), grated (423), apples (792), **fruit** (3,978), grapefruit (113), carrot (359)

Top 3 responses from EAT: water (33,449), tub (332), clean (6,599)
Stimulus word from EAT: **bath** (415)
Computed Stimuli: rinsed (177), **bath** (2,819), soak (315), rinse (288), wash (2,449), refill (138), rainwater (160), polluted (393), towels (421), sanitation (156)

Top 4 responses from EAT: grass (4,295), blue (9,986), red (13,528), yellow (4,432)
Stimulus word from EAT: **green** (10,606)
Computed Stimuli: **green** (10,606), jersey (359), ochre (124), bright (5,313), pale (3,583), violet (396), purple (1,262), greenish (136), stripe (191), veined (103)

Top 5 responses from EAT: drink (7,894), gin (507), bottle (4,299), soda (356), Scotch (621)
Stimulus word from EAT: **whiskey** (129)
Computed Stimuli: whisky (1,451), **whiskey** (129), tonic (511), vodka (303), brandy (848), Whisky (276), scotch (151), lemonade (229), poured (1,793), gulp (196)

Table 4.4. *Top 10 computed stimuli for various numbers of given responses. Numbers in brackets refer to corpus frequencies in the BNC*

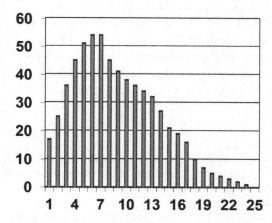

Figure 4.3. *Percentage of correctly predicted stimuli (vertical axis) depending on the number of given words (horizontal axis)*

Let us mention that we were positively surprised by the 54% performance figure, which is about three times as good as for forward associations (first column in Figure 4.3). On the one hand, in the reverse association task, there are several clues pointing to the same stimulus word. On the other hand, the task seems non-trivial for humans, and typically there are several plausible options how the given words can disambiguate each other. For example, given *apple* and *juice* (see Table 4.4), the solution our system came up with, namely *orange*, seems quite as plausible as the expected solution *fruit*. However, in our evaluation, *orange* is counted as wrong, and this is true for many others of the 46% incorrect results.

4.4. Human performance

Now that we have investigated machine performance on the reverse association task, the next question is how humans perform on this task, and how the results of a human and a machine compare. For this purpose, we conducted an experiment with the aim of collecting human reverse associations, and later on compared its results with those obtained in a simulation.

4.4.1. *Dataset*

As our dataset we used the test set which – as a follow-up activity of [RAP 13] – we had prepared for the *CogALex-IV Shared Task on the Lexical Access Problem* [RAP 14]. The aim of this shared task had been to compare different automatic methods for computing reverse associations[11]. In contrast, here we investigate human performance on this task. Therefore, to compare human with machine associations, it is good that both studies use a dataset based on the same source, i.e. on the EAT[12].

---

11 The best performing systems in this shared tasks all used methodologies similar to the system described in this paper (i.e. based on word co-occurrences in large corpora) and their results were of similar quality. As they have been discussed in detail in [RAP 13], we do not repeat this discussion here.
12 Note that, in addition to the dataset described here, the shared task at CogALex-IV used an additional so-called "training set", which was intended for the development and optimization of the automatic algorithms. This training set was produced in exactly the same way, just using a different selection of 2,000 items.

The dataset has been produced by conducting the following steps:

1) take the EAT as the basis;

2) modify capitalization;

3) extract 2,000-item subset (mainly at random);

4) retain only the top five associations for each stimulus word;

5) remove stimulus words.

Let us now look a bit closer at this procedure. The EAT lists for each of the 8,400 stimulus words the associative responses as obtained from about 100 test persons[13] who were asked to produce the word coming spontaneously to their mind.

As, given its origin in the 1970s, the EAT uses uppercase characters only, we decided to modify its capitalization. For this purpose, for each word occurring in the EAT, we looked up which form of capitalization showed the highest occurrence frequency in the *British National Corpus* [BUR 98]. By this form, we replaced the respective word, e.g. *DOOR* was replaced by *door*, and *GOD* was replaced by *God*. This way we hoped to come close to what might have been produced during compilation of the EAT if case distinctions had been taken into account. Since this method is not perfect, e.g. words often occurring in the initial position of a sentence might be falsely capitalized, we did some manual checking, but cannot claim to have achieved perfection.

For each stimulus word, only the top five associations (i.e. the associations produced by the largest numbers of test persons) were retained, and all other associations were discarded. The decision to keep only a small number of associations was motivated by the results shown in Figure 4.3, which indicate that associations produced by only very few test persons tend to be of an arbitrary nature. We also wanted to avoid unnecessary complications, which is why we decided on a fixed number, although the choice of exactly five associations is somewhat arbitrary.

---

13 To distribute work, different groups of test persons operated on the stimulus words.

Given words	Target words
able incapable brown clever good	capable
able knowledge skill clever can	ability
about near nearly almost roughly	approximately
above earth clouds God skies	heavens
above meditation crosses passes rises	transcends
abuse wrong bad destroy use	misuse
accusative calling case Latin nominative	vocative
ache courage blood stomach intestine	guts
ache nail dentist pick paste	tooth
aches hurt agony stomach period	pains
action arc knee reaction jerk	reflex
actor theatre door coach act	stage

Table 4.5. *Top 12 items of the dataset. The target words were of course undisclosed to the test takers*

From the remaining dataset, we removed all items that contained non-alphabetical characters. We also removed items that contained words that did not occur in the BNC. The reason for this is that quite a few of them are misspellings. By these measures, the number of EAT items was reduced from initially 8,400 to 7,416. From these, we randomly selected 2,000 items that were used as our dataset. Table 4.5 shows the (alphabetically) first 12 items in this dataset[14].

4.4.2. Test procedure

As, according to comments from our test persons, producing multi-stimulus associations is subjectively hard, in order not to overload the test takers, we divided the dataset into 40 sections, each comprising 50 items. A total of 40 test sheets were printed, each showing the test items from one section. The test takers were instructed to produce for each test item of five words the spontaneous association that first came to mind. It was also

14 The full dataset can be downloaded from http://pageperso.lif.univ-mrs.fr/~michael.zock/ ColingWorkshops/CogALex-4/co-galex-webpage/pst.html

mentioned that there are no "right" or "wrong" answers, but that we were only surveying human word associations[15].

As an example, the list *"work desk bureau secretary post"* was provided, together with the response *"office"*. As the experiment was conducted at the University of Mainz, Faculty of Translation Studies, Linguistics and Cultural Studies (FTSK) in Germersheim (Germany), all participants were non-native English speakers. For this reason, they were asked to rank their English proficiency on the following scale: *native / very good / good / satisfactory / basic knowledge / no or very little knowledge*. Note, however, that because the FTSK is specialized in training translators and interpreters, of whom many focus on English, as expected the majority of students indicated "very good" English proficiency, and none indicated a proficiency below "satisfactory". The experiments were conducted in the winter term of 2014/15 and in the summer term of 2015 in the courses given by the author. Course topics were *Translation Systems, Language and Cognition, Computational Linguistics, Electronic Dictionaries, Corpus Linguistics, Machine Translation, Computer Aided Translation* and *Translation Memories* (mostly seminars at either undergraduate or graduate level)[16]. Participation in the association test was voluntary. As the task was perceived to be tedious, many questionnaires showed some omissions. Altogether, 66 questionnaires were filled out, which implies that some of the 40 sections of the dataset were dealt with by more than one student.

4.4.3. Evaluation

For evaluation, we simply compared the associations produced by the test persons with the expected results as taken from the EAT (see Table 4.2). In a first round of evaluation, we only considered exact matches as correct, with the only flexibility that word capitalization was not taken into account. However, in a second round, we also counted inflectional variants and derived forms of the expected word as correct. For example, for the given words *"car round cart spoke bicycle"*, a test person responded with *wheels*, whereas the expected solution was *wheel*. This was now also considered to be a match.

---

15 Note that we did not inform the participants about the nature of the test data, i.e. that the underlying idea is based on the reverse association task.

16 For the purpose of this paper, German course titles were translated into English.

English proficiency	exact matches	tolerant matches
very good (48 subjects)	4.38 %	7.42 %
good (14 subjects)	2.14 %	4.29 %
satisfactory (4 subjects)	4.50 %	7.50 %
All 66 subjects	3.91 %	6.76 %

Table 4.6. *Results by language proficiency*
for the two modes of evaluation

The overall results are shown in Table 4.6. For each proficiency group, the percentage of correctly predicted words is specified for the two modes of evaluation. All accuracies are in a range between 2 and 8%. As can be seen, the students who rated their language proficiency to be very good did considerably better than the ones who rated themselves to be only good; but surprisingly, the students with an only satisfactory self-rating did even better. Note, however, that this is a small group of only four subjects who cannot be taken as representative (and whose self-ratings might simply reflect modesty).

4.5. Performance by machine

Although we have presented simulation results concerning the reverse association task in section 4.3, these results were obtained using a much smaller test set than the one used for the human survey. We therefore conducted an additional system run using the human test data. The parameters were the same as described previously (section 4.2). However, as our corpus we did not use the BNC but rather ukWaC, a web-derived corpus of about 2 billion words[17]. This has the advantage that our results can also be compared with those of the CogALex shared task [RAP 14], where the ukWaC corpus has also been used.

As done previously for the BNC, we lemmatized the ukWaC corpus and removed stop words. As our vocabulary we used a list of words which in the BNC had an occurrence frequency of 100 or higher. According to our definition of a word (string of alpha or of non-alpha characters), this was the case for 36,097 words. However, like the ukWaC corpus, we also

---

17 http://wacky.sslmit.unibo.it/doku.php?id=corpora

lemmatized this list and removed stop words. We also eliminated any strings containing non-alpha characters. This reduced the vocabulary to 22,578 words.

Using this list, by counting word co-occurrences in the ukWaC-corpus, we built up a co-occurrence matrix, thereby considering a window size of plus and minus two words around a given word. The resulting co-occurrence matrix was converted into a weight matrix by applying the log-likelihood ratio [DUN 93] to each value in the matrix. Finally, the product-of-ranks algorithm was applied to compute the results for each item of five words in the test set.

Of the 2,000 items, the system got 613 right, i.e. the word on rank 1 of the computed list was exactly identical to the expected word as provided in the test set. This corresponds to an accuracy of 30.65%. Despite the larger corpus (ukWaC is about 20 times larger than the BNC), this accuracy is considerably lower than the top performance figure of 54% reported in section 4.3. However, the discrepancy is not surprising as, in section 4.3, the test was conducted with the 100 words from the Kent–Rosanoff test. This test contains mostly very common words well known for their salient associations, which are comparatively easy to predict. Also, in section 4.3, the words in the test set had been lemmatized, which is not the case here.

Therefore, let us compare our results with those of the CogALex shared task [RAP 14]. There, on exactly the same test set and based on the same corpus, the best system showed a performance of 30.45%, which almost exactly matches the performance of our system. However, while this system used sophisticated technology involving word embeddings (neural network technology), we achieved a similar result using very simple technology. It should also be noted that we did not do any parameter optimization and simply used the parameters from a previous paper [RAP 13], i.e. to obtain the current results, we did not even look at the training set, which was also provided for the CogALex shared task.

It should also be noted that our selection of vocabulary was completely unrelated to the current task. Our word list was derived from the BNC rather than from ukWaC, and the fact that the test set was derived from the EAT had no influence on our choice of words[18]. We also used an arbitrary

---

18 Would the vocabulary be derived from the EAT, better results could be expected.

frequency threshold (BNC frequency of 100) for the selection of our vocabulary, rather than trying to optimize this threshold using the training set.

It should be noted that the choice of vocabulary is very important for this task, and much better results could be achieved by applying "informed guesses" on this issue. Let us mention that of the 2,000 unique solutions from the test set[19], in our vocabulary of 22,755 words, only 1,482 occurred, i.e. for 518 words, our system had no chance to come up with the correct solution.

4.6. Discussion, conclusions and outlook

4.6.1. *Reverse associations by a human*

Associating with several given words, as required in the reverse association task, is not easy. This has been remarked by several test persons and is also confirmed by a considerable number of omissions in the test sheets. For several reasons, it is difficult to come up with an expected word:

1) In many cases, the given words might almost quite as strongly point to other target words. For example, when given the words *gin*, *drink*, *scotch*, *bottle* and *soda*, instead of the target word *whisky*, the alternative spelling *whiskey* should also be fine, and possibly some other alcoholic beverages, such as *rum* or *vodka*, might also be acceptable[20].

2) The target vocabulary was not restricted in any way, so in principle hundred thousands of words had to be considered as candidates.

3) Although most of the target words were base forms, the dataset also contains a good number of cases where the target words were inflected forms. Of course, it is hard to get these inflected forms exactly right.

Owing to these difficulties, we expected low-performance figures, and this expectation was confirmed.

---

19 Each solution can occur just once because they are derived from the EAT stimuli.

20 As our data source (the EAT) did not provide any, it was not practical for us to try to come up with alternative solutions in the chosen reverse association framework. However, we think that doing so would mainly affect absolute but not relative performance (i.e. the ranking between different systems should remain similar).

As mentioned in section 4.4.2, we had not disclosed the nature of the dataset to the test subjects, i.e. did not tell them that the underlying idea is the reverse association task. Alternatively, we could have informed people about this and asked them to come up with a word with which they would associate each of the five given words. But this would probably have been conceived as an even more sophisticated task, potentially blocking spontaneity. We also tend to think that, although associations can occasionally be asymmetric[21], assymetry is likely not to have a decisive effect in our scenario. However, this is certainly a question which requires further investigation

While in previous studies involving multi-stimulus associations, human performance had always been much better than what simulation programs produced (see e.g. [RAP 08]), this is not the case here. In the CogALex shared task, a number of teams had tested their algorithms, which were mostly based on the analysis of word co-occurrences in large text corpora, on the very same dataset, and achieved performances of up to 30.45%. In the current paper, on this dataset, we presented a very similar result. These results are much better than the human performance of our non-native speakers shown in Table 4.6. Although native speakers can be expected to do better, we think that it will be challenging to outperform the automatic results. This might well mean that in one of its core disciplines, namely association, human intelligence is not better than a machine, although of course further investigation is required to confirm this finding.

4.6.2. *Reverse associations by a machine*

We introduced the product-of-ranks algorithm and showed that it can be successfully applied to the problem of computing associations if several words are given. To evaluate the algorithm, we used the EAT as our gold standard, but assumed that it makes sense to look at this data in the reverse direction, i.e. to predict the EAT stimuli from the EAT responses.

Although this is a task even difficult for humans, and although we applied a conservative evaluation measure that insists on exact string matches between a predicted and a gold standard association, our algorithm was able to do so with a success rate of approximately 30% (54% for the

---

21 For example, *flower pot* → *soil* are associated strongly, but *soil* → *flower pot* not to the same extent.

Kent–Rosanoff vocabulary). We also showed that, up to a certain limit, with increasing numbers of given words, the performance of the algorithm improves, and only thereafter degrades. The degradation is in line with our expectations because associative responses produced by only one or very few persons are often of almost arbitrary nature and therefore not helpful for predicting the stimulus word[22].

Given the notorious difficulty to predict experimental human data, we think that the performance of approximately 30% is quite good, especially in comparison to the human results shown in Table 4.6, but also in comparison to the related work mentioned in the introduction (11.54%), and to the results on single stimuli (17%). However, there is of course still room for improvement, even without moving to more sophisticated (but also more controversial) evaluation methods that allow alternative solutions. We intend to advance from the product-of-rank algorithm to a product-of-weights algorithm. But, this requires that we have a high-quality association measure with an appropriate value characteristic. One idea is to replace the log-likelihood scores by their significance levels. Another is to abandon conventional association measures and move on to empirical association measures as described in Tamir and Rapp [TAM 03]. These do not make any presuppositions on the distribution of words, but determine this distribution from the corpus. In any case, the current framework is well suited for measuring and comparing the suitability of any association measure. Further improvements might be possible by using neural vector space models (word embeddings), as investigated by some of the participants of the CogALex-IV shared task [RAP 14].

Concerning applications, we see a number of possibilities: one is the tip-of-the-tongue problem, where a person cannot recall a particular word but can nevertheless think of some of its properties and associations. In this case, descriptors for the properties and associations could be fed into the system in the hope that the target word comes up as one of the top associations, from which the person can choose.

Another application is in information retrieval, where the system can help to sensibly expand a given list of search words, which is in turn used to conduct a search. A more ambitious (but computationally expensive) approach would be to consider the (salient words in the) documents to be

---

22 Such associations might reflect very specific experiences of a test person.

retrieved as our lists of given words, and to predict the search words from these using the product-of-ranks algorithm.

A further application is in multiword semantics. Here, a fundamental question is whether a particular multiword expression is of compositional or of contextual nature. The current system could possibly help to provide a number of quantitative measures relevant for answering the following questions:

1) Can the components of a multiword unit predict each other?

2) Can each component of a multiword unit be predicted from its surrounding content words?

3) Can the full multiword unit be predicted from its surrounding content words?

The results of these questions might help us to answer the question regarding a multiword unit's compositional or contextual nature, and to classify various types of multiword units.

The last application we would like to propose here is natural language generation (or any application that requires it, e.g. machine translation or speech recognition). If in a sentence, one word is missing or uncertain, we can try to predict this word by considering all other content words in the sentence (or a somewhat wider context) as our input to the product-of-ranks algorithm.

From a cognitive perspective, the hope is that such experiments might lead to some progress in finding an answer concerning a fundamental question: is human language generation governed by associations, i.e. can the next content word of an utterance be considered as an association with the representations of the content words already activated in the speaker's memory?

4.7. Acknowledgments

This research was supported by a Marie Curie Career Integration Grant within the 7th European Community Framework Programme. Many thanks to the students who participated in the association experiments.

4.8. Bibliography

[BUR 98] BURNARD L., ASTON G., *The BNC Handbook: Exploring the British National Corpus*, Edinburgh University Press, Edinburgh, 1998.

[CHU 90] CHURCH K.W., HANKS P., "Word association norms, mutual information, and lexicography", *Computational Linguistics*, vol. 16, no. 1, pp. 22–29, 1990.

[DE 96] DE SAUSSURE F., *Cours de linguistique générale*, Payot, Paris, 1996.

[DUN 93] DUNNING T., "Accurate methods for the statistics of surprise and coincidence", *Computational Linguistics*, vol. 19, no. 1, pp. 61–74, 1993.

[GRI 07] GRIFFITHS T.L., STEYVERS M., TENENBAUM J.B., "Topics in semantic representation", *Psychological Review*, vol. 114, no. 2, pp. 211–244, 2007.

[JAM 90] JAMES W., *The Principles of Psychology*, Holt, New York, 1890.

[JEN 70] JENKINS J., "The 1952 Minnesota word association norms", in POSTMAN L., KEPPEL G. (eds), *Norms of Word Association*, Academic Press, New York, 1970.

[KAR 92] KARP D., SCHABES Y., ZAIDEL M. *et al.*, "A freely available wide coverage morphological analyzer for English", *Proceedings of the 14th International Conference on Computational Linguistics*, Nantes, pp. 950–955, 1992.

[KEN 10] KENT G.H., ROSANOFF A.J., "A study of association in insanity", *American Journal of Psychiatry*, vol. 67, pp. 317–390, 1910.

[KIS 73] KISS G.R., ARMSTRONG C., MILROY R. *et al.*, "An associative thesaurus of English and its computer analysis", in AITKEN A., BAILEY R., HAMILTON-SMITH N. (eds), *The Computer and Literary Studies*, Edinburgh University Press, 1973.

[KIS 75] KISS G.R., "An associative thesaurus of English: structural analysis of a large relevance network", in KENNEDY A., WILKES A. (eds), *Studies in Long Term Memory*, Wiley, London, 1975.

[NEL 98] NELSON D., MCEVOY C., SCHREIBER T.A., "The University of South Florida word association, rhyme, and word fragment norms", available at: http://www.usf.edu/FreeAssociation, 1998.

[PAL 64] PALERMO D.S., JENKINS J.J., *Word Association Norms: Grade School Through College*, University of Minnesota Press, Minneapolis, 1964.

[RAP 96] RAPP R., *Die Berechnung von Assoziationen*, Olms, Hildesheim, 1996.

[RAP 02] RAPP R., "The computation of word associations: comparing syntagmatic and paradigmatic approaches", *Proceedings of the 19th International Conference on Computational Linguistics*, Taipeh, vol. 2, pp. 821–827, 2002.

[RAP 08] RAPP R., "The computation of associative responses to multiword stimuli", *Proceedings of the Workshop on Cognitive Aspects of the Lexicon*, Manchester, pp. 102–109, 2008.

[RAP 13] RAPP R., "From stimulus to associations and back", *Proceedings of the 10th Workshop on Natural Language Processing and Cognitive Science*, pp. 78–91, Marseille, France, 2013.

[RAP 14] RAPP R., ZOCK M., "The CogALex-IV shared task on the lexical access problem", *Proceedings of the 4th Workshop on Cognitive Aspects of the Lexicon*, Dublin, Ireland, pp. 1–14, 2014.

[RUS 96] RUSSELL W.A., MESECK O.R., "Der Einfluß der Assoziation auf das Erinnern von Worten in der deutschen, französischen und englischen Sprache", *Zeitschrift für experimentelle und angewandte Psychologie*, vol. 6, pp. 191–211, 1959.

[SCH 89] SCHVANEVELDT R.W., DURSO F.T., DEARHOLT D.W., "Network structures in proxymity data", in BOWER G. (ed.), *The Psychology of Learning and Motivation: Advances in Research and Theory*, vol. 24, Academic Press, New York, 1989.

[SMI 13] SMITH K.A., HUBER D.E., VUL E., "Multiply-constrained semantic search in the Remote Associates Test", *Cognition*, vol. 128, pp. 64–75, 2013.

[TAM 03] TAMIR R., RAPP R., "Mining the web to discover the meanings of an ambiguous word", *Proceedings of the 3rd IEEE International Conference on Data Mining*, Melbourne, FL, pp. 645–648, 2003.

[WET 89] WETTLER M., RAPP R., "A connectionist system to simulate lexical decisions in information retrieval", in PFEIFER R., SCHRETER Z., FOGELMAN F. *et al.* (eds), *Connectionism in Perspective*, Elsevier, Amsterdam, 1989.

[WET 05] WETTLER M., RAPP R., SEDLMEIER P., "Free word associations correspond to contiguities between words in texts", *Journal of Quantitative Linguistics*, vol. 12, no. 2, pp. 111–122, 2005.

Hidden Structure and Function in the Lexicon

How many words (and which ones) are needed to define all the rest of the words in a dictionary? We applied graph theoretic analysis to the Wordsmyth suite of dictionaries. By recursively removing every word that is defined but defines no further words, every dictionary can be reduced to a small subset of words, the *Kernel*, which can define all the rest of the words in the dictionary, including one another; but the Kernel, though unique, is not the smallest subset that can define all the rest. The Kernel consists of one huge strongly connected component (SCC), the *Core*, about three quarters the size of the Kernel, surrounded by many tiny SCCs, the Satellites. Core words can define one another but cannot define the rest of the dictionary. The smallest number of words that can define all the rest of the dictionary is the "minimum feedback vertex set" or *Minimal Grounding Set (MinSet)*. MinSets are not unique. Each dictionary's Kernel contains many overlapping MinSets. About a fifth of the size of the Kernel, each MinSet is part-Core, part-Satellites; its words can define all the rest of the dictionary, but not one another. The Core words are more frequent, learned earlier and less concrete than the Satellite words, which are in turn more frequent and learned earlier but more concrete than the rest of the dictionary. We will discuss implications for language learning, language evolution and the representation of word meaning in the mental lexicon.

5.1. Introduction

Dictionaries catalogue and define the words of a language[1]. In principle, since every word in a dictionary is defined, it should be possible to learn the

---

Chapter written by Philippe Vincent-Lamarre, Mélanie Lord, Alexandre Blondin-Massé, Odile Marcotte, Marcos Lopes, Stevan Harnad.

1 Almost all the words in a dictionary (whether nouns, verbs, adjectives or adverbs) are "content" words, i.e. they are the names of categories [HAR 05]. Categories are kinds of things, both concrete and abstract (objects, properties, actions, events, states). The only words that are not the names of categories are logical and grammatical "function" words such as if, is, the, and, not. Our analysis is based solely on the content words; function ("stop") words are omitted.

meaning of any word through verbal definitions alone [BLO 13]. However, in order to understand the meaning of the word that is being defined, we have to understand the meaning of the words used to define it. If not, we have to look up the definition of those words too. However, if we have to keep looking up the definition of each of the words used to define a word, and then the definition of each of the words that define the words that define the words, and so on, we will eventually come full circle, never having learned a meaning at all.

This is the symbol grounding problem: the meanings of all words cannot be learned through definitions alone [HAR 90]. The meanings of some words, at least, have to be "grounded" by some means other than verbal definitions. That other means is direct sensorimotor experience with the referent of the word [HAR 10, PÉR 16], but the learning of categories from sensorimotor experience is not the subject of this paper. Here, we ask only how many words need to be grounded by some means other than verbal definition such that all the rest can be learned via definitions composed out of only those already grounded words -- and how do those grounding words differ from the rest?

5.2. Methods

5.2.1. *Dictionary graphs*

To answer this question, dictionaries can be analyzed using graph theory. We have previously analyzed several English dictionaries (including Longman's, Cambridge, Merriam-Webster and WordNet; [VIN 16]). In the present paper, we replicate and extend these results for the Wordsmyth Advanced Dictionary-Thesaurus (WADT, 70K words) as well as for three reduced versions of it [Wordsmyth Children's Dictionary-Thesaurus (WCDT, 20K words), Wordsmyth Beginner's Dictionary-Thesaurus (WBDT, 6K words), and Wordsmyth Illustrated Learner's Dictionary (WILD, 4K words)], modified for specific purposes by using fewer and more frequent words [PAR 98]: http://www.wordsmyth.net.

Apart from proper names, all words used in a dictionary are defined. In a dictionary graph, there is a directional link from each defining word to each defined word. Our analyses of dictionary graphs have revealed a hidden structure in dictionaries, which has not been observed or reported previously (see Figure 5.1). If we remove recursively all those words that are defined but that do not go on to define any further word, every dictionary graph can be reduced by about 85% (the percentage reducibility becomes less, the smaller the dictionary) to a unique set of words (which we have called the Kernel, K), out of which all the words in the dictionary can be defined [BLO 08]. There is only one such Kernel in any dictionary, but the Kernel is not the *smallest* number of words, *M*, out of which all the words in the dictionary can be defined. That smallest number of words is the dictionary graph's "minimum feedback vertex set" (see, e.g., [FOM 08, KAR 72, LAP 12], which we will call the dictionary's *MinSet (Minimal Grounding Set)*.

For the Wordsmyth dictionaries analyzed in this paper, the relative size of the Kernel varies from 17% of the whole dictionary in the largest dictionary (WADT) to 49% in the smallest (WILD). In all four dictionaries, the MinSet size is about a fifth of the Kernel (Table 5.1). Unlike the Kernel, however, the MinSet is not unique: There are a huge number of (overlapping) MinSets in every dictionary, each of the same minimal size, M. Each is a subset of the Kernel and any one of them, if it were grounded, would then ground the entire dictionary[2].

	WADT	WCDT	WBDT	WILD
Total word meanings	73,158	20,129	6,038	4,244
First word meanings	43,363	11,626	4,456	3,159
Rest	36,196 (83%)	8,617 (74%)	3,254 (73%)	1,608 (51%)
Kernel	7,167 (17%)	3,009 (26%)	1,202 (27%)	1,551 (49%)
Satellites	1,999 (5%)	609 (5%)	258 (6%)	159 (5%)
Core	5,168 (12%)	2,400 (21%)	944 (21%)	1,392 (44%)
MinSets	1,335 (3%)	548 (4.7%)	234 (5.3%)	330 (10.4%)
Satellite-MinSets	569 (1.3%)	160 (1.4%)	67 (1.5%)	39 (1.2%)
Core-MinSets	766 (1.7%)	388 (3.3%)	167 (3.8%)	291 (9.2%)

Table 5.1. *The Wordsmyth dictionaries. The Kernel's Core (its biggest SCC) is always much larger than the Satellites (small SCCs), and each MinSet (part-Core, part-Satellites) is about a fifth of the size of the Kernel. [Wordsmyth Advanced Dictionary-Thesaurus (WADT, 70K words), Wordsmyth Children's Dictionary-Thesaurus (WCDT, 20K words), Wordsmyth Beginner's Dictionary-Thesaurus (WBDT, 6K words) and Wordsmyth Illustrated Learner's Dictionary (WILD, 4K words)]; Parks et al. [PAR 98])*

---

2 There may be something informative in an analogy between all the MinSets as potential bases for all of semantic space and the infinite number of different sets of M linearly independent vectors that can serve as the potential bases for M-dimensional vector space.

The Kernel, however, is not just a large number of overlapping MinSets. It has structure too. It consists of a large number of strongly connected components (SCCs). (A directed graph -- in which a directional link indicates that word W_1 is part of the definition of word W_2 -- is "strongly connected" if any word in the graph can be reached by a chain of definitional links from any other word in the graph.) Most of the SCCs of the Dictionary's Kernel are small, but in every dictionary, we have analyzed so far that there also turns out to be one very large SCC, about half the size of the Kernel. We call this the Kernel's Core $(C)^3$.

The Kernel itself is a self-contained dictionary, just as is the dictionary as a whole (D). K is a sub-dictionary of D. Every word in the Kernel can be fully defined using only words in the Kernel. The Core is likewise a self-contained dictionary; but, in all the full-size dictionaries of natural languages that we have so far examined,[4] the Kernel is not an SCC: Within the Core (but not the Kernel), every word can be reached by a chain of definitions from any other word in the Core.

The Kernel is a Grounding Set for the dictionary as a whole, but it is not a *Minimal* Grounding Set (MinSet), i.e. not the smallest number of words capable of defining all the rest of the dictionary. The Kernel's Core is not only a MinSet for the dictionary as a whole: it is not even a Grounding Set at all. The words in the Core alone are not enough to define all the rest of the words in the dictionary, outside the Core.

In contrast, all the MinSets are, like the Core, contained entirely within the Kernel, but no MinSet is completely contained within the Core: each straddles the Core and the Satellites. Each MinSet can define all the rest of the words in the dictionary outside the MinSet, but no MinSet is an SCC: its words are not even connected (Table 5.2). Indeed, the MinSet cannot define any of the words *within* the MinSet: only the words *outside* the MinSet.

---

3 Formally, the Core is defined as the union of all the strongly connected components (SCCs) of the Kernel that do not receive any incoming definitional links from outside themselves. (In graph theoretic language: there is no incoming link into the Core, i.e. no definitional link from a word not in the Core to a word in the Core.) It turns out to be an empirical fact about all the full-sized dictionaries we have analyzed so far, however, that their Core is itself always an SCC, and also by far the largest of the SCCs in the Kernel, the rest of which look like many small satellites surrounding one big planet (Figures 5.1 and 5.2).

4 In some of the mini dictionaries generated in our online dictionary game, however, the Core is not an SCC, but a disjoint union of SCCs (Figures 5.5 and 5.6).

The MinSets of Dictionaries hence turn out empirically[5] to consist of two parts: words in the Core (C) (which is entirely within the Kernel) and words in the remainder of the Kernel (K); the Satellites (S). The MinSet S/C ratio varies across dictionaries, but, within a given dictionary, it is the same for all of its MinSets. The MinSets, the smallest subsets capable of defining all the rest of the dictionary, are hence part-Core and part-Satellite. The natural question, then, is: *What, if anything, is the difference between the kinds of words that are in the various components of this hidden structure of the dictionary: the MinSets, the Core, the Satellites, the Kernel, and the rest of the dictionary outside the Kernel?*

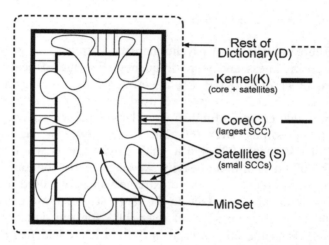

Figure 5.1. *Schematic diagram of the hidden structure of the Wordsmyth dictionaries: the Rest of the Dictionary, Kernel, Core and MinSets (only one shown). The words in these different structural components of the dictionary graph tend to differ in their psycholinguistic properties (Figures 5.2 and 5.3). Satellite words are more frequent than the words in the Rest of the Dictionary, more concrete and learned at a significantly younger age. The Core words, in turn, are even younger and more frequent than the Satellite words, but they are the Satellite words that are most concrete ones than the Core and the Rest of the Dictionary (Figure 5.2). (Note that the diagram is not drawn to scale: the relative size of the Kernel varies from 17% of the whole dictionary for the largest Wordsmyth dictionary (WADT) to 49% for the smallest (WILD))*

---

5 Most of the properties described here are empirically observed properties of dictionary graphs, and not necessary properties of directed graphs in general.

Is a column entry (right) necessarily a row entry (below)?	Dict	Kern	GS	SCC	Core	MinSet
Dictionary (D)	yes	yes	–	–	yes	–
Grounding Set (GS)	yes	yes	yes	–	–	yes
Strongly Connected Component (SCC)	–	–	–	yes	yes	–
Minimal Grounding Set (MinSet)	–	–	–	–	–	yes

Table 5.2. *Table indicating which of the hidden structures are Dictionaries, Grounding Sets, Strongly Connected Components and Minimal Grounding Sets*

5.2.2. Psycholinguistic variables

We used three databases of psycholinguistic correlates of words: age of acquisition, concreteness and word frequency. For age of acquisition, we used Kuperman *et al.*'s [KUP 12] age-of-acquisition ratings for 30,000 English words. For concreteness, we used Brysbaert *et al's.* [BRY 14] database for 40,000 common English word lemmas. For frequency, we used the SUBTLEX$_{US}$ Corpus [BRY 09]. The vast majority of these words had frequencies of less than 1,000, but a small percentage of word frequencies ranged from 5,000 to 2 million, heavily skewing the distribution. To avoid a disproportionate influence from these extreme values, we used the log base 10 of the raw frequency + 1. Due to the large size of these databases, we were able to achieve a coverage of over 90% for age and concreteness. Words missing from the corpus used in SUBTLEX$_{US}$ had frequency zero, which implies that we had frequency coverage for all words.

5.2.3. Data analysis

We did not apply statistical tests to the comparison between structures because we have access to all the words in each dictionary. Hence, our datasets are not samples from each dictionary: they are the entire population, making statistical estimation supererogatory. In addition, the number of observations for each hidden structure is so large that almost any marginal difference would yield statistically significant results. Thus, we rely on the replication of our observed pattern of results across all the individual dictionaries we have studied as confirmation of the generality of the findings.

5.3. Psycholinguistic properties of Kernel, Satellites, Core, MinSets and the rest of each dictionary

The words in the Kernel (K) differ significantly from words in the rest of the Dictionary (D) for all three psycholinguistic variables: Kernel words are learned significantly younger, more concrete and more frequent than the words in the rest of the dictionary. The same effect was found in comparing Core (C) words with D words for age and frequency, but not for concreteness, where only the Satellite words (S) are more concrete than C and D. Hence, the effects get stronger as we move inward from the rest of the Dictionary to the Kernel to the Core for age and frequency (but not concreteness), as schematized in the left part of Figure 5.2. There were only two small exceptions to the overall pattern in the two smallest dictionaries: for WCDT, S was not more concrete than C and D as in the other three dictionaries, and for WILD, D was not less frequent than S. Apart from these two exceptions, our results replicated the patterns observed previously for four other dictionaries [VIN 16].

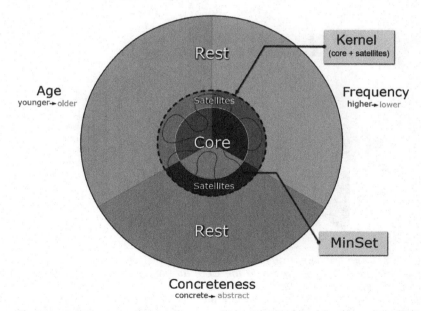

Figure 5.2. *Moving outward from the Core (C) to the Satellite (S) layer of the Kernel (K) to the rest of the Dictionary (D, about 80%), words are increasingly frequent and younger (as indicated by darkness) (see Figure 5.1 for arrows identifying each of these structures). For a color version of this figure, see www.iste.co.uk/sharp/cognitive.zip*

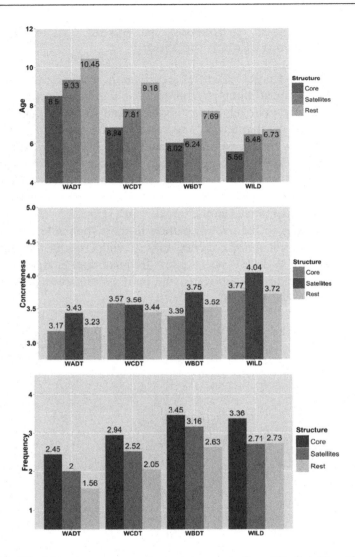

Figure 5.3. *Word frequency, age and concreteness in Core (C), Satellites (S) and rest of the Dictionary (D). Top: words in the Core are learned earlier than the Satellites, which are in turn learned earlier than the rest of the Dictionary. The pattern is the same in all four dictionaries. Middle: In three of the four dictionaries, the Satellite words are more concrete than the Core words and the rest of the Dictionary (WADT, WBDT, WILD); no pattern for concreteness in WCDT. Bottom: Core words are more frequent than the Satellites, which are more frequent than the rest of the Dictionary. The pattern is the same across the four dictionaries for word frequency except that, in the smallest, WILD, there is no significant difference between Satellites and the rest of the Dictionary. For a color version of this figure, see www.iste.co.uk/sharp/cognitive.zip*

The Kernel contains a very large number of word combinations that can form a MinSet, but each dictionary's MinSet is composed of the same proportion of C and S words (from 57% C in the largest, WADT, to 88% C in the smallest, WILD). What is the complementary role of the Core and the Satellites in generating a MinSet?

To get a better idea of which Kernel words are included in MinSets, we compared their frequency, age and concreteness with those of randomly selected Kernel words. All MinSets are part-Core and part-Satellite. For a given dictionary, some Kernel words are in all MinSets, some in no MinSet, and the rest are in between. However, the number of MinSets for a given dictionary is too large to enumerate them all and calculate the percentage of MinSets in which each Satellite and Core word appears.[6] We therefore used a Monte Carlo simulation to compare each dictionary's real MinSets with randomly selected pseudo-MinSets of the same size and Core/Satellite ratio.

---

6 A graph D is a dictionary graph (or dictionary for short) if each of its nodes (words), w, has at least one predecessor (defining word), w'. The set of all words of D is denoted by W and its set of arcs (the directed definitional links from each defining word w' to each defined word w) is denoted by L.

A sub-dictionary of D is a subset D' of W with the property that for any word w in W, each predecessor w' of w in D also belongs to W. Mathematically speaking, (word w belongs to W) and (link(w',w) belongs to L) together imply that (w' belongs to W).

If W is a strongly connected component (SCC) of D, then no proper subset W' of W (i.e. no subset W' such that there is at least one word w belonging to W but not to W') is a sub-dictionary of D.

Hence, a sub-dictionary is always the union of a collection of SCCs: By contracting each SCC (i.e. replacing each SCC by one supernode), we obtain an acyclic graph. We can thus speak of the SCCs that precede a given SCC.

An acyclic graph is a graph with no cycles, i.e. no sequence of the form (w1, w2, w3,..., wn, w1), in which each word in the sequence is used in the definition of the word that follows it in the sequence.

The union of a collection of SCCs is a sub-dictionary if and only if each predecessor of an SCC in the collection also belongs to the collection.

To find a MinSet for the Core C, we first apply certain reductions (see [LEV 88, LIN 00] to obtain a smaller graph called the reduced Core and denoted by C'. Finding a MinSet for C' is less difficult than finding one for C itself, although it remains a non-trivial task. Once a MinSet has been found for C', it is easy to transform it into a MinSet for C.

For the C' of each of our dictionaries, we have been able to determine that they contain three classes of words: essential words (a word is essential if it belongs to every MinSet), superfluous words (a word is superfluous if it does not belong to any MinSet) and ordinary words (a word is ordinary if it is neither essential nor superfluous).

We still have to extend these results to the Core itself and to the full dictionary, but it is likely that we will find essential words and superfluous words for these sets as well.

We then computed the average value of each psycholinguistic variable for each part-core and part-satellite independently between the MinSets and the pseudo-MinSets (see Figure 5.4).

With a few exceptions, the same pattern was observed for all four dictionaries. The Core words in the real MinSets are younger, more concrete and more frequent than those in the random pseudo-MinSets. The Satellite words in the real MinSets are less young and less frequent than their pseudo-MinSet counterparts; no clear pattern was observed for concreteness. T-tests for all paired comparisons between the MinSets and random samples were all highly significant ($p < 0.01$) with the exception of one comparison (concreteness for the Satellite part of WBDT MinSets).

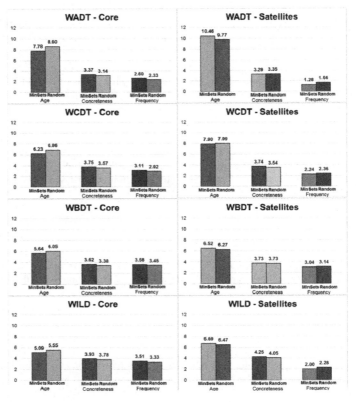

Figure 5.4. *Word frequency, age and concreteness within real MinSets versus random Pseudo-MinSets. Core/Satellite differences are magnified in MinSets compared with random samples of Core and Satellite words in the same proportion. For a color version of this figure, see www.iste.co.uk/sharp/cognitive.zip*

5.4. Discussion

Our findings suggest that in addition to the overall tendency for words to be younger and more frequent as we move from the outer rest of the dictionary to the Satellites to the Core (an effect that is accentuated in the MinSets), something different is happening at the Satellite layer, which is more concrete than the deeper Core as well as the rest of the dictionary. The Satellite words are more concrete and they are needed in every MinSet. We still do not know what functional role is played by the S words and the C words in forming a MinSet. Nor do we know what the differences are between MinSets. We doubt that anyone ever learns one MinSet from direct experience and then learns everything else through verbal definition, but our findings do suggest that it is in the nature of lexical representation that this is possible in principle.

In this work, we extended our previous analyses to four new dictionaries not included in our prior studies. Our objective was twofold. First, we wanted to replicate our findings and show that the hidden structures we had discovered were present in any dictionary. Second, the three smaller dictionaries of the Wordsmyth suite are purpose-designed for learners at different levels, which makes them interesting from the standpoint of language acquisition. As we move from the advanced (WADT) to the simplest dictionary (WILD), the relative proportion of the Kernel increases markedly and so does the relative size of the structures within it (S, C, M). One conclusion from this might have been that smaller dictionaries simply have larger Kernels. However, this was not true in the dictionaries we had studied previously, where the Kernel remained between 8 and 12% of the dictionary despite huge variation in dictionary size [VIN 16]. One possible explanation is that it is precisely because the words in the three smaller Wordsmyth dictionaries with their larger Kernels were all deliberately selected in order to teach the English language that they contain a greater proportion of "grounding words".

These results have implications for the understanding of symbol grounding and the learning and mental representation of word meaning. For language users to be able to learn and understand the meaning of words from verbal definitions, they have to have the vocabulary to understand the words in those definitions, or at least to understand the definitions of the words in the definitions, and so on. They need an already grounded set of word meanings that is sufficient to carry them, verbally, to the meaning of any other word in the language, if they are to be able to learn its meaning through words alone. A Grounding Set clearly has to be acquired before it is true that all further words can be acquired verbally; hence, we would expect the Grounding Set to be acquired earlier in life. It also makes sense that the words in the Grounding Set are used more frequently, perhaps partly because they are used more often to define or explain later words.

Grounding words being more concrete is also to be expected, because word meanings that do not come from verbal definitions have to be acquired by nonverbal means, and those nonverbal means are likely to be the learning of categories through direct sensorimotor experience: learning what to do and not do with what kind of thing [HAR 10, PÉR 16]. It is easy, then, to associate a category that we have already learned non-verbally with the (arbitrary) name that a language community agrees to call it [BLO 13]. The words denoting sensorimotor categories are hence likely to be more concrete.

Categorization itself, however, is by its nature also abstraction: to abstract is to single out some features of a thing, and ignore others. The way we learn what *kinds* of things there are, and what to do and not do with them (including what to call them), is not by simply memorizing raw sensorimotor experiences by rote. To be able to do the right thing with the right kind of thing, we learn through trial-and-error sensorimotor interactions to detect and abstract the features that distinguish the members of a category from the non-members and to ignore the rest of the features as irrelevant. The process of abstraction in the service of categorization leads in turn to higher-order categories, which are hence more likely to be verbal categories rather than purely sensorimotor ones. For example, we can have a preverbal category for "bananas" and "apples", based on their sensory projections and the differing sensorimotor actions needed to eat them; but the higher-order category "fruit" is not as evident at a non-verbal level, being more abstract. It is also likely that having abstracted the sensorimotor features that distinguish the members and non-members of a concrete category nonverbally, we will not just give the members of the category a name, but we may go on to abstract and name their features (yellow, red, round, elongated) too. It may be that some names for more abstract, higher-order categories are as essential in forming a Grounding Set as the more concrete categories and their names and that this may have something to do with the complementary functional role played by the Satellites in the make-up of a MinSet.

Finally, the lexicon of the language – our repertoire of categories – is open-ended and always growing. To understand the grounding of meaning, it will be necessary not only to look at the growth across time of the vocabulary (both receptive and productive) of the child, adolescent and adult, but also the growth across time of the vocabulary of the language itself (diachronic linguistics), to understand which words are necessary, and when, in order to have the full lexical power to define all the rest [LEV 12]. We have discussed Minimal Grounding Sets (MinSets), and it is clear that there are potentially very many of these; but it is not clear that anyone uses just one MinSet, or could actually manage to learn everything verbally knowing just one MinSet. It is almost certain that we need some redundancy in our Grounding Sets. The Kernel, after all, is only five times as big as a MinSet. Perhaps we do not even need a full Grounding Set in order to get by, verbally; maybe we can manage with gaps, (certainly, the child must, at least initially.) Nor is it clear – even if we have full mastery of enough MinSets or even a Kernel – that the best way to learn the meaning of all subsequent words is from verbal definitions alone. Language may well have evolved in order to make something like that possible in principle: to acquire new categories through recombinatory subject/predicate propositions, purely by verbal "telling", without sensorimotor "showing" [BLO 13]. However, in practice, the learning of new word meanings may still draw on some hybrid show-and-telling.

5.4.1. *Limitations*

Many approximations and simplifications have to be taken into account in interpreting these findings. We are treating a definition as an unordered string of words, excluding functional (stop) words and not making use of any syntactic structure. Many words have multiple meanings, and we are using only the first meaning of each word. The problem of extracting MinSets is NP-hard. In the special case of dictionary graphs -- and thanks also to the empirical fact that the Core turns out to be so big, and surrounded by small Satellites -- we have been able, using the algorithm of Lin and Jou [LIN 00] and techniques from integer linear programming (e.g. [NEM 99]), to extract a number of MinSets for the Wordsmyth suite of dictionaries, whose results we are reporting here as well as for other English dictionaries [VIN 16], such as Merriam-Webster and WordNet [FEL 10]. The analysis is now being extended to dictionaries in other languages.

5.5. Future work

In order to compare the emerging hidden structure of dictionaries with the way word meaning is represented in the mind (the "mental lexicon"), we have also created an online dictionary game in which the player is given a word to define; they must then define the words they used to define the word, and so on, until they have defined all the words they have used. This generates a mini-dictionary of a much more tractable size (usually less than 200 words; Figures 5.5 and 5.6)[7]. http://lexis.uqam.ca:8080/dictGame

We are currently performing the same analyses on these much smaller mini-dictionaries, to derive their Kernel, Core, Satellites and MinSets, and

---

7 The 37-word mini-dictionary in Figures 5.5 and 5.6 is displayed because it is small enough to illustrate the hidden dictionary graph structure at a glance. It was generated before we had added a new rule that a definition is not allowed to be just a synonym: in the more recent version of the game, a definition has to be at least two content words (and we may eventually also rule out second-order circularity [A = B + C, B = C + not-A, C = A + not-B]. However, it has to be borne in mind that (because of the symbol grounding problem) every dictionary is necessarily approximate and (at some level) circular (much the way all SCCs [other than trivial single-word SCCs] are circular). This is true whether it is a full dictionary or a game mini-dictionary generated by one player. Definitions can only convey new meanings if the mind already has enough old meanings, grounded by some means other than definition.

their psycholinguistic correlates (age, concreteness, frequency), to determine whether these inner "mental" dictionaries share the hidden structure and function that we are discovering in the formal external lexicon (see Figures 5.5 and 5.6). These mini-dictionaries will also allow us to analyze the difference in the functional role between the Satellite words and the Core words that make up a MinSet, which is much more difficult to do with full-size dictionaries.

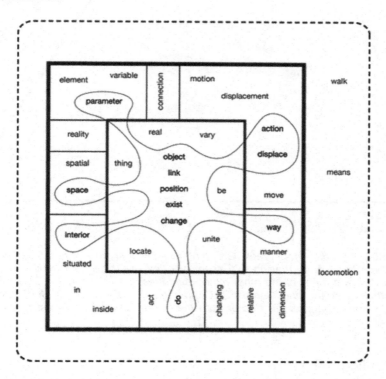

Figure 5.5. *Mini-dictionary diagram. The diagram is the same as Figure 5.1, but with real words to provide a concrete example. This 37-word mini-dictionary was generated by a player of our online dictionary game. The player is given a word and must define that word, as well as all the words used to define it, and so on, until all the words used are defined. The smallest resulting dictionary so far (37 words) is used here to illustrate the mini-dictionary's Kernel and Core plus one of its MinSets. Note that all the words in this mini-dictionary are in the Kernel except the start word, "walk", plus "locomotion" and "means". Figure 5.6 displays the graph for this mini-dictionary*

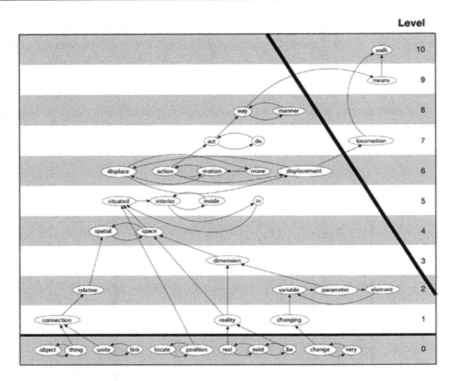

Figure 5.6. *Mini-dictionary graph. Graph of mini-dictionary in Figure 5.5, showing the definitional links. Note that, in this especially tiny mini-dictionary, unlike in the full dictionaries and many of the other mini-dictionaries, the words in the Core (level 0), rather than being the single largest SCC, are the union of multiple SCCs. The oblique boldface line separates the Kernel from the (three) words in the rest of this mini-dictionary*

5.6. Bibliography

[BLO 08] BLONDIN-MASSÉ A., CHICOISNE G., GARGOURI Y. *et al.*, "How is meaning grounded in dictionary definitions?" *TextGraphs-3 Workshop – 22nd International Conference on Computational Linguistics*, available at: http://www.archipel.uqam.ca/657/, 2008.

[BLO 13] BLONDIN-MASSÉ A., HARNAD S., PICARD, O. *et al.*, "Symbol grounding and the origin of language: from show to tell", in LEFEBVRE C., COMRIE B., COHEN H. (eds), *Current Perspective on the Origins of Language*, John Benjamins Publishing Company, Amsterdam, 2013.

[BRY 09] BRYSBAERT M., NEW B., "Moving beyond Kučera and Francis: a critical evaluation of current word frequency norms and the introduction of a new and improved word frequency measure for American English", *Behavior Research Methods*, vol. 41, no. 4, pp. 977–990, 2009.

[BRY 14] BRYSBAERT M., WARRINER, A.B., KUPERMAN V., "Concreteness ratings for 40 thousand generally known English word lemmas", *Behavior Research Methods*, vol. 46, no. 3, pp. 904–911, 2014.

[FEL 10] FELLBAUM C., *WordNet*, Springer, Netherlands, 2010.

[FOM 08] FOMIN F.V., GASPERS S., PYATKIN A.V. *et al.*, "On the minimum feedback vertex set problem: exact and enumeration algorithms", *Algorithmica*, vol. 52, no. 2, pp. 293–307, 2008.

[HAR 90] HARNAD S., "The symbol grounding problem", *Physica D*, vol. 42, pp. 335–346, 1990.

[HAR 05] HARNAD S., "To cognize is to categorize: cognition is categorization", in LEFEBVRE C., COHEN H. (eds), *Handbook of Categorization*, Elsevier, Amsterdam, 2005.

[HAR 10] HARNAD S., "From sensorimotor categories and pantomime to grounded symbols and propositions", in TALLERMAN M., GIBSON K.R. (eds), *The Oxford Handbook of Language Evolution*, Oxford University Press, Oxford, 2010.

[KAR 72] KARP R.M., "Reducibility among combinatorial problems", in MILLER R.E., THATCHER J.W., BOHLINGER J.D. (eds), *Proceedings of a Symposium on the Complexity of Computer Computations*, IBM Thomas J. Watson Research Center, New York, 1972.

[KUP 12] KUPERMAN V., STADTHAGEN-GONZALEZ H., BRYSBAERT M., "Age-of-acquisition ratings for 30,000 English words", *Behavior Research Methods*, vol. 44, no. 4, pp. 978–990, 2012.

[LAP 12] LAPOINTE M., BLONDIN-MASSÉ A, GALINIER P. *et al.*, "Enumerating minimum feedback vertex sets in directed graphs", *Bordeaux Graph Workshop*, Bordeaux, 2012.

[LEV 12] LEVARY D., ECKMANN J.P., MOSES E. *et al.*, "Loops and self-reference in the construction of dictionaries", *Physical Review X*, vol. 2, no. 3, pp. 031018, 2012.

[LEV 88] LEVY H., LOW, D.W., "A contraction algorithm for finding small cycle cutsets", *Journal of Algorithms*, vol. 9, no. 4, pp. 470–493, 1988.

[LIN 00] LIN H.M., JOU J.Y., "On computing the minimum feedback vertex set of a directed graph by contraction operations", *IEEE Transactions on CAD of Integrated Circuits and Systems*, vol. 19, no. 3, pp. 295–307, 2000.

[NEM 99] NEMHAUSER G., WOLSEY L., *Integer and Combinatorial Optimization*, John Wiley & Sons, New York, 1999.

[PAR 98] PARKS R., RAY J., BLAND S., *Wordsmyth English Dictionary/Thesaurus [WEDT]*, University of Chicago, Chicago, 1998.

[PÉR 16] PÉREZ-GAY F, SABRI H., RIVAS D. *et al.*, "Perceptual changes induced by category learning", *23rd Annual Meeting Cognitive Neuroscience Society*, New York, April 2016.

[VIN 16] VINCENT-LAMARRE P., BLONDIN M.A., LOPES M. *et al.*, "The latent structure of dictionaries", *Topics in Cognitive Science*, 2016.

6

Transductive Learning Games
for Word Sense Disambiguation

This chapter presents a semi-supervised approach to word sense disambiguation, formulated in terms of evolutionary game theory, where each word to be disambiguated is represented as a player and each sense as a strategy. The players receive a reward for interacting with other players, which gives them an incentive to select strategies with a higher payoff. The interactions among players are modeled with a weighted graph and it is assumed that some players have a defined strategy (labeled players) and others have to choose their strategy (unlabeled players). The information is propagated over the graph from labeled players to unlabeled, exploiting information from different sources: word similarity that weights the importance of the interactions and sense similarity that determines the payoffs of the games. In this way, similar players influence each other, selecting correlated senses. The method has been tested on four datasets with different numbers of labeled words. The experimental results demonstrate that the proposed approach performs well compared with state-of-the-art algorithms.

6.1. Introduction

Word Sense Disambiguation (WSD) is the task of identifying the intended sense of a word in a computational manner based on the context in which it appears [NAV 09]. Understanding the ambiguity of natural languages is considered an AI-hard problem [MAL 88]. Computational problems like this are the central objectives of Artificial Intelligence (AI) and Natural Language Processing (NLP) because they aim to solve the epistemological question of how the mind works. It has been studied since the beginning of NLP [WEA 55], and today is a central topic of this discipline.

Chapter written by Rocco TRIPODI and Marcello PELILLO.

The identification of the intended meaning of the words in a sentence is a hard task and humans have problems with it. This happens because people use words in a different manner from their literal meaning, and misinterpretation is a consequence of this habit. To solve this task, it is not only required to have a deep knowledge of the language, but also to be an active speaker of the language. In addition, languages change over time, in an evolutionary process, by virtue of the use of language by speakers, which leads to the formation of new words and new meanings which could be understood only by active speakers of the language. In fact, active speakers can identify the intended meanings of the words according to the knowledge of the communication context in which the words are expressed.

WSD is also a central topic in applications such as text entailment [DAG 04], machine translation [VIC 05], opinion mining [SMR 06] and sentiment analysis [REN 09]. These applications require the disambiguation of ambiguous words, as preliminary process; otherwise, they remain on the surface of the word [PAN 02], compromising the coherence of the data to be analyzed.

Our approach to WSD is a graph based on the way the data are represented, similarity based on the way the senses and the words of a sentence are compared, and is formulated in game theoretic terms to combine this information and to find consistent labeling of the data. It is aimed at maximizing the textual coherence imposing that the meaning of each word in a text must be related to the meaning of the other words in it. To do this, we exploited distributional and proximity information to weight the influence that each word has on the others. We also exploited semantic similarity information to weight the strengths of compatibility among senses and use transductive learning principles to propagate the information over the graph.

The rest of this chapter is structured as follows: section 6.2 presents graph-based algorithms for WSD. Section 6.3 introduces our approach to semi-supervised learning and some notions of game theory. This chapter continues with the presentation of our methods in section 6.4 and concludes with section 6.5 where we present the evaluation of our system and the comparison of our approach with state-of-the-art algorithms.

6.2. Graph-based word sense disambiguation

Graph-based WSD algorithms try to identify the actual sense of a word in a determined phrase, exploiting the information derived from its context. They are gaining much attention in the NLP community. This is because graph theory is a powerful tool that can be employed for different purposes, from the organization of the contextual information to the computation of the relations among word senses. These kind of approaches construct a graph collecting all the possible senses of the words in a text and represent them as nodes. Then, using connectivity measures can identify the most relevant word senses in the graph [SIN 07, NAV 07a].

Navigli and Lapata [NAV 07a] conducted an extensive analysis of graph connectivity measures for unsupervised WSD. Their approach uses a knowledge base, such as WordNet, to collect and organize all the possible senses of the words to be disambiguated in a graph structure, then uses the knowledge base to search for a path between each pair of senses in the graph and if it exists, it adds all the nodes and edges on this path to the graph. These measures analyze local and global properties of the graph. The results of the study indicate that local measures outperform global measure and, in particular, degree centrality and PageRank [PAG 99] achieve the best results.

PageRank [PAG 99] is one of the most popular algorithms for WSD; in fact, it has been implemented in several different ways by the research community [MIH 04, HAV 02, AGI 14, DEC 10]. It represents the senses of the words in a text as nodes of a graph. It uses a knowledge base to collect the senses of the words in a text and represents them as nodes of a graph. The structure of the knowledge base is used to connect each node with its related senses in a directed graph. The main idea of this algorithm is that whenever a link from one node to another node exists, a vote is produced, increasing the rank of the voted node. It works by counting the number and quality of links to a node to determine an estimation of how important the node is in the network. The underlying assumption is that more important nodes are likely to receive more links from other nodes [PAG 99]. Exploiting this idea, the ranking of the nodes in the graph can be computed iteratively with the following equation:

$$Pr = c\,M\,Pr + (1 - c)\,v$$

where M is the transition matrix of the graph, v is a $N \times 1$ vector representing a probability distribution and c is the so-called damping factor that represents the chance that the process stops, restarting from a random node. At the end of the process, each word is associated with the most important concept related to it. One problem of this framework is that the labeling process is not assumed to be consistent.

An algorithm, which tries to improve centrality algorithms, is SUDOKU, introduced by Manion and Sainudiin [MAN 14]. It is an iterative approach that simultaneously constructs the graph and disambiguates the words using a centrality function. It starts inserting in the graph the nodes corresponding to the senses of the words with low polysemy. The advantages of this method are that it uses small graphs at the beginning of the process, reducing the complexity of the problem; furthermore, it can be used with different centrality measures.

Recently, a model based on an undirected graphical model has been introduced by [CHA 15]. This method interprets the WSD problem as a maximum a posteriori query on a Markov random field [JOR 02]. The graph is constructed using the content words of a sentence as nodes and connecting them with edges if they share a relation, determined using a dependency parser. The values that each node in the graphical model can take include the senses of the corresponding word. The senses are collected using a knowledge base and weighted using a probability distribution based on the frequency of the senses in the knowledge base. Furthermore, the senses between two related words are weighted using a similarity measure. The goal of this approach is to maximize the joint probability of the senses of all the words in the sentence, given the dependency structure of the sentence, the frequency of the senses and the similarity among them.

A new graph-based, semi-supervised approach, introduced to deal with multilingual WSD [NAV 12b] and entity linking problems, is Babelfy [MOR 14]. Multilingual WSD is an important task because traditional WSD algorithms and resources are mainly focused on English language. It exploits the information from large multilingual knowledge bases, such as BabelNet [NAV 12a] to perform this task. Babelfy creates the semantic signature of each word to be disambiguated, that consists of collecting, from a semantic network, all the nodes related to a particular concept, exploiting the global structure of the network. This process leads to the construction of a graph-based representation of the whole text. Then, it applies Random Walk with

Restart [TON 06] to find the most important nodes in the network, solving the WSD problem.

Approaches that are more similar to ours in the formulation of the problem have been described by Araujo [ARA 07] and a specific evolutionary approach to WSD has been introduced by Menai [MEN 14]. It uses genetic algorithms [HOL 75] and memetic algorithms [MOS 89] to improve the performances of a gloss-based method. It assumes that there is a population of individuals, represented by all the senses of the words to be disambiguated and that there is a selection process that chooses the best candidates in the population. The selection process is defined as a sense similarity function that gives a higher score to candidates with specific features, increasing their fitness. This process is repeated until the fitness level of the population regularizes and, at the end, the candidates with higher fitness are selected as solutions of the problem.

6.3. Our approach to semi-supervised learning

Our approach uses some labeled nodes to propagate the information over the graph disambiguating unlabeled nodes in a consistent way. This process is based on two fundamental principles: the homophily principle, borrowed from social network analysis, and the transductive learning. The former simply states that similar objects are expected to have the same class [EAS 10]. We extended this principle assuming that objects, which are similar, are expected to have a similar class; an idea used also by [KLE 02], within a Markov random field framework. The latter is a case of semi-supervised learning [SAM 11] particularly suitable for relational data (see section 6.3.1).

In our system, we used a graph to model the geometry of the data and an evolutionary process to propagate the information over it. The graph construction method is described in section 6.4.1 and the evolutionary process in section 6.4.4. This work extends our previous works on unsupervised and semi-supervised WSD [TRI 15a, TRI 17].

6.3.1. Graph-based semi-supervised learning

Transductive learning was introduced by Vladimir Vapnik [VAP 98]. It was motivated by the fact that it is easier than inductive learning, given the fact that inductive learning tries to learn a general function to solve a specific problem, while transductive learning tries to learn a specific function for the problem at hand.

It consists of a set of labeled objects (x_i, y_i) $(i = 1, 2, ..., 1)$, where $x_i \in \mathbb{R}^n$ are objects represented by real-valued attributes and $y_i \in (1, 2, ..., m)$ are the possible labels of these objects. Together with the labeled objects, there is also a set of k unlabeled objects $(x_{l+1}, ..., x_{l+k})$. Rather than finding a general rule for classifying future examples, transductive learning aims at classifying only (the k) unlabeled objects exploiting the information derived from labeled ones.

Within this framework, it is common to represent the geometry of the data as a weighted graph. For a detailed description of algorithms and applications on this field of research, named graph transduction, we refer to [ZHU 05]. The purpose of this method is to transfer the information given by labeled nodes to unlabeled ones, exploiting the graph structure. Formally, we have a graph $G = (V, E, w)$ in which V is the set of nodes representing both labeled and unlabeled points, $V = \{v_l, v_u\}$, E is the set of edges $E \subseteq V \times V$ connecting the nodes of the graph and $w : \varepsilon \rightarrow \mathbb{R}^+$ is a weight function assigning a similarity value to each edge $\varepsilon \in E$. The task of transduction learning is to estimate the labels of the unlabeled points, given the pairwise similarity among the data points and a set of possible labels $\varphi = \{1, ..., c\}$.

6.3.2. *Game theory and game dynamics*

Game theory was introduced by Von Neumann and Morgenstern [VON 44] in order to model the essentials of decision-making in interactive situations. In its normal-form representation, it consists of a finite set of players $I = \{1, .., n\}$, a set of pure strategies for each player $S_i = \{s_1, ..., s_m\}$ and a utility function $S_1 \times ... \times S_n \rightarrow \mathbb{R}$, which associates strategies with payoffs. Each player can adopt a strategy in order to play a game and the obtained payoff depends on the combination of strategies played at the same time by two players (strategy profile). In non-cooperative games, the players choose their strategies independently, considering what other players can play and trying to find the best strategy to employ in a game.

The players can play mixed strategies, which are probability distributions over pure strategies. A mixed strategy can be defined as a vector $x = \{x_1,...,x_m\}$, where m is the number of pure strategies and each component x^h denotes the probability that a player chooses its h-th pure strategy. Each mixed strategy corresponds to a point on the simplex, whose corners correspond to pure strategies.

A strategy profile can be defined as a pair (p,q), where $p \in \Delta_i$ and $q \in \Delta_j$. The expected payoff for this strategy profile is computed as:

$$u_i(p,q) = p \cdot A_i q$$

and

$$u_j(p,q) = q \cdot A_j p$$

where A_i and A_j are the payoff matrices of players i and j respectively.

In evolutionary game theory, we have a population of agents that play games repeatedly with their neighbors and update their beliefs on the state of the system, choosing their strategy according to what action has been effective and what has not in previous games, until the system converges. The strategy space of each player i is defined as a mixed strategy x_i, as defined above. The payoff corresponding to a single strategy can be computed as:

$$u(x_i^h) = \sum_{j=1}^{n}(A_{ij}x_j)^h$$

and the average payoff is:

$$u(x_i) = \sum_{j=1}^{n}x_i^T A_{ij} x_j$$

where n is the number of players with whom the games are played and A_{ij} is the payoff matrix among players i and j. The replicator dynamic equation [TAY 78] is used to find those states of the system that correspond to the Nash equilibria of the games:

$$x_i^h(t+1) = x_i^h(t)\frac{u\left(x_i^h(t)\right)}{u\left(x_i(t)\right)} \forall h \in x_i$$

This equation allows better than average strategies to grow at each iteration. Each iteration of the dynamics can be considered as an instance of an inductive learning process, in which the players learn from the others how to play their best strategy in a determined context. When equilibrium is reached, we assign to each player the label corresponding to the strategy with the highest value, which is computed with the following equation:

$$\phi_i = \underset{h=1,\dots,m}{\operatorname{argmax}}\, x_i^h.$$

Experimentally, we noticed that the selected values are always close to 1.

6.4. Word sense disambiguation games

In this section, we describe how we formulate the WSD problem in game theoretic terms, extending our previous works [TRI 15a, TRI 15b, TRI 17] with the use of transductive learning principles. In the next section, we describe how the interaction graph is constructed, in section 6.4.2 we describe how the strategy space of the players is initialized, then we introduce the payoff matrices of the games and the system dynamics.

6.4.1. *Graph construction*

The graph is constructed selecting from a text all the words that have an entry in a knowledge base such as WordNet [FEL 98], denoted by $I = \{1,\dots,N\}$, where N is the number of target words. From I, we constructed the $N \times N$ similarity matrix W where each element w_{ij} is the similarity among words i and j. W can be exploited as a useful tool for graph-based algorithms, since it is treatable as a weighted adjacency matrix of a weighted graph.

A crucial factor for the graph construction is the choice of the similarity measure, $sim(\cdot, \cdot) \rightarrow \mathbb{R}$ to weight the edges of the graph. For our experiments, we used the Dice coefficient [DIC 45], since it has performed well on different datasets [TRI 15a, TRI 17]. This measure determines the strength of co-occurrence between two words and is computed as follows:

$$Dice(i, j) = \frac{2c(i, j)}{c(i) + c(j)}$$

where $c(i)$ is the total number of occurrences of word i in a large corpus and $c(i, j)$ is the co-occurrence of the words i and j in the same corpus. This formulation is particularly useful to decrease the ranking of words that tend to co-occur frequently with many other words. For the experiments in this work, we used as corpus the British National Corpus [LEE 92]. The similarity graph W encodes the information of how two target words are similar, in a distributional semantics perspective [HAR 54].

6.4.2. Strategy space

The strategy space of the players is created using a knowledge base to collect the sense inventories $S_i = \{1, \ldots, m_i\}$ of each word in the text, where m_i is the number of senses associated with word i. Then, it creates the list $C = (1, \ldots, c)$ of the unique senses in all inventories, which corresponds to the space of the game.

With this information, it is possible to initialize the mixed strategy space x of each player. It can be initialized using a uniform distribution or considering information from sense labeled corpora, allocating more mass to frequent senses. In the former case, we initialize the strategy spaces of each player with the following equation:

$$x_i^h = \begin{cases} m_i^{-1}, & \text{if sense } h \text{ is in } S_i \\ 0, & \text{otherwise} \end{cases}, \forall h \in C$$

In the latter case, we assign to each sense a probability according to its rank, assigning higher probabilities to senses with a high frequency. To model this scenario, we used a geometric distribution that produces a decreasing probability distribution. This initialization is defined as follows:

$$x_i^h = \begin{cases} p(1-p)^{r^h}, & \text{if sense } h \text{ is in } S_i \\ 0, & \text{otherwise} \end{cases}, \forall h \in C,$$

where p is the parameter of the geometric distribution and determines the scale or statistical dispersion of the probability distribution, and r^h is the rank of sense h, which ranges from 1, the rank of the most common sense for word i, to m_i, the rank of the least frequent sense. These values are divided by $\sum_{h \in C} x^h$ to make them add up to 1. In our experiments, we used the ranked system provided by the Natural Language Toolkit (version 3.0) [BIR 06] to rank the senses associated with each word to be disambiguated.

6.4.3. The payoff matrix

We encoded the payoff matrix of the games as a sense similarity matrix among all the senses in the strategy spaces of the game. In this way, the higher the similarity between the senses of two words, the higher the incentive for a player to select that sense, and play the strategy associated with it.

The $c \times c$ sense similarity matrix A is defined in the following equation:

$$a_{ij} = sim(a_i, a_j) \forall i, j \in C : i \neq j$$

This matrix can be computed using the information derived from the same knowledge base used to construct the strategy space of the game. It is used to extract the partial payoff matrix A_{ij} for all the single games played between two players i and j. This operation is performed by extracting from A the entries relative to the indices of the senses in the sense inventories S_i and S_j. It produces an $m_i \times m_j$ payoff matrix, where m_i and m_j are the numbers of senses in S_i and S_j, respectively.

The semantic measure that we used in this work is the Gloss Vector measure [PAT 06], since it has been demonstrated to have stable performances in different datasets [TRI 17]. It is based on the computation of the similarity between the definitions of two concepts in a lexical database. They are used to construct a co-occurrence vector $v_i = (v^1, v^2, \ldots, v^n)$ for each concept i, with a bag-of-words approach, where v^h represents the number of times word v occurs in the gloss and n is the total number of different words in the corpus. From this representation, it is possible to compute the similarity between two vectors using the cosine distance:

$$cos\theta \frac{v_i.v_j}{\|v_i\|\|v_j\|}$$

The vectors are constructed using the concept of super-gloss introduced by [PAT 06]. It is the concatenation of the gloss of the synset and the glosses of the synsets connected to it with any relation in the knowledge base.

6.4.4. System dynamics

At each iteration of the system, each player plays a game with its neighbors N_i according to the graph W. The payoff of the h-th strategy is calculated as:

$$u\left(x^h\right) = \sum_{j \in N_i} (w_{ij} A_{ij} x_j)^h$$

and the player's payoff as:

$$u(x) = \sum_{j \in N_i} x_i^T (w_{ij} A_{ij} x_j)$$

where N_i represents the neighbors of player i. We assume that the payoff of word i depends on the similarity that it has with word j, w_{ij}, the similarities among its senses and those of word j, A_{ij}, and the sense preference of word j, (x_j), that can be unambiguous if j is a labeled player.

We use the replicator dynamics equation (see section 6.3.2) to find the Nash equilibria of the games. During each phase of the dynamics, a process of selection allows strategies with a higher payoff to emerge and at the end of the process each player chooses its sense according to these constraints.

6.5. Evaluation

In this section, we present the experimental setting for the evaluation and comparison of our system with state-of-the-art algorithms.

6.5.1. *Experimental setting*

We evaluated our algorithm with three fine-grained datasets: Senseval-2 English all-words[1] (S2) [PAL 01], Senseval-3 English all-words[2] (S3) [SNY 04], SemEval-2007 all-words[3] (S7) [PRA 07] and one coarse-grained dataset, SemEval-2007 English all-words[4] (S7CG) [NAV 07b], using WordNet as a knowledge base. The descriptions of the datasets are presented in Table 6.1.

The results of the evaluation are presented as $F1$, which is calculated as:

$$F1 = 2 \cdot \frac{precision\ recall}{precision + recall}.$$

This measure determines the weighted harmonic mean of precision and recall. Precision is defined as the number of correct answers divided by the number of provided answers and recall is defined as the number of correct

---

1 www.hipposmond.com/senseval2
2 http://www.senseval.org/senseval3
3 http://nlp.cs.swarthmore.edu/semeval/tasks/index.php
4 http://lcl.uniroma1.it/coarse-grained-aw

answers divided by the total number of answers to be provided. In our evaluation, we excluded labeled points in this calculation. Experimentally we noticed that precision is always equal to recall, since the system is always able to provide an answer.

We evaluated two different versions of the system, one using a uniform probability distribution to initialize the strategy space of the games and the other using information from sense labeled corpora (see section 6.4.2). Furthermore, to make the evaluation unbiased, we present the mean and standard deviation results of our system over 25 trials with different sizes of randomly selected labeled points.

Dataset	Text	N	C	Tot. N
S2	1	670	2195	
S2	2	997	1836	2387
S2	3	720	1916	
S3	1	783	2472	
S3	2	633	1426	2007
S3	3	591	1881	
S7	1	111	593	
S7	2	150	798	455
S7	3	194	1035	
S7CG	1	368	1287	
S7CG	2	379	1473	
S7CG	3	499	1926	2268
S7CG	4	677	1666	
S7CG	5	345	1410	

Table 6.1. *Number of target words and senses for each text of the datasets*

6.5.2. *Evaluation results*

The results of the evaluation are presented in Figures 6.1 and 6.2, where the results with the two initializations described in section 6.4.2 are shown: uniform (Figure 6.1) and geometric (Figure 6.2).

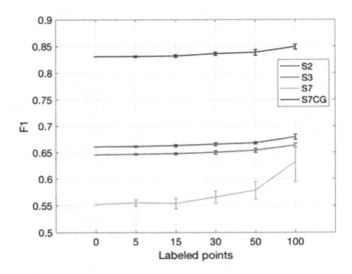

Figure 6.1. *Uniform distribution. Results as F1 varying the number of labeled nodes. For a color version of this figure, see www.iste.co.uk/sharp/cognitive.zip*

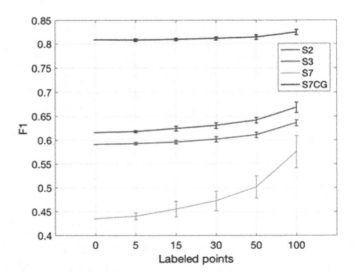

Figure 6.2. *Geometric distribution. Results as F1 varying the number of labeled nodes. For a color version of this figure, see www.iste.co.uk/sharp/cognitive.zip*

As we can see from the plots, the performance of our system on S7CG is very different from the others. This is because this dataset is coarse grained which means that the disambiguation of each word is not restricted to just one sense, as in the fine-grained datasets, but to a set of similar senses.

An important aspect to note is that the performance of the system is always increasing with the increasing labeled points. This is particularly evident on S7, where the performance passes from 0.43 to 0.57 using the uniform distribution and from 0.55 to 0.63 using the geometric distribution. For the other datasets, the improvements given by the labeled point are in the range of $3-5\%$.

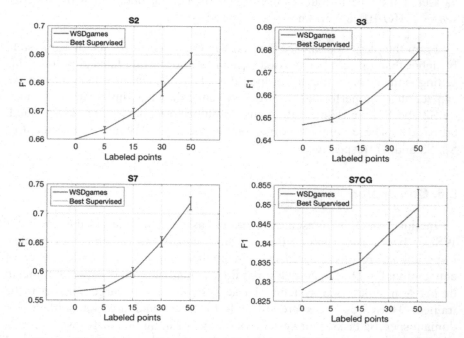

Figure 6.3. *Results as F1 using the geometric distribution and considering as correct the labeled nodes. The results are compared with the best supervised system on each dataset. For a color version of this figure, see www.iste.co.uk/sharp/cognitive.zip*

The information given by labeled points is more effective when we use a uniform distribution to initialize the strategy space of the system. This can be explained considering that with this initialization, we use less information, and for this reason, the presence of labeled points can balance this lack.

6.5.3. *Comparison with state-of-the-art algorithms*

The comparison of our system using a geometric distribution to initialize the strategy space of the games is presented in Figure 6.3. We compared our results with the best system that participated in each competition on each dataset if their performances are higher than those obtained with *It makes sense*[5] [ZHO 10], a well-known supervised system.

From the plots, we can see that, on S7CG, the performance of our system is higher than those of supervised systems without using labeled points. This setting is the same as the one proposed in [TRI 17]. On the other datasets, we can see that the performance of our system follows a similar trend. In fact, on S2 and S3, we require 50 points to outperform supervised systems and, on S7, 15. These numbers correspond to the 2.09, 249 and 3.29 percent of S2, S3 and S7, respectively.

6.6. Conclusion

In this work, we presented a graph-based semi-supervised system for WSD, based on game theory and consistent labeling principles. Experimental results showed that our method improves the performance of conventional methods and that it requires a small amount of labeled points to outperform supervised systems. These systems require large corpora to be trained. These resources are difficult to create and are not suitable for domain specific tasks. Our system infers the meaning of a target word from a small amount of labeled data exploiting relational and contextual information. In fact, the information of a labeled point is not only used locally by near words but also propagated over the graph and used globally by the dynamical system obtained with our game theoretic framework.

---

5 This system achieves higher results on S7CG and S3.

6.7. Bibliography

[AGI 14] AGIRRE E., DE LACALLE O.L., SOROA A., "Random walks for knowledge-based word sense disambiguation", *Computational Linguistics*, vol. 40, no. 1, pp. 57–84, 2014.

[ARA 07] ARAUJO L., "How evolutionary algorithms are applied to statistical natural language processing", *Artificial Intelligence Review*, vol. 28, no. 4, pp. 275–303, 2007.

[BIR 06] BIRD S., "NLTK: the natural language toolkit", *Proceedings of the COLING/ACL on Interactive Presentation Sessions*, pp. 69–72, 2006.

[CHA 15] CHAPLOT D.S., BHATTACHARYYA P., PARANJAPE A., "Unsupervised word sense disambiguation using Markov random field and dependency parser", *AAAI*, pp. 2217–2223, 2015.

[DAG 04] DAGAN I., GLICKMAN O., "Probabilistic textual entailment: generic applied modeling of language variability", *Proceeding of Learning Methods for Text Understanding and Mining*, pp. 26–29, 2004.

[DEC 10] DE CAO D., BASILI R., LUCIANI M. *et al.*, "Robust and efficient page rank for word sense disambiguation", *Proceedings of the 2010 Workshop on Graph-based Methods for Natural Language Processing*, pp. 24–32, 2010.

[DIC 45] DICE L.R., "Measures of the amount of ecologic association between species", *Ecology*, vol. 26, no. 3, pp. 297–302, 1945.

[EAS 10] EASLEY D., KLEINBERG J., *Networks, Crowds, and Markets*, Cambridge University Press, Cambridge, 2010.

[FEL 98] FELLBAUM C., *WordNet*, Wiley Online Library, 1998.

[HAR 54] HARRIS Z.S., "Distributional structure", *Word*, vol. 10, nos. 2–3, pp. 146–162, 1954.

[HAV 02] HAVELIWALA T.H., "Topic-sensitive PageRank", *Proceedings of the 11th International Conference on World Wide Web*, pp. 517–526, 2002.

[HOL 75] HOLLAND J.H., *Adaptation in Natural and Artificial Systems: an Introductory Analysis with Applications to Biology, Control, and Artificial Intelligence*, University of Michigan Press, Ann Arbor, 1975.

[JOR 02] JORDAN M.I., WEISS Y., "Graphical models: probabilistic inference", in ARBIB M.A. (ed.), *The Handbook of Brain Theory and Neural Networks*, MIT Press, Cambridge, 2002.

[KLE 02] KLEINBERG J., TARDOS E., "Approximation algorithms for classification problems with pairwise relationships: metric labeling and Markov random fields", *Journal of the ACM (JACM)*, vol. 49, no. 5, pp. 616–639, 2002.

[LEE 92] LEECH G., "100 million words of English: The British National Corpus (BNC)", *Language Research*, vol. 28, no. 1, pp. 1–13, 1992.

[MAL 88] MALLERY J.C., Thinking about foreign policy: finding an appropriate role for artificially intelligent computers, Masters Thesis, MIT, 1988.

[MAN 14] MANION S.L., SAINUDIIN R., "An iterative sudoku style approach to subgraph-based word sense disambiguation", *Proceedings of the Third Joint Conference on Lexical and Computational Semantics (* SEM 2014)*, pp. 40–50, 2014.

[MEN 14] MENAI M., "Word sense disambiguation using evolutionary algorithms – application to Arabic language", *Computers in Human Behavior*, vol. 41, pp. 92–103, 2014.

[MIH 04] MIHALCEA R., TARAU P., FIGA E., "PageRank on semantic networks, with application to word sense disambiguation", *Proceedings of the 20th International Conference on Computational Linguistics, Association for Computational Linguistics*, p. 1126, 2004.

[MOR 14] MORO A., RAGANATO A., NAVIGLI R., "Entity linking meets word sense disambiguation: a unified approach", *Transactions of the Association for Computational Linguistics*, vol. 2, pp. 231–244, 2014.

[MOS 89] MOSCATO P., "On evolution, search, optimization, genetic algorithms and martial arts: towards memetic algorithms", *Caltech Concurrent Computation Program, C3P Report*, vol. 826, p. 1989, 1989.

[NAV 07a] NAVIGLI R., LAPATA M., "Graph connectivity measures for unsupervised word sense disambiguation", *International Joint Conference on Artificial Intelligence*, pp. 1683–1688, 2007.

[NAV 07b] NAVIGLI R., LITKOWSKI K.C., HARGRAVES O., "SemEval-2007 task 07: coarse-grained English all-words task", *Proceedings of the 4th International Workshop on Semantic Evaluations, Association for Computational Linguistics*, pp. 30–35, 2007.

[NAV 09] NAVIGLI R., "Word sense disambiguation: a survey", *ACM Computing Surveys (CSUR)*, vol. 41, no. 2, p. 10, 2009.

[NAV 12a] NAVIGLI R., PONZETTO S., "BabelNet: the automatic construction, evaluation and application of a wide-coverage multilingual semantic network", *Artificial Intelligence*, vol. 193, pp. 217–250, 2012.

[NAV 12b] NAVIGLI R., PONZETTO S., "Joining forces pays off: multilingual joint word sense disambiguation", *Proceedings of the 2012 Joint Conference on Empirical Methods in Natural Language Processing and Computational Natural Language Learning*, pp. 1399–1410, 2012.

[PAG 99] PAGE L., BRIN S., MOTWANI R. *et al.*, The PageRank citation ranking: bringing order to the web, Technical report, Stanford InfoLab, 1999.

[PAL 01] PALMER M., FELLBAUM C., COTTON S. *et al.*, "English tasks: all-words and verb lexical sample", *The Proceedings of the Second International Workshop on Evaluating Word Sense Disambiguation Systems*, pp. 21–24, 2001.

[PAN 02] PANTEL P., LIN D., "Discovering word senses from text", *Proceedings of the Eighth ACM SIGKDD International Conference on Knowledge Discovery and Data Mining*, pp. 613–619, 2002.

[PAT 06] PATWARDHAN S., PEDERSEN T., "Using WordNet-based context vectors to estimate the semantic relatedness of concepts", *Proceedings of the EACL 2006 Workshop Making Sense of Sense-Bringing Computational Linguistics and Psycholinguistics Together*, vol. 1501, pp. 1–8, 2006.

[PRA 07] PRADHAN S.S., LOPER E., DLIGACH D. *et al.*, "SemEval-2007 task 17: English lexical sample, SRL and all words", *Proceedings of the 4th International Workshop on Semantic Evaluations*, pp. 87–92, 2007.

[REN 09] RENTOUMI V., GIANNAKOPOULOS G., KARKALETSIS V. *et al.*, "Sentiment analysis of figurative language using a word sense disambiguation approach", *RANLP*, pp. 370–375, 2009.

[SAM 11] SAMMUT C., WEBB G.I., *Encyclopedia of Machine Learning*, Springer, Berlin, 2011.

[SIN 07] SINHAR S., MIHALCEA R., "Unsupervised graph-based word sense disambiguation using measures of word semantic similarity", *ICSC*, vol. 7, pp. 363–369, 2007.

[SMR 06] SMRŽ P., "Using WordNet for opinion mining", *Proceedings of the Third International WordNet Conference*, pp. 333–335, 2006.

[SNY 04] SNYDER B., PALMER M., "The English all-words task", *Senseval-3: Third International Workshop on the Evaluation of Systems for the Semantic Analysis of Text*, pp. 41–43, 2004.

[TAY 78] TAYLOR P.D., JONKER L.B., "Evolutionary stable strategies and game dynamics", *Mathematical Biosciences*, vol. 40, no. 1, pp. 145–156, 1978.

[TON 06] TONG H., FALOUTSOS C., PAN J., "Fast random walk with restart and its applications", *Proceedings of the Sixth International Conference on Data Mining*, pp. 613–622, 2006.

[TRI 15a] TRIPODI R., PELILLO M., "WSD-games: a game-theoretic algorithm for unsupervised word sense disambiguation", *Proceedings of SemEval-2015*, pp. 329–334, 2015.

[TRI 15b] TRIPODI R., PELILLO M., DELMONTE R., "An evolutionary game theoretic approach to word sense disambiguation", *Proceedings of Natural Language Processing and Cognitive Science 2014*, pp. 39–48, 2015.

[TRI 17] TRIPODI R., MARCELLO P., "A Game-Theoretic Approach to Word Sense Disambiguation", *Computational Linguistics*, vol. 1, p. 43, 2017.

[VAP 98] VAPNIK V.N., *Statistical Learning Theory*, Wiley-Interscience, Hoboken, 1998.

[VIC 05] VICKREY D., BIEWALD L., TEYSSIER M. *et al.*, "Word-sense disambiguation for machine translation", *Proceedings of the conference on Human Language Technology and Empirical Methods in Natural Language Processing, Association for Computational Linguistics*, pp. 771–778, 2005.

[VON 44] VON NEUMANN J., MORGENSTERN O., *Theory of Games and Economic Behavior*, Princeton University Press, Princeton, 1944.

[WEA 55] WEAVER W., "Translation", in LOCKE W., BOOTH D. (eds), *Machine Translation of Languages*, vol. 14, Technology Press, MIT, Cambridge, 1955.

[ZHO 10] ZHONG Z., NG H.T., "It makes sense: a wide-coverage word sense disambiguation system for free text", *Proceedings of the ACL 2010 System Demonstrations, Association for Computational Linguistics*, pp. 78–83, 2010.

[ZHU 05] ZHU X., LAFFERTY J., ROSENFELD R., Semi-Supervised Learning with Graphs, Language Technologies Institute, School of Computer Science, Carnegie Mellon University, 2005.

7

Use Your Mind and Learn to Write:
The Problem of Producing Coherent Text

To produce written text can be a daunting task, presenting a challenge not only for high school students or second-language learners, but actually for most of us, including scientists and PhD students writing in their mother tongue. Text production involves several tasks: *ideation* (what to say?), *text structuring* (message grouping and linearization), *expression* (mapping of content onto linguistic forms) and *revision*. We will address here only one of them, *text structuring*, which is probably the most challenging task as it implies the grouping (chunking), ordering and linking of messages, which at the end of conceptual input lack this kind of information. Our goal is to find out whether part of this task can be automatized, the user providing a set of inputs (messages to be conveyed) and the computer then building automatically one or several *topic trees* from which the user will choose. While these trees still lack rhetorical information, functionally speaking they have a similar role as an outline: reduce the cognitive load of the writer and the reader. They help the writer to get some control over the information glut, telling him when to "say" what (order of sentences and paragraphs), and they help the reader to understand the functions of the different parts, i.e. how do the different parts relate to each other? As we can see, this is a very complex task. Having just begun to work on it, we will present here only preliminary results based on a very simple example and confined to a specific text type, *descriptions*. Yet, if ever this method works well for this type, it should also work, be it only partially, for other text types.

7.1. The problem

Spontaneous speech is a cyclic process involving a loosely ordered set of tasks: conceptual preparation, formulation, articulation [LEV 89, REI 00]. Given a goal, we have to decide *what to say* (conceptualization) and *how to say it* (formulation), making sure that the chosen elements, words, can be

---

Chapter written by Michael ZOCK and Debela Tesfaye GEMECHU.

integrated into a coherent whole (sentence frame) and do conform to the grammar rules of the language (syntax, morphology). During vocal delivery, in itself already quite a demanding task, the speaker may decide to initiate the next cycle, namely starting to plan the subsequent ideational fragment. In sum, speaking or acquiring this skill is a daunting task requiring the planning and execution of a number of subtasks. Given some goal, a speaker must plan *what to say* and *how to say* it, i.e. (a) *find* the right *words*; (b) determine an appropriate *sentence frame*; (c) put the chosen lemma in the right place; (d) add *function words*; (e) perform morphological adjustments; (f) articulate.

If speaking is difficult, writing is even a greater challenge, despite the huge amount of extra time. An author must not only know how to carry out most of the operations mentioned, but also be able to perform some additional operations which are not trivial at all. Some of them are at the *linguistic level* (cohesive devices: links, pronouns, choice of adequate determiner, etc.), and others are at the *conceptual level*: analysis and synthesis of knowledge[1], determination of information to provide and ensure reference (Henry Pu Yi, the last emperor, he), grouping, ordering and linking of messages, aggregation, i.e. merging syntactic constituents, etc. These last four operations are fundamental, as otherwise the reader may misunderstand or not understand at all. Being unable to see the connection between the parts, s/he cannot make sense of the whole. The document is perceived as a set of unrelated, i.e. incoherent segments. Yet, texts are characterized by the fact that goals, ideas and expressions are linked via a set of *rhetorical* (concession, rebuttal, etc.), *conceptual* (tense, cause–effect, set inclusion, etc.) and *linguistic* relations (anaphora, reference chains). Indeed, it is quite rare to see "texts" whose elements (propositions or sentences) are related only on the basis of statistical considerations (weight, frequency, etc.).

---

1 Reading the following sentence: "While there are many similarities between Japan and Germany there are also quite a few differences", it would be a mistake to believe that the expressed "facts" are stored like that in our memory. Indeed, what is expressed is probably the result of an analysis/synthesis of a large set of data concerning these two countries. Once we have performed this task, we may well conclude that despite the number of *commonalities* (discipline, work ethic, clean, well organized), there are also quite a few *differences* between the two countries: geographical location, religion, food (rice/potatoes, fish/meat), behavior (individualism vs. collective behavior), etc.

Text structuring is a particularly challenging task, because the ideas to be conveyed generally lack the links needed to build a coherent topic map, i.e. a tree showing which ideas go together and how the different chunks are related. Moreover, ideas tend to come to our mind in any order, i.e. via association [IYE 09]. Hence, in this case, "order of conceptual fragments" is a by-product of priming. It depends only on the relative associative strength between two items: a prime (doctor) and a probe (target, e.g. nurse). Obviously, this kind of order is very different from the one we see in ordinary texts, where the author guides the reader from some starting point (problem) to the conclusion (end point, solution). In conclusion, the order in which ideas come to our mind is fundamentally different from the order in which they will be conveyed in the final document, a well-structured text. Obviously, this transformation is not an easy task, yet, what makes things worse is the fact that the information needed to impose order on this data is generally absent in the conceptual input, i.e. the messages to be conveyed. This information needs to be inferred. This is probably the reason why writing is so much harder than speaking. Let us illustrate this via some concrete problems.

7.2. Suboptimal texts and some of the reasons

Texts are somehow like movies; they introduce and develop some objects over time. Having set the stage (context), the director introduces a topic, which s/he describes then from a chosen point of view, to move then on in various directions. To enable the onlooker to understand the movie (What is the point? How did we find the solution? What caused the coming about of some event?), the film director has to provide cues allowing the person watching the movie to grasp the topic, to see the details and to realize the evolution of the topic (topic changes, or return to the initial topic). In discourse, this is done via language, though the hardest parts are done in the brain. They are being taken care of by the reasoning and conceptual component: choice of the topic, relative order (development), type of connections, importance (focus) of the various elements at a given moment, etc. If sentences are like snapshots, texts are more like movies. They both have a framework, but the slides of a movie evolve over time. Hence, producing a good movie is, cognitively speaking, more demanding than taking a good snapshot.

While it is easy to tell whether a sentence is correct or not, it is not easy at all to do the same for a text. We may even wonder whether it makes sense to use the notion of correctness in this case. Texts span a wide spectrum, ranging from very well written to hardly understandable, with various stages in between. There are many reasons why a text may not read well: lack of coherence or cohesion, faulty reference, inadequate choice of a linguistic resource, etc., just a few examples to illustrate our point.

7.2.1. Lack of coherence or cohesion

Hearing someone say "John left for Taiwan to practice his French" may make us wonder about the connection between the two clauses, while the following, actually very similar sentence "John left for Taiwan to practice his Chinese" seems to be clear. We all understand that the second clause explains the reason of the first. The discourse is now coherent. Coherence is probably the single most important feature of text. Yet, even if texts lack coherence, they are hardly ever entirely incoherent. Readers sometimes even do not realize this, though it affects readability. Let us illustrate this via an example, where a cohesive document lacks *coherence*:

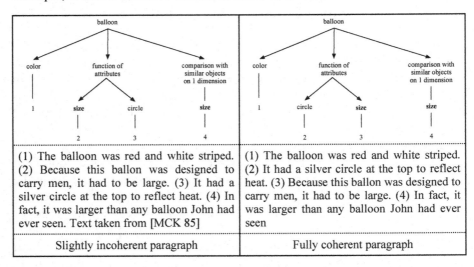

Figure 7.1. *Two paragraphs varying in terms of coherence*

Both paragraphs are composed of the same messages, but in slightly different orders, affecting readability. In the text of the left-hand side, there is a conceptual disruption. The author starts by providing information concerning the "color" of the described object. Next, he mentions its "size" and "form", to make then again a comment concerning "size", a feature mentioned already earlier on. This caveat has been avoided in the text of the right-hand side which for these reasons reads better. It is more coherent.

Texts may be coherent, but lack *cohesion* which also makes them suboptimal. *Cohesion* is generally achieved via specific linguistic means: hypernyms, pronouns, rhetorical relations, etc. The text of Figure 7.2 here below was generated by a computer to answer the following question: "What is the location of Uruguay?" The text on the left-hand side is coherent as it clusters information concerning various features: (**a**) *position* (a-c: latitude and longitude); (**b**) *neighboring countries* (d-e: northeast and west) and (**c**) *natural borders* (f-h: rivers and coasts). Despite this fact, the text does not read well. It lacks integration. There are many repetitions, which could have been avoided by using cohesive devices (pronouns, links). This has been done in the text on the right-hand side, which for this reason is more fluent than its counterpart.

(a) The location of Uruguay is South America. (b) The latitude ranges from -30 to -35 degrees. (c) The longitude ranges from -53 to -58 degrees. (d) The northern and eastern bordering country is Brazil. (e) The western bordering country is Argentina. (f) The boundary is the Uruguay river. (g) The southeastern coast is the Atlantic ocean. (h) The southern coast is the Rio de la Plata. (Example taken from [CAR 70])	(1a-1c) Uruguay lies in South America between -30 to -35 degrees latitude and -53 to 58 degrees longitude. (2d-2e) Its neighbour countries are Brazil in the north and the east, and Argentina in the west. (3f) The boundary between the two of them is the Uruguay river. (4g-4h) Uruguay's natural borders are the Atlantic Ocean in the south-east and the Rio de la Plata in the south
Coherent but incohesive paragraph	Coherent and cohesive paragraph

Figure 7.2. *Two paragraphs varying in terms of cohesion*

7.2.2. *Faulty reference*

One of the first things children learn are referring expressions [MAT 07]: cat, dog, mouse. Yet, the production of these forms is not a simple task, and there are various reasons for this. While the objects we talk about hardly ever change during our conversation, the linguistic means used for describing them do. In fact, they vary considerably, depending on whether we refer to an object for the first time or not: Once upon a time, there was a king named Henry. He had three daughters. Yet, the form depends not only on the referent's intrinsic features (gender: he/she), but also on competing elements. Imagine the scene here below, with the goal to refer to e_2.

Entity	type	feature$_1$	feature$_2$
e_1	cat	size: small	color: white
e_2	dog	size: big	color: black
e_3	mouse	size: small	color: gray

Table 7.1. *Description of the elements composing the scene*

In theory, we could use any of the following six forms: (1) a minimal description and basic-level term (dog), (2) a more specific word (Doberman), (3) a relational description (the dog *on* the lawn), (4) the referent's role (shepherd dog), (5) his proper noun (Fido) or (6) simply a pointer, i.e. a pronoun (it). While all these forms are correct, not all of them suit equally well the situation. The first example, using a basic level term (dog) and the definite article, fits best. Examples 5 and 6 (Fido, it) are underspecified and appropriate only under very specific circumstances: knowledge of the object's name, object currently in focus of attention, second mention, etc. Examples 2 and 3 are overspecified, as they provide more information than needed for identifying the intended object. For more details, see [ZOC 15b].

7.2.3. *Unmotivated topic shift*

Texts have both a hierarchical and linear structure. Entities (objects, topics) are introduced, and then developed. Since objects can be viewed from many perspectives, it is important to signal the viewpoint. The first entity is generally the perspective from which the described topic or scene is viewed. Topics are generally confined to a paragraph, and unless being "broadcasted", they remain stable. Consider the following paragraph from one of O'Connor's novels [OCO 71]

> "Mrs. Shortley was watching a black car turn through the gate from the highway. Over by the toolshed, about fifteen feet away, the two Negroes, Astor and Sulk, had stopped work to watch. They **were hidden** by a mulberry tree but Mrs. Shortley knew they were there."

If in this last sentence the author had used the *active voice*, writing, "The mulberry tree **hid** them but...", the text would still be correct, though much less fluent. Indeed, the use of this particular grammatical option would introduce an *unmotivated topic shift*[2], changing the perspective from "Astor and Sulk" to the mulberry tree.

7.3. How to deal with the complexity of the task?

Language production can be an overwhelming task. To deal with its complexity, people have conceived a wide range of strategies and tools: decomposition of the whole task, incremental processing[3], use of external resources (encyclopedias, dictionaries) or material support (text editors, pen and paper), allowing them to jot down ideas, make an outline, write a draft, etc.

Writing is not only re-writing but also thinking. We cannot just dump our ideas onto the world; we must give them a certain structure and form. Scardamalia and Bereiter [SCA 87] introduced the important distinction between *knowledge-telling* and *knowledge-transforming*. Novices use this first strategy, expressing ideas basically in the order in which they come to their mind, making text production look like a linear process, while in reality it is a hierarchical one [MAN 87]. There is a main point, there are subsidiary points, transitions from one level to the next, etc. What the knowledge-

---

2 Note that topic shifts can be triggered by various linguistic devices: voice (active/passive) or verb choice (buy vs. sell). Proper control of topics is important, as the reader interprets the clause accordingly.

3 Psychologists [KEM 87] have studied *incremental processing* noting that speakers start expression (articulation) before having fully encoded, i.e. completely specified all the details of the message. What holds for speaking holds, of course, also for writing.

teller's strategy is crucially lacking is a purposeful reflection concerning the content, role and form of the elements to be used. The author seems to be a prisoner of his associations. Rather than stepping back and deciding from there "what" to say, "when" and "how", s/he jumps too quickly to verbal delivery. Moreover, the fact that this behavior is largely controlled by local associations (*what to say next?*) makes everything look as if it were on the same plane.

The second production mode, *knowledge-transforming,* is generally used by mature writers who develop far more elaborated networks (more connections, better integration among goals) than novices. *Knowledge-transforming* can be characterized by an inclusion in the writing process of reflective operations that transform intentional, structural and conceptual representations (gist). Analysis of thinking-aloud protocols [FLO 80] has shown that mature writers plan by globally working through a writing task at an abstract level before working through it at a more concrete level. During the text production process, problems are tackled both at the level of content (*what do I mean?*) and at the level of form (*how do I say it?*). Reflection on both levels during composition leads to the transformation of content and form, giving rise to new thoughts.

In sum, there is a fundamental difference between expert and novice writing with respect to the nature of planning. Planning by beginners is opportunistic and driven by local constraints, while the planning by experts is strategic: the writer plans much carefully *what* she wants to say and *how* to express "things". It is mainly the authors' goals that determine content, structure and form. As we can see, writing is not an easy task and we are wondering whether and to what extent computers could help. We will focus here only on one task, *coherence,* i.e. grouping messages by category. Yet before doing so, let us take a look at the work done by computational linguists, cognitive psychologists and rhetoricians on whose theories any good application rests.

7.4. Related work

Let us start by taking a look at the work done by computational linguists working on *text* generation [REI 00, BAT 16]. Their ambition consists of the automatic production of texts based on messages and goals[4]. Since everyone seems to agree with the fact that texts are structured [MAN 87], this seems the right place to go. Alas, even there one will be disappointed. To avoid misunderstandings, the work produced by this community is important and impressive in many ways. Nevertheless, it seems to be based on assumptions incompatible with respect to our goal, which is to assist a writer in text production, i.e. help her or him to organize a set of ideas that prior to that point were a more or less random bunch of thoughts (at least for the reader).

Here are some of the reasons why we believe that this kind of work is not compatible with our goal. First of all, interactive generation (our case) is quite different from automatic text generation. Next, most text generators are based on assumptions that hardly apply in normal writing: (a) *all the messages* to be included in the final document are available at the very moment of building the text plan [HOV 91]; (b) ideas are retrieved *after* a text plan has been determined [MCK 85], or the two are done more or less in parallel [MOO 93]; (c) the links between ideas (messages) or the topics to be addressed are all known at the onset of building the text plan. This last point applies both to Marcu's work [MAR 97] and to data-based generators [REI 95].

Practically all these premises can be challenged, and none of them accounts for the psycholinguistic reality of composition, i.e. text production by human beings [DEB 84, BER 87, AND 96]. For example, authors often do not know the kind of links holding between ideas[5], neither do they always know the topical category of a given message[6]. Both have to be inferred. Authors have to discover the link(s) between messages and the nature of the topical category to which a message or a set of messages belongs. Both tasks are complex, requiring a lot of practice before leading to the skill of good writing (coherent and cohesive discourse).

---

4 This is often seen as a top-down process : goals triggering ideas, i.e. messages, which trigger words, which are inserted in some sentence frame, to be adjusted morphologically, etc.
5 The following two messages [(a) get married (x), (b) become pregnant (x)] could be considered as a *cause*, *consequence* or as a natural *sequence*.
6 What we mean by *topic* is the following. Suppose you were to write "foxes hide underground". In this case, a reader may conclude that you try to convey something concerning the foxes' "habits" (hide) or "habitat" (underground).

The above-mentioned work also fails to model the dynamic interaction between idea generation (messages) and text structure, [SIM 88] being arguably an exception. Indeed, a topic may trigger a set of ideas (top-down generation), just as ideas may evoke a certain topic (bottom-up), and of course, the two can be combined, a bottom-up strategy being followed by a top-down strategy (see Figure 7.3). This kind of interaction often occurs in spontaneous writing where ideas lead to the recognition of a topical category, which in turn leads to the generation of new data of the same kind. Hence, ideas or messages may have to be dropped. Not having enough conceptual material, the author may decide either not to mention a given fragment, to put it in a footnote, or to continue searching for additional material.

Another community interested in writing is that of psychologists. Clearly, a lot has been written on this subject[7]. Yet, despite the vast literature on composition and despite the recognition of the paramount role played by idea structuring (outlining) for yielding readable prose, little has been produced to clarify what it takes concretely speaking to achieve this goal (i.e. to help authors). Even the book series "Studies in Writing" [RIJ 96b][8] will tell you next to nothing concerning the topic we are interested in: how to find commonalities between conceptual fragments (ideas) to group them into chunks, or, how to "see" the hidden links between ideas.

In the remainder of this paper, we will present a small prototype trying to emulate the first strategy mentioned here above: to structure data, or discover potential structures in data (messages). Yet before doing so, we would like to spell out in more detail some of the assumptions underlying our work and show how they relate to what is known about the natural writing process.

7.5. Assumptions concerning the building of a tool assisting the writing process

As mentioned already, authors tend to use different strategies when writing: they start from topics or goals (top-down), from initially unrelated data or ideas (bottom-up), or they combine these two strategies. Bottom-up

---

7 Among others: [ALA 01, BER 87, FLO 80, KEL 99, LEV 13, MAT 87, OLI 01, RIJ 96a, RIJ 96b, TOR 99]. For more pointers, see: http://www.writingpro.eu/references.php
8 http://www.emeraldinsight.com/products/books/series.htm?id=1572-6304

activated ideas lead to the recognition of a subsumption category (Figure 7.3(b)), which in turn causes the activation of more data (top-down again, see Figure 7.3(c)).

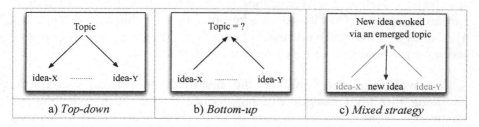

a) *Top-down*   b) *Bottom-up*   c) *Mixed strategy*

Figure 7.3. *Three strategies of discourse planning: top-down, bottom-up or both*

The first strategy is probably the most frequent one. Starting from a goal, authors seek relevant content (messages), organize it according to topical and rhetorical criteria, translate it into language and revise it. This is known as top-down planning. Note that revision can take place at any stage, and that, during the initial stage of conceptual planning (brainstorming), little filtering takes place. Authors are mainly keen on potentially interesting ideas. (Incidentally, this is why the term "brainstorming" better captures the reality of this situation than "idea planning".) It is only at the next step that contents are thoroughly examined. This may lead to a modification of the message base: some ideas will be dropped, others added. The result of this is generally a so-called outline or text plan, i.e. a definition of *what* is to be expressed *when*, i.e. in what *order*.

Another strategy involves going the opposite way. Starting from the ideas coming spontaneously to the authors' mind (bottom-up planning), s/he will try to group them into sets (topical trees) and to link these clusters. In this kind of bottom-up planning, the structure or topic emerges from the data. These topics may act as seeds, eventually triggering additional material (mixed strategy). Bottom-up planning is a very difficult problem (even for people). Yet, this is the one we are interested in. A question remains on the basis of what knowledge writers know which ideas cohere, i.e. what goes with what and in what specific way? Suppose you have an assignment asking you to write a small document about "foxes" and their similarities and differences compared with "wolves" or "coyotes", two animals with which they are sometimes confused. This might trigger search for information concerning "foxes", possibly yielding a set of messages like the one shown in Figure 7.3(a). For the time being, it does not really matter where these ideas come from (author's brain, external resources or others), what we are interested in here is to find an answer to the following questions: (a) How does the author group these messages or ideas into topical categories? (b) How does s/he order them within each group? (c) How does s/he link and name these chunks? (d) How does s/he discover and name the relations between each sentence or chunk?

We will focus here only on the first question (topical clustering), assuming that (a) messages will be grouped if they have something in common, and (b) messages or message elements do indeed have something in common. The question is how to show this. Actually, this can be either hard or fairly trivial, as in the case of term identity. Imagine the following inputs: (a) give (I, dog$_1$, my_son) and (b) like_to_chase (dog$_1$, milk-man). Since these two propositions share an argument (dog$_1$), they can be clustered, yielding two independent clauses (I've given my son a dog. He likes to chase the milk-man.), or a subordinate, i.e. relative clause (I've given my son a dog who likes to chase the milkman). Of course, you could also topicalize the "dog", producing the following sentence: "The dog I've given to my son likes to chase the milkman". Which of these forms is the most adequate one depends, among other things, on the context (surrounding sentences) and the discourse goal: what do you want to stress or highlight?

The question of how to reveal commonalities or links between data in the non-obvious cases remains. We can think of several methods. For example, we could try to enrich input elements by adding information (features, attribute-values, etc.) coming from external knowledge sources: corpora (co-occurrence data, word associations), dictionaries (definitions), etc. Another method could consist of determining similarities between message elements (words). This is the one we have used, and we will explain it in more depth here below (section 7.6). Once such a method has been applied, we should be able to cluster messages by category, even though we may not be able to give it a name. The name may be implicit, and name-giving may require other methods.

The result of this will be one or several topic trees, grouping (ideally) all inputs. While different trees may achieve different rhetorical goals (the focus being different), all of them ensure coherent discourse. The effect of these variances can probably only be judged by a human user, who shall pick the one best fitting his or her needs. While our developed software will not be able to achieve this goal, i.e. build a structure that conceptually and rhetorically matches the authors' goals, it should nevertheless be able to help the user perceive conceptual coherence, hence allow him to create a structure (topic tree) where all messages cohere, something that not all grown-up human beings are able to do. Concerning goals and bottom-up planning, consider the following.

Goals can be of various sorts. They can be coarse grained ("convince your father to lend you his car") or more fine-grained, relating to a specific topic: describe an animal and show how it differs from another one with which it is often confused (alligator-crocodile; fox-coyote/wolf). *Messages* may feed back on the conceptual component, altering messages or goals (addition, deletion, modification). This cyclic process between top-down and bottom-up processing is a very frequent case in human writing. We will focus here only on the latter, confining ourselves to propositions composed of two place predicates. These will be the inputs for which we try to check whether there is a commonality or link between them. Of course, even the linking of simple propositions may be a very complex problem. Think of causal relations that can be viewed as a systematic correlation between two events, or a state and an event[9]. Since these cases require a special approach, we will not deal with them here.

7.6. Methodology

We present in this section a description of the method used to allow for the kind of grouping mentioned here above. Messages can be organized on various dimensions and according to various viewpoints: *conceptual* relations (taxonomic, i.e. set inclusion, causal, temporal, etc.), *rhetorical* relations (concession, disagreement), etc. We will focus here only on the former, assuming that messages can at least to some extent be organized via the semantics[10] of their respective constituent elements.

---

9 The perception of the causal relationship between the underlined elements in – "Be careful, the road may be *dangerous*. They've just announced a *Typhoon*." – supposes that we know that Typhoons are dangerous.

10 Of course, the term semantics can mean many things (association, shared elements between a set of words, etc.), and which of them an author is referring to needs to be made explicit.

Put differently, in order to reveal the relative proximity or relation between a set of messages, we may consider the similarity of some of their constituent elements. Summing similarity values is a typical component of a vector space model and has been well described by [WID 04, MAN 08]. Concerning "similarity", we need to be careful though, as the words' similarity does not guarantee "relatedness"; it may even be one of its preconditions. Indeed, many researchers have used this feature for sentence similarity detection [BUL 07, TUR 06], but most of them based their analysis on the surface form, which may lead to erroneous results, because similar meanings can be expressed via very different syntactic categories (e.g. "use for" vs. "instrument", "have" vs. "her"). Likewise, a given form or linguistic resource, say the possessive adjective, may encode very different meanings. Compare – his car versus his father versus his toe – which express quite different relations: ownership, family relationship, inalienable part of the human body.

What we present here is a very preliminary work. Hence, our method is designed to address only very simple cases, two-place predicates, i.e. sentences composed of two nouns (a subject and an object) and a (linking) predicate. Given a set of these kind of inputs, our program determines their proximity regardless of their surface forms. The sentences will be clustered on the basis of semantic similarity between the constituent words. This yields a tree whose nodes are categories (whose type should ideally be expressed explicitly, e.g. food, color, etc.) and whose leaves are the messages or propositions given as input.

In the following sections, we will explain in more detail our approach by taking the inputs shown in Figure 7.4(a) to illustrate our purpose. The goal is to cluster these messages by topic to create a kind of outline or topical tree. Indeed, {1, 4} address *physical features* (appearance), {2, 6} provide spatial information, the place where foxes live or hide (*habitat*), while {3, 5, 7} deal with their *habits*. This last category can be split into two subtopics, in our case, "theft" {3} and "consumption" {5, 7}. The result of this analysis can be displayed in the form of a tree (Figure 7.4(b))[11].

---

11 Note that generally we can come up with more than one tree. Any set of data allowing for multiple analyses (depending on the point of view), and multiple rhetorical effects.

1° resemble (foxes, dog)
2° live_in (foxes, woods)
3° steal (foxes, chicken)
4° be (foxes, red)
5° eat (foxes, fruits)
6° hide (foxes, underground)
7° eat (foxes, eggs)

Figure 7.4(a) *Conceptual input (messages)*

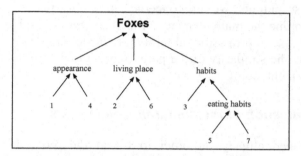

Figure 7.4(b) *Clustered output (topic tree)*

In order to achieve this result, we have defined an algorithm carrying out the steps referred to in Table 7.2. We will describe and explain them in more depth in the following sections. Note that what we called messages here above is now called sentence which is processed by a parser.

1) Determine the role of words, i.e. perform a syntactic analysis;
2) Find potential seed words;
3) Align words playing the same role in different sentences;
4) Determine the semantic proximity between the aligned words;
5) Determine the similarity between sentences;
6) Group sentences according to their semantic affinity (similarity).

Table 7.2. *Main steps for topic clustering*

7.6.1. *Identification of the syntactic structure*

The goal of this step is to identify the dependency structure of the sentence. This information will be used later on (**a**) to identify the semantic seeds (see section 7.6.2), (**b**) to align words playing a similar role and (**c**) to identify the role of the different elements of the underlying proposition, i.e. the respective predicate, subject or object. To obtain this information, we used the Stanford parser[12]. For example, the input "Foxes eat fruits" would yield the following output:

Tagging: **Foxes**/NNS **eat**/VBP **fruits**/NNS./.

Dependencies: N_{subj} (eat, foxes); D_{obj} (eat, fruits)

Of all these outputs, we are concerned here only with N_{subj} and D_{obj} in order to determine the main elements of the message: the subject, object and the main verb, or, in propositional terms [predicate (argument$_1$, argument$_2$)]. Next, we used the similarity of the parts (words) to determine the similarity of the wholes (sentences).

7.6.2. *Identification of the semantic seed words*

As mentioned already, in order to reveal the proximity or potential relation between two or more sentences, we can try to identify the similarity between the respective constituent words. We need to be careful though. If we do this by taking into account only the similarity values of the respective (pair of) words, we may bias the analysis and get incorrect results.

There are several problems at stake. For example, the number of identical words does not necessarily imply relatedness or similarity. Actually, two sentences may be composed of exactly the same words, and still mean quite different things, compare: "Women without their men are helpless" versus "Men without their women are helpless". Given the fact that such cases are quite frequent in natural language, we decided not to rely on (all) the words occurring in a sentence, or to use a "bag of words" approach (sentence without stop words). We preferred to rely only on specific words, called seeds, to compare the similarity of different sentences. We consider seed words to be elements conveying the core meaning of a sentence. For example, for the two sentences here above we could get the following seeds: (a) without (man, women); (b) without (women, men), which reveal quite readily their difference.

Our idea of choosing seed words seems all the more justified as different kinds of words (lexical categories) have different statuses: some words

---

12 http://nlp.stanford.edu/software/lex-parser.shtml

conveying more vital information than others. Nouns and verbs are generally more important than adjectives and adverbs, and each one of them normally conveys more vital information than any of the other parts of speech[13]. We assumed here that the core information of our sentences is presented via the nouns (playing different roles: subject, object) and the verb linking them (predicate). We further assumed that dependency information was necessary in order to be able to carry out the next steps. In the following two sentences, (a) "Foxes hide underground" and (b) "Foxes hide *their prey* underground", a "bag of word" method or a simple surface analysis would not do, as neither of them reveals the fact that the object of hiding ("fox" vs. "prey") is different in each sentence, a fact that needs to be made explicit.

To avoid this problem, we used the dependency information produced by the parser, which allowed us to determine the role of the nouns (deep-*subject*, deep-*object*) and the predicate (verb) linking the two. For example, this reveals the fact that the following two sentences are somehow connected: "*Foxes eat eggs*" and "*Foxes eat fruits*". In both cases, the concept "fox" is connected to some object ("egg" vs. "fruit") via some predicate, the verb "*eat*". The core of these two sentences is identical. Both of them tell us something about the foxes' diet or eating habits (egg, fruits). Note that while this method does not reveal the nature of the link (diet, food), it does suggest that there is some kind of link (both sentences talk about the very same topic: food). Hence, syntactic information (part of speech, dependency structure) is precious as it allows us to identify potential seed words that will be useful for subsequent operations.

7.6.3. *Word alignment*

In order to compare sentences in terms of similarity, we not only need a method for doing so, but also need the data to be in a comparable form. Hence, we need the input to be given in a standardized form or we need to carry out some normalization. The latter can be accomplished to some extent via a dependency parser that reveals the roles played by different words. We can now align the words of the various sentences and compare those playing the same semantic role.

Word alignment consists not simply in finding identical words in different sentences, but rather in finding and aligning words playing the

---

13 Note that we do not consider "connectors" (yet, despite, because) here, as they are not known at this stage.

same role in these sentences. This means in our case that we have to compare, say, the subject of one sentence with the subject of another, and do the same for the other syntactic categories or semantic roles (verbs, deep-objects, etc.). To allow for this, we rely again on the dependency information produced by the parser (section 7.6.1). Note that our example showed only surface relations (subject, object, etc.), while ideally we need information in terms of deep-case roles: agent, beneficiary, etc. [FIL 68]. Applied to our examples, "Foxes eat fruit" and "Foxes eat eggs", it is clear that "fruit" can be aligned with "eggs", since both nouns play the same role.

Note that we also need to be able to detect synonyms or semantic equivalences: "instrument ≡ is used for"; "resemble ≡ be alike", "for example ≡ somehow ≡ like", etc. These words are very useful and could be used as topic-signatures [LIN 00], hence seed words. Note that such information is obtained indirectly in our approach via the *vector space model* which is briefly described in the next section.

7.6.4. *Determination of the similarity values of the aligned words*

While there are various ways to detect links between sentences or words (e.g. shared features or associations), two obvious ones are coreference and class-membership. See our example in Figure 7.4(a), where the two sentences – ("Foxes eat eggs"; "Foxes eat fruits") – have an identical referent, the actor "fox", and two different instances of the same class, the generic element "food".

As mentioned already, in order to compare sentences in terms of their meaning, the compared structures must have a common format. Similarity of meaning supposes, of course, that we are able to extract somehow the meaning of the analyzed objects (sentences, words). Yet, word meanings depend on the context in which a word occurs. Words occurring in similar contexts tend to have similar meanings. This idea known as the "distributional hypothesis" has been proposed by several scholars, e.g. [FIR 57, HAR 54, WIT 22]. For surveys, see [SAH 08, DAG 99], or (http://en.wikipedia.org/wiki/Distributional_semantics).

Since we try to capture meaning via word similarity, the question of how to operationalize this notion arises. One way of doing so is to create a vector space composed of the target word and its neighbors [LUN 96]. The meaning of a word is represented as a vector based on the co-occurring words. In the following two sentences – "Foxes eat eggs" and "Foxes eat fruits" – we have four distinct tokens or words: foxes, eat, fruits and eggs. Hence, we constructed a vector for each one of them by considering their co-occurrences in COHA (Corpus of Historical American English), a part of speech tagged n-gram corpus of 400 million words [MAR 11]. This allowed us to apply the vector space model in order to compute the degree of similarity between a set of words. To this end, we computed the distance (cosine) of the respective vectors. Let us suppose that there are only four words co-occurring with "fruit" and "egg" ("juice, vitamin, price, eat" and "chicken, protein, eat and oval"), then the vector for fruit would be "juice, vitamin, price, eat" and the vector for egg would be chicken, protein, eat, oval.

Note that we will also count the frequency of the co-occurrence. To compute the similarity between two vectors, we computed the cosine of their angle. Hence, we constructed such vectors for all major words of our sentences. For the example here above, we have four vectors, one for each of the words occurring in both sentences: fox, eat, fruit and eggs. Note that the words in the vectors are replaced by their weight, i.e. a numerical value representing the meaning of the respective concepts. For instance, for the *fruit* vector, all of the following words, "juice, vitamin, price, eat" are replaced by a numerical value (weight). For details, see step 1 here below.

Consider the following example that measures the proximity between *fruit* and *eggs* by using the co-occurrence information gleaned from COHA, yielding the matrix shown here below.

I	word space	$\vec{T}_{1\,fruit}$	$\vec{T}_{2\,eggs}$
1	bird	0	1
2	hummus	0	1
3	food	1	1
4	incubator	0	1
5	banana	0	1
6	store	1	0
7	gene	1	1

Table 7.3. *Sample word space matrix*

The vectors are built by carrying out the following four steps:

1) Extract words co-occurring with *fruit* in a predefined window. In our experiments, we considered the window to be six content-bearing words of the sentence containing the target, hence, three terms preceding and following the term *fruit* that would yield in this case the following list of words: banana, food, gene and store.

2) Do the same for *eggs*, that would yield: bird, hummus, food, incubator and gene.

3) Build the corresponding vectors for each word (*fruit, egg*) based on its co-occurrences (see the table here above). In its simplest form, the vectors for each word are built on the basis of their co-occurrences. A term (*fruit, eggs*) receives frequency values depending on the number of times it co-occurs with another term within the defined window, which would yield the following vectors for *fruit* [0,0,1,0,0,1,1] and *vehicle* [1,1,1,1,1,0,1]. Note that the words here are referred to via their index, e.g. their position in the word space. Hence, the first position refers to the first term, the second to the second, and the last to the last one in the matrix. Unlike in the example given here above, we use in our experiments the weighted frequency of the words' co-occurrences rather than binary values.

4) Measure the distance between the vectors. In order to quantify the similarity between two words, we have to compute the cosine similarity of their respective vector representations. This is done in the following way:

Take a pair of words, say *fruit* (T_1) and *eggs* (T_2), with their respective vectors and weights, and carry out the steps described here below. To determine the similarity between "fruit" and "eggs" we compute their cosine similarity, i.e. the angle between their vectors. The cosine similarity value is computed in the following way:

1) sum the product of the weight of the terms in the respective vectors ($\vec{T}_1$ and $\vec{T}_2$);

2) sum the square of the weights of each vector term of $\vec{T}_1$ and $\vec{T}_2$;

3) take the square root of the product of the results of step 2;

4) divide the result of step 1 by the result of step 3.

We have used this method already for another task, the automatic extraction of part–whole relations [ZOC 15a]. Since then, we have extended it to allow for the computation of similarity between words. The method consists basically of two operations: creating a vector for all words and identifying the similarity of the aligned words.

Step 1: *Creation of a vector for all words based on co-occurrence information*

The co-occurrence information is gleaned from COHA. The vector is built on the basis of a word's co-occurrence within a defined window (phrase, sentence and paragraph). Since different words make different contributions, we assign weights to reflect their relative contribution in terms of relevance allowing determination of the meaning of a sentence or word. Hence, meanings are expressed via a weight, which may depend on the context of a given term, a factor that needs to be taken into account. The formula used to assign the weight is very similar to the TF-IDF. Since our objective is determining the similarity between words, the document–term matrix in the conventional TF-IDF is adopted to term–term matrix in our case. In order to determine the weight (W), we use the percentage of the term's co-occurrence frequency (TCF) with respect to a given concept out of the **total number** of co-occurrences (TOTNBC) of the term with any other term in the corpus. For example, in order to determine the weight of the term *egg* (one of the words in the "chicken vector") for the vector of "chicken", we count the co-occurrence frequency of *chicken* with *egg* and then divide this result by the total number of co-occurrences of the term *egg* with any other term in the corpus. We did the same for all relevant terms. The above-described operations can be captured via the following formula, which is used to determine the weight (W) of a given word for determining the meaning of another co-occurring term:

W = TCF-yz/TOTNBC-xy, where

TCF-y is the frequency, i.e. the number of times y co-occurs with z;

TF-y is the total frequency y in the corpus.

To build a vector for a given concept, we use the weighted value of all co-occurring words. Hence, we calculated the relative weight of each word of the vectors in defining the meaning of the term for which the vector was built.

Step 2: *Identification of the cosine similarity between vectors of the aligned words*

The similarities between words are computed on the basis of the cosine of the words' vectors. Note that the similarity value is calculated only for the aligned words.

7.6.5. *Determination of the similarity between sentences*

The meaning of a sentence can (at least to some extent) be obtained via the combined meanings of the constituent elements, words. Having identified in section 7.6.4 the similarity values between the aligned words, we now build a matrix showing their respective similarity values. The rows and columns of the vectors are built on the basis of co-occurring words and the cells contain their similarity values. In order to identify the similarity between two sentences, we add the similarity values of the component words and then compute the average to derive a single similarity value between the sentences. Hence, in order to identify the degree of similarity between a pair of sentences, we sum up the respective similarity values of their subjects, verbs and objects, and then divide the result by 3 in order to get the average.

	resemble	eat	are	live	steal	hide
resemble						
Eat	0.293					
Are	0.550	0.152				
Live	0.210	0.365	0.139			
Steal	0.428	0.392	0.210	0.306		
Hide	0.527	0.430	0.240	0.631	0.450	

Table 7.4. *Sample word similarity matrix of co-occurrences*

With respect to our fox example (Figure 7.4), all messages apart from the third one are clustered this way. Message 3 is clustered with 5 and 7 according to the algorithm presented in the next section.

7.6.6. *Sentence clustering based on their similarity values*

As mentioned already, our strategy consists of the creation of a tree based on the similarity values of the sentences given as input. Sentences are clustered in three steps on the basis of the similarity value of the subjects, the verb and the objects. Accordingly, sentences talking about different topics, say "foxes" and "fruits", are placed in different clusters. At the next cycle, each group is further clustered depending on the topic (habit, physical appearance, etc.), which may be signaled via the verb or the object. The clustering algorithm is given below.

---

1) Take any sentence of the considered pool of inputs and search for another one whose topic similarity (i.e. value of the subject) is closer to the target than any of its competitors to form a cluster. Similarity values are obtained via the word-similarity-matrix (see section 7.6.5 and table 7.4).

2) Continue to cluster the sentences of the groups obtained so far by using the similarity values of the verb linking the subject and the object.

3) Perform the same operation as in step 2 based on the similarity of the objects.

4) Repeat steps 1 to 3 until all the sentences, or the greatest possible number of sentences are clustered.

5) Create a link between the clusters based on the respective similarity values of the verb and the object.

Table 7.5. *The clustering algorithm*

7.7. Experiment and evaluation

In order to test our system, we used a text collection of 28 sentences. The test set contains four groups of sentences talking respectively about "foxes", "fruits", and "cars". The last set, called rag bag, is composed of topically unrelated sentences. It is used only for control purposes.

Topic₁ *Fox*	Topic₂ *Fruits*
1) Foxes resemble dogs.	8) An apple a day keeps the doctor away.
2) Foxes live in the woods.	9) Apples are expensive this year.
3) Foxes steal chicken.	10) Oranges are rich in vitamin C.
4) Foxes are red.	11) The kiwi fruit has a soft texture.
5) Foxes eat fruits.	12) Grapes can be eaten raw.
6) Foxes hide underground.	13) Grapes can be used for making wine.
7) Foxes eat eggs.	14) The strawberries are delicious.
Topic₃ *Cars*	Topic₄ *ragbag*
15) A car is a wheeled motor vehicle.	22) Olive oil is a fat obtained from olives.
16) Cars are used for transporting passengers.	23) Playboys usually have a lot of money.
17) Cars are mainly designed to carry people.	24) A finger is a limb of the human body.
18) The first racing cars amazed the automobile world.	25) Apple is the name of a software company.
19) Cars typically have four wheels.	26) Eau Sauvage is a famous perfume.
20) Cars are normally designed to run on roads.	27) Wine is an alcoholic beverage made from fermented grapes or other fruits.
21) Cars also carry their own engine.	28) IBM is an American corporation manufacturing computer hard ware.

Table 7.6. *Our test sentences*

The system's task is now to integrate as many messages as possible. This will yield a tree containing as many branches as there are topics, in our case four. At the next cycle, the system will try to create subcategories, i.e. branches are further divided, or, viewed in the opposite direction, messages are clustered in more specific categories (habits, living places, etc., in the fox group). Whether this is feasible depends, of course, on the message elements. Since the function of the control group (the set of topically unrelated sentences) is only to check the system's accuracy, none of its sentences should appear in any other group than the "control group".

Once this clustering is done, we can determine the system's performance by counting the number of sentences assigned properly in the tree. For the evaluation, we used the classical metrics, defining *precision, recall* and *F-measure* in the following way:

recall (Number of sentences correctly assigned to the valid cluster/Total number of sentences);

precision (Number of sentences correctly assigned to the valid cluster/ Number of sentences clustered);

f-measure $(2/ [(1/\text{precision}) + (1/\text{recall})])$.

We obtained the following results: 22 out of the 28 sentences are placed in the correct cluster, six occur in the wrong place. For example, the sentences 25 and 27 are both placed in the fruit cluster (topic 2), while they should not. This is due to the fact that our method does not take into account the semantics of the string "apple", which can refer both to the fruit and to the computer manufacturing company located at Cupertino. The same holds for "wine", which can be both an alcoholic drink and a fruit.

We have also evaluated our system by further clustering the sentences within each topic based on their similarity. All the sentences of topic 1 are clustered correctly, with sentences 1-4, 2-6 and 5-7 forming three clusters. Sentence 3 is more closely related to the clusters containing 5 and 7 than to any other cluster. All the sentences in group 2 are clustered and two of them (10 and 14) are grouped in a more specific category. However, we do have some problems as some items appear in the wrong place. Sentences 25 and 27 are intruders, dealing with a different topic. Hence, they should not be in this category. The same holds for sentence 9, which is put in the (sub)group of sentences 10 and 14. This is clearly wrong, since it is about the *computer company* and not about "apple", the *fruit*. On the other hand, sentence 11 should be there, i.e. in the same cluster as 10 and 14. Yet it is not. It is placed elsewhere, standing like a loner, which, of course, is a mistake. In topic 3 (cars), all the sentences are clustered correctly, and two sentences (15, 16) are placed in a more specific category. However, the system failed to cluster 19 and 21 together.

Given the above described results, the system has a recall, precision and f-measure of 78.6%, if we consider only the sentences placed in the correct position in the tree. It is interesting to note that some of the sentences placed in the wrong category have a similarity value fairly close to the one of the correct cluster. Actually, most of them have the second highest similarity value. Note also that, at this point, the clusters, i.e. the nodes of the tree, are not labeled. Whether this could be achieved via topic signatures [LIN 00] remains to be shown and is clearly a work for the future.

7.8. Outlook and conclusion

Organizing thoughts and expressions in such a way that discourse flows is a challenge for everyone. The difficulties are of various sorts. We have to acquire knowledge concerning linguistic resources (lexicon, grammar, textual devices: anaphora, etc.) and learn how our mind interprets the use of each one of them or the combination of conceptual entities (propositions). Once we have acquired this, effects can be used as goals, the normal starting point in communication.

A lot of work has been done on language production, yet, most of it from an engineering perspective [REI 00, BAT 16], i.e. fully automated text generation. Unfortunately, next to nothing has been done to assist the writer (our goal), though there is clearly a need for it. Note that the conditions under which an author works are far more multifarious and open-ended than the ones computers have to struggle with [AND 96]. Hence, instead of presenting a fully automatic solution as most computational linguists do, we propose here an interactive one. More precisely, we try to build an authoring tool where the computer helps the writer to organize his/her thoughts. Given a set of messages (user's knowledge state) and possibly a goal, the system structures the data in such a way that, conceptually speaking, the output is coherent, yielding one or several conceptual trees. Since this is quite a complex task, we have started with a very simple set of data. Nevertheless, we have tried to go beyond the obvious (co-references), to reveal some of the hidden and distributed information.

Our next steps will involve dealing with other conceptual relations and with the labeling of topical categories. We would also like to explore the idea of the information associated with the seed words. This should allow us to see on what basis (origin of the information on which) topical clustering is performed. As data can be grouped in many ways (depending on the chosen criteria or view points), and as various orders are likely to yield different effects, it would be nice to show the relationship between ordering and rhetorical effects.

Concerning the method, we could have also considered the following strategy: take a set of well-written texts of a specific type (description), extract its sentences, normalize and scramble them and have the computer try to reorganize them to produce a coherent whole, matching, as well as possible, the initial document (gold standard). Yet, having thought about this strategy too late (see, e.g., [BAR 08, BOL 10, LAP 06, WAN 06]), we used a different approach.

7.9. Bibliography

[ALA 01] ALAMARGOT D., CHANQUOY L., *Through the Models of Writing*, Kluwer, Dordrecht, 2001.

[AND 96] ANDRIESSEN J., DE SMEDT K., ZOCK M., "Discourse planning: empirical research and computer models", in DIJKSTRA T., DE SMEDT K. (eds). *Computational Psycholinguistics: AI and Connectionist Models of Human Language Processing*, Taylor Francis, London, 1996.

[BAR 08] BARZILAY R., LAPATA M., "Modeling local coherence: an entity-based approach", *Computational Linguistics*, vol. 34, no. 1, pp. 1–34, 2008.

[BAT 16] BATEMAN J., ZOCK M., "Natural language generation", in MITKOV R. (ed.), *Handbook of Computational Linguistics*, Oxford University Press, London, 2016.

[BER 87] BEREITER C., SCARDAMALIA M., *The Psychology of Written Composition*, Erlbaum, Hillsdale, 1987.

[BOL 10] BOLLEGALA D., OKAZAKI N., ISHIZUKA M., "A bottom-up approach to sentence ordering for multi-document summarization", *Information Processing and Management*, vol. 46, no. 1, pp. 89–109, 2010.

[BUL 07] BULLINARIA J.A., LEVY J., "Extracting semantic representations from word co-occurrence statistics: a computational study", *Behavior Research Methods*, vol. 39, pp. 510–526. 2007.

[CAR 70] CARBONELL J.R., Mixed-initiative man-computer instructional dialogues, PhD Thesis, Massachusetts Institute of Technology, 1970.

[DAG 99] DAGAN I., LEE L., PEREIRA F., "Similarity-based models of co-occurrence probabilities", *Machine Learning*, vol. 34, no. 1–3 special issue on Natural Language Learning, pp. 43–69, 1999.

[DEB 84] DE BEAUGRANDE R., *Text Production: Towards a Science of Composition*, Ablex, Norwood, 1984.

[DEM 08] DE MARNEFFE M.C., MANNING C.D., Stanford typed dependencies manual, Technical report, Stanford University, 2008.

[FIL 68] FILLMORE C., "The case for case", in BACH E., HARMS R. (eds), *Universals in Linguistic Theory*, Holt, Rinehart and Winston, New York, 1968.

[FIR 57] FIRTH J.R., "A synopsis of linguistic theory 1930–1955", in *Studies in Linguistic Analysis*, Philological Society, Oxford, 1957.

[FLO 80] FLOWER L., HAYES J.R., "The dynamics of composing: making plans and juggling constraints", in GREGG L., STEINBERG E.R. (eds), *Cognitive Processes in Writing*, Erlbaum, Hillsdale, 1980.

[HAR 54] HARRIS Z., "Distributional structure", *Word*, vol. 10, no. 23, pp. 46–162, 1954.

[HOV 91] HOVY E.H., "Approaches to the planning of coherent text", in PARIS C.L., SWARTOUT W.R., MANN W.C. (eds), *Natural Language Generation in Artificial Intelligence and Computational Linguistics*, Kluwer Academic, 1991.

[IYE 09] IYER L.R., DOBOLI S., MINAI A.A. *et al.*, "Neural dynamics of idea generation and the effects of priming", *Neural Networks*, vol. 22, no. 5–6, pp. 674–686, 2009.

[KEL 99] KELLOGG R., *Psychology of Writing*, Oxford University Press, New York, 1999.

[KEM 87] KEMPEN G., HOENKAMP E., "An incremental procedural grammar for sentence formulation", *Cognitive Science*, vol. 11, pp. 201–258. 1987.

[LAP 06] LAPATA M., "Automatic evaluation of information ordering", *Computational Linguistics*, vol. 32, no. 4, 2006.

[LEV 89] LEVELT W., *Speaking: From Intention to Articulation*, MIT Press, Cambridge, 1989.

[LEV 13] LEVY C.M., RANSDELL S., *The Science of Writing. Theories, Methods, Individual Differences and Applications*, Routledge, Abingdon, 2013.

[LIN 00] LIN C.-Y., HOVY E., "The Automated Acquisition of Topic Signatures for Text Summarization", *Proceedings of the COLING Conference*, pp. 495–501, 2000.

[LUN 96] LUND K., BURGESS C., "Producing high-dimensional semantic spaces from lexical co-occurrence", *Behavior Research Methods, Instruments, and Computers*, vol. 28, no. 2, pp. 203–208, 1996.

[MAN 87] MANN W.C., THOMPSON S.A., "Rhetorical structure theory: a theory of text organization", in POLANYI L. (ed.), *The Structure of Discourse*, Ablex, Norwood, 1987.

[MAN 08] MANNING C.D., RAGHAVAN P., SCHÜTZE H., *Introduction to Information Retrieval*, Cambridge University Press, Cambridge, 2008.

[MAR 97] MARCU D., "From local to global coherence: a bottom-up approach to text planning", *Proceedings of the Fourteenth National Conference on Artificial Intelligence*, pp. 629–635, 1997.

[MAR 11] MARK D., "N-grams and word frequency data", Corpus of Historical American English (COHA), 2011.

[MAT 87] MATSUHASHI A., "Revising the plan and altering the text", in MATSUHASHI A. (ed.), *Writing in Real Time*, Ablex, Norwood, 1987.

[MAT 07] MATTHEWS D.E., LIEVEN E., TOMASELLO M., "How toddlers and preschoolers learn to uniquely identify referents for others: a training study", *Child Development*, vol. 78, no. 6, pp. 1744–1759, 2007.

[MCK 85] MCKEOWN K.R., *Text Generation: Using Discourse Strategies and Focus Constraints to Generate Natural Language Text*, Cambridge University Press, Cambridge, 1985.

[MOO 93] MOORE J.D., PARIS C.L., "Planning text for advisory dialogues: capturing intentional and rhetorical information", *Computational Linguistics*, vol. 19, no. 4, 1993.

[OCO 71] O'CONNOR F., "The displaced person", in *The Complete Stories*, Macmillan, London, 1971.

[OLI 01] OLIVE T., LEVY C.M., *Contemporary Tools and Techniques for Studying Writing*, Kluwer, Dordrecht, 2001.

[REI 95] REICHENBERGER K., RONDHUIS K.J., KLEINZ J. *et al.*, "Communicative goal-driven NL generation and data-driven graphics generation: an architectural synthesis for multimedia page generation", *9th International Workshop on Natural Language Generation*, Niagara on the Lake, 1995.

[REI 00] REITER E., DALE R., *Building Natural Language Generation Systems*, Cambridge University Press, Cambridge, 2000.

[RIJ 96a] RIJLAARSDAM G., VAN DEN BERGH H., "The dynamics of composing – an agenda for research into an interactive compensatory model of writing: many questions, some answers", in LEVY C.M., RANSDELL S. (eds), *The Science of Writing: Theories, Methods, Individual Differences and Applications*, Erlbaum, Hillsdale, 1996.

[RIJ 96b] RIJLAARSDAM G., VAN DEN BERGH H., COUZIJN M., *Effective Teaching and Learning of Writing*, Amsterdam University Press, Amsterdam, 1996.

[SAH 08] SAHLGREN M. "The distributional hypothesis", *Rivista di Linguistica*, vol. 20, no. 1, pp. 33–53, 2008.

[SCA 87] SCARDAMALIA M., BEREITER C., "Knowledge telling and knowledge transforming in written composition", in ROSENBERG S. (ed.), *Reading, Writing and Language Learning*, Cambridge University Press, Cambridge, 1987.

[SIM 88] SIMONIN N., "An approach for creating structured text", in ZOCK M., SABAH G. (eds), *Advances in Natural Language Generation: An Interdisciplinary Perspective*, Pinter, London and Ablex, Norwood, 1988.

[TOR 99] TORRANCE M., JEFFERY G., *The Cognitive Demands of Writing*, Amsterdam University Press, Amsterdam, 1999.

[TUR 06] TURNEY P.D., "Similarity of semantic relations", *Computational Linguistics*, vol. 32, no. 3, pp. 379–416, 2006.

[WAN 06] WANG Y.W., Sentence ordering for multi-document summarization in response to multiple queries, PhD Thesis, Simon Fraser University, 2006.

[WID 04] WIDDOWS D., *Geometry of Meaning*, University of Chicago Press, Chicago, 2004.

[WIT 22] WITTGENSTEIN L., *Tractatus Logico-Philosophicus*, Kegan Paul, London, 1922.

[ZOC 15a] ZOCK M., TESFAYE D., "Automatic creation of a semantic network encoding part_of relations", *Journal of Cognitive Science*, vol. 16, no. 4, pp. 431–491, 2015.

[ZOC 15b] ZOCK M., LAPALME G., YOUSFI-MONOD M., "Learn to describe objects the way 'ordinary' people do via a web-based application", *Journal of Cognitive Science*, vol. 16, no. 2, pp. 175–193, 2015.

8

Stylistic Features Based on Sequential Rule Mining for Authorship Attribution

Authorship attribution is the task of identifying the author of a given document. Various style markers have been proposed in the literature to deal with the authorship attribution task. Frequencies of function words and Part-Of-Speech n-grams have been shown to be very reliable and effective for this task. However, despite the fact that they are state of the art, they partly rely on the invalid bag-of-words assumption, which stipulates that text is a set of independent words or segments of words. In this chapter, we present a comparative study using two different types of style markers for authorship attribution. We compare the effectiveness of using sequential rules of function words and Part-Of-Speech tags as style markers that do not rely on the bag-of-words assumption, on the one hand, and their raw frequencies, on the other hand. Our results show that the frequencies of function words and Part-Of-Speech n-grams outperform the sequential rules.

8.1. Introduction and motivation

Authorship attribution is the task of identifying the author of a given document. The authorship attribution problem can typically be formulated as follows: given a set of candidate authors for whom samples of written text are available, the task is to assign a text of unknown authorship to one of these candidate authors [STA 09].

Chapter written by Mohamed Amine BOUKHALED and Jean-Gabriel GANASCIA.

This problem has been addressed mainly as a problem of multi-class discrimination, or as a text categorization task [SEB 02]. Text categorization is a useful way to organize large document collections. Authorship attribution, as a subtask of text categorization, assumes that the categorization scheme is based on the authorial information extracted from the documents. Authorship attribution is a relatively old research field. A first scientific approach to the problem was proposed in the late 19th Century, in the work of Mendenhall in 1887, who studied the authorship of texts attributed to Bacon, Marlowe and Shakespeare. More recently, the problem of authorship attribution gained greater importance due to new applications in forensic analysis and humanities scholarship [STA 09].

To achieve high authorship attribution accuracy, we should use features that are most likely to be independent of the topic of the text. There is an agreement among different researchers that function words are the most reliable indicator of authorship. There are two main reasons for using function words in lieu of other markers. First, because of their high frequency in a written text, function words are very difficult to consciously control, which minimizes the risk of false attribution. The second is that function words, unlike content words, are more independent of the topic or the genre of the text, and hence we should not expect to find great differences of frequencies across different texts written by the same authors on different topics [CHU 07]. The Part-Of-Speech-based markers are also shown to be very effective because they partly share the advantages of function words [STA 09].

Despite the fact that function word-based markers are state-of-the-art, they basically rely on the *bag of words* assumption, which stipulates that text is a set of independent tokens. This approach completely ignores the fact that there is a syntactic structure and latent sequential information in the text. This is partly true for Part-Of-Speech n-grams as well, since they are based on the underlying assumption stipulating that text is a set of independent n-tokens' segments. De Roeck *et al.* [DER 04] have shown that frequent words, including function words, are not distributed homogeneously over a text. This provides evidence of the fact that the bag of words assumption is invalid. In fact, critiques have been made in the field of authorship attribution charging that many works are based on invalid assumptions [RUD 97] and that researchers are focusing on attribution techniques rather than coming up with new style markers that are more precise and based on less strong assumptions.

In an effort to develop more complex yet computationally feasible stylistic features that are more linguistically motivated, Hoover [HOO 03] pointed out that exploiting the sequential information existing in the text could be a promising line of work. He proved that frequent word sequences and collocations can be used with high reliability for stylistic attribution. In another study, Quiniou *et al.* [QUI 12] showed the interest of sequential data mining methods for the stylistic analysis of large texts. They claimed that relevant and understandable patterns that may be characteristic of a specific type of text can be extracted using sequential data mining techniques.

In this line of thought, here we study the problem of authorship attribution in classic French literature. Our aim is to evaluate the effectiveness of style markers extracted using sequential data mining techniques for authorship attribution. In this contribution, we focus on extracting style markers using sequential rule mining. We compare results given by these new style markers with that of the state-of-the-art features like function words frequencies and Part-Of-Speech n-grams, and we assess whether this type of marker is sufficient for accurate identification of authors.

The rest of the chapter is organized as follows. In section 8.2, we give a theoretical overview of the computational authorship attribution process. Then, in section 8.3, we present our working hypothesis and its corresponding stylistic markers. In section 8.4, we make a projection of the sequential data mining problem in our context, and we explain how the sequential rule-based style markers are extracted. The experimental evaluation settings are presented in section 8.5 in which we describe the dataset used in the experiment, and then present the employed classification scheme and algorithm. The results and discussions are presented in section 8.6. Finally, section 8.7 concludes the chapter.

8.2. The authorship attribution process

Authorship attribution and stylometry, which refers to the statistical analysis of literary style, have always been closely related research fields. In fact, authorship analysis relies on the notion of style and on the process of drawing conclusions about authorship information of a document by analyzing and extracting its stylometric characteristics. This assumes that the author of a document has a specific style by which he/she can be completely or partly distinguished from another author. Following this idea, current authorship attribution methods have two key steps (see Figure 8.1):

1) an indexing step based on style markers is performed on the text using some natural language processing techniques, such as Part-Of-Speech tagging, parsing and morphological analysis;

2) an identification step is applied using the indexed markers to determine the most likely authorship.

An optional feature selection step can be employed between these two key steps to determine the most relevant markers. This selection step is done by performing some statistical measures such as mutual information or Chi-square testing [HOU 06].

The identification step involves using methods that fall mainly into two categories: the first category includes methods that are based on statistical analysis, such as principle component analysis [BUR 02] or linear discriminant analysis [STA 01]; the second category includes machine learning techniques, such as simple Markov chain [KHM 01], Bayesian networks, support vector machines (SVMs) [KOP 04] and neural networks [RAM 04]. SVMs, which have been used successfully in text categorization and in other classification tasks, have been shown to be the most effective attribution method [DIE 03]. This is due to the fact that SVMs are less sensitive to irrelevant features in terms of degradation in accuracy, and enable us to handle high-dimensional data instances more efficiently.

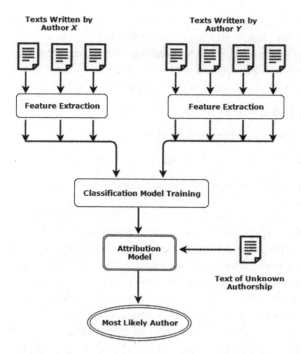

Figure 8.1. *Prototype of the process of authorship attribution process*

8.3. Stylistic features for authorship attribution

Many style markers have been used for the authorship attribution task, from early works based on features such as sentence length and vocabulary richness [YUL 44] to more recent and relevant works based on function words [HOL 01, ZHA 05], punctuation marks [BAA 02], Part-Of-Speech tags [KUK 01], parse trees [GAM 04] and character-based features [KEŠ 03]. As mentioned before, function words are shown to be a very reliable and effective indicator of authorship, and hence they are suitable to handle the task of authorship attribution or some other related tasks such as authorship verification [KOP 09]. In fact, function words have little lexical role to play, but instead they serve mostly a syntactic role by expressing grammatical relationships among words or collections of words within a sentence.

At this point, as an illustration of the idea prompted in the introduction of this chapter, we propose to explore the predictive property of stylistic features based on sequential rule mining. Therefore, as the main experiment in our work, we study the stylistic characterization of 10 classic French authors using different stylistic features ranging from a relatively low linguistic level to a higher and more complex one. We chose to focus on the syntactic aspect of style, so as stylistic features in this experiment, we took:

- frequency of function words;

- sequential rules of function words;

- tri-gram of Part-Of-Speech tags;

- sequential rules of Part-Of-Speech tags.

From the above list, the frequencies of function words are obviously the least complex linguistic features and subsequently the least relevant and interesting to characterize the style of authors. They neither offer explicit stylistic lexical preferences, nor an explicit stylistic syntactic trait. The other stylistic features are linguistically more complex and stylistically more interesting. For instance, the sequential rules of function words can capture the differences between the periodic and loose styles. While, the sequential rules of Part-Of-Speech tags can play an alternative role in the grammatical production rules used in formal grammar [O'NE 01]; except that in this case, these rules will give insights into the syntactic choices of an author rather than describing the grammar in a general way as is done using the production rules.

What we should expect from such a configuration is that the more relevant the feature is to describe the stylistic choices of a given author, the more it is able and suitable to distinguish their own writing from that of a different author. That is to say, for a stylistic characterization based on the classification approach, we would expect the sequential rules of function words to be more effective than the frequencies of function words since they are more stylistically relevant (they are able to tell us more about the writing style of an author, they are easier to interpret at the same time, and they are not based on invalid assumptions). We would expect the sequential rules of Part-Of-Speech tags to be the more effective of the two for the same reason.

Our aim in this experiment is to test the validity of this hypothesis by evaluating the effectiveness of stylistic features presented above in the context of authorship attribution. Well, it turns out that this hypothesis is not true, at least for the corpus that we have considered in this experiment. This can be considered as a clear argument, suggesting that less complex features, acting on a relatively low linguistic level and based on invalid assumptions, are more suitable for authorship studies from a classification point of view. Our experiment explores these issues.

8.4. Sequential data mining for stylistic analysis

Sequential data mining is a data mining subdomain introduced by Agrawal *et al.* [AGR 93], which is concerned with finding interesting characteristics and patterns in sequential databases. Sequential rule mining is one of the most important sequential data mining techniques used to extract rules describing a set of sequences. In what follows, for the sake of clarity, we will limit our definitions and annotations to those necessary to understand our experiment.

Considering a set of literals called items, denoted by $I = \{i_1, ..., i_n\}$, an itemset is a set of items $X \subseteq I$. A sequence S (single-item sequence) is an ordered list of items, denoted by $S = \langle i_1 \ ... i_n \rangle$ where i_1 to i_n are items.

Sequence ID	Sequence
1	< a, b, d, e >
2	< b, c, e >
3	< a, b, d, e >

Table 8.1. *Sequence database SDB*

A sequence database SDB is a set of tuples (id, S), where id is the sequence identifier and S a sequence. Interesting characteristics can be extracted from such databases using sequential rules and pattern mining. A sequential rule $R : X \Rightarrow Y$ is defined as a relationship between two itemsets X and Y, such that $X \cap Y = \varnothing$. This rule can be interpreted as follows: if the itemset X occurs in a sequence, the itemset Y will occur afterward in the same sequence. Several algorithms have been developed to efficiently extract this type of rule, such as Fournier-Viger and Tseng [FOU 11]. For example, if we run this algorithm on the SDB containing the three sequences presented in Table 8.1, we will get as a result sequential rules, such as "$a \Rightarrow d, e$" with support equal to 2, which means that this rule is respected by two sequences in the *SDB* (i.e. there exist two sequences of the *SDB* where we find the item a, and we also find d and e afterward in the same sequence).

In our study, the text is first segmented into a set of sentences, and then each sentence is mapped into two sequences: one for function words appearing in order in that sentence, and another sequence for the Part-Of-Speech tags resulting from its syntactic analysis. For example, the sentence "J'aime ma maison où j'ai grandi." will be mapped to < je,ma,où,je > as a sequence of French function words, and will be mapped to < PRO:PER, VER:pres, DET:POS, NOM, PRO:REL, PRO:PER, VER:pres, VER:pper, SENT> as a sequence of Part-Of-Speech tags. "je $\Rightarrow$ où", "ma $\Rightarrow$ où,je" or "DET:POS, NOM $\Rightarrow$ SENT" are examples of sequential rules respected by these sequences. The whole text will produce two sequential databases, one for the function words and another for the Part-Of-Speech tags. The rules extracted in our study represent the cadence authors follow when using function words in their writings for instance. This gives us more explanatory properties about the syntactic writing style of a given author than frequencies of function words or Part-Of-Speech n-grams could offer.

8.5. Experimental setup

In this section, we present the experimental setup of our approach. We first describe the dataset used in the experiment, and then present the classification scheme and algorithm employed for this experiment. The results and discussion are presented in the next section.

8.5.1. *Dataset*

To test the effectiveness of sequential rules over Part-Of-Speech tags and function words for authorship attribution, we used texts written by Balzac, Dumas, France, Gautier, Hugo, Maupassant, Proust, Sand, Sue and Zola. This choice was motivated by our special interest in studying the classic French literature of the 19th Century, and the availability of electronic texts from these authors on the Gutenberg project website[1] and in the Gallica

---

1 http://www.gutenberg.org/

electronic library[2]. Our choice of authors was also affected by the fact that we want to cover the most important writing styles and trends from this period. For each of the 10 authors mentioned above, we collected four novels, so that the total number of novels is 40. The next step was to divide these novels into smaller pieces of texts in order to have enough data instances to train the attribution algorithm. Researchers working on authorship attribution on literature data have been using different dividing strategies. For example, Hoover [HOO 03] decided to take just the first 10,000 words of each novel as a single text, while Argamon and Levitan [ARG 05] treated each chapter of each book as a separate text. In our experiment, we chose to slice novels by the size of the smallest one in the collection in terms of the number of sentences. This choice respects the condition proposed by Eder [EDE 13] that specifies the smallest reasonable text size to achieve good attribution; more information about the dataset used in the experiment is presented in Table 8.2.

Author Name	# of words	# of texts
Balzac, Honoré de	548778	20
Dumas, Alexandre	320263	26
France, Anatole	218499	21
Gautier, Théophile	325849	19
Hugo, Victor	584502	39
Maupassant, Guy de	186598	20
Proust, Marcel	700748	38
Sand, George	560365	51
Sue, Eugène	1076843	60
Zola, Émile	581613	67

Table 8.2. *Statistics for the dataset used in our experiment*

8.5.2. *Classification scheme*

In the current approach, each text was segmented into a set of sentences (sequences) based on splitting done using the punctuation marks of the set $\{`.', `!', `?', `:', `...'\}$, then the corpus was Part-Of-Speech tagged and function words were extracted. The algorithm described in Fournier-Viger and Tseng [FOU 11] was then used to extract sequential and association

---

2 http://gallica.bnf.fr/

rules over the function words and the Part-Of-Speech tag sequences from each text. These rules will help us gather not only sequential information from the data, but also structural information, due to the fact that a text characterized by long sentences will result in more frequencies of the rules.

Each text is then represented as a vector R_K of frequencies of occurrence of rules, such that $R_K = \{r_1, r_2, \ldots, r_K\}$ is the ordered set, by decreasing normalized frequency of occurrence of the top-K rules in terms of support in the training set. Each text is also represented by a vector of normalized frequencies of occurrence of function words and Part-Of-Speech tag 3-grams. The normalization of the vector of frequency representing a given text was done by the size of the text. Our aim is first to compare the classification performance of the top-K function word sequential rules (SR) with the function words frequencies. Second, to compare the classification performance of the top-K sequential rules of Part-Of-Speech tag with the 3-gram frequencies.

Given the classification scheme described above, we used SVMs classifier to derive a discriminative linear model from our data. To get a reasonable estimation of the expected generalization performance, we used 5-fold cross-validation. The dataset was split into five equal subsets; the classification was done five times by taking four subsets for training each time and leaving out the last one for testing. The overall classification performance is taken as the average performance over these five runs. In order to evaluate the attribution performance, we used the common measures used to evaluate supervised classification performance: we have calculated precision (P), recall (R) and F-measure F_β, where TP stands for true positive, TN for true negative, FP for false positive and FN for false negative:

$$P = \frac{TP}{TP + FP} \tag{8.1}$$

$$R = \frac{TP}{TP + FN} \tag{8.2}$$

$$F_\beta = \frac{\left(1 + \beta^2\right) RP}{\left(\beta^2 R\right) + P} \tag{8.3}$$

We consider that precision and recall have the same weight, and hence we set β equal to 1.

8.6. Results and discussion

The results of measuring the attribution performance for the different feature sets presented in our experiment setup are summarized in Table 8.3 for features derived from function words, and in Table 8.4 for those derived from Part-Of-Speech tags. These results show, in general, a better performance when using function words and Part-Of-Speech tag 3-gram frequencies, which achieved a nearly perfect attribution, over features based on sequential rules for our corpus.

Our study here shows that the SVMs classifier combined with features extracted using sequential data mining techniques can achieve a high attribution performance (e.g. F1 = 0.939 for Top 300 FW-SR). Until a certain limit, adding more rules increases the attribution performance (e.g. F1 = 0.733 for Top 100 POS-SR compared with F1 = 0.880 for Top 800 POS-SR).

Contrary to our hypothesis, function word frequency features, which fall under the bag-of-word assumption, known to be blind to sequential information, outperform features extracted using the sequential rule mining technique. The same thing can be said for the Part-Of-Speech tag 3-grams.

Feature set	P	R	F_1
Top 100 FW-SR	0.901	0.886	0.893
Top 200 FW-SR	0.942	0.933	0.937
Top 300 FW-SR	0.940	0.939	0.939
FW frequencies	**0.990**	**0.988**	**0.988**

Table 8.3. *Five-fold cross-validation for our dataset. SR refers to sequential rules and FW refers to function words*

By taking a closer look at the sequential rules extracted from the Part-Of-Speech tag sequences, we found that these rules, especially the most frequent ones, are more likely to be language-grammar dependent (e.g. ADJ NC,PONCT with sup = 63,569 and DET,NC,P ⇒ ADJ with sup = 63,370).

To reduce this effect, we added a $TF-IDF$-like heuristic that measures the overall discriminative power of each sequential rule. The $TF-IDF$-like weight of a sequential rule R_i present in a text t is calculated as follows:

$$TF-IDF_t(R_i) = \left(1 + supp_t(R_i)\right) * \log\left(\frac{N}{N_t}\right)$$ [8.4]

where $supp_t(R_i)$ is the support of the rule R_i in the text t, N is the total support of all rules in the corpus and N_t is the total support of all rules in the text t.

Results given by this $TF-IDF$ weighting in Table 8.5 are better than the original ones, but they still cannot reach the performance given by the state-of-the-art style markers. This suggests that in future studies, we should add an adequate feature selection method that will filter the rules to capture the most relevant ones.

By analyzing the individual attribution performance for each author separately, we notice a significant variance between the attribution performance of one author and that of another (e.g. $F_1 = 1$ for Proust compared with $F_1 = 0.673$ for Dumas); some individual results are presented in Table 8.5. This particularity is due to the fact that some authors have more characterizing style than others in the works used for the experiment. This property can be clearly visualized by carrying out the principal components analysis (see Figure 8.2) on the 40 books used in the dataset.

Feature set	P	R	F_1
Top 200 POS-SR	0.72	0.70	0.71
Top 300 POS-SR	0.83	0.81	0.82
Top 400 POS-SR	0.84	0.83	0.83
Top 500 POS-SR	0.85	0.84	0.84
Top 600 POS-SR	0.87	0.85	0.86
Top 700 POS-SR	0.88	0.86	0.87
Top 800 POS-SR	0.88	0.87	0.88
POS 3-gram frequencies	**0.99**	**0.99**	**0.99**

Table 8.4. *Five-fold cross-validation results for our dataset. SR refers to sequential rules and POS refers to Part-Of-Speech*

Feature set	P^*	R^*	F^*_1
Top 200 POS-SR	0.82	0.79	0.81
Top 300 POS-SR	0.86	0.84	0.85
Top 400 POS-SR	0.87	0.86	0.87
Top 500 POS-SR	0.89	0.88	0.88
Top 600 POS-SR	0.89	0.88	0.88
Top 700 POS-SR	0.91	0.90	0.90
Top 800 POS-SR	**0.92**	**0.91**	**0.91**

Table 8.5. *Five-fold cross-validation results given by considering the TF-IDF-like weighting for our dataset. SR refers to sequential rules and POS refers to Part-Of-Speech*

Even if these results are in line with previous works that claimed that bag-of-words-based features are more relevant than sequence-based features for stylistic attribution [ARG 05], they show that style markers extracted using sequential rule mining techniques can be valuable for authorship attribution. We believe that our results open the door to a promising line of research by integrating and using sequential data mining techniques to extract more linguistically motivated style markers for computational, stylistic and authorship attribution.

Author Name	P	R	F_1
Balzac	0.88	0.75	0.80
Dumas	0.65	0.69	0.67
France	0.92	0.96	0.93
Gautier	0.95	0.85	0.89
Hugo	0.88	0.95	0.91
Maupassant	1.00	0.85	0.91
Proust	1.00	1.00	1.00
Sand	0.92	0.90	0.91
Sue	0.86	0.86	0.86
Zola	0.98	1.00	0.99

Table 8.6. *Individual 5-fold cross-validation results for each author evaluated for the Top 700 Part-Of-Speech tag sequential rules*

Actually, despite the fact that function words are not very relevant features to describe the stylistic characterization, they are a reliable indicator of authorship. Owing to their high frequency in a written text, function words are very difficult to consciously and voluntarily control, which makes them a more inherent trait and consequently minimizes the risk of false attribution. Moreover, unlike content words, they are more independent of the topic or the genre of the text, and therefore we should not expect to find great differences of frequencies across different texts written by the same authors on different topics [CHU 07]. Yet, they basically rely on the bag-of-words assumption, which stipulates that text is a set of independent words.

As we have seen, it turns out that the hypothesis, stated as a basis for the experiment, is not true, at least for the corpus that we have considered in this experiment. This can be considered as a clear argument, suggesting that complex features such as sequential rules are not suitable for authorship attribution studies. In fact, there is a difference between the characterizing ability of a stylistic feature, on the one hand, and its discriminant power, on the other. The most relevant and suitable stylistic features to perform a discriminant task such as stylistic classification are the ones that operate on the low linguistic levels as function words do. These are subsequently more difficult to linguistically interpret and understand and do not necessarily enhance the knowledge concerning the style of the text from which they were extracted.

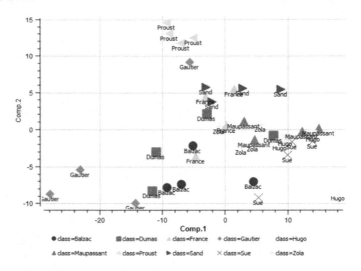

Figure 8.2. *Principal components analysis of the 40 books (four books per author) in the dataset, Top 200 SR analyzed. For a color version of this figure, see www.iste.co.uk/sharp/cognitive.zip*

8.7. Conclusion

In this chapter, we have presented a first study on using style markers extracted using sequential data mining techniques for authorship attribution. We have considered extracting linguistically motivated markers using a sequential rule mining technique based on function word and Part-Of-Speech tags. To evaluate the effectiveness of these markers, we conducted experiments on a classic French corpus. Our preliminary results show that sequential rules can achieve a high attribution performance that can reach an F_1 score of 93%. Yet, they still do not outperform low-level features, such as frequencies of function words.

Based on the current study, we have identified several future research directions. First, we will explore the effectiveness of using probabilistic heuristics to find a minimal feature set that still allows good attribution performance, which would be very helpful for stylistic and literary analysis. Second, this study will be expanded to include sequential patterns (n-gram with gaps) as style markers. Third, we intend to experiment with this new type of style markers for other languages and text sizes using standard corpora employed in the wider field.

8.8. Bibliography

[AGR 93] AGRAWAL R., IMIELIŃSKI T., SWAMI A., "Mining association rules between sets of items in large databases", *ACM SIGMOD*, vol. 22, no. 2, pp. 207–216, 1993.

[ARG 05] ARGAMON S., LEVITAN S., "Measuring the usefulness of function words for authorship attribution", *Proceedings of the Joint Conference of the Association for Computers and the Humanities and the Association for Literary and Linguistic Computing*, pp. 3–7, 2005.

[BAA 02] BAAYEN H., VAN HALTEREN H., NEIJT A. *et al.*, "An experiment in authorship attribution", *Proceedings of 6th International Conference on the Statistical Analysis of Textual Data*, pp. 29–37, 2002.

[BUR 02] BURROWS J., "'Delta': a measure of stylistic difference and a guide to likely authorship", *Literary and Linguistic Computing*, vol. 17, no. 3, pp. 267–287, 2002.

[CHU 07] CHUNG C., PENNEBAKER J., "The psychological functions of function words", *Social Communication*, pp. 343–359, 2007.

[DER 04] DE ROECK A., SARKAR A., GARTHWAITE P., "Defeating the homogeneity assumption", *Proceedings of 7th International Conference on the Statistical Analysis of Textual Data*, pp. 282–294, 2004.

[DIE 03] DIEDERICH J., KINDERMANN J., LEOPOLD E. *et al.*, "Authorship attribution with support vector machines", *Applied Intelligence*, vol. 19, pp. 109–123, 2003.

[EDE 13] EDER M., "Does size matter? Authorship attribution, small samples, big problem", *Digital Scholarship in the Humanities*, 2013.

[FOU 11] FOURNIER-VIGER P., TSENG V., "Mining top-k sequential rules", *Advanced Data Mining and Applications*, Springer, pp. 180–194, 2011.

[GAM 04] GAMON M., "Linguistic correlates of style: authorship classification with deep linguistic analysis features", *Proceedings of the 20th International Conference on Computational Linguistics*, pp. 611–617, 2004.

[HOL 01] HOLMES D., ROBERTSON M., PAEZ R., "Stephen Crane and the New-York Tribune: a case study in traditional and non-traditional authorship attribution", *Computers and the Humanities*, vol. 35, no. 3, pp. 315–331, 2001.

[HOO 03] HOOVER D., "Frequent collocations and authorial style", *Literary and Linguistic Computing*, vol. 18, no. 3, pp. 261–286, 2003.

[HOU 06] HOUVARDAS J., STAMATATOS E., "N-gram feature selection for authorship identification", *Artificial Intelligence: Methodology, Systems, and Applications*, Springer, pp. 77–86, 2006.

[KEŠ 03] KEŠELJ V., PENG F., CERCONE N. *et al.*, "N-gram-based author profiles for authorship attribution", *Proceedings of the Conference Pacific Association for Computational Linguistics*, PACLING, vol. 3, pp. 255–264, 2003.

[KHM 01] KHMELEV D., TWEEDIE F., "Using Markov chains for identification of writer", *Literary and Linguistic Computing*, vol. 16, no. 3, pp. 299–307, 2001.

[KOP 04] KOPPEL M., SCHLER J., "Authorship verification as a one-class classification problem", *Proceedings of the Twenty-First International Conference on Machine Learning*, pp. 62–67, 2004.

[KOP 09] KOPPEL M., SCHLER J., ARGAMON S., "Computational methods in authorship attribution", *Journal of the American Society for Information Science and Technology*, vol. 60, no. 1, pp. 9–26, 2009.

[KUK 01] KUKUSHKINA O., POLIKARPOV A., KHMELEV D., "Using literal and grammatical statistics for authorship attribution", *Problems of Information Transmission*, vol. 37, no. 2, pp. 172–184, 2001.

[O'NE 01] O'NEILL M., RYAN C., "Grammatical evolution", *IEEE Transactions on Evolutionary Computation*, vol. 10, no. 6, pp. 349–358, 2001.

[QUI 12] QUINIOU S., CELLIER P., CHARNOIS T. *et al.*, "What about sequential data mining techniques to identify linguistic patterns for stylistics?", *Computational Linguistics and Intelligent Text Processing*, Springer, pp. 166–177, 2012.

[RAM 04] RAMYAA C.H., RASHEED K., "Using machine learning techniques for stylometry", *Proceedings of International Conference on Machine Learning*, 2004.

[RUD 97] RUDMAN J., "The state of authorship attribution studies: some problems and solutions", *Computers and the Humanities*, vol. 34, no. 4, pp. 351–365, 1997.

[SEB 02] SEBASTIANI F., "Machine learning in automated text categorization", *ACM Computing Surveys*, vol. 34, no. 1, pp. 1–47, 2002.

[STA 01] STAMATATOS E., FAKOTAKIS N., KOKKINAKIS G., "Computer-based authorship attribution without lexical measures", *Computers and the Humanities*, vol. 35, pp. 193–214, 2001.

[STA 09] STAMATATOS E., "A survey of modern authorship attribution methods", *Journal of the American Society for Information Science and Technology*, vol. 60, no. 3, pp. 538–556, 2009.

[YUL 44] YULE G., *The Statistical Study of Literary Vocabulary*, CUP Archive, 1944.

[ZHA 05] ZHAO Y., ZOBEL J., "Effective and scalable authorship attribution using function words", *Information Retrieval Technology*, Springer, pp. 174–189, 2005.

A Parallel, Cognition-oriented
Fundamental Frequency
Estimation Algorithm

9.1. Introduction

The fundamental frequency F0 plays an important role in human speech perception and is used in all fields of speech research. For instance, humans identify emotional states based on a few features, one of which is F0 [ROD 11]. For speech synthesis, accurate estimates of F0 are a prerequisite for prosody control in concatenative speech synthesis [EWE 10].

Fundamental frequency detection has been an active field of research for more than 40 years. Early methods used the autocorrelation function and inverse filtering techniques [MAR 72, RAB 76]. In most of these approaches, threshold values are used to decide whether a frame is assumed to be voiced or unvoiced. More advanced methods incorporate a dynamic programming stage to calculate the F0 contour based on frame-level F0 estimates gained from either a conditioned linear prediction residual or a normalized cross correlation function [SEC 83, TAL 95]. The normalized cross correlation-based RAPT algorithm is also known as getf0. Praat's well-known pitch detection algorithm calculates cross-correlation or autocorrelation functions and considers local maxima as F0 hypotheses

---

Chapter written by Ulrike GLAVITSCH.

[BOE 01]. The fundamental frequency estimator for speech and music YIN with no upper limit on the frequency search range uses the autocorrelation function and a number of modifications to prevent errors [DEC 02]. In the last decade, techniques like pitch-scaled harmonic filtering (PSHF), non-negative matrix factorization (NMF) as well as time-domain probabilistic approaches have been proposed for F0 estimation [ACH 05, ROA 07, SHA 05, PEH 11]. In SAFE, F0 estimates are inferred from prominent signal-to-noise ratio (SNR) peaks in the speech spectra [CHU 12]. Pitch and probability-of-voicing estimates gained from a highly modified version of the getf0 (RAPT) algorithm are used in an automatic speech recognition system for tonal languages [GHA 14]. These recent methods achieve low error rates and high accuracies but at a high computational cost – either at run-time or during model training. These calculative approaches generally disregard the principles of human cognition and the question is whether F0 estimation can be performed equally well or better by considering them.

In this chapter, we propose an F0 estimation algorithm based on the elementary appearance and inherent structure of the human speech signal. A period, i.e. the inverse of F0, is primarily defined as the time distance between two maximum and two minimum peaks, and we use the same term to refer to the speech section between two such peaks. Human speech is a sequence of alternating speech and pause segments. Speech segments are word flows uttered in one breath of air. The speech segments are usually much longer than the pause segments. In speech segments, we distinguish voiced and unvoiced parts. The speech signal is periodic in voiced regions, whereas it is aperiodic in unvoiced regions. The voiced regions can be further divided into stable and unstable intervals [GLA 15]. Stable intervals show a quasi-constant energy or a quasi-flat envelope, whereas unstable intervals exhibit significant energy rises or decays. On stable intervals, the F0 periods are mostly regular, i.e. the sequence of maximum or minimum peaks is more or less equidistant, whereas the F0 periods in unstable regions are often shortened, elongated, doubled, or may show little similarity with their neighboring periods. Speech signals are highly variable and such special cases occur relatively often. Thus, it makes sense to compute F0 estimates in stable intervals first and use this knowledge to find F0 of unstable intervals in a second step. The F0 estimation method for stable intervals is straightforward as regular F0 periods are expected. The F0 estimation approach for unstable intervals computes variants of possible F0 continuation sequences and evaluates them for highest plausibility. The variants reflect the regular and all the irregular period cases and are calculated using a peak look-ahead strategy. We denote this F0 estimation method for unstable intervals as F0 propagation, since it computes and verifies F0 estimates by considering previously computed ones.

It turns out that the whole F0 estimation can be performed in parallel on the different speech segments of a recording. The speech segments can be considered as separable units of speech that can be treated as computationally independent entities.

We consider the proposed algorithm as cognition oriented inasmuch as it incorporates several principles of human cognition. First, human hearing is also a two-stage process. The inner ear performs a spectral analysis of a speech section, i.e. different frequencies excite different locations along the basilar membrane and as a result different neurons with characteristic frequencies [MOO 08]. This spectral analysis delivers the fundamental frequency and the harmonics. The brain, however, then checks the information delivered by the neurons, interpolating and correcting it where necessary. Our proposed F0 estimation algorithm performs in a similar way, in that the F0 propagation step proceeds from regions with reliable F0 estimates to those where F0 is not clearly known yet. We observed that F0 is very reliably estimated on high-energy stable intervals, which typically represent vowels. Thus, we always compute F0 for unstable intervals by propagation from high-energy stable intervals to lower energy regions. Second, we have adopted the hypothesis-testing principle of human thinking for generating variants of possible F0 sequences and testing them for the detection of F0 in unstable intervals [KAH 11]. Next, human cognition uses context to decide a situation. For instance, in speech perception humans bear the left and right context of a word in mind if its meaning is ambiguous. In an analogous way, our algorithm looks two or three peaks ahead to find the next valid maximum or minimum peak for a given F0 hypothesis. Special cases in unstable intervals can very rarely be disambiguated by just looking a single peak ahead. Finally, performing the tasks of the F0 estimation algorithm in parallel on different speech segments is also adopted from human cognition. The human brain is able to process a huge number of tasks in parallel.

The resulting algorithm is very efficient, thoroughly extensible, easy to understand and has been evaluated on a clean speech database. Recognition rates are better than those of a reference method that uses cross-correlation functions and dynamic programming. In addition, our algorithm structures the speech signal in spoken and pause segments, voiced and unvoiced regions, and stable and unstable intervals. This structure may be useful for further speech processing, such as automatic text-to-speech alignment, automatic speech or speaker recognition.

9.2. Segmentation of the speech signal

As mentioned in the introduction, a speech signal consists of speech units separated by pauses. The speech units contain voiced and unvoiced regions and on the voiced parts, we distinguish stable and unstable intervals. The algorithms and criteria to detect these different structures are described in the following sections.

9.2.1. Speech and pause segments

We use the algorithm to determine the endpoints of isolated utterances by Rabiner and extend it by a heuristics to find the pauses between the spoken segments in a speech signal [RAB 75]. We refer to this combined algorithm as a pause-finding algorithm. Rabiner's algorithm decides whether a signal frame, i.e. a small 10 ms long section of the signal, is characterized as speech or pause based on its energy and the silence energy. The silence energy is the mean energy of an interval that contains silence or signal noise. The silence or noise in our algorithm is expected at the beginning of the speech signal. Users may configure the length over which the silence energy is computed; the default value is 100 ms. First, the pause-finding algorithm calculates an initial segment list, where each segment is characterized by its start and end sample positions and the segment type – either SPEECH or PAUSE. Second, the algorithm merges pause segments that are too short with their neighboring speech segments. In a similar way, it merges speech segments that are too short with their neighboring pause segments. Some of the segments in the initial segment list are too short to form a true speech or pause segment. For instance, a glottal stop before a plosive or a low-energy speech segment is often identified as a pause segment. The minimum lengths of both pause and speech segments are configurable. Finally, the algorithm extends the speech segments by a certain small length. This is necessary since the ends of speech segments may be low-energy phonemes. These phonemes are automatically included by extending the speech segments by a configurable length. The pause-finding algorithm consists of six steps that we present in the following:

1) *Energies, peak energy and silence energy*: the energies E(k) are computed at discrete points k every 10 ms each over a window of 10 ms in the speech signal, k = 0, ..., n − 1. The peak energy Emax is the maximum energy of all energies E(k). Emin is the mean energy of the initial silence that is supposed to occur at the beginning of the speech signal.

2) *Threshold ITL for speech/pause decision*: the threshold ITL is computed as in [RAB 75]:

$$I1 = 0.03 * (Emax - Emin) + Emin \qquad [9.1]$$

$$I2 = 4 * Emin \qquad [9.2]$$

$$ITL = min(I1, I2) \qquad [9.3]$$

3) *Initial segment list*: each frame is classified as either speech or pause frame, comparing its energy with ITL. It is a speech frame if its energy is larger than ITL, a pause frame otherwise. Consecutive speech frames form a speech segment, and consecutive pause frames form a pause segment of the initial segment list.

4) *Merging of too short segments of type PAUSE*: pause segments shorter than the configurable minimum pause length are merged with their neighboring speech segments. The default minimum pause length is 200 ms.

5) *Merging of too short segments of type SPEECH*: speech segments shorter than the configurable minimum speech length are merged with their neighboring pause segments. The default minimum speech length is 150 ms.

6) *Extension of segments of type SPEECH*: all segments of type SPEECH are extended by the length given by the configurable maximum speech segment extension (default value 50 ms). At the same time, the pause segments to the left and right of each speech segments are reduced by that amount.

Figure 9.1 shows the result of the pause-finding algorithm for a speech recording, where a female speaker reads the beginning of the story "The north wind and the sun" of the Keele pitch database [PLA 95].

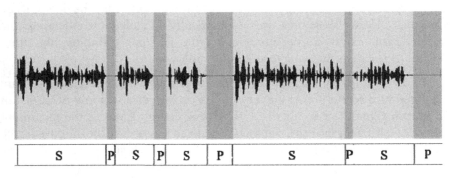

Figure 9.1. *Segmentation of a speech signal into speech (S) and pause (P) segments. For a color version of this figure, see www.iste.co.uk/sharp/cognitive.zip*

9.2.2. *Voiced and unvoiced regions*

Voiced and unvoiced regions are detected in the speech segments only. The pause segments are unvoiced by default. To detect voiced and unvoiced regions, the speech segment is again subdivided into a sequence of frames. However, the frames are longer than that for the pause-finding algorithms, i.e. 20 ms, and they overlap by half of the frame length, namely 10 ms. We define a frame to be voiced if its mean energy exceeds a certain threshold, the absolute height of the frame's maximum or minimum peak is above a given level and the number of counted zero crossings in the frame is lower than a certain number. A voiced region is a sequence of voiced frames and similarly, an unvoiced region contains only unvoiced frames.

A zero crossing is the location in the speech signal where there is change from a positive sample value to a negative value or vice versa. Voiced regions such as vowels, nasals, etc. exhibit a low number of zero crossings, whereas unvoiced regions, e.g. fricatives, usually have a rather high number of zero crossings.

For computation of the mean energy of a frame, we use a more elaborate method than the standard approach, i.e. computing the sum of the squares of the frame's samples and dividing it by the frame length. This standard approach is not precise if the F0 of a frame is not an integer multiple of the frame length [GLA 15]. It may falsify the voiced/unvoiced decision and the stable/unstable classification of a frame in a later step (see section 9.2.3) that is based on the mean energy, too. However, as the period of a frame – the inverse of F0 – is not known at this stage of processing, we compute the mean energy on a scale of window lengths, each of which corresponds to a different period length. An optimization step then finds the best window length for each frame. This procedure is similar to pitch-scaled harmonic filtering (PSHF) [ROA 07], where an optimal window length is calculated for finding harmonic and non-harmonic spectra. The window lengths are selected such that periods of F0 between 50 and 500 Hz roughly fit in one of the selected windows a small number of times at least. The selected window lengths correspond to fundamental frequencies of 50, 55, 60,..., 95 Hz. Each window length is centered around a frame's center position. The optimal window length is the one where the mean energies of a small number of frames around the frame's middle position show the least variation [GLA 15].

9.2.3. *Stable and unstable intervals*

We further segment voiced regions into stable and unstable intervals. As mentioned in the introduction, stable intervals have a quasi-constant energy, whereas in unstable intervals, the energy rises or falls significantly. Given the mean energy of a frame as computed in section 9.2.2, a frame is defined as stable in the following way: its mean energy must not deviate by more than 50% from the mean energy of the previous frame and also not by more than 50% from the mean energy of the next frame. By setting the threshold for the relative mean energy difference to 50%, we allow some tolerance for the energy differences between frames of a stable interval. This is justified since speech signals show high variations.

Figure 9.2 shows a voiced region of a speech segment with three stable intervals S_1, S_2 and S_3. The figure also depicts the series of overlapping speech frames for processing.

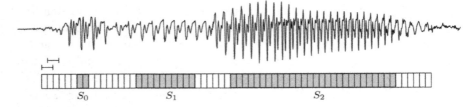

Figure 9.2. *Overlapping frames of a voiced region of a speech segment that contains the words "the north" uttered by a male speaker. Three stable intervals S_1, S_2 and S_3 of lengths 1, 9 and 27 are identified.*

9.3. F0 estimation for stable intervals

The F0 estimation method for stable intervals finds a quadruple of signal peaks $P = (p_L, p_0, p_1, p_R)$ of either maximum or minimum peaks p_i, $i = L$, 0, 1, R, such that the center position of the frame is between p_0 and p_1. A peak is defined as either a local minimum or a local maximum in the sequence of signal samples. For each peak, p_k, $k = 0, ..., n - 1$, in a speech segment, we maintain a triple of values (x_k, y_k, c_k), where x_k and y_k are the peak coordinates and c_k is the peak classification – either a minimum or a maximum. The F0 estimate is the inverse of the mean of the period lengths found in P, i.e. the mean of the distances between peaks p_L and p_0, p_0 and p_1 as well as p_1 and p_R. The tuple P is selected among a series of possible candidate tuples according to a similarity score. Furthermore, it is checked whether the peak tuple is not a multiple of the supposedly true F0 period, otherwise a different peak tuple is selected. In the following, we describe the algorithm to find such a peak tuple P for each stable frame.

We start by finding the peak in the frame that has the highest absolute value. We then look for candidate peaks that have a similar absolute height and whose distance from the highest peak is within the permissible range of period lengths. The search for candidate peaks is performed in the direction of the center position of the frame. Given a peak pair p_0 and p_1 – one of them with the highest absolute value and a candidate peak – the algorithm looks for peaks to the left and the right to complete the quadruple. We select those peaks with the highest absolute values in about the same distance to the left and to the right of p_0 and p_1 as the distance between the two peaks. The peak quadruple may reduce to a triple peak sequence if such a peak at one side of the middle peak pair cannot be found. Each such candidate peak quadruple or peak triple is scored and the tuple with the highest score is selected as the tentatively best candidate.

The proposed score measures the equality of peak distances and absolute peak heights of a peak tuple. The score s for peak tuple $P = (p_L, p_0, p_1, p_R)$ is the product of partial scores s_x and s_y. The value s_x measures the equality of the peak intervals, whereas s_y is a measure for the similarity of the absolute peak heights. The partial score s_x is defined as $1 - a$, where a is the root of the mean squared relative differences between the peak distances at the edges and the distance between the middle peaks. The partial score s_y is given as $1 - b$, where b is the root of the mean squared difference of the absolute peak heights from the maximum absolute peak height in the given peak tuple. The equations below show how the score s is computed for a peak quadruple in detail. The formulas are easily adapted for tuples with only three peaks:

$$s = s_x s_y \tag{9.4}$$

The partial score s_x is defined as follows:

$$s_x = 1 - a \tag{9.5}$$

$$a = \sqrt{(b_0^2 + b_1^2)/2} \tag{9.6}$$

$$b_0 = \frac{d_1 - d_0}{d_1}, b_1 = \frac{d_1 - d_2}{d_1} \tag{9.7}$$

$$d_0 = x_0 - x_L, d_1 = x_1 - x_0, d_2 = x_R - x_1 \tag{9.8}$$

The value x_i, i = L, 0, 1, R, refers to the x-coordinate of peak p_i as mentioned above.

The partial score s_y is given by:

$$s_y = 1 - b \tag{9.9}$$

$$b = \sqrt{\frac{1}{4}(g_L^2 + g_0^2 + g_1^2 + g_R^2)}$$ [9.10]

$$g_i = (|y_i| - y_{max}) / y_{max}, i = L, 0, 1, R$$ [9.11]

$$y_{max} = \max(|y_i|, i = L, 0, 1, R)$$ [9.12]

The value y_i denotes the peak height of peak p_i, $i = L, 0, 1, R$. The score s delivers exactly 1 if the peak heights and peak intervals are equal and less than 1 if they differ.

The peak tuple with the highest score may be a multiple of the true period. Thus, we check for the existence of equidistant partial peaks within the peak pair p_0 and p_1. Such partial peaks must have about the same absolute height as the original candidate peaks p_0 and p_1. If such partial peaks on both sides of the x-axis are found, we look for a candidate peak tuple with the partial peak distance and install it as the current best candidate.

Next, we find the peak tuple in the center of the frame that has the same period length as the best candidate tuple. This is achieved by looking for peaks to either the left or the right side of the best candidate in the distance of the period length until a peak tuple is found, where the frame's center is between the two middle peaks p_0 and p_1.

Finally, we detect sequences of roughly equal F0 estimates in a stable interval. These sequences are referred to as equal sections. The F0 estimates of the frames in an equal section must not deviate from the mean F0 in the equal section by more than a given percentage that is currently set to 10%. The longest such equal section with a minimum length of 3 is set as *the* equal section of the stable interval. The remaining equal sections are maintained in a list and may be used during F0 propagation (see section 9.4).

9.4. F0 propagation

The F0 propagation is the second major stage of the proposed F0 estimation algorithm. Its purpose is to calculate and check F0 estimates in regions where no reliable F0 estimates exist. This mainly affects unstable intervals and also portions of stable intervals where, for example, the F0 estimates do not belong to an equal section. The main idea is that the F0 propagation starts at the stable interval with the highest energy from where it proceeds to the regions to both its left and right side. It always progresses from higher energy to lower energy regions. Once a local energy minimum is reached, it continues with the next stable interval in propagation direction, i.e. a local energy maximum. For the verification and correction of calculated F0 estimates, we developed a peak propagation procedure that computes the most plausible peak continuation sequence given the peak tuple of the previous frame. The most plausible peak continuation sequence is found by considering several variants of peak sequences that reflect the regular and irregular period cases. In the following, we describe the control flow of the F0 propagation and explain the particular peak propagation procedure.

9.4.1. *Control flow*

The propagation of F0 estimates is performed separately for each voiced region. The first step in this procedure is the definition of the propagation order and the propagation end points. The propagation starts with the stable interval that contains the frame with the highest mean energy in the equal section. From this equal section, the propagation flows first to the left side and then to the right side. For each stable interval containing an equal section, we define the right and the left propagation end points. They are the start and end frames of the voiced region if there is only one stable interval in the voiced region. The propagation end points are the local energy minimum frame and its direct neighbor if there is a local energy minimum between two stable intervals S_1 and S_2. They are the start frame and the preceding frame of S_2 if there is no local energy minimum between S_1 and S_2 and S_2 has lower energy than S_1. In a similar way, the propagation end points are the end frame and its successor frame of S_1 if there is no local energy minimum between S_1 and S_2 and S_1 has lower energy than S_2. Figure 9.3 shows the propagation directions, order and end points of a voiced segment that contains two stable intervals with equal sections E_0 and E_1. For simplicity, the stable interval that contains E_0 and the stable interval that contains E_1 are not shown in the figure.

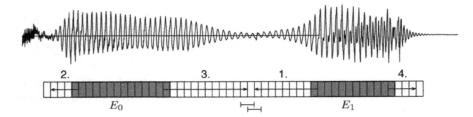

Figure 9.3. *Propagation order, directions and end points of a voiced region. Propagation start and end points are marked at the center positions of frames. Propagation starts from equal section E_1 as it has frames with higher energies than E_0*

Next, we compute candidate F0 values for the unstable frames using the method presented in section 9.3, but with a restricted range of tolerable F0. We calculate the mean F0 of the equal section where the propagation starts. The lower and upper bounds of the tolerable F0 range are an octave lower than this mean F0 value and two-thirds of an octave higher than it. In contrast to the F0 estimation method for stable intervals, the check for multiple periods is omitted since it would hardly work in unstable intervals with potentially strongly varying peak heights.

The main part of the F0 propagation stage is to check whether the F0 estimate of a frame is in accordance with the F0 of its previous frame and if not, to perform a peak propagation step (see section 9.4.2) to find the most plausible peak continuation sequence. The frame's actual F0 is derived from the detected peak continuation sequence. The peak continuation sequence may be regular but may also contain elongated or shortened periods or octave jumps. As soon as a propagation end point is reached, we check whether the mean F0 of the equal section of the next stable interval is similar to the mean F0 of the most recently calculated values. Propagation continues normally from the next stable interval if this condition holds. Otherwise, the list of equal sections in a stable interval is inspected for a better fitting equal section and the algorithm uses this as the new propagation starting point if such an equal section is found.

9.4.2. *Peak propagation*

The peak propagation step computes a set of peak sequence variants that may follow the peak tuple of the previous frame and evaluates each of them for plausibility. Each peak sequence is computed by a look-ahead strategy for the next peak. In general, we look two peaks ahead before deciding on the next one.

The following peak sequence variants are considered:

– V1 (regular case): the peaks continue at about the same distance as the peaks in the previous frame;

– V2 (elongated periods): the periods are elongated and the peak distances become larger;

– V3 (octave jump down): the peaks follow at double distance as in the previous frame;

– V4 (octave jump up): the peaks follow at half the distance as in the previous frame.

The peak sequence variants are computed depending on the octave jump state of the previous frame. The octave jump state is maintained for each frame and its default value is "none". There are two additional values, "down" and "up", for the state of F0 that is an octave higher than normal, and the state of an F0 estimate that has fallen by an octave. Variant V2 is used to detect extended periods that may not be captured by V1. V3 is necessary to test the case of a sudden octave jump down but is only calculated in the case of an octave jump "none". V4 is considered only in an octave jump down state to check whether such a phase has ended. Currently, our algorithm detects neither repeated octave jumps down nor a sudden octave jump up, but the recognition of these cases can be implemented in the future.

For each of the variants V1–V4, we define interval ranges where subsequent peaks are expected, check which peaks occur in these intervals and try to find an optimal peak sequence. The interval ranges are defined relative to the last peak distance D, i.e. D is the distance between the last two peaks in propagation direction of the previously computed peak sequence. The continued peak sequence starts with the peak tuple for the previous frame and adds peaks to the left side if the peak propagation is to the left or to the right side, otherwise. Each new peak to be added is searched for in the expected interval, while at the same time checking whether a similar peak exists in the interval that follows. Therefore, we have a peak look-ahead of 2. Each such peak pair – the two peaks looked ahead – is scored by computing their mean absolute height. The first peak of the pair that achieves the highest such score is installed as the definite next peak in the peak sequence. This peak propagation stops as soon as the center position of the addressed frame has passed by two peaks or if no further peak is found. It is deliberate that the score to evaluate peaks in unstable intervals considers only the absolute peak heights. A measure that accounted also for the peak distances would deliver false peak sequences, owing to the irregular peak distances that we expect in unstable intervals. Figure 9.4 illustrates the look-ahead strategy for successor peaks in left propagation direction, starting at peak p_0 that is part of peak tuple (p_0, p_1, p_R) of the previous frame. It shows the case of elongated periods V2. The interval where a first peak is expected is denoted by I_0. We find two possible candidate peaks in I_0, namely $p_{k(1)}$ and $p_{k(2)}$. For both candidate peaks, we look for possible look-ahead peaks. In Figure 9.4, the look-ahead peaks for $p_{k(2)}$ are shown, they are $p_{k(2,1)}$ and $p_{k(2,2)}$ in interval $I_{k(2)}$ that depends on the position of $p_{k(2)}$. The peak pair $p_{k(2)}$ and $p_{k(2,2)}$ achieves the highest score, and thus $p_{k(2)}$ is installed as the next valid peak in the peak propagation.

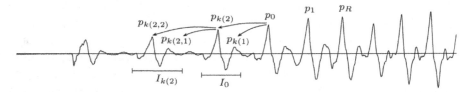

Figure 9.4. *Peak look-ahead strategy of 2 for the extended period case V2 that starts with tuple P = (p_1, p_1, p_R) in propagation direction to the left. Peaks $p_{k(1)}$ and $p_{k(2)}$ are inspected from p_0 in interval I_0, and peaks $p_{k(2,1)}$ and $p_{k(2,2)}$ are the look-ahead peaks found in expected interval $I_{k(2)}$ starting from $p_{k(2)}$.*

The final step in the peak propagation stage is the evaluation of the various peak sequence variants. In general, the variant with the highest score, i.e. with the highest mean absolute peak height, is the best peak continuation sequence. However, some additional checks are needed to verify it. Here, we describe the evaluation procedure for the case where the previous frame has no octave jump. A similar procedure is applied if the previous frame is in an octave jump down state. In the case of currently no octave jump, we first check whether V1 and V2 deliver the same peak sequence. If so, we keep V1 and discard V2. Otherwise, an additional peak propagation step for the next frame is performed to see whether V2 diverges in a double period case. If this is the case, V2 is discarded and V1 is kept. In all other cases, we keep the variant of either V1 or V2 with the larger score, i.e. the higher absolute mean peak height, in V1. Then, we evaluate V1 against the double period variant V3 if V3 has a score greater than or equal to V1. V3 is installed and the frame's octave jump state is set to "down" only if V3 has no middle peaks of sufficient heights, i.e. if the absolute height of a potential middle peak is smaller than a given percentage of the minimum of the absolute heights of the enclosing peaks. Otherwise, the peak tuple from the peak sequence of V1 is installed for the current frame.

9.5. Unstable voiced regions

Voiced regions without stable intervals or voiced regions that have no sufficiently large subsequences of equal F0 are treated in a separate post-processing step. Basically, the same propagation procedure is applied but the propagation starting or anchor point is found using looser conditions and additional information.

First, we compute the mean F0 of the last second of speech considering only frames with a verified F0, i.e. frames in equal sections of stable intervals used as starting points for F0 propagation or propagated frames in unstable intervals. We then compute candidate F0 values for all unstable frames of the voiced regions in the range of the mean F0. The permissible F0 range is the same as described in section 9.4.1. The anchor point for propagation is found by inspecting the equal section list of the stable intervals in the voiced regions, or a small section around the highest energy frame if no stable interval in the voiced region exists. The propagation starts from such a section if the mean of the F0 estimates does not deviate too largely from the last second's mean F0. If no such section takes place, we leave the F0 estimate unchanged. In this case, no propagation takes place.

9.6. Parallelization

The proposed F0 estimation algorithm can be parallelized in different ways. The algorithm delivers a segmentation of the audio signal into speech and pause segments that represent quasi-independent units. On the one hand, the whole F0 estimation algorithm may run on the speech segments in parallel. On the other hand, the different tasks of the F0 estimation algorithm may run in parallel across the speech segments but sequentially within each speech segment. The three tasks of our algorithm are: (1) the preprocessing of computing the energies and the peaks, (2) the F0 estimation on stable intervals and (3) the peak propagation. A list of all signal peaks – both local minima and maxima – is maintained for each speech segment and computed at an early stage of processing. This second way of parallelized computation of F0 is implemented in our algorithm, as illustrated in Figure 9.5. The figure shows a series of speech and pause segments denoted by S and P. The three tasks (1)–(3) of the algorithm are depicted as blocks for each speech segment. These tasks are processed in different parallel processes T1, T2 and T3 across all speech segments. Of course, T2 has to wait until T1 has finished, and T3 waits until T2 has finished for the same speech segment.

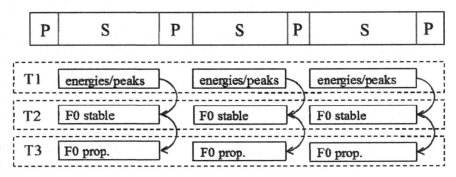

Figure 9.5. *Parallelized computation of F0 estimation. The three tasks are run in parallel across all speech segments but sequentially within each speech segment*

9.7. Experiments and results

The presented F0 estimation algorithm was evaluated on the Keele pitch reference database for clean speech. We measured the voiced error rate (VE), the unvoiced error rate (UE) and the gross pitch error rate (GPE). A voiced error is present if a voiced frame is recognized as unvoiced, an unvoiced error exists if an unvoiced frame is identified as voiced and a gross pitch error is counted if the estimated F0 differs by more than 20% from the reference pitch. The precision is given by the root-mean-square error (RMSE) in Hz for all frames classified as correct. The results for our parallel, cognition-oriented F0 estimation algorithm (PCO) are given in

Table 9.1. We also cite the results of other state-of-the-art F0 estimation or pitch detection algorithms where these figures were available: RAPT, PSHF and non-negative matrix factorization (NMF) [ROA 07, SHA 05, TAL 95]. RAPT is one of the best time-domain algorithms based on cross-correlation and dynamic programming.

	VE (%)	UE (%)	GPE (%)	RMSE (Hz)
PCO	*4.84*	*4.12*	*1.96*	*5.89*
RAPT	3.2	6.8	2.2	4.4
PSHF	4.51	5.06	0.61	2.46
NMF	7.7	4.6	0.9	4.3

Table 9.1. *Results of the parallel, cognition-oriented F0 estimation (PCO) in comparison with other state-of-the-art algorithms on the Keele pitch reference database*

The results show that the voiced and unvoiced error rates of PCO are comparable to those of the other state-of-the-art algorithms. In fact, the sum of both voiced and unvoiced error rates is the smallest for PCO, namely 8.96%, whereas it is 10% for RAPT, 9.57% for PSHF and 12.3% for NMF. The gross pitch error rate (GPE) of 1.96 is lower than that for RAPT but clearly not as low as that for the frequency-domain algorithms PSHF and NMF. Our algorithm is a pure time-domain method, and it performs better than RAPT which operates on the time domain also, but uses the normalized cross-correlation function. However, the gross pitch error (GPE) of our algorithm PCO is far lower than those of Praat, YIN and SAFE, as given in Table 9.2. The GPE of the cited algorithms are reported in [CHU 12]. For clarity, Table 9.2 also gives the GPE of the algorithms PSHF, NMF and RAPT, cited in Table 9.1.

	PSHF	NMF	*PCO*	RAPT	Praat	YIN	SAFE
GPE (%)	0.61	0.9	*1.96*	2.2	3.22	2.94	2.98

Table 9.2. *Gross pitch error rates alone of method PCO compared with standard F0 estimation algorithms on the Keele pitch reference database*

The root-mean-square error (RMSE), at 5.89 Hz – a measure for the preciseness of the correctly calculated F0 estimates – is higher than that for the other algorithms, as given in Table 9.1. This can be explained as follows. The fundamental frequency F0 is defined as the inverse of the time between two minimum signal peaks or two maximum peaks. However, maximum or minimum peaks may have an inclination – either to the left or to the right – and often, there is a set of close peaks around the maximum or minimum peak, so that F0 is not as accurately calculated as with other methods. However, the accuracy of F0 estimates can certainly be improved by adjustment methods.

9.8. Conclusions

We have presented an F0 detection algorithm as an approximate model of the human cognitive process. It purely operates in the time domain, achieves very low error rates and outperforms the state-of-the-art correlation-based method RAPT in this respect. These results are achieved with little resources in terms of memory and computing power. Obviously, the strengths and the potential of the algorithm lie in the concepts that simulate human recognition of F0.

The question asked in the introduction whether F0 estimation using principles of human cognition can be performed equally well or better than the state-of-the-art F0 detection algorithms can be answered with a partial "yes" for clean speech. The gross pitch error rates of our algorithm are among the lowest of the standard F0 estimation algorithms. However, the accuracy in terms of root-mean-square error is still higher than that of other algorithms.

The presented algorithm is thoroughly extensible, as new special cases are easily implemented. In this sense, the algorithm can also be applied to other tasks, e.g. spontaneous or noisy speech, by analyzing the new cases and modeling them. In this way, it will become more and more generic. This procedure closely reflects human learning, which is said to function by adopting examples and building patterns independently of the frequency or probability of their occurrence [KUH 77]. For this reason, we have refrained from using weights or probabilities to favor one or other cases but instead to look ahead and evaluate until the case is decided.

A major strength of the algorithm is the segmentation of the speech signal into various structures additionally to the F0 contour. The recognition of speech and pause segments makes a parallelization of the algorithm possible. The classification into stable and unstable intervals may be used for automatic speech recognition. Similarly to the presented F0 estimation, automatic speech recognition may first recognize the phonemes in stable intervals before detecting the phonemes in the unstable intervals. Spectra are more reliably computed on stable than on unstable intervals.

Future work will focus on extending the algorithm for both spontaneous and particularly noisy speech data and improving the accuracy of the F0 estimates.

9.9. Acknowledgments

The author wishes to thank Prof. Jozsef Szakos from the Hong Kong Polytechnic University for valuable comments and Prof. Guy Aston from the University of Bologna, Italy, for his careful proof-reading. She is also very grateful to Christian Singer who implemented the basic version of the pause-finding algorithm during his diploma thesis.

9.10. Bibliography

[ACH 05] ACHAN K., ROWEIS S., HERTZMANN A. et al., "A segment-based probabilistic generative model of speech", Proceedings of ICASSP, pp. 221–224, 2005.

[BOE 01] BOERSMA P., "PRAAT, a System for doing phonetics by computer", Glot International, vol. 5, no. 9/10, pp. 341–345, 2001.

[CHU 12] CHU W., ALWAN A., "SAFE: a statistical approach to F0 estimation under clean and noisy conditions", IEEE Transactions on Audio, Speech and Language Processing, vol. 20, no. 3, pp. 933–944, 2012.

[DEC 02] DE CHEVEIGNÉ A., KAWAHARA A., "YIN, a fundamental frequency estimator for speech and music", Journal of the Acoustical Society of America, vol. 111, no. 4, pp. 1917–1930, 2002.

[EWE 10] EWENDER T., PFISTER B., "Accurate pitch marking for prosodic modification of speech segments", Proceedings of INTERSPEECH, pp. 178–181, 2010.

[GHA 14] GHAHREMANI P., BABA ALI B., POVEY D. et al., "A pitch extraction algorithm tuned for automatic speech recognition", Proceedings of INTERSPEECH, pp. 2494–2498, 2014.

[GLA 15] GLAVITSCH U., HE L., DELLWO V., "Stable and unstable intervals as a basic segmentation procedure of the speech signal", *Proceedings of INTERSPEECH*, pp. 31–35, 2015.

[KAH 11] KAHNEMAN D., *Thinking, Fast and Slow*, Farrar, Straus and Giroux, New York, 2011.

[KUH 77] KUHN T.S., "Second thoughts on paradigms", *The Essential Tension, Selected Studies in Scientific Tradition and Change*, The University of Chicago Press, Chicago, pp. 837–840, 1977.

[MAR 72] MARKEL J.D., "The SIFT algorithm for fundamental frequency estimation", *IEEE Transactions on Audio and Electroacoustics*, vol. 20, no. 5, pp. 367–377, 1972.

[MOO 08] MOORE B.C.J., *An Introduction to the Psychology of Hearing*, Emerald, Bingley, 2008.

[PEH 11] PEHARZ R., WOHLMAYR M., PERNKOPF F., "Gain-robust multi-pitch tracking using sparse nonnegative matrix factorization", *Proceedings of ICASSP*, pp. 5416–5419, 2011.

[PLA 95] PLANTE F., MEYER G.F., AINSWORTH W.A., "A pitch extraction reference database", *Proceedings of Eurospeech*, pp. 837–840, 1995.

[RAB 75] RABINER L.R., SAMBUR M.R., "An algorithm for detecting the endpoints of isolated utterances", *Bell System Technical Journal*, vol. 54, no. 2, 1975.

[RAB 76] RABINER L.R., CHENG M.J., ROSENBERG A.E. *et al.*, "A comparative performance study of several pitch detection algorithms", *IEEE Transactions on Acoustics, Speech, and Signal Processing*, vol. 24, no. 5, pp. 399–418, 1976.

[ROA 07] ROA S., BENNEWITZ M., BEHNKE S., "Fundamental frequency based on pitch-scaled harmonic filtering", *Proceedings of ICASSP*, pp. 397–400, 2007.

[ROD 11] RODERO E., "Intonation and emotion: influence of pitch levels and contour type on creating emotions", *Journal of Voice*, vol. 25, no. 1, pp. e25–e34, 2011.

[SEC 83] SECREST B.G., DODDINGTON G.R., "An integrated pitch tracking algorithm for speech systems", *Proceedings of ICASSP*, pp. 1352–1355, 1983.

[SHA 05] SHA F., SAUL L. K., "Real-time pitch determination of one or more voices by nonnegative matrix factorization", *Advances in Neural Information Processing Systems*, MIT Press, vol. 17, pp. 1233–1240, 2005.

[TAL 95] TALKIN D., *A Robust Algorithm for Pitch Tracking (RAPT)*, Speech Coding and Synthesis, Elsevier Science B.V., Amsterdam, 1995.

Benchmarking n-grams, Topic Models and Recurrent Neural Networks by Cloze Completions, EEGs and Eye Movements

Previous neurocognitive approaches to word predictability from sentence context in electroencephalographic (EEG) and eye movement (EM) data relied on cloze completion probability (CCP) data effortfully collected from up to 100 human participants. Here, we test whether three well-established language models can predict these data. Together with baseline predictors of word frequency and position in sentence, we found that the syntactic and short-range semantic processes of n-gram language models and recurrent neural networks (RNN) perform about equally well when directly accounting CCP, EEG and EM data. In contrast, a low amount of variance explained by a topic model suggests that there is no strong impact on the CCP and the N400 component of EEG data, at least in our Potsdam Sentence Corpus dataset. For the single-fixation durations of the EM data, however, topic models accounted for more variance, suggesting that long-range semantics may play a greater role in this earlier neurocognitive process. Though the language models were not significantly inferior to CCP in accounting for these EEG and EM data, CCP always provided a descriptive increase in explained variance for the three corpora we used. However, n-gram and RNN models can account for about half of the variance of the CCP-based predictability estimates, and the largest part of the variance that CCPs explain in EEG and EM data. Thus, our approaches may help to generalize neurocognitive models to all possible novel word combinations, and we propose to use the same benchmarks for language models as for models of visual word recognition.

Chapter written by Markus J. HOFMANN, Chris BIEMANN and Steffen REMUS.

10.1. Introduction

In neurocognitive psychology, manually collected cloze completion probabilities (CCPs) are the standard approach to quantifying a word's predictability from sentence context [KLI 04, KUT 84, REI 03]. Here, we test a series of language models in accounting for CCPs, as well as the data they typically account for, i.e. electroencephalographic (EEG) and eye movement (EM) data. With this, we hope to render time-consuming CCP procedures unnecessary. We test a statistical n-gram language model [KNE 95], a Latent Dirichlet Allocation (LDA) topic model [BLE 03], as well as a recurrent neural network (RNN) language model [BEN 03, ELM 90] for correlation with the neurocognitive data.

CCPs have been traditionally used to account for N400 responses as an EEG signature of a word's contextual integration into sentence context [DAM 06, KUT 84]. Moreover, they were used to quantify the concept of word predictability from sentence context in models of eye movement control [ENG 05, REI 03]. However, as CCPs are effortfully collected from samples of up to 100 participants [KLI 04], they provide a severe challenge to the ability of a model to be generalized across all novel stimuli [HOF 14], which also prevents their ubiquitous use in technical applications.

To quantify how well computational models of word recognition can account for human performance, Spieler and Balota [SPI 97] proposed that a model should explain variance at the item-level, i.e. latencies averaged across a number of participants. Therefore, a predictor variable is fitted to the mean word naming latency as a function of $y = f(x) = \sum a_n x_n + b + error$ for a number of n predictor variables x that are scaled by a slope factor a, an intercept of b, and an error term. The Pearson correlation coefficient r is calculated, and squared to determine the amount of explained variance r^2. Models with a larger number of n free parameters are more likely to (over-)fit error variance, and thus fewer free parameters are preferred (e.g. [HOF 14]).

While the best cognitive process models can account for 40–50% of variance in behavioral naming data [PER 10], neurocognitive data are noisier. The only interactive activation model that gives an amount of explained variance in EEG data [BAR 07, MCC 81] was that of Hofmann *et al.* [HOF 08], who account for 12% of the N400 variance. Though models of eye movement control use item-level CCPs as predictor variables [ENG 05, REI 03], computational models of eye movement control have hardly been benchmarked at the item-level, to our knowledge [DAM 07].

While using CCP-data increases the comparability of many studies, the creation of such information is expensive and they only exist for a few languages [KLI 04, REI 03]. If it were possible to use (large) natural language corpora and derive the information leveraged from such resources automatically, this would considerably expedite the process of experimentation for under-resourced languages. Comparability would not be compromised when using standard corpora, such as that available through Goldhahn *et al.* [GOL 12] in many languages. However, it is not yet clear what kind of corpus is most appropriate for this enterprise, and whether there are differences in explaining human performance data.

10.2. Related work

Taylor [TAY 53] was the first to instruct participants to fill a cloze with an appropriate word. The percentage of participants who fill in the respective word serves as cloze completion probability. For instance, when exposed to the sentence fragment "He mailed the letter without a ___", 99% of the participants complete the cloze by "stamp", thus CCP equals 0.99 [BLO 80]. Kliegl *et al.* [KLI 04] logit-transformed CCPs to obtain *pred* = *ln(CCP/(1−CCP))*.

Event-related potentials are computed from human EEG data. For the case of the N400, words are often presented word-by-word, and the EEG waves are averaged across a number of participants relative to the event of word presentation. As brain-electric potentials are labeled by their polarity and latency, the term N400 refers to a negative deflection around 400 ms after the presentation of a target word.

After Kutas and Hillyard [KUT 84] discovered the sensitivity of the N400 to cloze completion probabilities, they suggested that it reflects the semantic relationship between a word and the context in which it occurs. However, there are several other factors that determine the amplitude of the N400 [KUT 11]. For instance, Dambacher *et al.* [DAM 06] found that word frequency (*freq*), the position of a word in a sentence (*pos*), as well as predictability (*pred*) affect the N400.

While the eyes remain relatively still during fixations, readers make fitful eye movements called saccades [RAD 12]. When successfully recognizing a word in a stream of forward eye movements, no second saccade to or within the word is required. The time the eyes remain on that word is called single-fixation duration (SFD), which shows a strong correlation with word predictability from sentence context (e.g. [ENG 05]).

10.3. Methodology

10.3.1. *Human performance measures*

This study proposes that language models can be benchmarked by item-level performance on three datasets that are openly available in online databases. Predictability was taken from the Potsdam Sentence Corpus[1], first published by Kliegl *et al.* [KLI 04]. The 144 sentences consist of 1,138 tokens, available in Appendix A of [DAM 09], and the logit-transformed CCP measures of word predictability were retrieved from Ralf Engbert's homepage1 [ENG 05]. For instance, in the sentence "Manchmal sagen Opfer vor Gericht nicht die volle Wahrheit" [Before the court, victims tell not always the truth.], the last word has a CCP of 1. N400 amplitudes were taken from the 343 open-class words published in Dambacher and Kliegl [DAM 07]. These are available from the Potsdam Mind Research Repository[2]. The EEG data published there are based on a previous study (see [DAM 06] for method details). The voltage of 10 centroparietal electrodes was averaged across up to 48 artifact-free participants from 300 to 500 ms after word presentation for quantifying the N400. SFD are based on the same 343 words from Dambacher and Kliegl [DAM 07], available from the same source URL. Data were included when this word was only fixated for

1 http://mbd.unipotsdam.de/EngbertLab/Software.html
2 http://read.psych.unipotsdam.de

one time, and these SFDs ranged from 50 to 750 ms. The SFD was averaged across up to 125 German native speakers [DAM 07].

10.3.2. *Three flavors of language models*

Language models are based on a probabilistic description of language phenomena. Probabilities are used to pick the most fluent of several alternatives, e.g. in machine translation or speech recognition. Word **n-gram models** are defined by a Markov chain of order $n-1$, where the probability of the following word only depends on previous $n-1$ words. In statistical models, the probability distribution of the vocabulary, given a history of $n-1$ words, is estimated based on n-gram counts from (large) natural language corpora. There exist a range of n-gram language models (see, e.g., Chapter 3 in [MAN 99], which are differentiated by the way they handle unseen events and perform probability smoothing). Here, we use a Kneser–Ney [KNE 95] 5-gram model[3]. For each word in the sequence, the language model computes a probability p in $]0; 1[$. We use the logarithm *log(p)* of this probability as a predictor. We used all words in their full form, i.e. did not filter for specific word classes and did not perform lemmatization. N-gram language models are known to model local syntactic structure very well. Since only n-gram models use the most recent history for predicting the next token, they fail to account for long-range phenomena and semantic coherence (see [BIE 12]).

Latent Dirichlet Allocation (LDA) topic models [BLE 03] are generative probabilistic models representing documents as a mixture of a fixed number of N topics, which are defined as unigram probability distributions over the vocabulary. Through a sampling process like Gibbs sampling, topic distributions are inferred. Words frequently co-occurring in the same documents receive a high probability in the same topics. When sampling the topic distribution for a sequence of text, each word is randomly assigned to a topic according to the document-topic distribution and the topic-word distribution. We use Phan and Nguyen's [PHA 07] GibbsLDA implementation for training an LDA model with 200 topics (default values for $\alpha = 0.25$ and $\beta = 0.001$) on a background corpus. Words occurring in too many documents (a.k.a. stopwords) or too few documents (mistyped or rare words) were removed from the LDA vocabulary. Then, retain the per

---

3 https://code.google.com/p/berkeleylm/

document topic distribution $p(z|d)$ and the per topic word distribution $p(w|z)$, where z is the latent variable representing the topic, d refers to a full document during training – during testing d refers to the history of the current sentence – and w is a word. In contrast to our earlier approach using only the top three topics [BIE 15], we here computed the probability of the current word w given its history d as a mixture of its topical components $p(w|d) = p(w|z)p(z|d)$. We hypothesize that topic models account for some long-range semantic aspects missing in n-gram models. While Bayesian topic models are probably the most widespread approach to semantics in psychology (e.g. [GRI 07]), latent semantic analysis (LSA) is not applicable in our setting [LAN 97]: we use the capability of LDA to account for yet unseen documents, whereas LSA assumes a fixed vocabulary and it is not trivial to fold new documents into LSA's fixed document space.

While Jeff Elman's [ELM 90] seminal work suggested early on that semantic and also syntactic structure automatically emerges from a set of simple recurrent units, such an approach has received little attention in language modeling for a long time, but is currently of interest to many computational studies. In brief, such **Neural Network Language Models** are based on the optimization probability of the occurrence of a word, given its history using neural units linking back to themselves, much as the neurons in the CA3 region of the human hippocampus [MAR 71, NOR 03]. The task of language modeling using neural networks was first introduced by Bengio *et al.* [BEN 03] and received at that point only little attention because of computational challenges regarding space and time complexity. Due to recent advancement in the field of neural networks – for an overview, see [MIK 12] – neural language models gained more popularity, particularly because of the so-called neural word embeddings as a side product. The language model implementation we use in this work is a recurrent neural network architecture[4] similar to the one used by Mikolov's Word2Vec[5] toolkit [MIK 13]. We trained a model with 400 hidden layers and hierarchical softmax. For testing, we used the complete history of a sentence up to the current word.

---

4 FasterRNN: https://github.com/yandex/faster-rnnlm
5 Word2Vec: https://code.google.com/archive/p/word2vec/

10.4. Experiment setup

Engbert *et al.*'s [ENG 05] data are organized in 144 short German sentences with an average length of 7.9 tokens, and provide features, such as *freq* as corpus frequency in occurrences per million [BAA 95], *pos* and *pred*. We test whether two corpus-based predictors can account for predictability, and compare the capability of both approaches in accounting for EEG and EM data. For training n-gram and topic models, we used three different corpora differing in size and covering different aspects of language. Further, the units for computing topic models differ in size.

NEWS: a large corpus of German online newswire from 2009, as collected by LCC [GOL 12], of 3.4 million documents/30 million sentences/ 540 million tokens. This corpus is not balanced, i.e. important events in the news are covered better than other themes. The topic model was trained on the document level.

WIKI: a recent German Wikipedia dump of 114,000 articles/7.7 million sentences/180 million tokens. This corpus is rather balanced, as concepts or entities are described in a single article each, independent of their popularity, and spans all sorts of topics. The topic model was trained on the article level.

SUB: German subtitles from a recent dump of opensubtitles.org, containing 7,420 movies/7.3 million utterances/54 million tokens. While this corpus is much smaller than the others, it is closer to a colloquial use of language. Brysbaert *et al.* [BRY 11] showed that word frequency measures of subtitles provide numerically greater correlations with word recognition speed than larger corpora of written language. The topic model was trained on the movie level.

Pearson's product-moment correlation coefficient was calculated (e.g. [COO 10, p. 293]), and squared for the N = 1,138 predictability scores [ENG 05] or N = 343 N400 amplitudes or SFD [DAM 07]. To address overfitting, we randomly split the material into two halves, and test how much variance can be reproducibly predicted on two subsets of 569 items. For N400 amplitude and SFD, we used the full set, because one half was too small for reproducible predictions. The correlations between all predictor variables can be examined in Table 10.1. We observe very high correlations between the n-gram and the RNN predictions within and across corpora. The correlations involving topic-based predictions are smaller, supporting our hypothesis that they reflect a somewhat different neurocognitive process.

		1.	2.	3.	4.	5.	6.	7.	8.	9.
NEWS	1. n-gram		0.65	0.87	0.87	0.56	0.84	0.83	0.59	0.80
	2. topic	0.65		0.68	0.66	0.78	0.70	0.61	0.77	0.61
	3. neural	0.87	0.68		0.84	0.59	0.88	0.77	0.62	0.79
WIKI	4. n-gram	0.87	0.66	0.84		0.61	0.90	0.79	0.59	0.78
	5. topic	0.56	0.78	0.59	0.61		0.65	0.55	0.75	0.55
	6. neural	0.84	0.70	0.88	0.90	0.65		0.76	0.64	0.79
SUB	7. n-gram	0.83	0.61	0.77	0.79	0.55	0.76		0.61	0.85
	8. topic	0.59	0.77	0.62	0.59	0.75	0.64	0.61		0.61
	9. neural	0.80	0.61	0.79	0.78	0.55	0.79	0.85	0.61	

Table 10.1. *Correlations between the language model predictors*

10.5. Results

10.5.1. *Predictability results*

In the first series of results, we examine the prediction of manually obtained CCP-derived predictability with corpus-based methods. A large amount of explained variance would indicate that predictability could be replaced by automatic methods. As a set of baseline predictors, we use *pos* and *freq*, which explains 0.243/0.288 of the variance for the first and the second half of the dataset, respectively. We report results in Table 10.2 for all single corpus-based predictors alone and in combination with the baseline, all combinations of the baseline with n-gram, topics and neural models from the same corpus.

Predictors	NEWS	WIKI	SUB
n-gram	0.262/0.294	0.226/0.253	0.268/0.272
topic	0.063/0.061	0.042/0.040	0.040/0.034
neural	0.229/0.226	0.211/0.226	0.255/0.219
base+n-gram	**0.462/0.490**	0.423/0.458	0.448/0.459
base+topic	0.348/0.375	0.333/0.357	0.325/0.355
base+neural	**0.434/0.441**	0.418/0.433	**0.447/0.418**
base+n-gram+topic	0.462/0.493	0.427/0.464	0.447/0.458
base+n-gram+neural	0.466/0.492	0.431/0.461	0.467/0.461
base+neural+topic	0.438/0.445	0.421/0.436	0.446/0.423
base+n-gram+topic+neural	0.466/0.493	0.433/0.465	0.467/0.460

Table 10.2. r^2 *explained variance of predictability, given for two halves of the dataset, for various combinations of baseline and corpus-based predictors*

It is apparent that the n-gram scores best, and also the neural model alone reaches r^2 levels that approach the baseline. In contrast, much as our earlier top-three topics approach [BIE 15], the mixture of all topics explains only a relatively low amount of variance. Combining the baseline with the n-gram predictor already reaches a level very close to the combination of all predictors, thus it may provide the best compromise between parsimony and explained variance. Again, this model performance is closely followed by the recurrent neural network (see Figure 10.1).

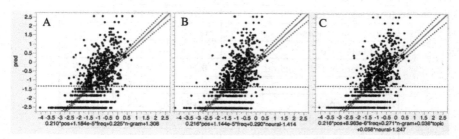

Figure 10.1. *Prediction models exemplified for the NEWS corpus in the x-axes and the N = 1,138 predictability scores on the y-axes. A) Prediction by baseline + n-gram (r^2 = 0.475), B) a recurrent neural network (r^2 = 0.437) and C) a model containing all predictors (r^2 = 0.478). The three pairwise Fisher's r-to-z tests revealed no significant differences in explained variance (Ps > 0.18). For a color version of this figure, see www.iste.co.uk/sharp/cognitive.zip*

We also fitted a model based on all corpus-based predictors from all corpora, which achieved the overall highest r^2 (0.490/0.507). In summary, it becomes clear that about half of the empirical predictability variance can be explained by a combination of positional and frequency features combined with either a word n-gram language model, or a recurrent neural network.

10.5.2. N400 amplitude results

For modeling N400, we have even more combinations at our disposal, since we can combine corpus-based measures with the baseline, the predictability performance and with both. We evaluate on all 343 data points for N400 amplitude fitting. Without using corpus-based predictors, the baseline predicts a mere 0.032 of variance, predictability alone explains 0.192 of variance and their combination explains 0.193 of variance – i.e. the baseline is almost entirely subsumed by CCP-based predictability. As can be observed from Table 10.3, this is a score that is not yet reached by the language models, even when combining all of them.

Predictors	NEWS	WIKI	SUB
n-gram	0.141	0.140	0.126
topic	0.039	**0.055**	0.025
neural	0.108	0.098	0.100
base+n-gram	**0.161**	0.153	0.135
base+topic	0.063	0.079	0.055
base+neural	0.133	0.116	0.114
base+n-gram+topic	0.161	0.158	0.132
base+n-gram+neural	0.167	0.153	0.141
base+neural+topic	0.133	0.123	0.112
base+n-gram+topic+neural	**0.167**	0.158	0.137
base+n-gram+pred	**0.223**	0.226	0.206
base+topic+pred	0.193	0.204	0.191
base+neural+pred	0.221	0.212	0.206
base+n-gram+topic+pred	0.225	0.228	0.203
base+n-gram+neural+pred	0.228	0.226	0.209
base+neural+topic+pred	0.224	0.215	0.203
base+n-gram+topic+neural+pred	**0.232**	0.228	0.206

Table 10.3. r^2 explained variance of the N400 for various combinations of the corpus-based predictors, in combination with the baseline, and with the empirical predictability

When comparing the performance of the computationally defined predictors, a picture similar to the prediction of the empirical predictability emerges. The n-gram model scores best, particularly for the larger NEWS and WIKI corpora. This confirms a generally accepted hypothesis that larger training data trumps smaller, more focused training data, see e.g. [BAN 01] and others. The n-gram model is, however, immediately followed by the neural model, and again, the topic predictor provides the poorest performance in explaining N400 amplitude variance, which suggests that the N400 does not reflect long-range semantic processes. The best combination without predictability, with a score of r^2 = 0.167, approaches the performance of the predictability and baseline (see Figure 10.2).

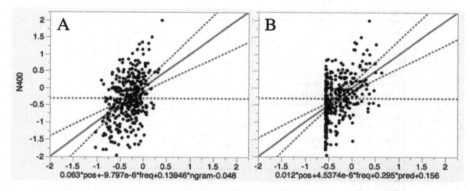

Figure 10.2. *Prediction models exemplified for the NEWS corpus in the x-axes and the N = 334 mean N400 amplitudes on the y-axes. A) Prediction by baseline + n-gram (r^2 = 0.161), and B) a standard approach to N400 data, consisting of the baseline of position and frequency, as well as the empirical predictability (r^2 = 0.193; e.g. [DAM 06]). Fisher's r-to-z tests revealed no significant differences in explained variance (P = 0.55). For a color version of this figure, see www.iste.co.uk/ sharp/cognitive.zip*

The experiments with predictability as an additional predictor confirm the results from the previous section: n-grams + baseline and predictability capture slightly different aspects of human reading performance, thus their combination explains up to 6% more net variance than predictability alone.

10.5.3. *Single-fixation duration (SFD) results*

Finally, we examine the corpus-based predictors for modeling the mean single fixations duration for 343 words. For this target, the *pos+freq* baseline explains $r^2 = 0.021$, whereas predictability, alone or combined with the baseline, explains $r^2 = 0.184$.

Predictors	NEWS	WIKI	SUB
n-gram	0.225	0.140	0.126
topic	0.135	0.140	0.100
neural	0.242	0.190	**0.272**
base+n-gram	0.239	0.226	0.226
base+topic	0.152	0.154	0.127
base+neural	0.265	0.204	**0.284**
base+n-gram+topic	0.260	0.262	0.246
base+n-gram+neural	0.287	0.238	0.297
base+neural+topic	0.279	0.235	0.298
base+n-gram+topic+neural	0.295	0.265	**0.307**
base+n-gram+pred	0.273	0.274	0.258
base+topic+pred	0.235	0.250	0.229
base+neural+pred	0.314	0.267	0.320
base+n-gram+topic+pred	0.297	0.301	0.275
base+n-gram+neural+pred	0.319	0.283	0.322
base+neural+topic+pred	0.319	0.289	0.329
base+n-gram+topic+neural+pred	0.323	0.304	**0.330**

Table. 10.4. *Explained variance of the single-fixation durations, for various combinations of baseline, predictability and corpus-based predictors*

The experiments confirm the utility of n-gram models in accounting for eye movement data. The n-gram model alone explains even more variance than predictability – however, the difference is not significant ($P > 0.46$).

In contrast to the previous approaches to predictability and N400 amplitudes, however, the recurrent neural network outperformed the n-gram model at a descriptive level, as it accounted for up to 3% more of the variance than the n-gram model. This performance was not reached at the largest NEWS corpus, but at the smaller SUB corpus. This suggests that – for SFD data – the dimension reduction seems to compensate for the larger amount of the noise in the smaller training dataset (see [BUL 07, GAM 16, HOF 14]). Therefore, the neural model may provide a better fit for such early neurocognitive processes when it is trained by colloquial language [BRY 11].

The topics model seems to have a stronger impact on SFDs than on the other neurocognitive benchmark variables, suggesting a greater influence of long-range semantics on SFDs than on predictability or the N400. Taken together, these findings suggest that SFDs reflect different cognitive processes than the N400 (see [DAM 07]).

Last but not least, though again adding predictability increased the total amount of explained variance by 2%, the language models did an excellent job in accounting for SFD data. When taking all language model-based predictors together, this accounts for significantly more variance than the standard model using predictability (see Figure 10.3).

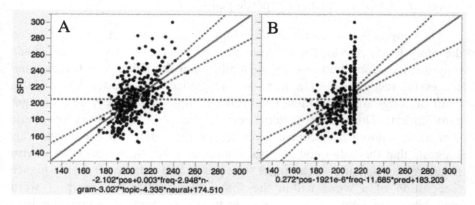

Figure 10.3. *Prediction models exemplified for the SUB corpus in the x-axes and the N = 334 mean SFD scores on the y-axes. A) Prediction by baseline + all three language models (r^2 = 0.295), and B) a standard approach to SFD data, using the baseline and predictability as predictors of SFDs (r^2 = 0.184). Fisher's r-to-z test revealed a significant difference in explained variance (z = 1.95; P = 0.05). For a color version of this figure, see www.iste.co.uk/sharp/cognitive.zip*

10.6. Discussion and conclusion

We have examined the utility of three corpus-based predictors to account for word predictability from sentence context, as well as the EEG signals and EM-based reading performance elicited by it. Our hypothesis was that word n-gram models, topic models and recurrent neural network models would account for the predictability of a token, given the preceding tokens in the sentence, as perceived by humans, as well as some electroencephalographic and eye movement data that are typically explained by it. Therefore, we used the amount of explained item-level variance as a benchmark, which has been established as a standard evaluation criterion for neurocognitive models of visual word recognition (e.g. [HOF 08, PER 10, SPI 97]).

Our hypothesis was at least partially confirmed: n-gram models and RNNs, sometimes in combination with a frequency-based and positional baseline, are highly correlated with human predictability scores and in fact explain variance of human reading performance to an extent comparable to predictability – slightly less on the N400 but slightly more on the SFD. This, however, might at least in part be explainable by a larger amount of noise in the EEG data with fewer participants when compared with the eye movement data with much more participants.

The long-range semantic relationships as captured by topic models, on the other hand, provided a different picture. If any, the topic model made only a minor contribution to predictability and the N400. For the fast and successful recognition of a word at one glance, as reflected by SFDs in contrast, long-range document level relationships seem to provide a stronger contribution. This result pattern occurs even in the context of single sentences, without a discourse level setting the topic of a document. This suggests that the colloquial and taxonomic far-reaching semantic long-term structure particularly determines the fast and effective single-glance recognition of a word within the first 300 ms after the onset of word recognition. In contrast, topic models hardly account for somewhat later processes around 400 ms in the brain-electric data and the time-consuming, probably late, processes being contained in the predictability scores.

For predicting the empirical word predictability from sentence context as well as the N400, recurrent neural network models often performed somewhat worse than the n-gram approach. For predicting SFDs, however, the neural model was superior. Most interestingly, the neural network model performs best when it is trained on a small but probably more representative sample of everyday language. Therefore, size probably does not trump everything and in any model [BAN 01]. It also hints at the generalization properties of its dimensionality reduction, which are more important for smaller training data [BUL 07, GAM 16, HOF 14], but probably leads to imprecise modeling when more data are available.

Can we now safely replace human predictability scores with n-gram statistics? Given the high correlation between predictability and the combination of n-grams with frequency and positional information, and given that n-gram-based predictors achieve similar levels of explained variance to predictability, the answer seems to be positive. However, though our corpus-based approaches explain most of the variances that manually collected CCP scores also account for, adding predictability always accounts for more variance – though this difference is not significant (see Figure 10.2; cf. Figures in [BIE 15]).

When contrasting the standard predictors of position, frequency and predictability used in eye tracking and EEG research (e.g. [DAM 06, REI 03]), only for the SFDs, all three corpus-based predictors did a better job than the standard model. However, with this approach, it is apparent that many more predictors are needed, and thus the probability for fitting error variance is much larger than that for the standard model. Thus, we think that much more evidence is required, before we dare to state this as a firm conclusion. Also for this three-predictor model, adding the empirical predictability provides a net gain of 2% explained variance.

As n-gram or neural models together with word frequency and position captured about half of the predictability variance, and most of the N400 and SFD variance elicited by it, we propose that it can be used to replace tediously collected CCPs. This not only saves a lot of pre-experimental work, but also opens the possibility to apply (neuro-) cognitive models in technical applications. For instance, n-gram models, topic models and neural models can be used to generalize computational models of eye movement control to novel sentences [ENG 05, REI 03].

In the end, language models can also improve our understanding of the cognitive processes underlying predictability, EEG and EM measures. While it is not clear what exactly determines human CCP-based predictability performance, the different language models provide differential grain size levels using their training data, thus paving the way for the question as to which neurocognitive measures of "word predictability" are affected by sentence- or document-level semantic knowledge. While Ziegler and Goswami [ZIE 05] discussed the optimal grain size of language learning at the word-level and sub-word-level grain sizes, recent evidence of a severe decline of comprehension abilities since the 1960s suggests the necessity to continue with that discussion at the level of supralexical semantic integration [SPI 16].

10.7. Acknowledgments

The "Deutsche Forschungsgemeinschaft" (MJH; HO 5139/2-1), the German Institute for Educational Research in the Knowledge Discovery in Scientific Literature (SR) program and the LOEWE Center for Digital Humanities (CB) supported this work.

10.8. Bibliography

[BAA 95] BAAYEN H.R., PIEPENBROCK R., GULIKERS L., The CELEX Lexical Database. Release 2 (CD-ROM), LDC, University of Pennsylvania, Philadelphia, 1995.

[BAN 01] BANKO M., BRILL E., "Scaling to very very large corpora for natural language disambiguation", *Proceedings of ACL '01*, Toulouse, pp. 26–33, 2001.

[BAR 07] BARBER H.A., KUTAS M., "Interplay between computational models and cognitive electrophysiology in visual word recognition", *Brain Research Reviews*, vol. 53, no. 1, pp. 98–123, 2007.

[BEN 03] BENGIO Y., DUCHARME R., VINCENT P. *et al.*, "A neural probabilistic language model", *Journal of Machine Learning Research*, vol. 3, no. 6, pp. 1137–1155, 2003.

[BIE 12] BIEMANN C., ROOS S., WEIHE K., "Quantifying semantics using complex network analysis", *Proceedings of COLING 2012*, Mumbai, pp. 263–278, 2012.

[BIE 15] BIEMANN C., REMUS S., HOFMANN M.J., "Predicting word 'predictability' in cloze completion, electroencephalographic and eye movement data", *Proceedings of the 12th International Workshop on Natural Language Processing and Cognitive Science*, Krakow, pp. 83–94, 2015.

[BLE 03] BLEI D.M., NG A.Y., JORDAN M.I. "Latent Dirichlet Allocation", *Journal of Machine Learning Research*, vol. 3, pp. 993–1022, 2003.

[BLO 80] BLOOM P.A., FISCHLER I., "Completion norms for 329 sentence contexts", *Memory & cognition*, vol. 8, no. 6, pp. 631–642, 1980.

[BUL 07] BULLINARIA J.A., LEVY J.P., "Extracting semantic representations from word co-occurrence statistics: a computational study", *Behavior Research Methods*, vol. 39, no. 3, pp. 510–526, 2007.

[BRY 11] BRYSBAERT M., BUCHMEIER M., CONRAD M. *et al.*, "A review of recent developments and implications for the choice of frequency estimates in German", *Experimental psychology*, vol. 58, pp. 412–424, 2011.

[COO 10] COOLICAN H., *Research Methods and Statistics in Psychology*, Hodder & Stoughton, London, 2010.

[DAM 06] DAMBACHER M., KLIEGL R., HOFMANN M.J. *et al.*, "Frequency and predictability effects on event-related potentials during reading", *Brain Research*, vol. 1084, no. 1, pp. 89–103, 2006.

[DAM 07] DAMBACHER M., KLIEGL R., "Synchronizing timelines: relations between fixation durations and N400 amplitudes during sentence reading", *Brain research*, vol. 1155, pp. 147–162, 2007.

[DAM 09] DAMBACHER M., *Bottom-up and Top-down Processes in Reading*, Potsdam University Press, Potsdam, 2009.

[ELM 90] ELMAN J.L., "Finding structure in time," *Cognitive Science*, vol. 211, pp. 1–28, 1990.

[ENG 05] ENGBERT R., NUTHMANN A., RICHTER E.M. *et al.*, "SWIFT: a dynamical model of saccade generation during reading", *Psychological Review*, vol. 112, no. 4, pp. 777–813, 2005.

[GAM 16] GAMALLO P., "Comparing explicit and predictive distributional semantic models endowed with syntactic contexts," *Language Resources and Evaluation*, pp. 1–17, doi:10.1007/s10579-016-9357-4, 2016.

[GOL 12] GOLDHAHN D., ECKART T., QUASTHOFF U., "Building large monolingual dictionaries at the Leipzig Corpora Collection: From 100 to 200 languages", *Proceedings of LREC 2012*, Istanbul, pp. 759–765, 2012.

[GRI 07] GRIFFITHS T.L., STEYVERS M., TENENBAUM J.B., "Topics in semantic representation", *Psychological Review*, vol. 114, no. 2, pp. 211–244, 2007.

[HOF 14] HOFMANN M.J., JACOBS A.M., "Interactive activation and competition models and semantic context: from behavioral to brain data", *Neuroscience & Biobehavioral Reviews*, vol 46, pp. 85–104, 2014.

[HOF 08] HOFMANN M.J., TAMM S., BRAUN M.M. *et al.*, "Conflict monitoring engages the mediofrontal cortex during nonword processing", *Neuroreport*, vol. 19, no. 1, pp. 25–29, 2008.

[KLI 04] KLIEGL R., GRABNER E., ROLFS M. *et al.*, "Length, frequency, and predictability effects of words on eye movements in reading", *European Journal of Cognitive Psychology*, vol. 16, no. 12, pp. 262–284, 2004.

[KNE 95] KNESER R., NEY H., "Improved backing-off for m-gram language modeling", *Proceedings of IEEE Int'l Conference on Acoustics, Speech and Signal Processing*, Detroit, pp. 181–184, 1995.

[KUT 11] KUTAS M., FEDERMEIER K.D., "Thirty years and counting: finding meaning in the N400 component of the event-related brain potential (ERP)", *Annual Review of Psychology*, vol. 62, pp. 621–647, 2011.

[KUT 84] KUTAS M., HILLYARD S.A., "Brain potentials during reading reflect word expectancy and semantic association", *Nature*, vol. 307, no. 5947, pp. 161–163, 1984.

[LAN 97] LANDAUER T.K., DUMAIS S.T., "A solution to Plato's problem: the latent semantic analysis theory of acquisition, induction, and representation of knowledge", *Psychological Review*, vol. 104, no. 2, pp. 211–240, 1997.

[MAN 99] MANNING C.D., SCHÜTZE H., *Foundations of Statistical Natural Language Processing*, MIT Press, Cambridge, 1999.

[MAR 71] MARR D., "Simple memory: a theory", *Philosophical transactions of the Royal Society of London. Series B, Biological sciences*, vol. 262, no. 841, pp. 23–81, 1971.

[MCC 81] MCCLELLAND J.L., RUMELHART D.E., "An interactive activation model of context effects in letter perception: part 1", *Psychological Review*, vol. 5, pp. 375–407, 1981.

[MIK 12] MIKOLOV T., Statistical language models based on neural networks, PhD Thesis, Brno University of Technology, 2012.

[MIK 13] MIKOLOV T., YIH W., ZWEIG G., "Linguistic regularities in continuous space word representations", *Proceedings of NAACL-HLT*, Atlanta, pp. 746–751, 2013.

[NOR 03] NORMAN K.A., O'REILLY R.C., "Modeling hippocampal and neocortical contributions to recognition memory: a complementary-learning-systems approach", *Psychological Review*, vol. 110, no. 4, pp. 611–646, 2003.

[PER 10] PERRY C., ZIEGLER J.C., ZORZI M., "Beyond single syllables: large-scale modeling of reading aloud with the Connectionist Dual Process (CDP++) model", *Cognitive Psychology*, vol. 61, no. 2, pp. 106–151, 2010.

[PHA 07] PHAN X.-H., NGUYEN C.-T., "GibbsLDA++: A C/C++ Implementation of Latent Dirichlet Allocation (LDA)", available at: http://gibbslda.sourceforge.net/, 2007.

[RAD 12] RADACH R., GÜNTHER T., HUESTEGGE L., "Blickbewegungen beim Lesen, Leseentwicklung und Legasthenie", *Lernen und Lernstoerungen*, vol. 1, no. 3, pp. 185–204, 2012.

[REI 03] REICHLE E.D., RAYNER K., POLLATSEK A., "The E-Z reader model of eye-movement control in reading: comparisons to other models", *The Behavioral and Brain Sciences*, vol. 26, no. 4, pp. 445–476, 2003.

[SPI 16] SPICHTIG A., HIEBERT H., VORSTIUS C. *et al.*, "The decline of comprehension-based silent reading efficiency in the U.S.: a comparison of current data with performance in 1960", *Reading Research Quarterly*, vol. 51, no. 2, pp. 239–259, 2016.

[SPI 97] SPIELER D.H., BALOTA D.A., "Bringing computational models of word naming down to the item level", *Psychological Science*, vol. 8, no. 6, pp. 411–416, 1997.

[TAY 53] TAYLOR W.L., "'Cloze' procedure: a new tool for measuring readability", *Journalism Quarterly*, vol. 30, p. 415, 1953.

[ZIE 05] ZIEGLER J.C., GOSWAMI U., "Reading acquisition, developmental dyslexia, and skilled reading across languages : a psycholinguistic grain size theory", *Psychological Bulletin*, vol. 131, no. 1, pp. 3–29, 2005.

List of Authors

Chris BIEMANN
University of Hamburg
Germany

Philippe BLACHE
Brain and Language Research
Institute
CNRS–University of Provence
Aix-en-Provence
France

Alexandre BLONDIN-MASSÉ
University of Quebec at Montreal
Canada

Mohamed Amine BOUKHALED
Paris VI University
France

Jean-Gabriel GANASCIA
Paris VI University
France

Izabela GATKOWSKA
Jagiellonian University
Kraków
Poland

Debela Tesfaye GEMECHU
Addis Ababa University
Ethiopia

Ulrike GLAVITSCH
Swiss Federal Laboratories for
Materials Science and
Technology (EMPA)
Zurich
Switzerland

Maciej GODNY
Jagiellonian University
Kraków
Poland

Stevan HARNAD
University of Quebec at Montreal
Canada
and
University of Southampton
UK

Markus J. HOFMANN
University of Wuppertal
Germany

Michał KORZYCKI
AGH
University of Technology
Kraków
Poland

Marcos LOPES
University of São Paolo
Brazil

Mélanie LORD
University of Quebec at Montreal
Canada

Wiesław LUBASZEWSKI
Jagiellonian University
Kraków
Poland

Odile MARCOTTE
University of Quebec at Montreal
Canada

Marcello PELILLO
European Centre for Living
Technology (ECLT)
Ca' Foscari University
Venice
Italy

Reinhard RAPP
University of Mainz
and
University of Applied Sciences
Magdeburg-Stendal
Germany

Steffen REMUS
University of Hamburg
Germany

Florence SÈDES
Paul Sabatier University
Toulouse
France

Bernadette SHARP
Staffordshire University
Stoke-On-Trent
England

Rocco TRIPODI
European Centre for Living
Technology (ECLT)
Ca' Foscari University
Venice
Italy

Philippe VINCENT-LAMARRE
Ottawa University
Canada

Michael ZOCK
Traitement Automatique du
Langage Ecrit et Parlé (TALEP)
Laboratoire d'Informatique
Fondamentale (LIF)
CNRS
Marseille
France

Index

Printed in the United States
By Bookmasters